IET ENERGY ENGINEERING 110

High Voltage Power Network Construction

Other volumes in this series:

Volume 1	**Power Circuit Breaker Theory and Design** C.H. Flurscheim (Editor)
Volume 4	**Industrial Microwave Heating** A.C. Metaxas and R.J. Meredith
Volume 7	**Insulators for High Voltages** J.S.T. Looms
Volume 8	**Variable Frequency AC Motor Drive Systems** D. Finney
Volume 10	**SF_6 Switchgear** H.M. Ryan and G.R. Jones
Volume 11	**Conduction and Induction Heating** E.J. Davies
Volume 13	**Statistical Techniques for High Voltage Engineering** W. Hauschild and W. Mosch
Volume 14	**Uninterruptible Power Supplies** J. Platts and J.D. St Aubyn (Editors)
Volume 15	**Digital Protection for Power Systems** A.T. Johns and S.K. Salman
Volume 16	**Electricity Economics and Planning** T.W. Berrie
Volume 18	**Vacuum Switchgear** A. Greenwood
Volume 19	**Electrical Safety: A guide to causes and prevention of hazards** J. Maxwell Adams
Volume 21	**Electricity Distribution Network Design, 2nd Edition** E. Lakervi and E.J. Holmes
Volume 22	**Artificial Intelligence Techniques in Power Systems** K. Warwick, A.O. Ekwue and R. Aggarwal (Editors)
Volume 24	**Power System Commissioning and Maintenance Practice** K. Harker
Volume 25	**Engineers' Handbook of Industrial Microwave Heating** R.J. Meredith
Volume 26	**Small Electric Motors** H. Moczala *et al.*
Volume 27	**AC–DC Power System Analysis** J. Arrillaga and B.C. Smith
Volume 29	**High Voltage Direct Current Transmission, 2nd Edition** J. Arrillaga
Volume 30	**Flexible AC Transmission Systems (FACTS)** Y.-H. Song (Editor)
Volume 31	**Embedded Generation** N. Jenkins *et al.*
Volume 32	**High Voltage Engineering and Testing, 2nd Edition** H.M. Ryan (Editor)
Volume 33	**Overvoltage Protection of Low-Voltage Systems, Revised Edition** P. Hasse
Volume 36	**Voltage Quality in Electrical Power Systems** J. Schlabbach *et al.*
Volume 37	**Electrical Steels for Rotating Machines** P. Beckley
Volume 38	**The Electric Car: Development and future of battery, hybrid and fuel-cell cars** M. Westbrook
Volume 39	**Power Systems Electromagnetic Transients Simulation** J. Arrillaga and N. Watson
Volume 40	**Advances in High Voltage Engineering** M. Haddad and D. Warne
Volume 41	**Electrical Operation of Electrostatic Precipitators** K. Parker
Volume 43	**Thermal Power Plant Simulation and Control** D. Flynn
Volume 44	**Economic Evaluation of Projects in the Electricity Supply Industry** H. Khatib
Volume 45	**Propulsion Systems for Hybrid Vehicles** J. Miller
Volume 46	**Distribution Switchgear** S. Stewart
Volume 47	**Protection of Electricity Distribution Networks, 2nd Edition** J. Gers and E. Holmes
Volume 48	**Wood Pole Overhead Lines** B. Wareing
Volume 49	**Electric Fuses, 3rd Edition** A. Wright and G. Newbery
Volume 50	**Wind Power Integration: Connection and system operational aspects** B. Fox *et al.*
Volume 51	**Short Circuit Currents** J. Schlabbach
Volume 52	**Nuclear Power** J. Wood
Volume 53	**Condition Assessment of High Voltage Insulation in Power System Equipment** R.E. James and Q. Su
Volume 55	**Local Energy: Distributed generation of heat and power** J. Wood
Volume 56	**Condition Monitoring of Rotating Electrical Machines** P. Tavner, L. Ran, J. Penman and H. Sedding
Volume 57	**The Control Techniques Drives and Controls Handbook, 2nd Edition** B. Drury
Volume 58	**Lightning Protection** V. Cooray (Editor)
Volume 59	**Ultracapacitor Applications** J.M. Miller
Volume 62	**Lightning Electromagnetics** V. Cooray
Volume 63	**Energy Storage for Power Systems, 2nd Edition** A. Ter-Gazarian
Volume 65	**Protection of Electricity Distribution Networks, 3rd Edition** J. Gers
Volume 66	**High Voltage Engineering Testing, 3rd Edition** H. Ryan (Editor)
Volume 67	**Multicore Simulation of Power System Transients** F.M. Uriate
Volume 68	**Distribution System Analysis and Automation** J. Gers
Volume 69	**The Lightening Flash, 2nd Edition** V. Cooray (Editor)
Volume 70	**Economic Evaluation of Projects in the Electricity Supply Industry, 3rd Edition** H. Khatib
Volume 72	**Control Circuits in Power Electronics: Practical issues in design and implementation** M. Castilla (Editor)
Volume 73	**Wide Area Monitoring, Protection and Control Systems: The enabler for Smarter Grids** A. Vaccaro and A. Zobaa (Editors)
Volume 74	**Power Electronic Converters and Systems: Frontiers and applications** A.M. Trzynadlowski (Editor)
Volume 75	**Power Distribution Automation** B. Das (Editor)
Volume 76	**Power System Stability: Modelling, analysis and control** B. Om P. Malik
Volume 78	**Numerical Analysis of Power System Transients and Dynamics** A. Ametani (Editor)
Volume 79	**Vehicle-to-Grid: Linking electric vehicles to the smart grid** J. Lu and J. Hossain (Editors)
Volume 81	**Cyber-Physical-Social Systems and Constructs in Electric Power Engineering** Siddharth Suryanarayanan, Robin Roche and Timothy M. Hansen (Editors)
Volume 82	**Periodic Control of Power Electronic Converters** F. Blaabjerg, K. Zhou, D. Wang and Y. Yang
Volume 86	**Advances in Power System Modelling, Control and Stability Analysis** F. Milano (Editor)
Volume 87	**Cogeneration: Technologies, optimisation and implementation** C.A. Frangopoulos (Editor)
Volume 88	**Smarter Energy: From smart metering to the smart grid** H. Sun, N. Hatziargyriou, H.V. Poor, L. Carpanini and M.A. Sánchez Fornié (Editors)
Volume 89	**Hydrogen Production, Separation and Purification for Energy** A. Basile, F. Dalena, J. Tong, T.N. Veziroğlu (Editors)
Volume 90	**Clean Energy Microgrids** S. Obara and J. Morel (Editors)
Volume 91	**Fuzzy Logic Control in Energy Systems with Design Applications in MATLAB®/Simulink®** İ.H. Altaş
Volume 92	**Power Quality in Future Electrical Power Systems** A.F. Zobaa and S.H.E.A. Aleem (Editors)
Volume 93	**Cogeneration and District Energy Systems: Modelling, analysis and optimization** M.A. Rosen and S. Koohi-Fayegh
Volume 94	**Introduction to the Smart Grid: Concepts, technologies and evolution** Salman K. Salman
Volume 95	**Communication, Control and Security Challenges for the Smart Grid** S.M. Muyeen and S. Rahman (Editors)
Volume 97	**Synchronized Phasor Measurements for Smart Grids** M.J.B. Reddy and D.K. Mohanta (Editors)
Volume 98	**Large Scale Grid Integration of Renewable Energy Sources** Antonio Moreno-Munoz (Editor)
Volume 100	**Modeling and Dynamic Behaviour of Hydropower Plants** N. Kishor and J. Fraile-Ardanuy (Editors)
Volume 101	**Methane and Hydrogen for Energy Storage** R. Carriveau and David S.-K. Ting
Volume 108	**Fault Diagnosis of Induction Motors** Jawad Faiz, Vahid Ghorbanian and Gojko Joksimović
Volume 905	**Power System Protection, 4 volumes**

High Voltage Power Network Construction

Keith Harker

The Institution of Engineering and Technology

Published by The Institution of Engineering and Technology, London, United Kingdom

The Institution of Engineering and Technology is registered as a Charity in England & Wales (no. 211014) and Scotland (no. SC038698).

First published 2018

The Institution of Engineering and Technology
Michael Faraday House
Six Hills Way, Stevenage
Herts, SG1 2AY, United Kingdom

www.theiet.org

British Library Cataloguing in Publication Data
A catalogue record for this product is available from the British Library

ISBN 978-1-78561-423-1 (hardback)
ISBN 978-1-78561-424-8 (PDF)

Typeset in India by MPS Limited
Printed in the UK by CPI Group (UK) Ltd, Croydon

Contents

About the author

Keith Harker has spent virtually the whole of his career with the UK electrical supply industry – with experience spanning transmission, distribution and generation engineering. In more recent years, he has held a number of senior positions in power network construction with the National Grid Company with responsibilities that have included design, project management, health and safety, quality management, engineering competency and technical assurance. He has written numerous papers on power engineering and is the author of an earlier book on the subject of power system commissioning and maintenance practice. He is currently concerned with both consultancy and lecturing in power engineering and is a Fellow of the Institution of Engineering and Technology.

Foreword

Electrical power networks are like no other national infrastructure in the way they underpin modern society. This is because any loss of electrical power particularly on a large scale increasingly risks an inability of many societies to adequately function, especially those that are industrialised. Such societies invariably demand that those that own and operate power networks must deliver the highest standards of performance in ensuring continuity of electricity supply. This requires power networks that are constructed (i.e. planned and built) to achieve those high standards of performance. Within this context, this book examines the key requirements, considerations, complexities and constraints relevant to the task of power network construction.

The book is specifically concerned with high-voltage power network construction with the aim of providing an overview of the holistic end-to-end construction task in a single volume. It covers all stages and requirements of construction from an investment 'need case' to newly commissioned equipment entering commercial service. The text specifically targets the 400, 275, 132 and 33 kV networks. Best and common practice is presented although it is accepted that there often a variety of ways of achieving the same end. The book is written primarily from a UK electrical power industry perspective, but recognises that much is already harmonised with the rest of Europe and increasingly so the rest of the world.

A central theme of the book is that the construction of a high-voltage power network must result not only in the provision of a secure and reliable supply of electrical power, but also one that satisfies legal requirements and codes of acceptable and best available practice. At the same time, it must retain public acceptance both during construction and following completion. To achieve these requirements, it is reasoned that the resulting end-to-end construction task necessitates the implementation of three complimentary deliverables. First, the specification and implementation of a technical solution that results in the required power system performance characteristics. Second, the execution of appropriate quality management system procedural arrangements to ensure that all construction activities are undertaken in accordance with efficient and consistent pre-defined systems of work – and confirmed as achieved. Third, that all duty holders have demonstrated that they possess the requisite competencies i.e. knowledge, skills and experience to discharge their responsibilities to the required professional standard. The book is structured around these three key requirements.

The book essentially covers the broad span of requirements associated with the three high-level stages of design, site installation and commissioning. It additionally encompasses the criterion surrounding financial investment and the practicalities associated with engineering contracts, construction delivery models and project management. In totality, the power system covers not only the power networks but also sources of power generation and therefore the impact of, and interface with, both thermal power stations and renewable generation has been included and briefly covered.

The target audience is primarily those engineers with construction-related responsibilities: development engineers, project managers, project engineers, design engineers, site managers, site engineers, commissioning engineers and those with health and safety and safety rules responsibilities. It is aimed at those who are at the outset of their careers – with an objective of providing an easily accessible information source to assist in fast tracking their engineering knowledge and development. In doing so, the objective is not only to focus on the speciality area of each individual engineering discipline but also to provide a broad understanding of, and empathy with, the whole construction task. It also provides a reference source both for more experienced engineers, and those in related disciplines who want a greater insight into the complexities of power network construction.

To those commencing their career in the power supply industry, particularly in the field of construction I trust that you may have as much excitement and sense of achievement as I have had throughout the whole of my career in power engineering – not only as a result of delivering a complex engineering task to the highest possible standard but also actively contributing to a critically important objective – namely that of 'keeping the lights on'.

Keith Harker
January 2018

Acknowledgements

Although the content of most technical publications is in the first instance a reflection of the knowledge and experience of the author, it is in the second instance also a reflection of the professional engineers and managers who have educated, corrected and enthused the author – often over many years. Within this context, acknowledgement is made to the following:

1. Mr R. Inglesfield, Dr H. Habibollahi, Mr N. Tart and the late Mr J.C. Whittaker – for their in-depth command and understanding of protection and control systems – and in particular the complexities of protection settings.
2. Mr R. Carpenter, Mr M. A. Waldron, Mr P.L. Fletcher and the late Mr C. Morris – and their wealth of knowledge on substation engineering.
3. Dr A.I. Erinmez, Dr N. Tleis and Mr A. Hiorns – who could resolve almost any question or problem on power system design.
4. Mr K. Chitty and Mr F. Waite – for their expansive knowledge on, and passion for, HV cable engineering.
5. Mr P.N. Jarman – who became a centre of excellence for transformer engineering and static equipment containing windings.
6. Mr J. (Jack) Gibson, Mr J.A. Thetford, Mr J. Warburton and Ms M. Clarke – in recognition of their enthusiasm for, and conceptual understanding of, OHL engineering and the infectious impact it had on the author.
7. Mr N. Tobin – who opened the author's eyes to the complexities and importance of civil engineering.
8. Mr L. Adams and S. Hughes – for their comprehensive understanding of, and long standing commitment to, site installation and commissioning practice.
9. Dr I. Sokhey and Mr G. Aberwayerdene – highly competent power engineers with the added bonus of an understanding of the complexities of power generation.
10. Mr D. Barron and D. Monkhouse – who could hold court on HVDC engineering and FACTS technology.
11. Mr M.C. Hanlon – for many years of commitment to, and belief in, both quality and safety management systems together with a passion for standards of performance.
12. Mr D. Poulton and Mr P.J. Johnson – for their enthusiasm and commitment over many years to furthering engineering competency – and the concept of the all-round power engineer.

13. To those engineers too numerous to mention who over the years have provided the author with advice, information and direction.
14. To Mr M.C. Hanlon, Mr S. Hughes, Mr N. Tart and Mr N. Tobin for proof reading the first draft of this book.

Special acknowledgement is made to the late Mr L. (Les) Adams an engineering manager of distinction who not only had a passionate interest in, and actively promoted, all of the subject matter contained in this publication – but also fostered and encouraged the training and development of many young engineers and technicians, in the field of power engineering.

Symbols used in figures

Symbols currently used in UK power network construction (and in many countries world-wide) generally accord with BS ISO 14617, Graphical Symbols for Diagrams, or with IEC 60617 which is also has the title Graphical Symbols for Diagrams (and which is one and the same as BS EN 60617) – virtually all symbols are identical. However, many existing drawings will accord with an older drawing symbols standard such as BS 3939, and therefore, the practicing engineer will have to deal with all (if interfacing with older equipment). Within this context, the author, in the following chapters, has frequently reverted to older symbols as many are both more distinctive, therefore making the diagrams easier to interpret, and still commonly used in documents, reports, etc. This applies in particular to the circuit breaker where the older symbol of the square is often used (and still preferred by many practicing engineers). Instances also arise of the author simplifying some symbols and drawings to provide a more focused and explanatory presentation. The traditional practice of placing a circle around a circuit breaker or disconnector to indicate that it is in the open state has also been used.

It is worthy of note that the three phase HV power system in the United Kingdom is still denoted in terms of Red, Yellow and Blue phases (whereas consumer LV wiring is aligned with the European phase colours in accordance with IEC 60445).

It is also worthy of note that in North America (and in other parts of the world), ANSI symbols tend to predominate.

The following illustrates both current and older symbols as used in this publication.

Symbols used in figures

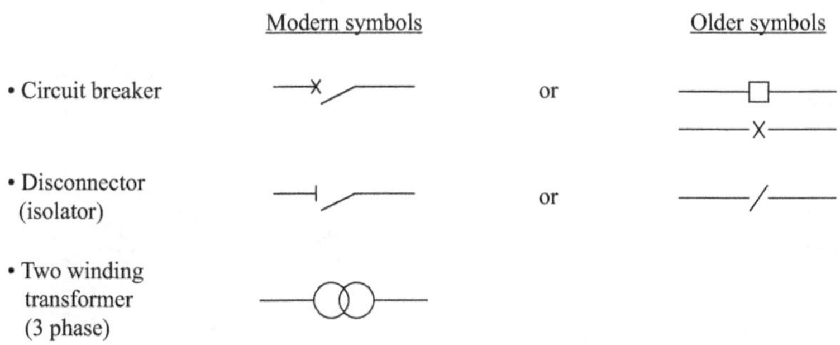

- Auto transformer (3 phase)

- AC generator

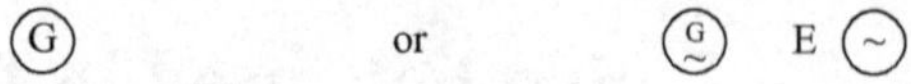

- Current transformer

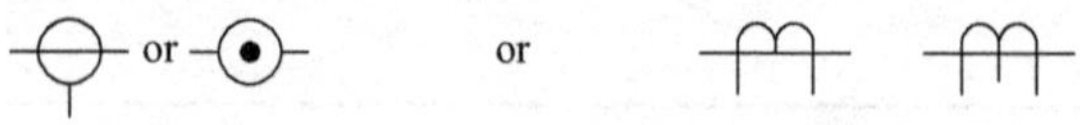

- Two winding transformer (single phase)

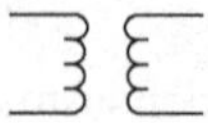

- Auto transformer (single phase)

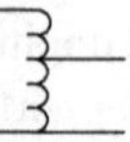

- Reactor

- Winding configurations

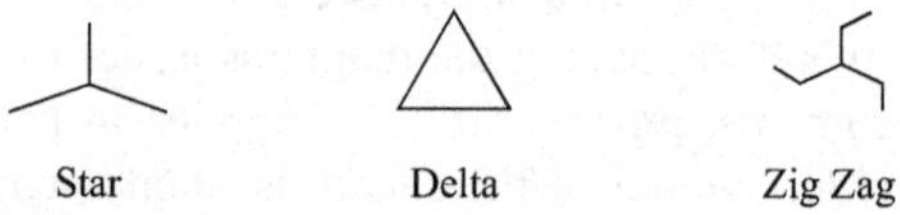

- Voltage transformer

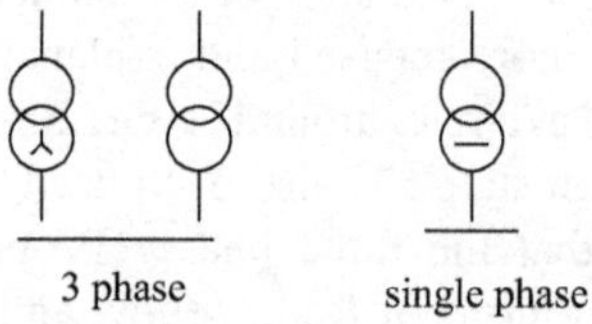

- Fuse

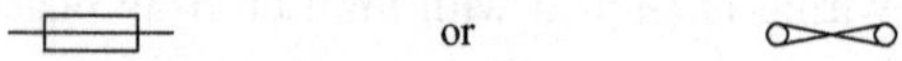

- Link

- Normally open contact

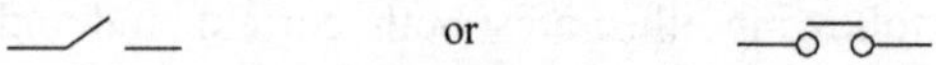

- Normally closed contact

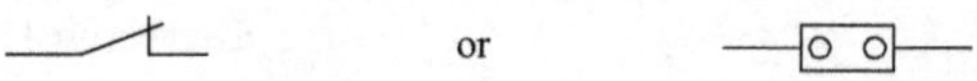

- Relay

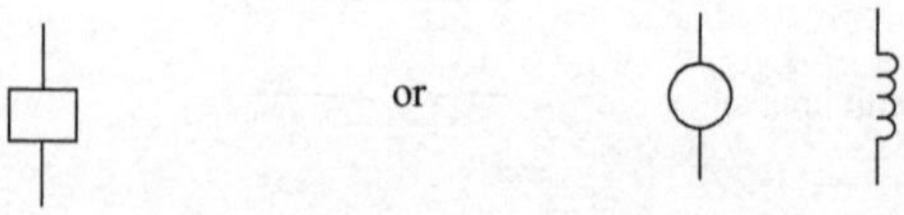

- IDMT overcurrent relay

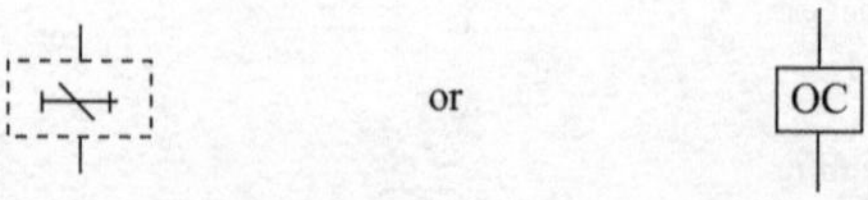

- Instantaneous overcurrent relay 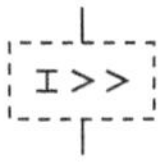

or 

- Impedance 
- Resistance 
- Inductance 
- Capacitance 
- Earth 
- Earth plane 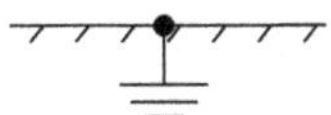
- Fault (short circuit) 
(any fault)

(fault to earth)
- Battery 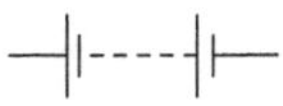
- Search voltage limiter (SVL) 
- HV cable 

Abbreviations and interpretations

The following provides a summary of the abbreviations used in the text together with an interpretation of some frequently used terms:

- **AC**
  Meaning alternating current
- **AIS**
  Meaning air insulated switchgear
- **CDM**
  Meaning the construction (design and management) regulations 2015
- **Circuit Earth**
  A protective earth capable of carrying network short-circuit currents
- **Contractor**
  Generally meaning the successful tenderer who has been contracted to deliver the project, and in the context of this publication generally responsible for the stages of detail design, manufacture/procure, site installation and commissioning
- **CT**
  Meaning current transformer
- **CVT**
  Meaning capacitor voltage transformer
- **DC**
  Meaning direct current
- **emf**
  Meaning electro-motive force
- **EPC**
  Meaning an engineer, procure and construct contract
- **DDS**
  Meaning 'Detail Design Specification'. This forms part of the project-specific design specification and is prepared post-sanction
- **Equipment**
  Within this publication, the term 'equipment' is used as a blanket term covering all electrical devices and conductors e.g. circuit breakers, OHL, HV cables, protection relays, etc. The term applies both before and after being subject to power network company safety rules. The terms 'plant' and 'apparatus' which are used in a variety of other publications are not used, but are encompassed by the term equipment
- **ERTS**
  Meaning emergency return to service

- **ESQCR**
  Meaning the electricity, safety, quality and continuity regulations 2002
- **FACTS**
  Meaning flexible alternating current transmission system
- **GIS**
  Meaning gas insulated switchgear
- **HASAWA**
  Meaning the Health and Safety at Work Act etc. 1974
- **HSE**
  Meaning the Health and Safety Executive
- **IDMTL**
  Meaning Inverse Definite Minimum Time Lag, (often abbreviated to IDMT)
- **Impressed Voltage Earth**
  An earth capable of protecting against impressed voltages
- **IV**
  Meaning impressed voltage, and covering capacitive coupling, inductive coupling, conductive coupling and trapped charge
- **MSC**
  Meaning manually switched capacitor
- **NPS**
  Meaning negative phase-sequence
- **OHL**
  Meaning overhead line
- **Power Network vs Power System**
  Within this publication, the term 'power network' refers to all transmission and distribution networks covering the UK voltages of 33 kV to 400 kV (and where stated in the text, lower voltage networks). The term 'power system' covers all networks together with sources of connected generation. It is recognised that in practice these terms are occasionally, and imprecisely, used interchangeably
- **Power Network Company**
  Meaning the electrical utility that instigates the construction work (and is usually one and the same as the client) and who usually own and operate one or more power networks
- **PPS**
  Meaning positive phase-sequence
- **Project-Specific Design Specification**
  A design specification unique to the project in question, which functions in conjunction with the project-generic design specification
- **Project-Generic Design Specification**
  A design specification common to most projects, which functions in conjunction with the project-specific design specification. The project-generic design specification usually comprises a number of 'technical specifications' which serve as a procurement specification e.g. circuit breaker technical specification. It may also include power network company technical policies

- **QMS**
  Meaning quality management system
- **RMS**
  Meaning root mean square (of a voltage or current wave-form)
- **Sanction**
  Financial and technical approval by nominated officers of a power network company that a scheme may be implemented and the necessary funds and resources committed, i.e. a project is commenced
- **Scheme vs Project**
  Within this publication, the term 'scheme' is interpreted as the end-to-end construction process from Need Case to formal closure, following commissioning. The term 'project' is interpreted as all those construction activities that take place from point of Sanction to completion of commissioning. A project therefore exists within a scheme, and a scheme may exist without a project taking place if terminated at the time of sanction. In practice, these terms are occasionally, and imprecisely, used interchangeably.

  NB: The term 'scheme' may also be used to describe an arrangement of protection or control equipment and associated circuitry, e.g. a busbar protection scheme
- **SCS**
  Meaning substation control system
- **SDS**
  Meaning 'scheme design specification'. This forms part of the project-specific design specification and is prepared pre-sanction
- **SMS**
  Meaning safety management system
- **SQP**
  Meaning site quality plan
- **SVC**
  Meaning static VAr compensator
- **S/S**
  Meaning substation
- **UK**
  Meaning the United Kingdom of Great Britain and Northern Ireland. The term 'UK' as used in the text is generally taken as a blanket term covering all member countries of the United Kingdom – whilst recognising that country-specific differences may occasionally arise, albeit usually small
- **VT**
  Meaning voltage transformer
- **ZPS**
  Meaning zero phase-sequence

Chapter 1
Construction execution model

1.1 Introduction

The term 'power network construction' encompasses the end-to-end activities required to bring into existence the extension, reinforcement, modification or replacement of an existing operational power network. With reference to this text, the term high voltage (HV) power network is primarily considered to include the main UK transmission and distribution voltages of 400, 275, 132 and 33 kV (and 66 kV where it remains). Although the 11 kV network is also an HV network (and the 22, 20, 6.6 and 3.3 kV networks where they remain), the construction activities are of a different scale/magnitude – and therefore not fully considered – although many of the same principles apply.

The end-to-end power network construction task essentially involves three high-level stages: design, site installation and commissioning. In the first instance, delivery of these three stages may be considered to be primarily a technical activity. However of necessity, construction must be undertaken within the wider environment constraints imposed by society. These constraints include not only technical performance requirements – but also those of health and safety, environmental pollution and the aesthetic/nuisance impact that a modern power network imposes on society in general – all of which must be positively managed.

1.2 Wider construction environment

With reference to Figure 1.1, construction invariably commences with a 'need case' (the input), i.e. the reason that justifies why power network construction needs to be undertaken – and ends with a commissioned economically viable and operational power network (the output). In progressing from input to output, the construction task is subject to a number of requirements imposed by the wider construction environment; the most notable of which are as follows:

- Legal requirements relating to technical performance standards.
- Legal requirements relating to health and safety and environmental pollution standards.
- Legal requirements relating to land ownership, access, aesthetic and amenity requirements.

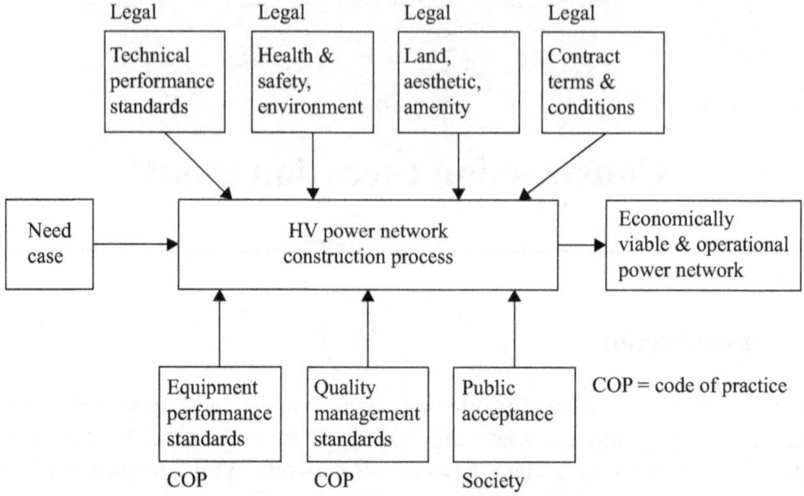

Figure 1.1 Wider construction environment

- Legal requirements relating to contractual terms and conditions whenever a contracting organisation is employed.
- Technical standards relating to equipment performance to which a power network company, contractor, consultant and manufacture are expected to adhere.
- Quality management system procedures when undertaking construction work – again to which professional/responsible power network companies, contractors, consultants and manufacturers are reasonably expected to adhere.
- Public acceptance both during on-site construction and following construction completion. This is an increasingly important consideration – since objections have the potential to significantly delay or de-rail construction. Thus, the management of public expectations is an integral and important aspect of the construction task.

Power networks operate for, and are fundamental to, the well-being of modern society. At the same time, society imposes requirements to which power network construction must adhere – as outlined above. These wider environment requirements must therefore be factored into any model used for the execution of HV power network construction.

1.3 Construction execution model – components

1.3.1 Technology, QMS and competency

With reference to Figure 1.2, the term 'construction execution model' relates to the key requirements necessary for the effective/efficient execution of the construction task. These are considered to comprise three complementary components:

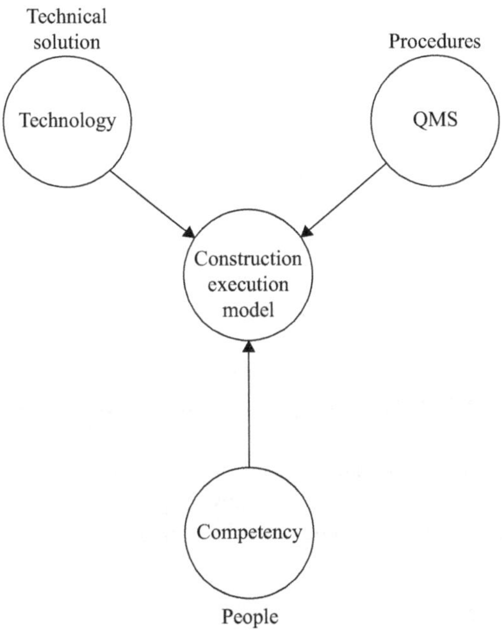

Figure 1.2 Construction execution model

technology (technical solution); quality management system (QMS) procedures including health, safety and environmental; and engineering competency. These are summarised in the following sections.

1.3.2 Technology

Technology refers to the technical component of the construction task, that is the preparation and implementation of an optimal engineering design solution based upon legal/technical standards requirements – resulting in operational equipment that has been subject to appropriate factory and commissioning tests – which have confirmed that the equipment performance characteristics satisfy both the need case and associated construction design specification.

1.3.3 QMS procedures

A procedure may succinctly be described as 'who does what and when'. It is primarily concerned with stipulating (a) the range of tasks to be undertaken, (b) the content of those tasks, (c) who is nominated to undertake those tasks and (d) when those tasks are to be executed.

Procedures are invariably defined in narrative form and/or process flow diagrams. They are typically underpinned by, and integral to, the requirements of a QMS such as ISO9001. Successful construction procedures must incorporate the key requirements of defined responsibilities, planning, step by step sequence of activities, interface management, essential documentation and essential meetings. Legal requirements must again be incorporated.

1.3.4 Engineering competency

Successful completion of any task can only be accomplished through individuals who possess the appropriated competence for the task in hand. Modern work methods have sought to de-skill many activities through automation or structured-process techniques – however, power network construction is still very dependent upon individuals with a high level of engineering competency. Within this context, removal of human error through the employment of a competent workforce remains a prime consideration in retaining high levels of power system reliability.

The construction execution model therefore requires engineers with appropriate knowledge/experience not only in technology-related requirements but also in the rigorous execution of the related construction QMS procedures (including health, safety and environmental). Criterion by which the relevant engineering duty-holders demonstrate the required knowledge/experience is a key factor in being assured that they are competent for the construction task to be undertaken.

1.4 Technology

An overview of the technology requirements of a power-network-construction project is summarised in the following sections.

1.4.1 Technology stages

The construction requirements falling under the heading of 'technology' may be subdivided into the following stages:

- Construction design specification
  - Project-specific design specification
  - Project-generic design specification
- Detail design
- Equipment manufacture/procure
- Equipment site installation
- Equipment commissioning.

The content of these requirements is summarised in the following sections.

1.4.2 Construction design specification

With reference to Figure 1.3, a construction design specification comprises a specification of design requirements in sufficient detail to enable the subsequent stages of detail design, manufacture/procure, site installation and commissioning to successfully take place. It is divided into two specifications: a project-specific design specification comprising those components of the design that are specific/unique to the construction work in question and a project-generic design specification comprising a suite of technical specifications (and network planning standards) which

	Spec. category	Document	Design task	
Construction design specification	Project-specific design specification	SDS	1. Power system electrical design	i) HV network design
				ii) HV equipment design
				iii) HV power system single line diagram
				iv) Power system protection design
				v) Power system control design
			2. Substation layout spatial design	
			3. Protection and control accommodation design	
			4. OHL route and electro-mechanical design	
			5. HV cable route and electro-mechanical design	
			6. Civil, structural and building engineering design	i) Equipment structure design
				ii) Ground design
				iii) Site infrastructure and building services design
			7. Temporary works design	
		DDS	Technical implementation. 1. Install/commission/operate & maintain/repair & replace/ decommission & remove – requirements 2. SDS detail/refine	
	Project-generic design specification	Technical policy and specifications	1. Technical policy (eg Network planning standards) 2. Equipment technical specifications	

Figure 1.3 Construction design specification model

are usually common to all projects undertaken by a specific power network company. The project-specific design specification is also usually divided into two, which are as follows:

1. **Scheme design specification (SDS)**
   The scheme design specification (SDS) (or other similar title) is prepared prior to the contract tender document release and comprises sufficient detail to cost the project for purposes of 'sanction', i.e. the authority to proceed which is usually given by the power network company (client). The SDS also forms the

technical section of the contract tender document (assuming the work is let to a contractor), see Section 1.4.3.

2. **Detail design specification (DDS)**
   The detail design specification (DDS) (or other similar title) is usually prepared after contract release and provides greater detail than that provided in the SDS, see Section 1.4.4.

1.4.3 Project-specific design specification – SDS

With reference to Figure 1.3, the composition of the SDS is typically as follows:

1. **Power system electrical design**
   This design is the core design task around which all other design and related tasks revolve. It consists of the following components:
   (i) **HV network design**
   This comprises a proposed design solution at the network level (i.e. new substations and connecting circuits including modifications to the existing network) which satisfies the need case. It is usually based upon HV network planning standards which define voltage levels, load and fault currents, security of supply and operational flexibility, etc. It is typically contained in a system development report (or other similar title) and referenced by the SDS. See Chapter 5.
   (ii) **HV equipment design**
   This comprises the proposed items of equipment (transformers, OHL, HV cables, circuit breakers, etc.) specified in terms of their project-specific ratings and electrical characteristics – which are specified in detail by the project-generic design specification.
   (iii) **HV power system single line diagram**
   This consists of a single line diagram (i.e. single phase representation) showing the proposed interconnection of substation equipment and connecting OHL and cable circuits. Points (i) and (iii) are interconnected activities and undertaken simultaneously.
   (iv) **Power system protection design**
   This comprises the proposed protection arrangements, as outlined in Chapter 10. Many power network companies have defined (standard) protection arrangements for specific HV equipment layouts – however, non-standard layouts will demand protection design tailored to the installation.
   (v) **Power system control design**
   This comprises the specification of a supervisory control and data acquisition system, and associated equipment controls, alarms, indications and metering requirements. It additionally includes auto-reclose, synchronising and interlocking systems. See Chapter 10.
2. **Substation spatial layout design**
   A substation spatial layout diagram is required to ensure that the basic design and access clearances for both personnel and equipment during maintenance,

etc., exceed minimum safety requirements. It is also required to determine the physical size of the substation, the land requirements and the visual profile. This diagram therefore shows three-phase physical dimensions. See Chapter 8.

3. **Protection and control accommodation design**
   The physical layout of protection and control panels/cubicles and associated kiosks, etc., needs to be specified – with an objective of both minimising the size of the buildings housing the panels – and ensuring that the arrangement is logical from an operational perspective (wherever possible it should mirror the substation physical layout).
4. **OHL route and electro-mechanical design**
   This usually comprises a single line diagram defining the geographic route of the OHL. The design must take into account: crossings with other utilities/services; impact on roads, houses, factories, etc.; access routes and minimising the impact on the general public. OHL tower, pole, conductor and spatial design need to be specified to a level sufficient for costing, sanction and entry into a contract tender document. See Chapter 6.
5. **Cable route and electro-mechanical design**
   Similar to an OHL, this comprises a single line diagram defining the geographic route of the cable. The design must take into account the impact on roads and crossings with other utilities/services and minimising the impact on the general public. Cable-design requirements, including terminations, laying mediums and bonding requirements, need to be specified, sufficient for costing, sanction and entry into a contract tender document. See Chapter 7.
6. **Civil, structural and building engineering design**
   Although power engineering is the prime construction activity – an operational power network cannot be constructed without a significant civil engineering contribution. The phrase 'Civil, Structural and Building Engineering' (CSBE) describes the wider civil engineering task, see Chapter 13. CSBE encompasses the following three key activities:
   (i) Equipment structure design
   (ii) Ground design
   (iii) Site infrastructure and building engineering design

   (i) **Equipment structure design**
   Power equipment such as circuit breakers, disconnectors, busbars, etc., are frequently mounted on structures. Such structures need to be designed to accommodate the forces (wind, short-circuit, etc.) to which they will be subject.

   (ii) **Ground design**
   Power system equipment and structures rest on the ground – as do transformer plinths, roadways, substation buildings, etc. Evaluation of the ground bearing capability must therefore be undertaken – with proposals for strengthening the ground where necessary.

   (iii) **Site infrastructure and building engineering design**
   Site infrastructure relates to the buildings, roadways, trenches, drainage systems, fences, etc., that comprise an operational substation. Building

engineering on the other hand describes the heating, lighting, ventilation, etc., requirements of buildings – including substation lighting.

7. **Temporary works design**
   The previous sections have focused upon the design requirements of the final, installed and operational equipment – and associated CSBE. However, additional design is frequently required relating to temporary installations which are essential for completion of the final design. Temporary design typically includes scaffolding, cable trench shuttering, temporary access roads, temporary protection systems for emergency return to service or commissioning purposes, etc. Detailed temporary works design need to be defined in the DDS.

1.4.4 Project-specific design specification – DDS

The preparation of a project-specific design specification must not only consider the performance characteristics of the constructed asset, as defined in a SDS – but also the detailed 'technical implementation' of the project, i.e. how the equipment will be: installed on site, including removal and re-use of existing equipment; commissioned; operated and maintained; repaired and replaced; decommissioned/removed from site; etc. The DDS specifies these requirements (see Chapter 20) together with a more detailed statement of the requirements of the SDS.

NB: The DDS should ideally be prepared prior to contract release – but time, resource and cost constraints usually result in preparation after contract release and immediately prior to detail design commencing. It is usually prepared by the successful contractor for agreement with the power network company.

1.4.5 Project-generic design specification

Project-generic design relates to network planning standards and equipment (generic) technical specifications, which are usually compiled or referenced by individual power network companies. They are employed and utilised on most projects, i.e. design standards. For example – whereas the project-specific design specification defines the requirements of, say, a circuit breaker operating at a specific network voltage, with a specific load carrying capability (i.e. requirements unique to the project) – the project-generic design specification defines a complete procurement specification for the circuit breaker (common to all projects). This would typically comprise:

- Electrical performance characteristics
- Pollution performance
- Type test requirements
- Phase–earth and phase–phase clearances
- Types of operating mechanism
- Maximum permissible equipment temperature range
- Access arrangements
- Lifetime performance capability
- Site-installation-specific requirements

- Commissioning test requirements
- Maintenance requirements and manuals
- Labelling:

Such specifications usually comprise a suite of documents (one for each type of equipment) and often require adherence to a defined international standard (e.g. IEC).

1.4.6 Detail design

Detail design comprises the production of all drawings, schedules and calculations that are required to be prepared to define the design in sufficient detail such that site installation and commissioning can be successfully carried out. Detail design is based upon both the project generic and project-specific design specifications – and usually takes place following award of contract – and therefore undertaken by the successful contractor. See Chapters 14 and 20.

1.4.7 Manufacture/procure

The manufacture/procure stage relates to the factory assembly of items of equipment, or the procurement of equipment, as specified in the construction design specification. The detail design and manufacture stages both take place in the same period of time – and are frequently interdependent tasks. For example – factory assembly may be dependent upon a drawing prepared as part of detail design; alternatively, a detail design drawing may require information about the manufactured equipment. The manufacture stage would usually include factory tests of the equipment to confirm key performance characteristics. See Chapter 14 and 21.

1.4.8 Equipment site installation

Site installation comprises the on-site building of equipment specified in the construction design specification. It is undertaken in accordance with both the detail design and manufacturer's instructions for equipment installation. It is worthy of note that the terms site installation and site construction tend to be alternative terms for site-based work. Temporary works design requirements would also be undertaken during site installation. See Chapters 14 and 22.

1.4.9 Equipment commissioning

Commissioning requires equipment tests and inspections that demonstrate that the installed equipment accords with both the construction design specification and the detail design (drawings, etc.) and is suitable for commercial operation. See Chapters 15 and 23.

1.4.10 Equipment and CSBE – terminology

In this text the term 'equipment' is taken to mean items which directly comprise the power network – both before and after entering service, i.e. circuit breakers, cables, OHL, protection systems, etc. The term 'CSBE' describes the non-power network items which are essential for the functioning of a power network, e.g. substation buildings, roads, drainage, fencing, etc.

However, there are no definitive rules relating to these terms and often, in practice, the word equipment is loosely taken as an all-embracing term to include all or part of CSBE. Some texts also use the term 'plant' or 'plant and apparatus' instead of equipment – and some utilities use the term 'plant' to describe an item prior to that item coming under the jurisdiction of the power network company's safety rules – and once under the safety rules that item is termed 'equipment'.

Rarely in practice does the use of these imprecise terms cause a problem – but the reader needs to be aware – since the terms may vary from company to company – and situations may arise when clarification is a necessary.

1.5 Construction QMS procedures

1.5.1 QMS scheme investment process

A QMS is a systematic approach to work, by an organisation, for the purposes of ensuring defined standards of work quality. It requires management commitment, appropriate organisational structures, defined responsibilities, procedural documentation, defined competencies and arrangements for review and improvement. The core of a QMS is a suite of interlocking procedures for planning and executing the work. An important component of a QMS is periodic audit of the procedures and their enactment, to confirm compliance.

A comprehensive suite of QMS procedures is essential to power network construction. Each of the parties involved, i.e. power network companies, consultants, manufacturers and contractors should have their own QMS. The QMS of the power network company and contractor must dove tail at the interfaces between the two organisations, to ensure that their joint endeavour is seamless.

The power network company (i.e. the client) will usually have a single overarching procedure (umbrella procedure) detailing the end-to-end process – covering the high-level requirements of the construction task. The title of this procedure may vary, but typically, it may be termed the 'scheme investment procedure' – and this title (see Chapter 17) will be used in this text. The scheme investment process typically comprises the sequence of the key stages in the scheme. Each stage in itself will comprise a step-by-step sequence of more detailed activities – sometimes defined in subordinate procedures.

Figure 1.4 illustrates a typical scheme investment process (sometimes termed a 'construction delivery model') in which a construction design specification is prepared by an power network company (the client) – and following a competitive tender and contract award process – the successful contractor undertakes detail design, site installation and commissioning. As discussed in Chapter 16, there are permutations of the model shown in Figure 1.4 – however, this model is very common and illustrates the key features.

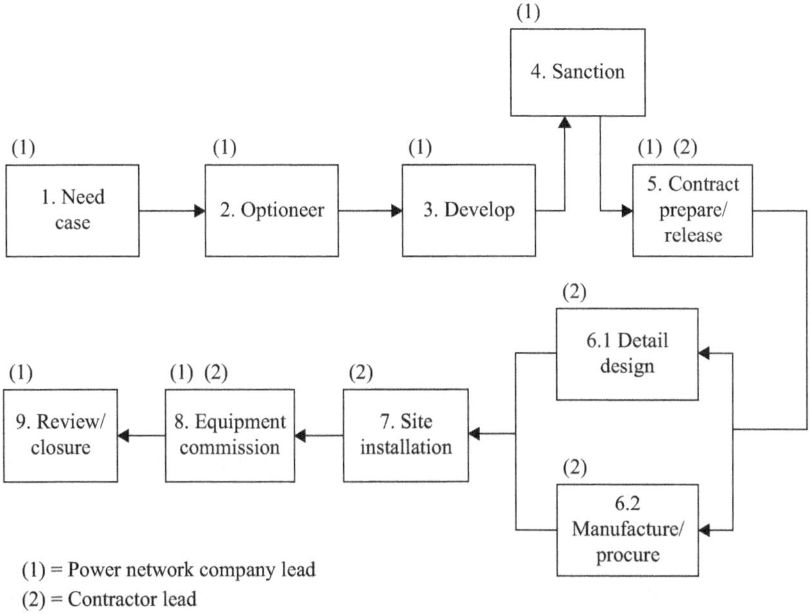

Figure 1.4 Scheme investment process (typical)

1.5.2 Scheme investment process – stages

With reference to Figure 1.4, the high-level stages of the scheme investment process comprise:

1. **Need case**
   The need case comprises the technical and economic reasons that justify power network construction. It is usually prepared by the power network company (client), see Chapter 17.
2. **Optioneering**
   Optioneering is the evaluation of a number of options which satisfy the need case – with justification for the chosen option. This is usually undertaken by the power network company (client), see Chapter 17.
3. **Development**
   Development comprises the preparation of a construction design specification for the chosen option – for subsequent inclusion in a contract tender document. A range of surveys (e.g. ground bearing conditions) may be required during this stage. Development is undertaken by the power network company (client) and is usually subject to a 'sanction' hold point where senior representatives of the power network company conclude whether the proposed investment is justified – and therefore should continue. Sanction is undertaken prior to release of the contract tender documents, see Chapter 17.

4. **Contract**
   The contract stage, see Chapters 16 and 19 comprises four sub-stages:
   (i) Preparation of a contract tender enquiry document by the power network company (incorporating the construction design specification) – and release of the same to competent contractors.
   (ii) Contract tender response from the contractors – proposing technical solutions with schedules of costs.
   (iii) Evaluation of the contractor responses by the power network company (client) – to determine the optimum technical/economic response – and subsequent selection of the successful contractor.
   (iv) Award of contract by the power network company to the successful contractor.
5. **Detail design**
   The detail design stage comprises procedural arrangements by which the detail design is undertaken by a contractor and confirmed as correct by the power network company, see Chapter 20.
6. **Manufacture/Procure**
   The manufacture/procure stage consists of the procedural arrangements by which equipment is either procured or factory assembled, tested and subject to a manufacturing surveillance/assurance process, see Chapters 19 and 21.
7. **Site installation**
   Site installation comprises procedural arrangements for both establishing and managing a construction site and constructing the equipment and associated site infrastructure (e.g. roads, buildings, power supplies, etc.; see Chapter 22).
8. **Commissioning**
   The commissioning stage consists of procedural arrangements for planning, managing and executing commissioning, see Chapter 23.
9. **Review**
   A review of the lessons learned and good/bad practice for improving/correcting both procedural documentation and performance on future projects completes the scheme investment process, see Chapter 17.

1.5.3 QMS procedures – format

Procedures are usually in text format supported by flow diagrams. However modern practice also utilises computer-based process flow diagrams. The structured contents of a QMS procedure usually comprise the following generic requirements:

- Roles, responsibilities and tasks of the named duty holders.
- Definitions relevant to the process (i.e. the meaning of specific terms)
- Step by step sequence of activities
- Documentation to be prepared
- Meetings to be convened with typical agendas
- Interaction with related processes
- Lessons learned.

QMS procedures are examined in more detail in Chapters 16–25.

1.5.4 Schemes and projects

With reference to Figure 1.4, the end-to-end construction process is often termed a 'scheme' and that part of the process that takes place after power network company sanction is termed a 'project'. When these terms are used in subsequent chapters, they shall have this meaning. It is worthy of note that a scheme may exist without a subsequent project – should the scheme be rejected at time of power network company (client) sanction.

NB: This use of the term 'scheme' must not be confused with the use of the same term in a technology sense. For example – the term power system protection scheme is frequently used to describe a protection equipment and associated connections as shown on a 'schematic diagram'.

1.6 Competency

1.6.1 Competency model

Figure 1.5 illustrates a simple competency model suitable for power network construction. Competency in technology, health and safety (including environmental), and QMS procedures delivers three key outputs:

1. Power network reliability. Keeping the lights on both during and following construction.
2. Health and safety (and environmental) performance excellence. No incidents, injuries, lost time or environmental incidents.
3. Construction efficiency. Providing the optimal solution efficiently and effectively – with no rework.

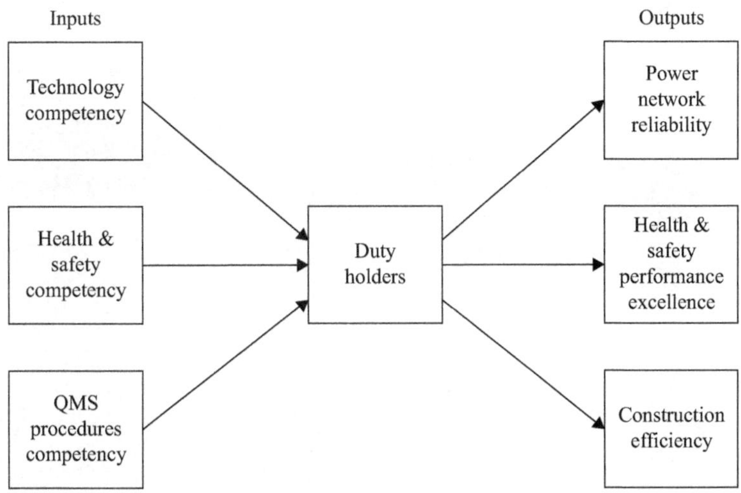

Figure 1.5 Competency input/output model

When considering competency and how it should be achieved, three prime factors need to be considered:

1. The content and level of competency necessary for the role to be undertaken.
2. How (1) above is tutored/lectured/provided.
3. How competency is to be assessed/demonstrated.

Competency will generally comprise a blend of a number of interlocking factors: academic qualifications relevant to the role to be undertaken; knowledge and practical application of both the technology and QMS procedures relevant to the role to be undertaken; and experience undertaking the role. A key decision is how (3) above is achieved – which in some critical instances may be through a formal assessment process. These considerations will be examined in more detail in Chapter 26.

Chapter 2
Legal and national/international standards

2.1 Introduction

Power network construction is undertaken in accordance with a wide range of legal and national/international standards requirements. The volume of documentation and the number of publishing organisations is extensive, and as such the requirement for adherence, and indeed which standards to use, may often be confusing. This chapter will identify some of the key documentation and explain its purpose and relevance. The main documentation categories are as follows:

- Legal
- National/international technical standards
- International organisation for standardisation (ISO) standards
- Publically available specifications (PAS)
- Occupational Health and Safety Assessment Series (OHSAS)
- Organisation-specific technical standards.

The term organisation specific as used above refers to power network companies, consultants, contractors and manufacturers. Figure 2.1 provides a documentation overview.

2.2 UK legal requirements

2.2.1 Key legal requirements

The primary legal documents relevant to power network construction comprise the following:

- UK Electricity Act 1989
- Health and Safety at Work etc. Act 1974
- Electricity Safety, Quality and Continuity Regulations (ESQCR) 2002
- Environmental regulations
- Planning acts.

NB: Although the above documents are specific to the United Kingdom, most have equivalents in other countries.

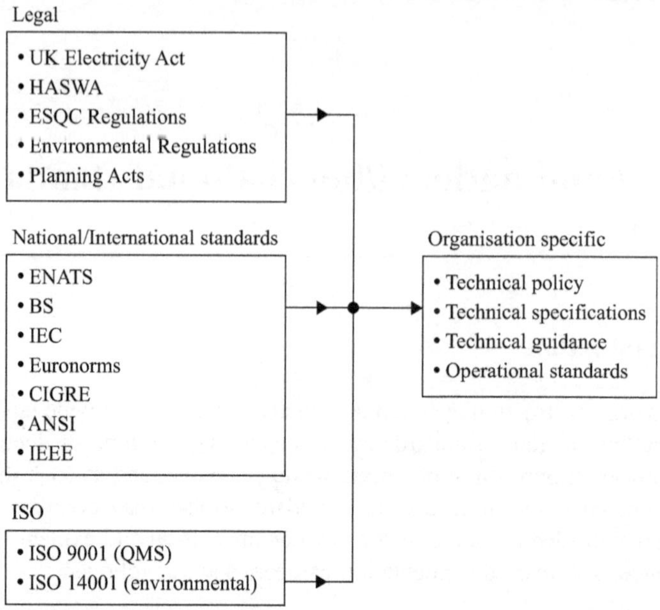

Figure 2.1 Legal and national/international standards documentation overview

2.2.2 UK Electricity Act 1989

The Electricity Act 1989 brought into existence the structural arrangements and requirements of the present UK electricity supply industry, including the granting of licences for transmitting and distributing electricity. The Act introduces the role and powers of a 'director of electricity supply', i.e. the regulator. The Act places duties on licence holders, for example:

1. Section 9 of the Act requires licence holders to develop and maintain an efficient, coordinated and economical system of electricity transmission and distribution.
2. Section 37 of the Act requires that the installation of an overhead line can only be undertaken with the consent of the UK Secretary of State.
3. Section 38 of the Act requires licence holders to:
   (i) Have regard to the desirability of preserving natural beauty, of conserving flora, fauna and geological or physiological features of special interest and of protecting buildings and objects of architectural, historic or archaeological interest.
   (ii) Do what is reasonable to mitigate any effect which his proposals would have on the natural beauty of the countryside or any such flora, fauna, features, sites, buildings or objects.

NB: National grid, the owner of the transmission licence, utilises the Holford and Horlock rules for delivering best practice in the siting and profile of overhead lines and substations, respectively, and these documents are equally used by many distribution companies, see Chapter 17.

2.2.3 The transmission and distribution licences

The transmission and distribution licences define obligations and rights of the companies to whom the licences have been awarded. For example, with reference to power network construction, the following is required:

- Preparation of, and compliance with, a grid code and distribution code for transmission and distribution licences, respectively.
- Compliance with defined security standards for planning and operation of the network.

2.2.4 The grid code

The grid code contains the minimum technical specification (TS) for the operation and development of the National Electricity Transmission System and the connection of equipment to that system. The grid code is currently under the stewardship of National Grid. It is to be noted that work is being undertaken on a European Network Code(s) and once developed it will become European Union (EU) Regulations and will have precedence over all electricity codes in individual EU member countries. The key contents of the grid code are summarised as follows:

1. **A planning code:** specifying technical and design criteria.
2. **Connection conditions:** specifying the minimum technical, design and operational criteria which must be complied with. It stipulates the standard levels of frequency and voltage and sets out criteria for quality and security of supply. It additionally stipulates fault clearance times.
3. **Compliance process:** which must be followed by any user (customer) to demonstrate compliance with the grid code in relation to new plant/apparatus being energised.
4. **Operating code:** defining requirements for safe operation of the transmission system.
5. **Balancing code:** comprising the balancing mechanism and the method of making and accepting offers for new connections, to ensure the electricity system remains balanced between supply and demand.
6. **Data registration code:** the data required from each user for grid technical performance evaluation.

2.2.5 The distribution code

The distribution code serves a similar purpose to the grid code for the 132 kV and lower voltage electrical networks. It is under the joint stewardship of each distribution company. It comprises the following:

1. **Distribution planning and connection code:** specifying technical and design criteria.

2. **Distribution operating code:** the provision of loading and generation output information by the users for purposes of evaluating system security and stability.
3. **Distribution data registration code:** i.e. data required from each user for assessment of distribution network technical performance.

2.3 UK health and safety legal requirements

2.3.1 The Health and Safety at Work etc. Act 1974

The United Kingdom in common with many other countries has a very high commitment to the health and safety of people at work. In the United Kingdom, the legal safeguards are enshrined in the Health and Safety at Work etc. Act (HASWA) 1974. The HASWA satisfies EU Regulations. The HASWA is referred to as an 'enabling' act in as much as it enables supporting regulations, codes of practice and guidance notes, the purpose of which is summarised as follows:

1. **Regulations:** These are laws made under the HASWA and enlarge upon the requirements of the HASWA in specific areas.
2. **Approved codes of practice (ACOP):** These offer practical examples of good practice and how to comply with the law. Failure to adhere to an ACOP or to discharge work to a similar standard can result in court of law finding the duty holder at fault. ACOPs are frequently concerned with specific regulations.
3. **Guidance:** The purpose of guidance is to interpret what the legal documents say, to help people comply with the law and to provide technical advice. Guidance is not mandatory – but failure to follow guidance will normally not be sufficient to comply with the law.

2.3.2 Health and safety documentation relevant to construction

With reference to power network construction, some of the main health and safety documentation comprises the following (the list is not exhaustive).

- Health and Safety at Work etc. Act 1974
- Construction (Design and Management Regulations) 2015
- Electricity at Work Regulations 1989
- Management of Health and Safety at Work Regulations 1999
- Control of Substances Hazardous to Health Regulations 2002
- Health and Safety (First Aid) Regulations 1981
- Pressure System Regulations 2000
- Manual Handling Regulations 1992
- Work at Height Regulations 2005
- Lifting Operations and Lifting Equipment Regulations 1998
- Personal Protective Equipment at work Regulations 1992
- Reporting of Injuries, Diseases and Dangerous Occurrences Regulations 2013
- HSG65 Managing for Health and Safety
- Managing Health and Safety in Construction (CDM 2015)
- ACOP L5 Control of Substances Hazardous to Health

- ACOP L22 Pressure Systems Safety Regulations
- HSG65 Managing for Health and Safety
- HSG47 Avoidance of Danger From Underground Services
- HSG85 Electricity at Work: Safe Working Practices
- L25 Personal Protective Equipment at Work
- HSG168 Fire Safety in Construction Work
- HSG150 Health and Safety in Construction
- L74 First aid at Work

2.3.3 Health and safety executive

Both the composition of the health and safety executive (HSE) and its function are defined in the HASWA. The HSE is a UK-independent watchdog for work-related health and safety and acts in the public interest to reduce work-related injuries.

The HSE can visit any workplace at any time to carry out health and safety inspections. They can investigate following a report of an injury or suspected unsafe working practices which may breach the HSAWA/supporting regulations. If a HSE inspector considers that health and safety law has been broken, or work activities give rise to serious risks, the options, depending upon the gravity of the situation, are as follows:

- Issue an informal warning.
- Issue an improvement notice or prohibition notice.
- Prosecute the company or individual.

2.3.4 HASWA – duties

The Act covers all people at work (with some exceptions) whether they be employers, employees or self-employed. It is aimed at people and their activities rather than premises and processes. Specific duties, with an emphasis on construction, are summarised as follows:

1. **Section 2: General duties of employers to their employees**
   This section is concerned with employers ensuring the health, safety and welfare at work of their employees. In particular, it requires the 'provision and maintenance of plant and systems of work that are, so far as reasonably practical, safe and without risks to health', a relevant requirement for a construction site. It also requires employers to provide 'information', 'instruction training' and 'supervision' – and that places of work are maintained 'in a condition that is safe and without risk to health'. Furthermore, employers must maintain 'working environments' that are 'safe, without risk to health, and adequate as far as arrangement for their welfare is concerned'.
2. **Section 3: General duties of employers and the self-employed to people other than their employees**
   This section is concerned with people who may be impacted upon by work carried out by the duty holders. It requires the duty holders to not

expose such people to 'risks to health and safety'. It also requires the duty holders to provide 'information' on the way their work might 'affect health and safety'.

3. **Section 4: General duties of people concerned with premises to people other than their employees**
   Section 4 is concerned with people who are in 'control of premises' and the requirement to provide a healthy and safe work environment to people other than their employees. It is particularly relevant to the instance of the duty of care owed to a contractor, working on a construction site which is under the health and safety control of another party.
4. **Section 5: Not discussed**
5. **Section 6: General duties of manufacturers and so on regarding articles and substances for use at work**
   This section is highly relevant to construction work. It specifies the duty of 'any person who designs, manufactures, imports or supplies articles for use at work'. In particular that the articles are 'designed and constructed to be safe and without risk to health', it also requires the articles to be tested, examined and accompanied by 'adequate information'. Very importantly, this section requires 'any person who erects or installs any article' to ensure 'that nothing about the way it is erected or installed makes it unsafe or a risk to health'.
6. **Section 7: General duties of employees at work**
   Section 7 places an obligation on employees to take 'reasonable care for the health and safety of themselves and others who may be affected by their acts'.
7. **Sections 8 and 9: Not discussed.**

The above-mentioned points summarise some of the key duties contained in the Act which have relevance to construction work in general. It is to be noted that some of the above requirements contain the caveat 'as is reasonably practical'. A full and complete understanding must be via a comprehensive examination of the Act itself supported by expert opinion. The requirements of the Act must be factored into TSs, quality management system (QMS) procedures and competency evaluation.

2.4 UK Electricity Safety, Quality and Continuity Regulations 2002/2009

2.4.1 Purpose of the regulations

The ESQCR became law in 2002 and subject to an amendment in 2009. They replaced the Electricity Supply Regulations 1988/1998. The regulations, initially under the stewardship of the Department of Trade and Industry (DTI), are now under the HSE.

The regulations are aimed at protecting the general public and consumers from danger. In addition, the regulations specify power quality and supply continuity requirements to achieve an efficient and economic electricity supply service for consumers.

2.4.2 The ESQC regulations

The following summarises some of the key requirements of the regulations relevant to power network construction which must be factored into UK technical standards and specifications.

1. **Part I: Introductory**
   The specified duty holders are generators and distributors (which includes transmission), and in some instances meter operators. The section on 'general adequacy of electrical equipment' requires duty holders to:
   (i) ensure metering equipment is sufficient for its purpose and constructed, installed, protected, used and maintained to prevent danger.
   (ii) assess overhead lines and substations for risks of danger from interference, vandalism and unauthorised access, to classify the degree of risk and record in a register and take safeguard measures commensurate with the risk.
   There is also a section on 'duty of cooperation' requiring duty holders and suppliers to disclose information and cooperate amongst themselves to ensure compliance with the regulations. There is also a section on 'inspection of networks' requiring a duty holder to inspect the network with sufficient frequency, so he is aware of the required action to ensure compliance with the regulations.
2. **Part II: Protection and earthing**
   This part commences with a section on the requirements of electrical protection, which states:

   'a generator or distributor shall be responsible for the application of such protective devices to his network as will, as far as reasonably practical, prevent any current, including any leakage current, from flowing in any part of his network for such a period that that part of his network can no longer carry that current without danger'.

   It is worthy of note that the above does not specifically stipulate that electrical protection shall operate quickly enough to avoid electrocution should a person come into contact the electrical network.
   The section on earthing generally comprises the following:
   (i) requirement for continuity of supply neutral conductors
   (ii) that high-voltage networks shall be connected with earth at or as near as practicable to the source of voltage
   (iii) requirements relating to protective multiple earthing
   (iv) requirements to earth metal work which encloses/supports equipment.
3. **Part III: Substations**
   This part is primarily concerned with requirements for enclosing substations, together with signs and notices warning of danger, and the name and telephone number of the duty holder responsible for the substation. A specific requirement for an air-insulated substation is a surrounding fence or wall not less than 2.4 m in height.

4. **Part IV: Underground cables**
   This part specifies the requirement for cable metallic screens to be connected to earth, and for cable metallic protection. It also specifies the requirement to keep the cable at a depth to prevent danger, and the need for cable protection and marking. It additionally specifies the requirement for up-to-date cable maps showing location and depth.
5. **Part V: Overhead lines**
   The part on overhead lines commences with the mandatory height of overhead lines as defined in Figure 2.2.
6. **Part VI: Generation**
   Not discussed.
7. **Part VII: Supplies to installations and to other networks**
   Two relevant areas from Part VII are as follows:
   (i) **Protection against supply failure:** This requires a distributor to ensure that his network is so arranged and so provided, where necessary with fuses or automatic switching devices, appropriately located and set, as to restrict, so far as reasonably practical, the number of consumers affected by any fault on his network. In addition, a distributor shall at all times take all reasonable practical steps to avoid interruption of supply resulting from his own acts (there are some caveats to this requirement).
   (ii) **Declaration of phases, frequency and voltage at supply terminals:** The regulations specify the following (unless otherwise agreed in writing):
      (a) the frequency of the network shall be 50 Hz with a permitted variation either above or below 1%.
      (b) the supply voltage of the network at the supply terminals shall be 230 V between phase and neutral, with a permitted variation of 10% above or 6% below.
      (c) a high-voltage supply operating at a voltage below 132 kV shall have a permitted variation of 6% above or below.
      (d) a high-voltage supply operating at a voltage of 132 kV or above shall have a permitted variation of 10% above or below.

Nominal voltage	Over road	Other locations
Up to 33 kV	5.8 m	5.2 m
Above 33 kV and up to 66 kV	6.0 m	6.0 m
Above 66 kV and up to 132 kV	6.7 m	6.7 m
Above 132 kV and up to 275 kV	7.0 m	7.0 m
Above 275 kV and up to 400 kV	7.3 m	7.3 m

Figure 2.2 Minimum height above ground of overhead lines

8. **Part VIII: Miscellaneous**
   This part is mostly about notifications to the secretary of state in the event of the following:
   (i) Injury, death, fire explosion, etc.
   (ii) Major interruption of supplies.

2.4.3 Application of ESQC regulations

The ESQC regulations comprise the top level of electrical network requirements imposed upon duty holders which must be factored into all related lower level standards. They are particularly significant to the activities of generators and distributors. The above has summarised key requirements relevant to the subject matter of this text. A complete understanding of requirements must again be through comprehensive examination of the regulations themselves.

2.5 UK environmental legislation

2.5.1 Environmental legislation categories

There is a very significant raft of environmental legislation relevant to power network construction. Unlike health and safety law, there is no over-arching document comparable to the HSAW Act. In general, the environmental legislation can be divided into seven categories which are as follows:

- **Waste:** The removal of hazardous and non-hazardous waste, excavated material and demolished equipment, etc.
- **Water:** This includes water pollution, site drainage, culverts and waterway requirements, etc.
- **Air:** Encompassing air pollution from emissions, vehicle discharges, greenhouse gas emissions (e.g. sulphur hexafluoride used in switchgear).
- **Land:** This comprises contaminated land and land pollution, rights of way, hedgerows and trees, heritage, ancient monuments and archaeological, etc.
- **Biodiversity:** Including ecological areas, wildlife protection, protected species, conservation of habitats, invasive plants and nature reserves, etc.
- **Nuisance:** Involving control of noise, road vehicles, control of dust, debris and maintenance of clean neighbourhoods, etc.
- **Other:** This covers a range of legislation which does not fall into the other categories – and mostly not applicable to construction.

2.5.2 Range of environmental legislation

The range of environmental legislation is huge and equals that of health and safety legislation. At first sight, it may seem a bewildering amount. Much is common with EU and international requirements. The following provides a list of some of the relevant UK legislation for purposes of providing an insight into the subject matter:

1. **Waste**
   - (i) Control Waste Regulations 1992/2012
   - (ii) Hazardous Waste Regulations 2005
   - (iii) Site Waste Management Plan Regulations 2007
   - (iv) Environmental Permitting Regulations 2007.
2. **Water**
   - (i) Groundwater Regulations 2009
   - (ii) Land Drainage Act 1991
   - (iii) Control of Pollution (Oil storage) Regulations 2001
   - (iv) Environmental Permitting Regulations 2010.
3. **Air**
   - (i) Road Traffic (vehicle emissions) Regulations 1997
   - (ii) Fluorinated Greenhouse Gas Regulations 2009
   - (iii) Environmental Protection Regulations 2011
   - (iv) Clean Air Act 1993
   - (v) Pollution Prevention and Control Act 1999.
4. **Land**
   - (i) Contaminated Land Regulations 2006
   - (ii) Building Regulations 2010
   - (iii) Energy Performance of Building Regulations 2007
   - (iv) Ancient Monuments and Archaeology Areas Act 1979
   - (v) Hedgerows Regulations 1997
   - (vi) Environmental Protection Act 1990
   - (vii) Town and Country Planning (environmental impact assessment) Regulations 1999.
5. **Biodiversity**
   - (i) Wildlife and Countryside Act 1971
   - (ii) Conservation of Habitats and Species Regulations 2010
   - (iii) Wild Mammals Protection Act 1996.
6. **Nuisance**
   - (i) Environmental Protection Act 1990
   - (ii) Road Traffic Act 1998
   - (iii) Noise and Statutory Nuisance Act
   - (iv) Control of Noise at Work Regulations 2005
   - (v) Road Vehicles (control and use) Regulations 1988
   - (vi) Clean Neighbourhood and Environment Act 2005.
7. **Other**
   - (i) None relevant.

2.5.3 UK Environment Agency

The UK Environment Agency (EA) in many ways parallels the HSE – but considers environmental legislation. In Scotland, the equivalent is the Scottish

Environment Protection Agency. The EA has responsibilities relating to the protection and enhancement of the environment. It is in effect the environmental watchdog.

The EA oversees the enactment of legislation on the environment. The EA also regulates a broad range of activities to minimise impact on the environment through granting of permits and licences – some of which are required during construction, e.g. waste disposal.

When an environmental breach has been committed and the delivery of advice and guidance has not, or will not, achieve a satisfactory outcome, the EA will progress to a form of sanction. Such sanctions may consist of the following:

1. Issue of a warning
2. Statutory enforcement notices
3. Prohibition notices
4. Fixed penalty notices
5. In the ultimate – prosecution.

Power network construction QMS systems must ensure the inclusion of environmental legislation – not only to satisfy legal requirements but as a measure of good practice.

2.6 UK planning acts

2.6.1 Planning Act 2008

The UK Planning Act 2008 is intended to speed up the process for approaching National Significant Infrastructure Projects such as those associated with major power network construction. The Act has created a new infrastructure planning unit with planning inspectors who will administer the process and recommend to the secretary of state (who will make the final decision), whether to grant/refuse a development consent to a developer. The process is required to follow a 12–15-month timescale as follows:

1. **Pre-application:** Developer carries out significant consultation and submits a development consent order (DCO) to the planning inspectorate.
2. **Acceptance:** Planning inspectorate has 28 days to determine whether the DCO application meets the requirements to proceed.
3. **Pre-examination:** Public can submit their views in writing.
4. **Examination:** Hearings undertaken by planning inspectorate to consider evidence submitted – this should be completed within 6 months.
5. **Decision:** Inspectorate considers evidence and, assisted by National Policy Statements, one of which is on power networks, has 3 months to prepare a report to the secretary of state recommending whether DCO should be granted or refused. Secretary of state has a further 3 months to accept or reject the recommendation.
6. **Post-decision legal challenge:** There is a 6-week period in which the decision can be challenged.

It is worthy of note that criterion for satisfying the Planning Act is specified in the Policy Statement on Electricity Network Infrastructure (EN-5). Major power network construction projects, particularly relating to electricity transmission infrastructure, i.e. concerning overhead lines, cables, substations, HV convertor stations, etc., must therefore be undertaken in accordance with the requirements of the 2008 Planning Act.

2.6.2 Town and Country Planning Order 1995

The Town and Country Planning Order 1995 grants planning permission for certain types of development. Schedule 2 of the order specifies the classes of development – which includes power network substations. Such development is referred to as 'permitted development'. Similar arrangements apply in Scotland. Thus, substation development in the United Kingdom cannot proceed without this requirement.

2.7 National/international technical standards

2.7.1 Relevant technical standards

Power network companies, consultants, contractors and manufacturers usually have a suite of internal technical policies, guidance notes and specifications that reference a wide range of national/international standards. These standards are very numerous which often causes ambiguity about their purpose and relative standing. This section will therefore identify: the main standards relevant to power network construction; that which the standards seek to achieve; and which organisations publish the standards. The following standards will be examined:

- Electricity Networks Association Technical Specifications (ENATS)
- British Standards (BS)
- International Electrotechnical Commission (IEC) standards
- European Committee for Electrotechnical Standardisation (CENELEC) standards
- Euronorm Standards
- International Council on Large Electric Systems (CIGRE) standards
- American National Standards Institute (ANSI) standards
- Institution of Electrical and Electronic Engineers (IEEE) standards.

2.7.2 Definition of a standard

Contemporary literature contains numerous definitions of a 'standard' albeit with a common theme. One typical definition is as follows:

'A standard is a published document that contains clear, precise criteria designed to be used consistently as a rule, guideline or definition'.

Standards are generally acknowledged as best practice compiled by experts in the field and as such drive product excellence and productivity. Often the terms 'standard' and 'code of practice' (COP) are used interchangeably – but some literature draws a distinction. This distinction defines two types of standard: a specification and a COP. A specification is taken to be a coherent set of absolute requirements for achieving specific outcomes which represent currently accepted good practice. In contrast, a COP is considered to be guidance and recommendations that reflect good practice. As such, a COP is less rigorous than a specification. In the following sections, the term 'standard' will generally be taken as meaning 'specification'.

2.7.3 Electricity networks association technical specifications

The UK Energy Networks Association (ENA) prepares and maintains a range of documents including the following: UK electrical industry developed technical specifications (TSs); engineering recommendations; and engineering technical reports. The TSs are largely applicable at distribution voltages, i.e. 132 kV and below in the United Kingdom. For example:

- ENATS 41-37 Part 2: Gas insulated switchgear (GIS) for use on 66 to 132 kV distribution systems
- ENATS 41-15 Part 1: Standard circuit diagrams for equipment in 132 kV substations (transformer feeder, transformer end).

The distribution network design in the United Kingdom is very much based on ENA TSs.

2.7.4 British Standards

British Standards have a long history dating back to 1901. British Standards are produced by the BSI Group which is recognised as the UK national standards body by the UK government. A BS is not a government regulation under the law, but compliance with a BS usually means compliance with relevant regulations, although other means may be equally acceptable without using a BS.

BSI is the UK member of the ISO and the 'Committee European de Normalisation' (CEN). CEN standards take precedence over national standards for those countries in the EU. CEN standards applicable in the United Kingdom would mostly be published through BSI and referenced as BS/ISO. Similarly, BSI represents the UK interests on the IEC and IEC standards available in the United Kingdom would mostly be published through BSI and referenced as BS/IEC.

Many British Standards are relevant to electrical power networks. For example:

- BS7354: Design of HV open terminal substations
- BS7671: Requirements for electrical installations (IET wiring regulations)
- BS/ISO/IEC 27001: Information technology.

2.7.5 International Electrotechnical Commission (IEC) standards

The IEC dates back to 1906 and was primarily brought about by the UK Institution of Electrical Engineers and the American Institute of Electrical Engineers. The IEC prepares and publishes international electrical standards. It is probably the leading international organisation in the electrical field. Today, the standards are prepared by experts of international standing drawn from across the world.

Power network construction in the United Kingdom and in many other countries draws heavily on IEC standards – the following of which are typical:

- IEC 60815: Specification of Insulators
- IEC 61936-1: Power installations exceeding 1 kV.

As previously noted, many of these standards are now published in the United Kingdom through BSI.

2.7.6 European Commission for Electrotechnical Standardisation (CENELEC) standards

CENELEC, although not an EU institution, is responsible for standardisation in the electrotechnical field. The commission has representation from across Europe and has cooperation agreements with similar bodies from across the world. From a power engineering perspective, CENELEC would mostly look to IEC for definition of standards.

2.7.7 Euronorm standards

The purpose of a Euronorm (EN) is to harmonise goods/services within the EU. EN standards have been ratified by a European Standards Organisation, e.g. CENELEC. Once ratified ENs take precedence over any national standards. With reference to power network construction, ENs greatest impact is probably concerned with structural design. For example:

- BS/EN 1990 (Eurocode 0): Basis of structural design.

2.7.8 International Council on Large Electric Systems (CIGRE) standards

Founded in 1921, CIGRE is an international association for promoting collaboration with experts from around the world to improve electrical power systems. CIGRE membership is open to individuals, companies or organisations. CIGRE publications fall within the general categories of standards, COP or discussion papers. A distinction between CIGRE and IEC is that CIGRE is more concerned with emerging practice, whereas IEC is concerned with specifications relating to current practice, as such CIGRE is a feeder organisation into IEC.

With reference to power network construction, CIGRE standards would usually be referenced in the absence of an IEC equivalent. For example:

- 283 working group B1.18: Special bonding of high-voltage power cables
- 44 working group 23.10: Earthing of GIS – an application guide.

2.7.9 American National Standards Institute (ANSI) standards

ANSI is an American (USA) body similar to BSI. However, unlike BSI, ANSI does not develop standards, but oversees the development of standards by other organisations. In particular, it promotes the use of American standards internationally – but at the same time interfaces and cooperates with other organisations such as IEC to provide international standards. ANSI standards are rarely referenced in UK power network engineering – but occasionally referenced by some manufacturers.

2.7.10 IEEE standards

The IEEE is an American standards organisation, combined with a professional organisation similar to the UK IET. It is not a formal body authorised by government – but produces highly respected standards with contributions from international experts. The IEEE liaises closely with IEC, and some standards have an IEC/IEEE dual logo. Use of IEEE standards for the UK power network is not common as the preference is for European standards.

2.8 International Organisation for Standardisation (ISO) standards

2.8.1 ISO – purpose

Founded in 1947, ISO is an international standards setting body comprising representatives from various national standards organisations e.g. BSI or IEC established for the purpose of harmonising international standards. Documents may typically be referenced BSI/ISO or ISO/IEC, etc.

Two ISO standards in particular have significant relevance to power network construction. These are as follows:

- BS EN ISO 9001: Quality management systems
- BS EN ISO 14001: Environmental management systems.

Many organisations, including those concerned with power network construction, will only work with other organisations that hold an ISO certification, for the above two standards. The award of certification follows independent evaluation and confirmation that the organisation in question meets the requirements of the above standards.

2.8.2 ISO 9001: Quality Management Systems

ISO 9001 specifies the requirements for a QMS, which may be defined as follows:

> A collection of business processes focused on achieving quality policy and quality objectives to meet customer requirements. It is expressed as the organisation's management, structure, policies and procedures needed to implement quality management.

A QMS system is focused on a continuous improvement cycle of 'plan/do/check/act' where act means to correct/improve. A QMS does not define neither the range of procedures to be prepared nor the precise content of those procedures – this is left to the judgement of the organisation in question.

An essential requirement for power network construction is a QMS (for each organisation participating in the work), comprising a comprehensive suite of procedures defining tasks, timing, responsibilities and the competency requirements for those duty holders who will discharge those responsibilities.

2.8.3 ISO 14001: Environmental Management Systems

ISO 14001 specifies the requirements for an environmental management system. It enables an organisation to develop and implement environmental policy and objectives which take into account legal requirements and information about significant environmental risks. The main requirements of this standard comprise:

- Environmental policy
- Planning
- Implementation including emergency response
- Checking performance
- Management review
- Continual improvement arrangements.

Similar to the QMS implementation, a procedure defining environmental requirements on a construction site must be prepared.

2.9 Publically available specification (PAS)

2.9.1 Publically available specification

A PAS is a standard document that closely resembles a formal standard such as a BS – but has a different consensus model particularly relating to content consensus. PASs are both issued with the agreement of, and published by, the BSI group. A PAS can be commissioned by any organisation who wishes to document and introduce best practice. PASs must neither contradict nor contravene any formal standards; they may have development time scales as short as 8 months, as such they are sometimes referred to as a fast track standard. Three PASs worthy of note are as follows:

- PAS 55 asset management (shortly to be ISO 5501)
- PAS 91 construction-related procurement
- PAS 99 specification of common management system requirements as a framework for integration.

The development of PAS 55 was heavily influenced by the UK power network companies and is relevant to the management of assets, i.e. post-construction.

PAS 91 sets out the content, format and use of questions that are applicable to pre-qualification for construction engineering. To be eligible for pre-qualification, it is necessary that suppliers demonstrate that they possess, and have access to, the governance, qualifications and references, expertise, competence and health and safety capabilities necessary for them to be considered competent to undertake the work. The means by which competency is demonstrated is by responding to a structured questionnaire. PAS 91 is very relevant to power network construction.

PAS 99 is the world's first specification for an integrated management system, i.e. a set of management procedures covering policies, procedures and processes, i.e. an umbrella approach encompassing such standards as: ISO 9001; ISO 14001; OHSAS 18001. PAS 99 is increasingly used by larger organisations.

2.10 Occupational Health and Safety Assessment Series

2.10.1 Occupational Health and Safety Assessment Series (OHSAS)

The OHSAS standards date from 1999 and were created through international collaboration to form a single unified approach to occupational health and safety management systems. UK BSI provides the secretariat. Two documents exist in the series, which are as follows:

- OHSAS 18001 – Occupational health and safety concerned with specifying the requirements for the system
- OHSAS 18002 – Guidelines for implementing OHSAS 18001.

Organisations are assessed and certified against these standards as recognition of a high standard of safety management. These standards are harmonised with ISO 9001 and ISO1401. The standards are based upon the requirements of HSG 65 (see Section 18.1) and the strategy of plan/do/check/act.

2.11 Organisation-specific technical standards

2.11.1 Range of technical documents

Organisations, i.e. mainly power network companies but also consultants, contractors and manufacturers, usually prepare their own suite of technical documents which fall under the umbrella of their QMS. These documents usually reference national/international standards to which may be added requirements particular to that organisation, and of course, they must embody relevant legal requirements. The range of documents usually includes the following:

1. **Technical policy**
   These are documents which are usually for internal use and stipulate an engineering decision or course of action, e.g. network planning standards or the criterion for constructing a GIS substation in preference to an AIS substation.

2. **Technical specifications (TS)**
   A TS is a procurement document (project generic design specification) for an item of equipment, e.g. a circuit breaker, HV cable, protection relay, etc.
3. **Technical guidance**
   This usually comprises supporting explanations/reasons behind a technical policy document or TS – or a model method of compliance when undertaking a specific task.
4. **Network operational standards**
   These define the requirements for operating the power network. This would include network outage planning, normal network running arrangements including routine switching, and network emergency response requirements. These are essentially in the domain of the power network company.

Part 1

Technology

Technology overview

This part of the book will examine the technology-related aspects of power network construction. The content will be broadly aligned to both the 'construction design specification', as outlined in Figure 1.3, and the 'technology stages', as specified in Section 1.4.1. The structure of the related chapters is as follows:

- Chapter 3 summarises the basic power-system theory which explains relevant aspects of both the power system, and associated electrical equipment. It underpins the technical content of most of the other chapters, especially Chapter 4 which examines the theory and practice of power system fault analysis.
- Chapter 5 examines design at an HV power network level, with much of the content having relevance to the preparation of a construction design specification.
- Chapters 6–13 are essentially concerned with design at the equipment (i.e. asset) level, including spatial/positioning requirements. It covers the following:
  - OHL design
  - Cable design
  - Substation design
  - Substation HV equipment design
  - Protection and control design
  - Earthing design
  - Civil, structural and building engineering design

  Much of the above form the basis of equipment (generic) technical specifications – which are used and referenced at all stages of construction.
- Chapter 11 is uniquely concerned with impressed voltages – which impacts both on personnel safety and equipment design. In the author's experience, this is a subject of critical importance, about which little is written, at least rarely in a fully holistic and practical way. As such an endeavour has been made to compile a joined up narrative, commencing with basic and fundamental concepts and progressing to site practical considerations. There is a very high and appropriate focus, in the United Kingdom in particular, on ensuring safety of personnel, in a work environment where IV is ever present.
- Chapters 14 and 15 examine the technical considerations relevant to the latter stages of construction, namely, detailed design, manufacture/procure, site installation and commissioning.

It is almost a self-evident observation to state that each of the technologies considered is worthy of a publication in its own right. As such, this text is targeted at providing the reader with a working understanding – covering the broad range of technical requirements and interlocking complexities – but with the additional objective of providing a 'spring board' for those who may require a more in depth understanding, which is readily provided and available in more specialist texts.

Chapter 3

Power system design fundamentals

3.1 Introduction

This chapter will examine fundamental aspects of electrical power system theory that form the basis of power system design and operation, and which in turn are essential to the preparation of a construction design specification. It will also briefly examine the fundamentals of the wider power system such as generator and high voltage direct current (HVDC) theory and operation. Although various parts of this chapter can be obtained from a wide range of contemporary technical publications, it is collated here to provide a concise and relevant reference source in a single publication for ease of assimilation. The following will be examined:

- Basic power system relationships
- Power system current flow analysis
- Transformer fundamentals
- Per cent impedance and fault level analysis
- Synchronous generator fundamentals
- Wind generation
- Power system transients
- HVDC transmission

3.2 Basic power system relationships

3.2.1 Polar notation

With reference to Figure 3.1:

$$V^* = a + jb = V\angle\Phi.$$

where V^* = vector quantity, V = modulus of V^* and $a = V\cos\Phi$, $b = V\sin\Phi$.

1. **Vector multiplication**
   Two vectors, I^* at an angle Φ and Z^* at an angle θ, may be multiplied as follows:

   $$I^*Z^* = I\angle\Phi \times Z\angle\theta = IZ\angle(\Phi + \theta)$$

   where I is the modulus of I^* and Z is the modulus of Z^*.
   This simple technique proves very useful in practice.

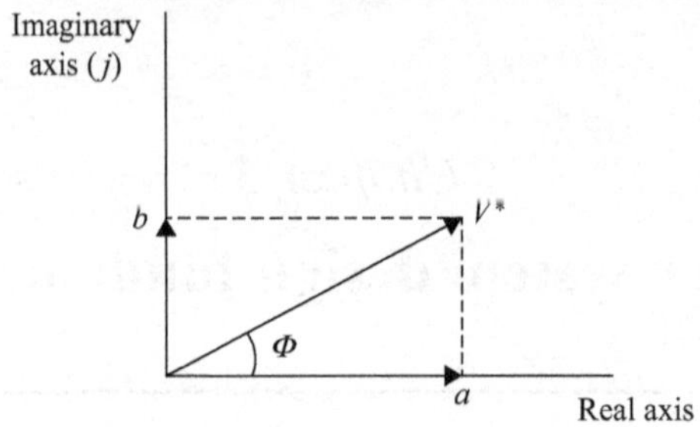

Figure 3.1 Polar notation

2. **Vector division**
   Similarly, two vectors, V^* at an angle Φ and Z^* at an angle θ, may be divided as follows:

$$\frac{V^*}{Z^*} = \frac{V}{Z}\angle(\Phi - \theta)$$

3.2.2 *The j operator*

With reference to Figure 3.2, the vector operator j has the following properties:

$$j = 1\angle 90^\circ$$
$$j^2 = 1\angle 180^\circ = -1$$
$$j^3 = 1\angle 270^\circ = 1\angle -90^\circ$$
$$j^4 = 1\angle 360^\circ = 1\angle 0^\circ.$$

3.2.3 *The* a *operator*

With reference to Figure 3.3, the operator a has the following properties:

$$a = 1\angle 120^\circ = -\frac{1}{2} + j\frac{\sqrt{3}}{2}$$
$$a^2 = 1\angle 240^\circ = -\frac{1}{2} - j\frac{\sqrt{3}}{2}$$
$$a^3 = 1\angle 0^\circ$$

and $1 + a + a^2 = 0$.

3.2.4 *Power flow relationships*

With reference to Figure 3.4:

$$\text{Power flow (real)} = \sqrt{3}\, V_L I_L \cos\, \Phi\, \text{W} \tag{3.1}$$

$$\text{Power flow (reactive)} = \sqrt{3}\, V_L I_L \sin\, \Phi\, \text{VAr} \tag{3.2}$$

$$\text{Power flow (total)} = \sqrt{3}\, V_L I_L\, \text{VA} \tag{3.3}$$

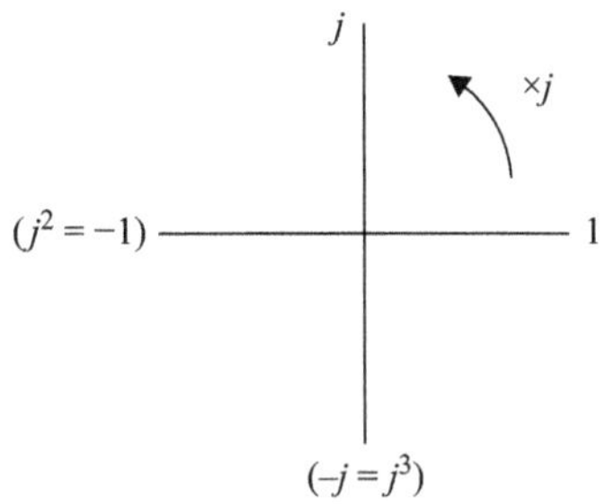

Figure 3.2 j Operator

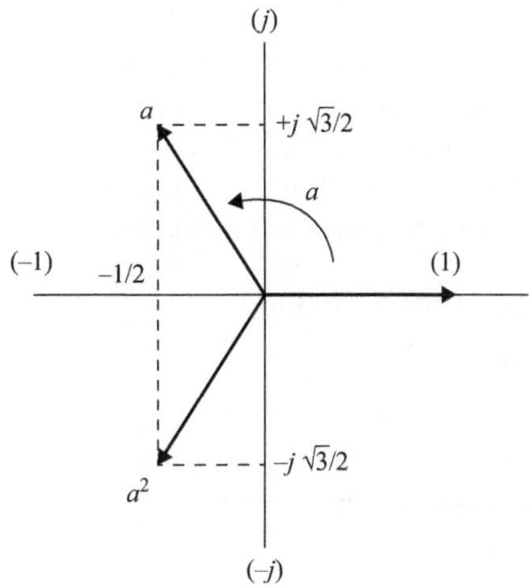

Figure 3.3 a Operator

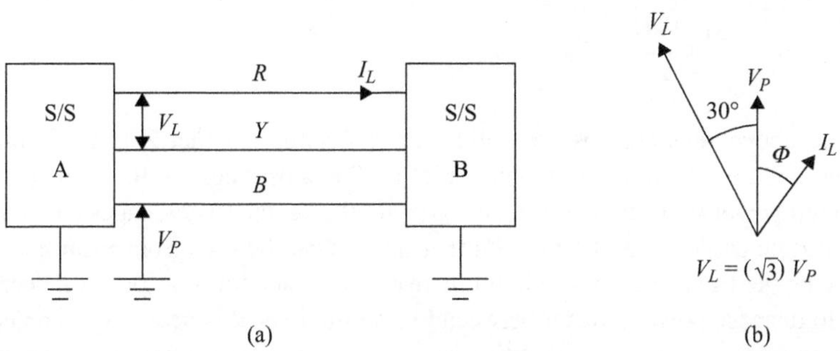

Figure 3.4 Power relationships: (a) power network and (b) vector diagram

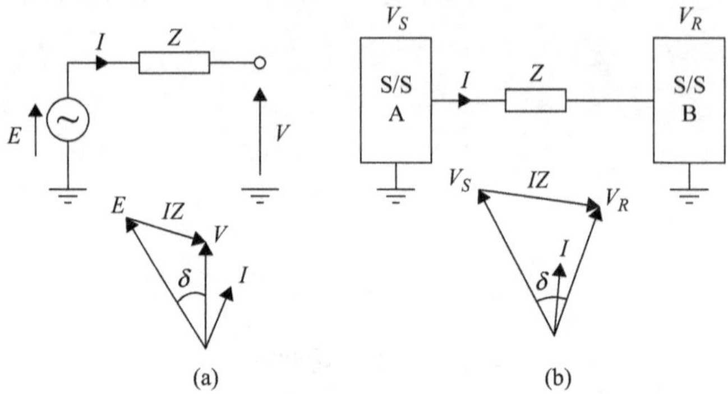

Figure 3.5 Power transfer relationships: (a) generator and (b) power network

Rule of thumb MVA to amperes conversion is provided as follows:

- At 11 kV – 1 MVA corresponds to 52.5 A/phase
- At 33 kV – 1 MVA corresponds to 17.5 A/phase
- At 132 kV – 1 MVA corresponds to 4.4 A/phase
- At 275 kV – 1 MVA corresponds to 2.1 A/phase
- At 400 kV – 1 MVA corresponds to 1.4 A/phase.

3.2.5 Power transfer relationship

With reference to Figure 3.5:

1. **Generator**
   With reference to Figure 3.5(a), the power flow from the generator to the rest of the power system is given by:

$$P = \frac{3EV \sin \delta}{Z} \text{W} \tag{3.4}$$

2. **Power network**
   With reference to Figure 3.5(b), the power flow from Substation A to Substation B is given by:

$$P = \frac{3V_S V_R \sin \delta}{Z} \text{W} \tag{3.5}$$

The above relationships are valid on the assumption that Z is predominately reactive. Figure 3.6 shows a graphical plot of P against angle δ. It can be seen that maximum power transfer occurs at an angle of 90°, i.e. the two voltages either side of Z are at right angles to each other. Furthermore, it can be seen from expression (3.5) that the power transfer across a circuit increases as the circuit impedance Z decreases. Thus, to increase power transfer between two substations, it is necessary to reduce the circuit impedance, e.g. by constructing parallel circuits or installing series capacitance compensation – a frequent requirement. Z is usually termed the 'transfer impedance'.

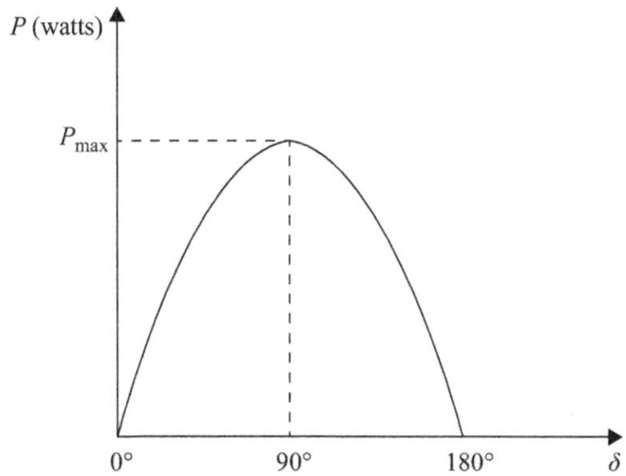

Figure 3.6 Power angle relationship

3.3 Power system current flow analysis

3.3.1 Network current flow analysis methods

There is a frequent requirement during the design process to determine levels of current flow – particularly fault current. Five methods for evaluating circuit currents will be briefly examined. These comprise:

- The Kirchhoff method
- The Maxwell mesh current method
- The superposition method
- The Thevenin method
- The extended superposition method.

Although network current flows can be determined by application of computer-based packages – there is no substitute both for understanding the underpinning theory and the competence to personally undertake the calculations.

3.3.1.1 The Kirchhoff method

The Kirchhoff method is based upon two laws: the current and the voltage.

1. **The current law**
   The current law states that the algebraic sum of the currents at a network junction is zero. Therefore, with reference to Figure 3.7(a):

   $$I_A + I_B + I_C + I_D = 0 \tag{3.6}$$

   where the currents above are vector quantities.
2. **The voltage law**
   The voltage law states that around any closed mesh, the algebraic sum of the emfs (electromotive force) is equal to the algebraic sum of the voltage drops.

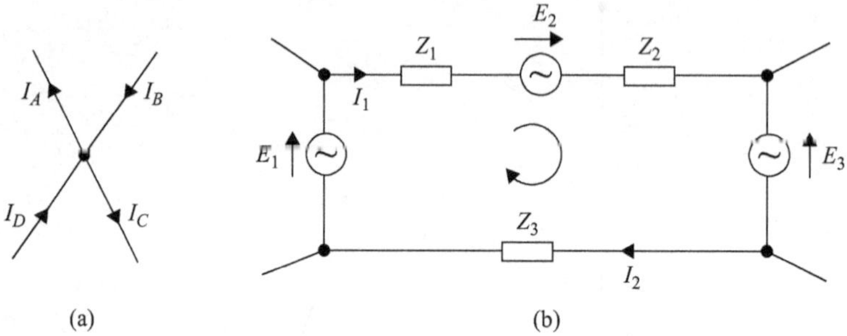

Figure 3.7 Kirchhoff's laws: (a) current law and (b) voltage law

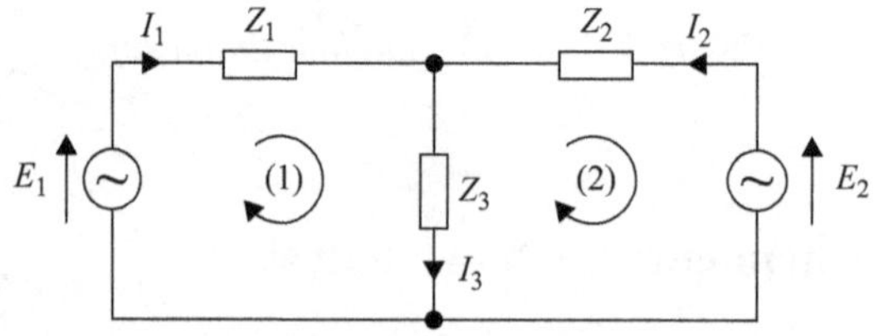

Figure 3.8 Kirchhoff method – example

Therefore, with reference to Figure 3.7(b), and arbitrarily taking a clockwise direction around the mesh:

$$E_1 + E_2 - E_3 = I_1(Z_1 + Z_2) + I_2 Z_3 \tag{3.7}$$

where the parameters above are vector quantities.

With reference to the simple network illustrated in Figure 3.8 (where the values of generated voltage and impedance are known), the Kirchhoff technique for determining the currents flowing within the network is to assign currents I_1, I_2 and I_3. The currents may be chosen in any arbitrary direction, and the mathematics will subsequently confirm a positive current (assumed current direction correct) or a negative current (current in reverse direction to that assumed). The following relationships apply:

Mesh (1)	$E_1 = I_1 Z_1 + I_3 Z_3$	(voltage law)	(1)
Mesh (2)	$-E_2 = -I_2 Z_2 - I_3 Z_3$	(voltage law)	(2)
Current junction	$I_1 + I_2 - I_3 = 0$	(current law)	(3)

From the above three simultaneous equations, and by process of substitution and elimination, the three currents I_1, I_2 and I_3 can be determined.

3.3.1.2 The Maxwell mesh current method

The Maxwell method employs Kirchhoff's laws but analyses the network from the perspective of mesh currents, i.e. a current of constant magnitude which circulates

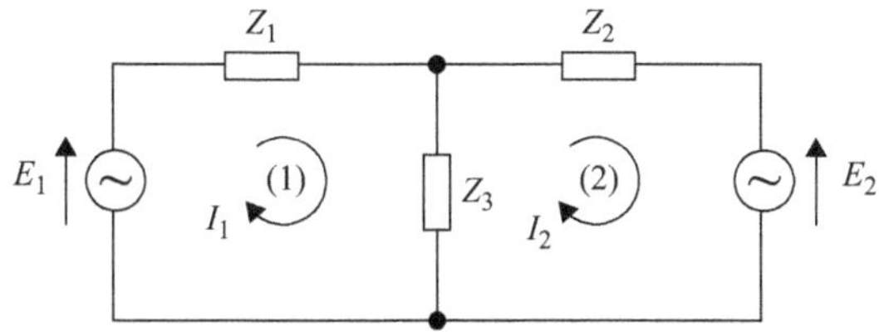

Figure 3.9 Maxwell method – example

around the mesh. This method can often reduce the number of equations to be solved. With reference to Figure 3.9 (where the values of generated voltage and impedance are known), the method for determining the current in each part of the network is as follows:

$$\text{Mesh 1} \quad E_1 = I_1(Z_1 + Z_3) - I_2(Z_3) \tag{1}$$

$$\text{Mesh 2} \quad -E_2 = I_2(Z_2 + Z_3) - I_1(Z_3) \tag{2}$$

From the above two simultaneous equations, the mesh currents I_1 and I_2, which are the currents through impedances Z_1 and Z_2, respectively, can be determined (via the process of substitute and elimination). The current through impedance Z_3 is given by $(I_1 - I_2)$.

3.3.1.3 The superposition method

The superposition method determines network currents by considering the currents that would arise due to each emf acting alone, and in turn, with the other emfs being supressed and replaced by their internal impedance. The resulting current is the integration (superposition) of each of the currents arising from the application of the individual emfs. This method is advantageous when analysing large networks. The currents in each limb of the network are determined by network reduction and back substitution, so avoiding the complexity of simultaneous equations, as with the Kirchhoff and Maxwell methods. Figure 3.10 illustrates the method.

- Figure 3.10(a) – for the network shown, currents I_1, I_2 and I_3 are required to be determined.
- Figure 3.10(b) – Step 1 is to supress E_B and determine the network currents due to E_A alone.
- Figure 3.10(c) – Step 2 is to supress E_A and determine the network currents due to E_B alone.

Step 3 is to determine the resulting currents by superposition, which are as follows:

$$I_1 = I_{1A} - I_{1B}$$

$$I_2 = I_{2B} - I_{2A}$$

$$I_3 = I_{3A} + I_{3B}$$

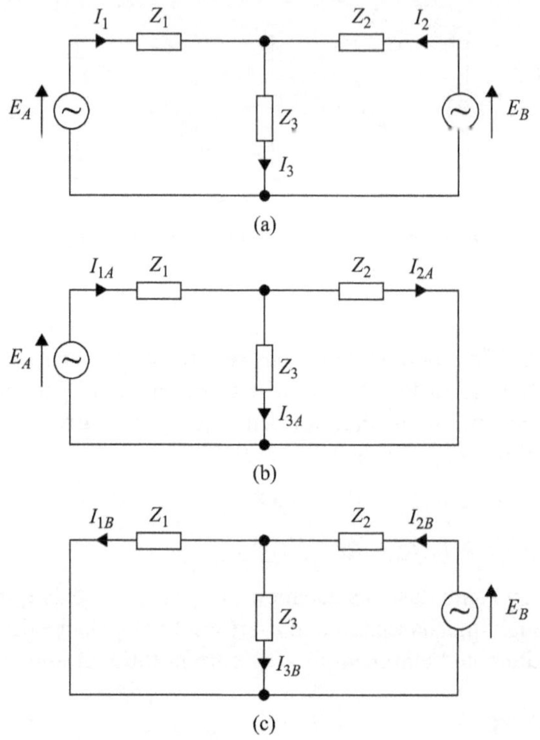

Figure 3.10 Superposition method – example: (a) network to be analysed, (b) step 1 = to supress E_B and (c) step 2 = to supress E_A

3.3.1.4 The Thevenin method

The Thevenin method determines the current in a single branch of a network and simplifies the analysis if only the current in the single branch is required – for example, the determination of current flowing in a fault. Figure 3.11 illustrates the method.

- Figure 3.11(a) – for the network shown, the current through impedance Z_3 is required to be determined.
- Figure 3.11(b) – Step 1: is to open circuit the branch whose current is to be determined, i.e. the branch containing impedance Z_3 and subsequently determine the open-circuit voltage V.
- Figure 3.11(c) – Step 2: is to connect the open-circuit voltage V in series with impedance Z_3 whilst supressing voltages E_1 and E_2. The current I_3 can then be determined as shown. It is necessary to reverse the polarity of V to obtain the correct direction of current flow.

3.3.1.5 The extended superposition method

As stated in Section 3.3.1.4, the Thevenin method only determines the current in the branch of the network under consideration. To obtain the current in the whole

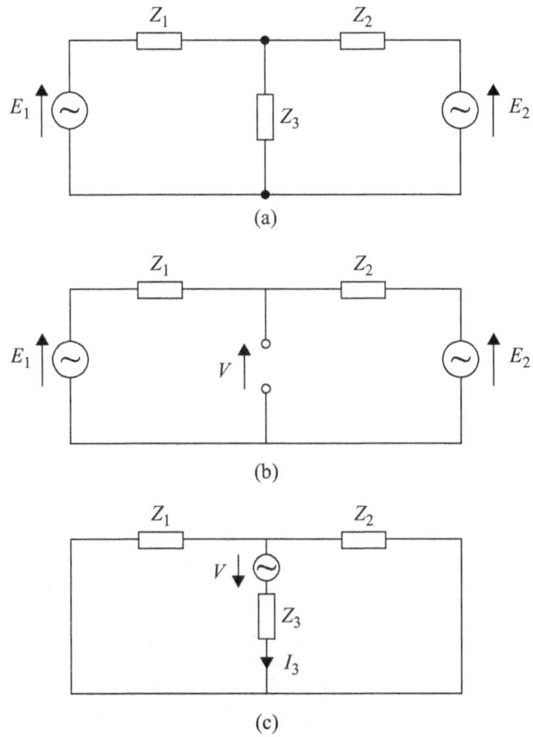

Figure 3.11 Thevenin method – example (a) network to be analysed, (b) determine the open-circuit voltage and (c) Thevenin equivalent circuit

network, a combination of Thevenin and the superposition method must be used – sometimes termed the extended superposition method. The technique is as follows:

- Figure 3.12(a) – a short-circuit fault takes place on the network and the current both in the short circuit and the rest of the network is required to be determined.
- Figure 3.12(b) – Step 1: is to open circuit the branch of the network subject to the short circuit to determine the Thevenin voltage V. The current flowing is equivalent to network load current, before the short circuit arises.
- Figure 3.12(c) – Step 2: using the Thevenin equivalent circuit determine the current flow in the short circuit and that around the remainder of the network.
- Figure 3.12(d) – Step 3: the resulting current in the short circuit and the rest of the network is obtained by integrating (superimposing) the currents determined in Steps 1 and 2.

It is worthy of note that with reference to all the above methods, engineers much prefer to determine network current flows using network reduction techniques (i.e. circuitry) rather than simultaneous equations. As such the concept of a single equivalent generator is very useful, see Section 4.3.6(5).

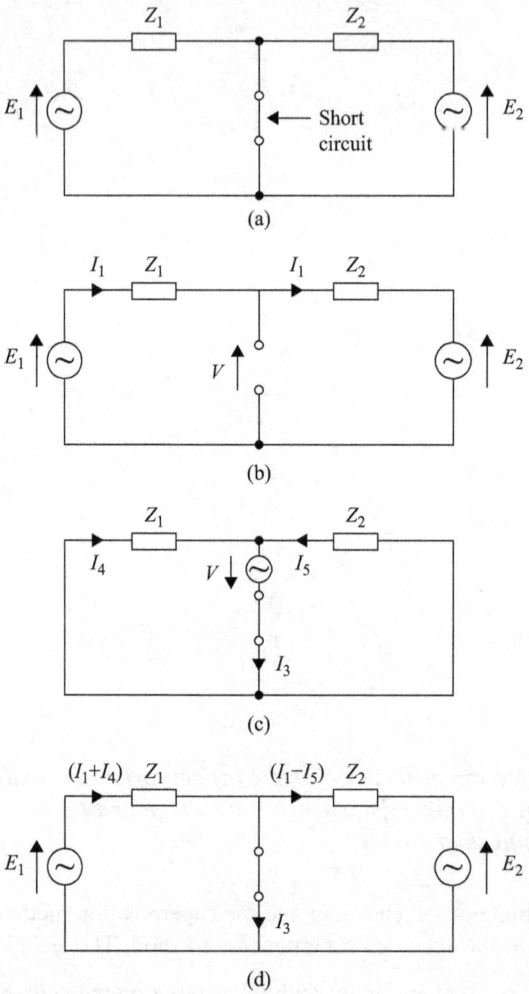

Figure 3.12 Extended superposition method: (a) network to be analysed, (b) obtain open-circuit voltage and current, (c) Thevenin equivalent circuit and (d) superimpose currents

3.4 Transformer fundamentals

3.4.1 Two-winding transformer – principles

With reference to Figures 3.13 and 3.14, essential features of a two-winding transformer are as follows:

- The windings comprise a primary winding (usually the higher voltage side) with turns of N_p and a secondary winding with turns of N_S.

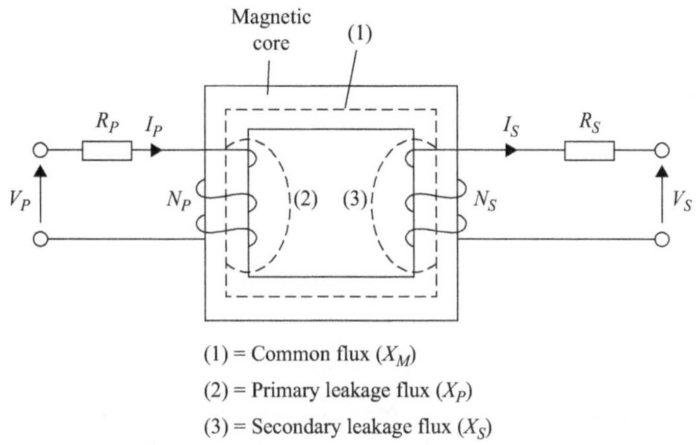

Figure 3.13 Transformer operation – basics

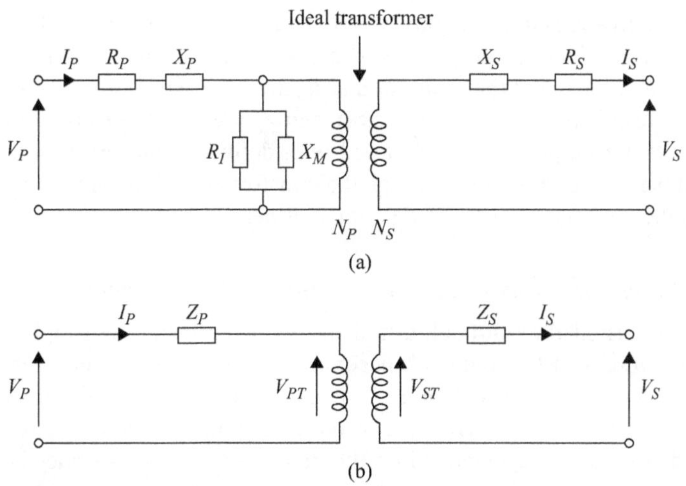

Figure 3.14 Transformer equivalent circuit: (a) equivalent circuit and (b) simplified equivalent circuit

- The core of the transformer contains three distinct areas of magnetic flux, which are as follows:
  - Flux common to the primary and secondary windings, represented by the magnetising impedance X_M.
  - The primary winding leakage flux, which links the primary winding only, represented by primary leakage reactance X_P.
  - The secondary winding leakage flux, which links the secondary winding only, represented by secondary leakage reactance X_S.
- The primary and secondary winding resistances R_P and R_S, respectively.

3.4.2 Transformer relationships

With reference to Figure 3.14, the following transformer relationships assume negligible losses in the transformer:

$$I_P N_P = I_S N_S \tag{3.8}$$

$$\frac{V_P}{N_P} = \frac{V_S}{N_S} \tag{3.9}$$

$$V_P I_P = V_S I_S \tag{3.10}$$

3.4.3 Transformer equivalent circuit

Figure 3.14(a) illustrates the transformer equivalent circuit which can be derived from basic electrical theory. The resistance R_I represents the iron losses comprising hysteresis loss and eddy current loss (in the steel core laminations). Both R_I and the magnetising impedance X_M are relatively large and, for most considerations relating to current flow through the transformer, can be ignored, as shown in the simplified equivalent circuit (Figure 3.14(b)). In addition, in Figure 3.14(b), R_P and X_P (as shown in Figure 3.14(a)) have been combined to form Z_P, and R_S and X_S to form Z_S. It is to be noted that both R_P and R_S are much smaller in magnitude than X_P and X_S, respectively, and therefore, Z_P and Z_S are mostly reactive. In practice, transformers are purposely manufactured with defined values of X_P and X_S since they limit the fault current flow through the transformer and therefore limit the fault levels on the busbars connected to the transformer.

3.4.4 Referred impedance – two-winding transformer

Calculating current flow through a transformer, even using the simplified circuit shown in Figure 3.14(b) would still necessitate the solution of a number of simultaneous equations. The calculation is greatly simplified if all the impedance is positioned either on the primary side or on the secondary side of the transformer. An impedance that is repositioned on the other side of a transformer is termed a 'referred' impedance and as such is defined in terms of a 'common base' voltage. Figure 3.15(a) shows the instance of the secondary impedance Z_S referred to the primary side of the transformer to become Z_{SR}, whose value is determined as follows:

$$Z_{SR} = Z_S \left(\frac{N_P}{N_S} \right)^2 \tag{3.11}$$

Alternatively, the primary impedance Z_P when referred to the secondary side of the transformer to become Z_{PR} is given by:

$$Z_{PR} = Z_P \left(\frac{N_S}{N_P} \right)^2 \tag{3.12}$$

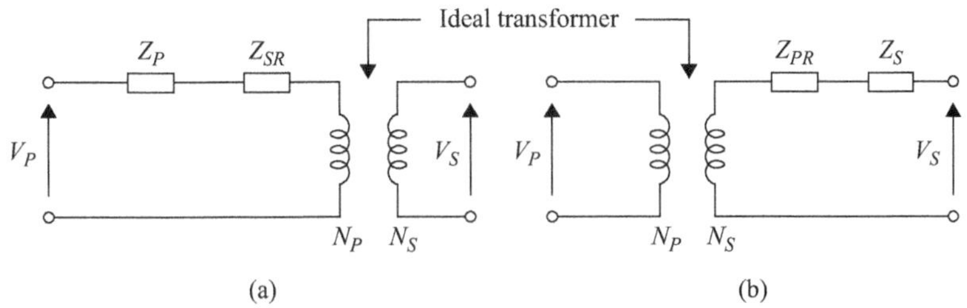

Figure 3.15 Impedance referred to primary and secondary: (a) secondary impedance referred to primary and (b) primary impedance referred to secondary

3.4.5 Autotransformer

In similar fashion to Figure 3.13, Figure 3.16 illustrates the basic features and operation of an autotransformer. The following is worthy of note:

- An autotransformer comprises a single winding which is tapped, with the whole of the winding (the primary) comprising turns N_P and the tapped portion of the winding (the secondary) comprising turns N_S.
- The winding, as shown in Figure 3.16(a), contains three distinct areas of magnetic flux, which are as follows:
  - Flux common to the whole winding, resulting in magnetising impedance X_M.
  - Flux linking that only part of the winding above the tapping point, i.e. flux resulting in leakage impedance Z_P (essentially reactive).
  - Flux linking only that part of the winding below the tapping point, i.e. flux resulting in leakage impedance Z_S (essentially reactive).
- Winding resistances are relatively small in value and are ignored and not shown.

Figure 3.16(b) shows the autotransformer equivalent circuit. Again, X_M, the magnetising impedance, and R_I, the iron loss, are relatively high in value and can usually be ignored. If I_P and I_S are taken to be Maxwell circulating currents, see Section 3.3.1.2, then the ampere turn balance relationship still holds, i.e.:

$$I_P N_P = I_S N_S.$$

In addition, expressions (3.9) and (3.10) are also valid.

3.4.6 Referred impedance – autotransformer

It can be shown that the referred autotransformer impedances are as given in Figure 3.17. Figure 3.17(a) shows all the impedance referred to the primary (HV side), and Figure 3.17(b) shows all the impedance referred to the secondary side. As can be seen, the composition of the impedances is more complex than for the two-winding transformer. It is worthy of note that impedance Z_S is frequently considered to be a common impedance (since both of the circulating currents I_P and I_S flow through Z_S) rather than a secondary impedance.

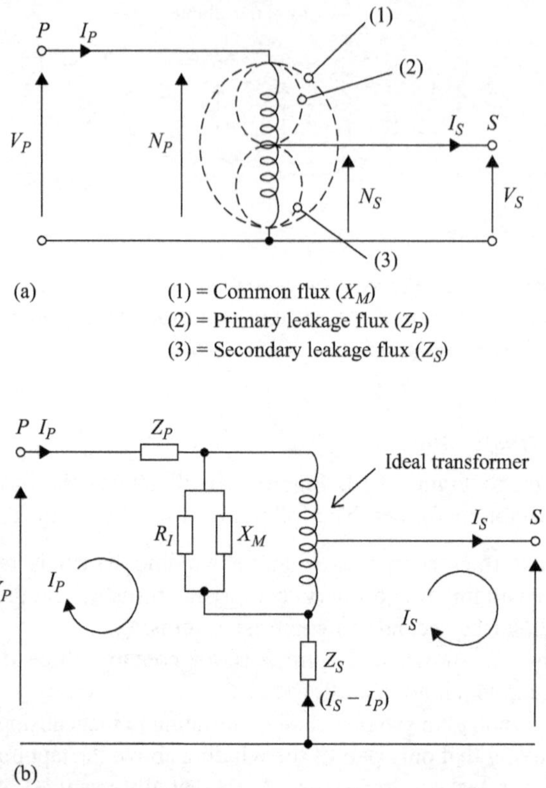

Figure 3.16 Autotransformer equivalent circuit: (a) winding and (b) equivalent circuit

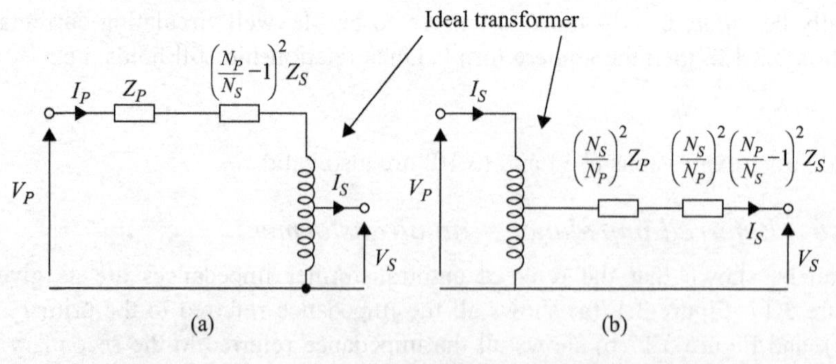

Figure 3.17 Autotransformer referred impedances: (a) all impedance referred to primary and (b) all impedance referred to secondary

3.4.7 Current transformers

Current transformers (CTs) are used to transform the high levels of power network current down to lower levels of current suitable for feeding protection relays and metering, etc. To achieve this, the CT requires a relatively high number of turns on the secondary winding, typically ranging from 250/1 (secondary to primary turns) at 33 kV to 2,000/1 at 400 kV.

The principal of operation is shown in Figure 3.18. Salient observations are as follows:

- Figure 3.18(a) shows a part of the 400-kV network supplying a load current of 1,000 A which passes through a CT of ratio 1,000/1, resulting in a secondary current of 1A flowing into a relay.
- Figure 3.18(b) shows a simplified equivalent circuit based upon Figure 3.14(a) which, in this instance, the leakage reactance and resistance is relatively small, and the iron loss relatively high, and hence both are ignored. Typical values for Z_M, the magnetising impedance (which is referred to the transformer secondary), and Z_B, the impedance of the relay, are shown. NB: The impedance associated with a relay is usually termed the 'burden'. The relative magnitudes of Z_M and Z_B are such that virtually all of the 1-A secondary current flows through the relay. The voltage across the relay by ohms law is therefore 2 V.
- Figure 3.18(b) shows both Z_M and Z_B referred to the primary to become Z_{MR} and Z_{BR}. This is achieved by multiplying the secondary impedances of the CT by the square of the turns ratio, i.e. $(1/1{,}000)^2$ – as defined in expression (3.11).
- It can be seen in Figure 3.18(b) that Z_{MR} and Z_{BR} are very small in value compared to the impedance of the load and as such have virtually zero impact on the magnitude of the 1,000-A load flow. In this, the CT differs markedly from the power transformer whose secondary impedance (i.e. essentially the load impedance) dictates the value of load current. CTs are therefore known as 'series connected' transformers.
- An implication of the CT secondary impedance having virtually no impact on the 1,000-A primary current is that the CT secondary current continues to be 1 A, no matter the burden of the relay. Therefore, should the burden Z_B become an open circuit, all of the 1 A flows through the magnetising impedance Z_M with a resulting secondary voltage of 10 kV. With a fault current of say 20 kA, the CT secondary voltage would theoretically rise to 200 kV. In practice, however, the CT magnetic core would saturate and limit the voltage rise, even though it would remain at a dangerously high level, i.e. it would risk both electrocution of personnel and damage to the CT and connected wiring. A critical requirement associated with CTs, therefore, is to keep the secondary impedance either a low value (as defined by CT manufacturers) or a short circuit when either not connected to a relay, etc., or during testing.

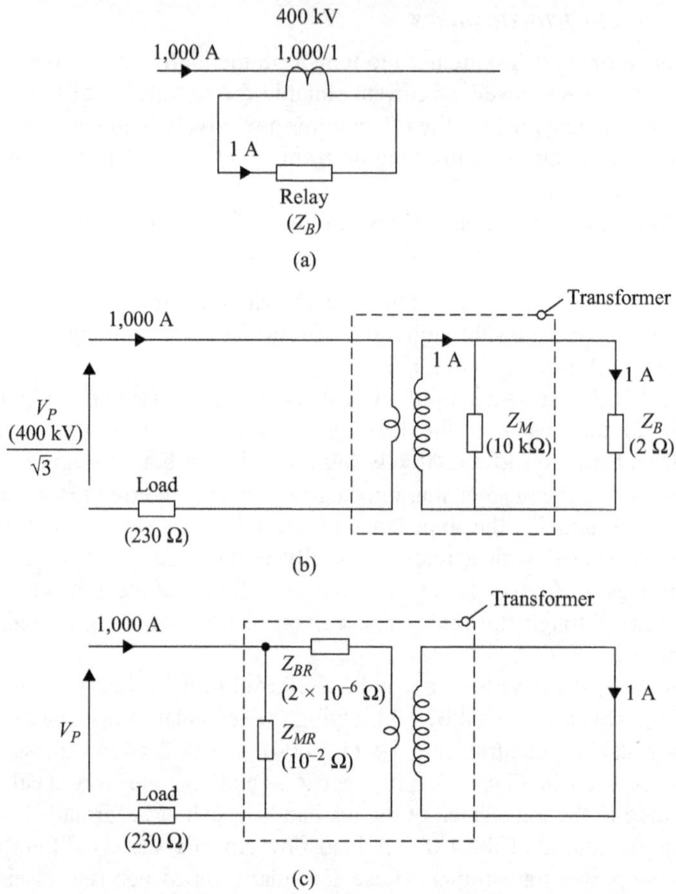

Figure 3.18 Current transformer – operation: (a) CT arrangement, (b) equivalent circuit and (c) impedance referred to primary

3.5 Per-cent impedance

3.5.1 Per-cent impedance – concept

Per-cent impedance is an extremely useful method of simplifying power system calculations. It is not actually a measure of impedance, as its name implies, but a measure of the per-cent voltage drop, of the network voltage, which occurs across an item of equipment, when a defined current flows through that equipment.

3.5.2 Per-cent impedance – basics

Consider a three-phase network with a nominal phase voltage of V_P, see Figure 3.19. Let a defined current I, known as the 'base current', flow through an item of equipment of impedance Z. A voltage drop IZ will arise across the

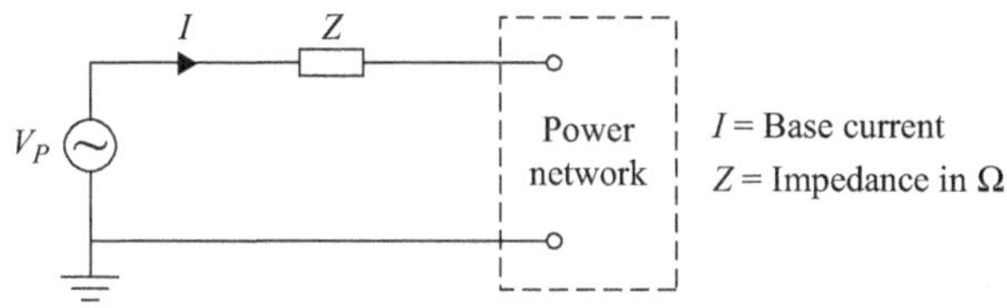

Figure 3.19 Per-cent impedance – basics

impedance. As a result, it can then be said that the per-cent impedance of the item of equipment is given by:

$$\text{Per-cent impedance} = (\%Z) = \frac{IZ}{V_P} \times 100\% \tag{3.13}$$

A 'base voltage' is also defined which is that of the line voltage, i.e. V_L:

$$\text{Now } V_P = \frac{V_L}{\sqrt{3}} \tag{3.14}$$

Inserting (3.14) into (3.13),

$$\text{then } (\%Z) = \sqrt{3}\frac{IZ}{V_L} \times 100\% \tag{3.15}$$

From the base voltage V_L and the base current I_{Base}, a base level of volt–amperes (VA) can be defined as follows:

$$\text{i.e. } \text{VA}_{\text{Base}} = \sqrt{3}\, V_L\, I_{\text{Base}}$$

In practice, it is usual to use and refer to a base MVA – which is given by:

$$\text{MVA}_{\text{Base}} = \sqrt{3}\, V_L I_{\text{Base}} \times 10^{-6} \tag{3.16}$$

Again, in practice, it is usual to quote (%*Z*) on an MVA_{Base} of 100 MVA.

NB: Transformer (%*Z*) is occasionally quoted on the MVA rating of the transformer.

Therefore, it can be said that if an item of equipment has a (%*Z*) of 2% on a 100-MVA_{Base}, then 2% of the phase to neutral voltage is dropped when a current equal to the base current I_{Base}, flows through the equipment. The base current obtained from expression (3.16) is as follows:

$$I_{\text{Base}} = \frac{100\text{ MVA}}{\sqrt{3}V_L} \tag{3.17}$$

3.5.3 Per-cent impedance – advantages of use

With reference to the network shown in Figure 3.20, if the impedances of all the items of equipment were stated in ohms, then to determine the fault current, all of

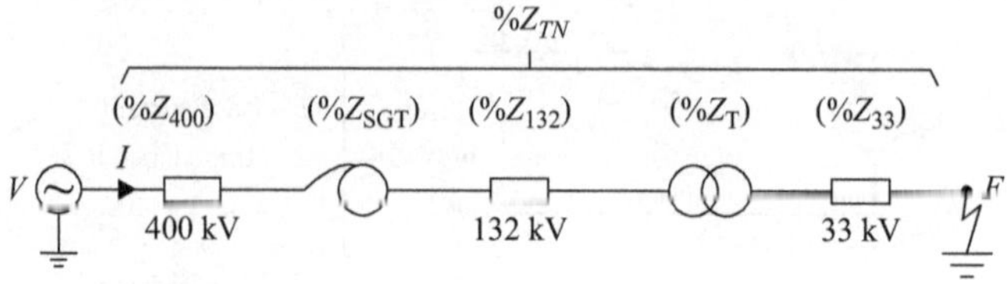

Figure 3.20 Per-cent impedance – across a network

the impedances would need to be referred to a common base voltage, see Section 3.4.4, before a calculation of fault current could be undertaken. This is very tedious. A far simpler solution arises if all impedances are stipulated in per-cent impedance.

Suppose, in the first instance that all impedances in Figure 3.20 are specified in per-cent impedance on 100 MVA base, and a current equivalent to 100 MVA flows through the network (i.e. prior to the fault F arising), then each impedance would have a percentage of the voltage V, dropped across it. The addition of all of those per-cent volt drops is one and the same as the total per-cent impedance $\%Z_{TN}$ of the whole network:

$$\text{i.e. } \%Z_{TN} = \%Z_{400} + \%Z_{SGT} + \%Z_{132} + \%Z_T + \%Z_{33}$$

It could then be stated that:

$$(\%Z_{TN})V \propto 100\,\text{MVA} \tag{3.18}$$

i.e. the per cent of voltage V dropped across the network is proportional to a flow of 100 MVA.

Now when fault F occurs, 100% of V is dropped across the network, and in this instance, a level of MVA termed $\text{MVA}_{\text{Fault}}$ flows across the network into the fault. Thus, similar to expression (3.18), it can be said that:

$$100\%\ V \propto \text{MVA}_{\text{Fault}} \tag{3.19}$$

Dividing expression (3.19) by (3.18):

$$\text{then } \frac{100\%\ V}{(\%Z_T)V} = \frac{\text{MVA}_{\text{Fault}}}{100\,\text{MVA}}$$

$$\text{and rearranging, then } \text{MVA}_{\text{Fault}} = \frac{100\%}{\%Z_T} \times 100\,\text{MVA} \tag{3.20}$$

And if, for example, $\%Z_T = 5\%$ and is inserted into expression (3.20):

$$\text{then } \text{MVA}_{\text{Fault}} = \frac{100\%}{5\%} \times 100\,\text{MVA}$$

$$\text{i.e. } \text{MVA}_{\text{Fault}} = 2{,}000\,\text{MVA}$$

From expression (3.20), the fault current I_F into the fault is simply derived from expression (3.3):

$$\text{i.e.}\ I_F = \frac{\text{MVA}_{\text{Fault}}}{\sqrt{3}\, V_L}$$

where $V_L = \sqrt{3}\, V$ (where V is the phase-neutral voltage).

3.5.4 *Ohms and per-cent impedance conversion*

It is frequently necessary, in practice, to convert from ohms to per-cent impedance and vice versa. The following well-known expressions provide the means for achieving this.

1. **Per-cent impedance converted to ohms**

$$Z(\Omega) = \frac{(\%Z)(V_L)^2}{100\% \times \text{VA}_{\text{Base}}} \tag{3.21}$$

As stated earlier, the VA base commonly used in practice is 100 MVA.

2. **Ohms converted to per-cent impedance**
Rearranging expression (3.21), then:

$$(\%Z) = \frac{Z(\Omega) \times \text{VA}_{\text{Base}}}{(V_L)^2} \times 100\% \tag{3.22}$$

Rule of thumb conversion from per-cent impedance to ohms on 100 MVA base is as follows:

- At 11 kV – 1% = 0.012 Ω
- At 33 kV – 1% = 0.11 Ω
- At 132 kV – 1% = 1.75 Ω
- At 275 kV – 1% = 7.5 Ω
- At 400 kV – 1% = 16 Ω.

3.6 Synchronous generator fundamentals

3.6.1 *Synchronous generator introduction*

Although this publication is essentially concerned with HV power networks, their performance is heavily influenced by those large and complex items of equipment that are connected to one end of the network, namely electrical generators. As such an understanding of HV power network design is incomplete without an appreciation of the performance and impact of generation. The bedrock of power system generation is the synchronous generator, usually powered by either nuclear, gas, coal, biomass, etc. (i.e. thermal power stations) or hydro (i.e. water power). The following relevant aspects of the synchronous generator will be summarised.

NB: The performance of generators under fault conditions is examined in Chapter 4.

- Physical arrangements
- Electrical characteristics
- Generator controls – thermal power station
- Generator performance chart
- Pole slipping.

3.6.1.1 Generator physical arrangements

Figure 3.21 illustrates the physical arrangements of a synchronous generator. The following is worthy of note:

- The generator comprises a rotating magnetic field, the rotor, which is encompassed by the three-phase stator windings (also termed the armature), which provide the electrical output. The stator comprises a cylindrical steel (magnetic) laminated case in which the windings are embedded.
- The rotor comprises field windings (which create the magnetic field) embedded in a magnetic core (usually a solid forging of alloy steel) which are usually supplied from a direct current generator, driven from the rotor shaft.
- The rotor is driven from the power source (e.g. steam turbine), and the rotating magnetic field induces an emf into the stator windings to create the three-phase voltages. Figure 3.21(b) shows the position of the red phase stator windings

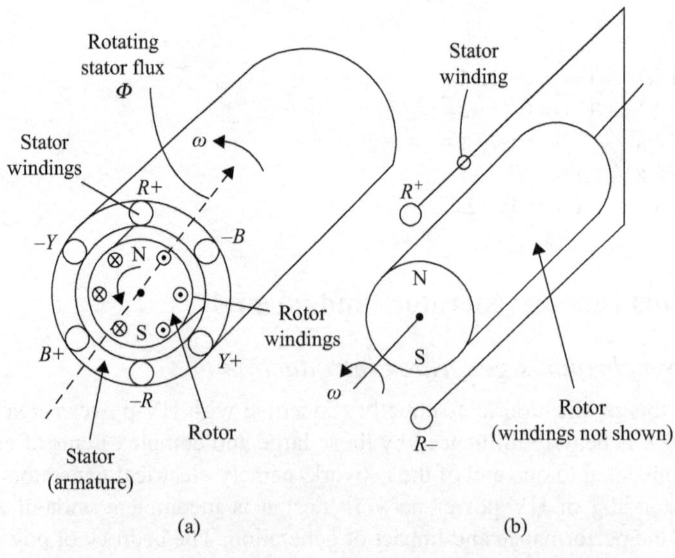

Figure 3.21 Synchronous generator physical arrangements – simplified: (a) generator cross-section and (b) position of stator winding

only. The yellow and blue phase windings (not shown) are physically displaced 120° either side of the red phase.

- When the stator windings are connected to a power network, a balanced three-phase current flows into the network. The three-phase currents each produce an associated magnetic field which collectively sum together to form a rotating magnetic field (similar to that in a three-phase induction machine), which rotates within the stator windings at the rotor speed of rotation ω (equivalent to a frequency of 50 Hz). Since the rotating stator field rotates at the same speed as the rotor, it induces no emf into the rotor.

3.6.1.2 Generator electrical characteristics

Figure 3.22(a) shows, electrically, how the three stator windings are connected together (usually in star formation) to form a three-phase generator. The three-phase equivalent circuit of the generator is shown in Figure 3.20(b) with the corresponding vector diagram shown in Figure 3.20(c). The following is worthy of note:

- The generated voltage E is the stator open-circuit voltage. Voltage V represents the 'infinite busbars' of the power system.
- The rotating stator magnetic field (i.e. flux) Φ shown in Figure 3.21(a) self-induces a voltage into the stator windings. This may be represented by an equivalent reactance in the generator equivalent circuit. It is helpful to divide Φ into two components, which are as follows:
  - A component that fully links with the rotor windings, represented by, and termed the 'armature reaction reactance' X_a.
  - A component that does not link the rotor, represented by, and termed the 'stator leakage reactance' X_L.

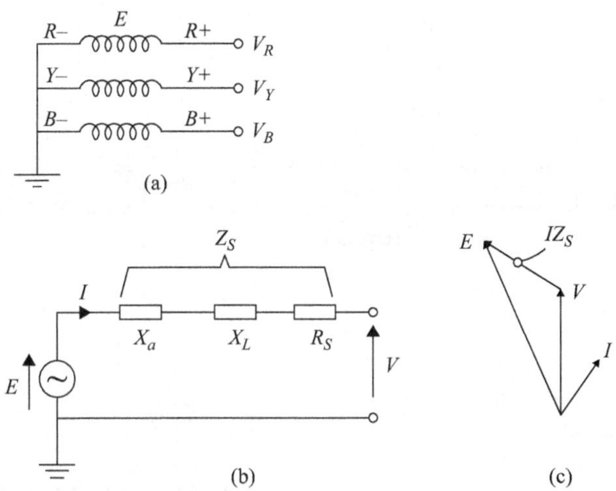

Figure 3.22 Synchronous generator – electrical characteristics: (a) winding arrangement, (b) equivalent circuit and (c) vector diagram

- The stator winding contains finite resistance R_S, and this together with X_a and X_L forms the synchronous impedance Z_S. In practice, R_S is relatively small in magnitude and therefore Z_S to good approximation is equal to X_a and X_L and predominately reactive.
- With reference to Figure 3.22(c), voltage V is taken as the fixed reference vector and the following vector relationship holds (NB: voltage drop IZ_S will always be at right angles – approximately – to current I):

$$E = V + IZ_S \tag{3.23}$$

3.6.1.3 Generator controls – thermal power stations

A thermal power station is one that converts heat energy into electrical energy and typically includes nuclear, gas and coal. Figure 3.23 shows a simplified diagram of the thermal/mechanical/electrical process and control system. The essential features of the generator control arrangements are as follows:

1. **The governor**
   The fuel is used to create steam in the boiler, which in turn flows into the steam turbine, the rotating shaft of which spins the rotor. A governor system measures the speed of rotation of the turbine shaft (i.e. the frequency) and adjusts the flow of steam into the turbine to maintain the frequency constant in accordance with the governor setting limits. Nominal frequency in the United Kingdom is 50 Hz.

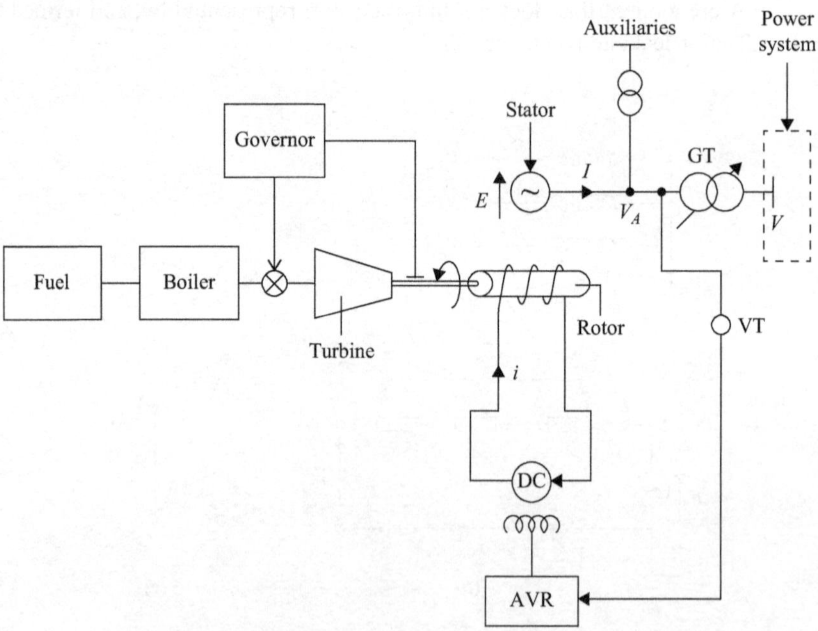

Figure 3.23 Generator controls – thermal power station

2. **The AVR**
The magnitude of the stator open-circuit voltage E is proportional to the rotor magnetic field and as such proportional the output current of the DC generator. This in turn is proportional to the magnitude of current in the DC generator field winding, as controlled by the automatic voltage regulator (AVR). The purpose of the AVR is generally to hold the voltage V_A constant to provide a stable voltage source for the generator auxiliaries (which will typically supply the fuel, boiler, turbine, AVR, etc.). This is achieved via a voltage transformer VT, as shown in Figure 3.23, which measures the magnitude of V_A, which is fed into the AVR, which via the control loop noted above modifies the magnitude of generated voltage E to hold V_A constant, as output from the generator varies.
3. **The generator transformer tap changer**
With reference to Figure 3.23, the generator transformer, GT, is usually fitted with a tap changer. Increasing or decreasing the tap position will change the magnitude of VArs (usually MVArs) flowing into the power system (to aid voltage control on the power system). A more specialist text will provide the mathematical explanation for this occurrence.

3.6.1.4 Generator performance chart

The output range of a generator is normally illustrated and defined on a generator performance chart, similar to that shown in Figure 3.24. Figure 3.24(a) shows the electrical circuit to be considered. Figure 3.24(b) provides the vector diagram which forms the basis of the performance chart. As can be seen the stator current I can be resolved into two components (axis), that of I_{MW} (real power) and I_{MVAr} (reactive power). These in turn results in voltage drops $I_{MW}\,Z_S$ and $I_{\mathrm{MVAr}}\,Z_S$ which vectorially add to become IZ_S. This analysis can be translated into the composition of the generator performance chart as illustrated in Figure 3.24(c) – salient points which are worthy of note are as follows:

- Voltage V is a fixed reference vector (i.e. the infinite busbar).
- A MW scale can be constructed, at right angles to V, which plots the locus of E (i.e. $V + I_{MW}\,Z_S$).
- A MVAr scale can similarly be constructed to plot the locus of E for both lagging and leading MVAr. This locus is in phase with V.
- In Section 3.2.5, it was stated that maximum power transfer arises when vectors V and E are at right angles, i.e. the theoretical stability limit. To obtain the practical stability limit, a factor of safety is added.
- The maximum allowable value of stator current, I_S, results in the maximum voltage drop I_SZ, which when vectorially added to V gives a locus of E equivalent to maximum MVA output.
- Maximum rotor current results in the maximum value of E, and a locus of E can be drawn accordingly.

The above boundary conditions determine the shape of the performance chart and the range of outputs from the generator. Power stations often have a ‘vector meter’

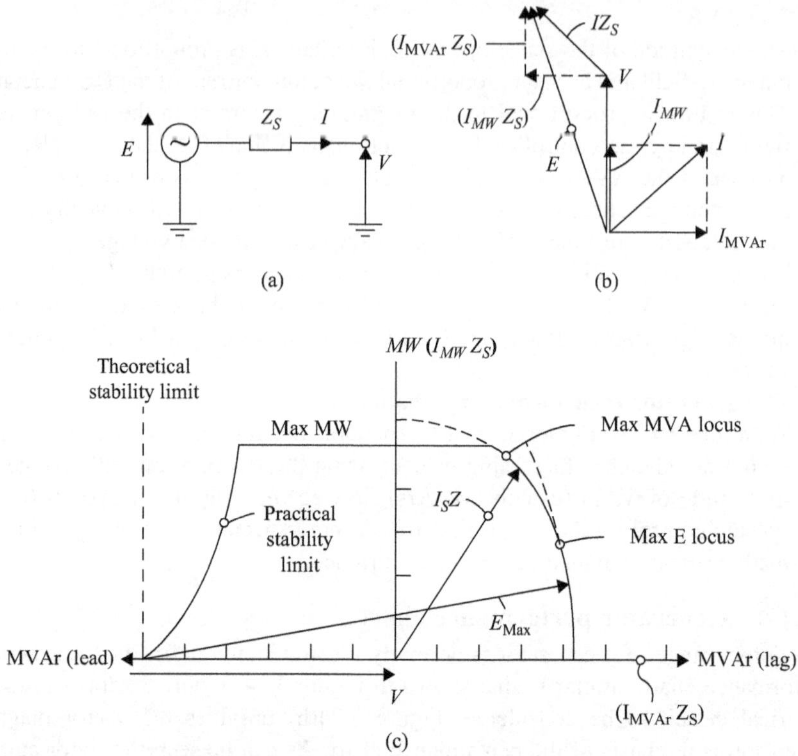

Figure 3.24 Generator performance chart: (a) equivalent circuit, (b) vector diagram and (c) derivation of performance chart (I_{MVAr}, Z_S)

or other similar software package which shows the performance chart and the instantaneous position of the generator output on the chart. This provides an indication of where the generator is operating and the capacity in hand.

3.6.1.5 Generator pole slipping

Pole slipping occurs when synchronism is lost between generators. This may be either an individual generator or between a group of generators and the rest of the power system. With reference to Figure 3.25, E_A is taken as the reference voltage and, as a result of loss of synchronism, E_B rotates (or slips) around E_A – resulting in pole slipping (NB: E_B could rotate in either direction). Characteristics of pole slipping include:

- Widely fluctuating voltages and currents across the power system as E_B rotates. Once during every rotation, a point on the power system drops to zero voltage (when E_A and E_B are anti-phase). This point is termed the system centre. It is the equivalent of a three-phase fault at that point.
- Damage to generators resulting from pulsating mechanical forces and overheating of the rotor.

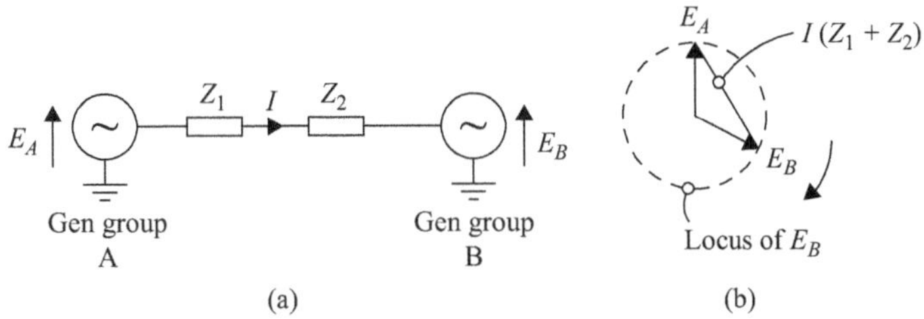

Figure 3.25 Pole slipping: (a) pole slip network and (b) vector diagram

- Protection operation, particularly that of distance and backup overcurrent protection, arising from the fluctuating voltages and currents. This is advantageous separating the slipping systems and preventing damage and limiting the possible loss of supply.

3.7 Wind generation

3.7.1 Induction generator – principles

The earliest wind generators operated on the three-phase induction generator principle. Such machines are termed asynchronous generators as they do not rotate at synchronous speed but, at a speed determined by the intermittent and varying speed of the wind. The principle of the induction generator is described with reference to Figure 3.26 (which shows a single-phase equivalent circuit). In this instance, the rotor windings are not fed from a direct current source, as with a synchronous machine but are shorted together. The circuit components of Figure 3.26 comprise the following:

- V = power system voltage applied to the stator
- Z_M = generator (stator) magnetising impedance
- X_S = stator leakage reactance
- R_S = stator resistance
- X_R = rotor impedance when rotor is stationary (based on 50 Hz)
- R_R = rotor resistance
- S = slip frequency
- N_1 = represents number of turns on stator winding
- N_2 = represents number of turns on rotor winding
- N_S = stator rotating magnetic field speed of rotation
- N_R = rotor speed of rotation.

The slip frequency is given by $S = N_S - N_R$
And the fractional slip is given by $S_F = \frac{(N_S - N_R)}{N_S}$.

The equivalent circuit as shown in Figure 3.26 is based upon the following:

- At the instant of energisation of the stator then:
  - The rotor speed $N_R = 0$
  - The rotor induced emf = E_R

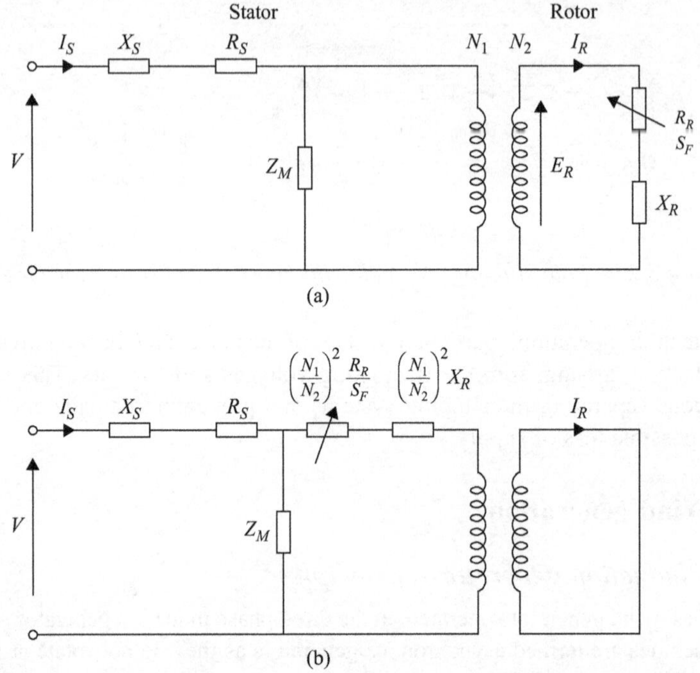

Figure 3.26 Induction generator – equivalent circuit: (a) single-phase equivalent circuit and (b) impedances referred to stator side

- Inductive reactance of the rotor $= X_R$
- Resistance of the rotor $= R_R$.

- When the rotor is rotating at a speed N_R then
  - Voltage induced in the rotor $= S_F\ E_R$ (i.e. a function of the frequency induced in the rotor)
  - Inductive reactance of the rotor $= S_F\ X_R$ (i.e. a function of the frequency induced in the rotor)
  - Resistance of the rotor $= R_R$ (i.e. independent of frequency).

Thus, it can be said that the current in the rotor windings is given by:

$$I_R = \frac{S_F E_R}{S_F X_R + R_R}$$

Or on rearranging:

$$I_R = \frac{E_R}{X_R + \frac{R_R}{S_F}}$$

NB: The parameters contained in the above expressions are vector quantities and are expressed as shown for reasons of simplicity.

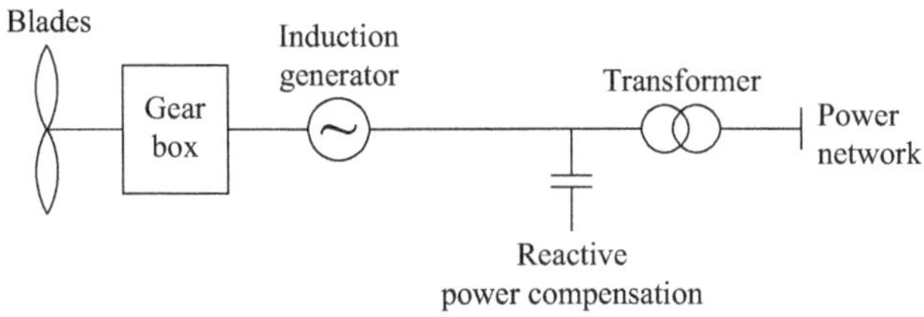

Figure 3.27 Early wind generator – arrangements

The above translate into the equivalent circuit shown in Figure 3.26(a) (similar to that of a transformer as described in Figure 3.14). And Figure 3.26(b) shows the same equivalent circuit but with the rotor impedances referred to the stator.

Therefore, with reference to Figure 3.26, when the stator is fed with a three-phase voltage whose phase to neutral voltage is V, a current is drawn into the stator resulting in a rotating magnetic field within the stator, whose speed of rotation is N_S (and equivalent to 50 Hz). The rotating magnetic field induces a voltage into the rotor causing a current to flow around the shorted rotor windings. This causes a motor action between stator and rotor, and the rotor commences to gather speed, rotating in the same direction as the stator magnetic field. If however the rotor is connected to a source of power such as the wind, and the speed N_R increases beyond that of N_S (i.e. super-synchronous), then the fractional slip S_F becomes negative. This means that in Figure 3.26 the value of R_R/S_F becomes negative, the implication of which is that (when the total apparent resistance of the generator becomes negative) power flows out of the machine, i.e. it becomes a generator.

3.7.2 Early wind generators

The arrangements associated with early wind generators are illustrated in Figure 3.27. Salient points are as follows:

- The blades usually have a pitch system to control rotor speed and keep them from turning in winds that are too high.
- The gear box connects a low blade speed to a faster connecting shaft speed (connected to the rotor) suitable for electricity generation.
- Maximum power output from an induction generator is achieved only a few per cent above synchronous speed.
- Induction generators draw significant reactive power from the power network, which may need to be compensated for by the installation of capacitor banks (i.e. power factor correction).
- The transformer connects the lower voltage system of the wind generator to the higher voltage of the power network.

3.7.3 Modern wind generators

Figure 3.28 illustrates the arrangements of a modern wind generator. In this instance the rotor is energised similar to that for a synchronous generator.

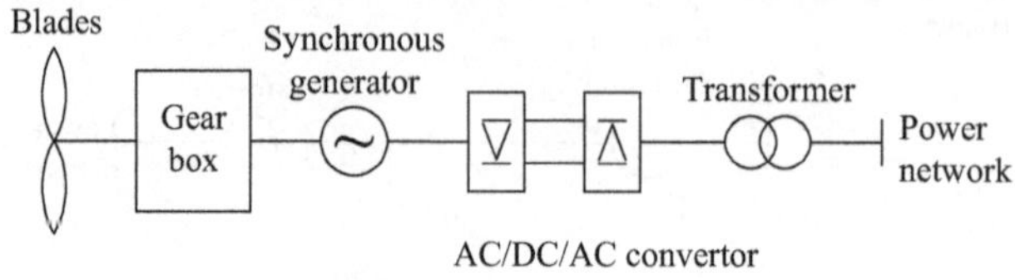

Figure 3.28 Modern wind generator – arrangement

Although the wind generator will be unable to rotate constantly at synchronous frequency (i.e. 50 Hz), due to wind speed variation, the output can be connected to the power network via an AC/DC/AC convertor system in a similar fashion to an HVDC transmission system. An advantage of this arrangement is that it is capable of generating VArs so dispensing with the requirement for reactive compensation, and contributing towards voltage control of the power network.

3.7.4 Wind generator – fault level contribution

A limitation of all wind generation systems is that they provide a relatively low contribution to power system fault current. This is an important consideration since a minimum level of fault current is required to ensure the operation of protection systems. With ever-increasing levels of wind generation, this is an ever-increasing problem – and needs to be identified and fully considered when connecting wind generation to the power network.

3.8 Power system transients

3.8.1 Types of transient

Power system impedance comprises series impedance which is predominately inductive and shunt impedance which is predominately capacitive. In addition to OHL and cables, the power system also comprises shunt reactors and capacitors. All of these give rise to transient currents and voltages on circuit energisation (and de-energisation). Such transients need to be taken into account when considering CT accuracy, protection design and circuit breaker interrupter design, etc.

3.8.2 Currents arising on energisation of an inductive circuit

This may typically apply to the energisation of a shunt reactor, or the application of a fault to an OHL, cable or busbar (since the network is predominately inductive). Figure 3.29 shows a simple inductive circuit, which on energisation is subject to the following expression:

$$E\sin(\omega t + \theta) = i_T R + L\frac{di_T}{dt}$$

where θ is the point on wave of switching.

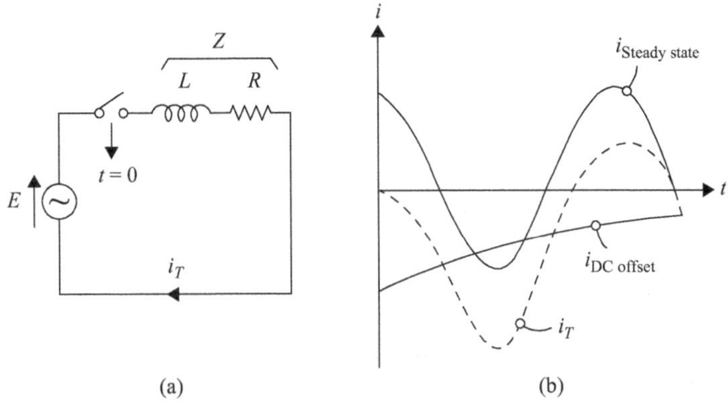

Figure 3.29 Transient current flow – inductive circuit: (a) circuit and (b) waveforms

The solution to the above equation is given by two components of current, an ongoing steady state sine wave component and a transient decaying DC offset component, as follows:

$$i_T = i_{\text{Steady state}} + i_{\text{DC offset}}$$

The solution of which is $i_T = i\sin(\omega t + \theta - \Phi) - i\sin(\theta - \Phi)e^{-\frac{Rt}{L}}$

where in the above $i = \dfrac{E_{\text{Max}}}{Z}$ and $\Phi = \tan^{-1}\left(\frac{2\Pi fL}{R}\right)$, where f is the frequency of E (i.e. 50 Hz).

With reference to the above and Figure 3.29, the following is worthy of note:

- The maximum value of DC offset (either positive or negative) occurs when: $t = 0$ and $(\theta - \Phi) = 90°$ (or 270°). Power networks are usually highly inductive, and as such Φ would be close to 90°, therefore θ, the point on wave at which the voltage is switched, needs to be close to 180° (or 0°).
- At the instant of energisation, i.e. $t = 0$, the steady state and DC offset components of current are equal and opposite; therefore, the total current always commences from zero.
- A DC offset with a long-time constant (i.e. a highly inductive circuit where the ratio *R/L* is small), resulting in a slow current decay, may result in the total current reaching a maximum of 2*i*.
- Fault current calculations in particular should take the impact of the transient current into account.

One instance of the switching of an inductive circuit is of particular significance, which is that of energisation of a transformer. If the transformer is energised (i.e. switched) at the instant of minimum voltage (on the sine wave), the resulting total flux in the transformer core may theoretically approach a peak value of twice the steady state peak flux – as a result of the transformer core having a high inductive reactance to resistance ratio. However, the transformer core usually saturates before reaching such a high flux which in turn results in a large transient increase in current – this is termed the 'transformer magnetising inrush current'. The magnitude of the inrush current may be up seven times RMS full load current, which may last for several seconds after switching, at which point it will have decayed to normal steady state magnetising current.

3.8.3 Currents arising on energisation of a capacitive circuit

This may typically apply to the energisation of a shunt capacitance bank, or a length of cable. Figure 3.30(a) shows a capacitive circuit, which on energisation is subject to the following expression:

$$E\sin(\omega t+\theta)=i_T R+\frac{1}{C}\int i_T dt$$

where θ is the point on wave of switching.

The solution to the above is once again given by two components of current, an ongoing steady state sine wave component and transient decaying DC offset component, as follows:

$$i_T=i_{\text{Steady state}}+i_{\text{DC offset}}$$

The solution of which is:

$$i_T=i\sin(\omega t+\theta+\Phi)+[i_M\sin\Phi-i\sin(\theta+\Phi)]e^{-\frac{t}{RC}} \tag{3.24}$$

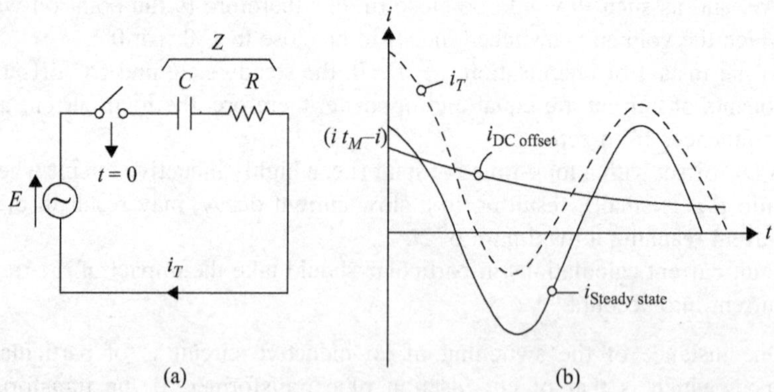

Figure 3.30 Transient current flow – capacitive circuit: (a) circuit and (b) waveform

where in the above $i_M = \frac{E_{Max}}{R}$, $i = \frac{E_{Max}}{\sqrt{R^2 + \frac{1}{(2\Pi fC)^2}}}$ and $\Phi = \tan^{-1}\left(\frac{1}{2\Pi fCR}\right)$.

And where f is the frequency of E (i.e. 50 Hz). With reference to the above, the following is worthy of note:

- For the instance of $1/(2\Pi fC)$ (i.e. capacitive reactance) being much greater than the value of R, then Φ becomes approximately 90°, and at time $t = 0$, and with reference to expression (3.24):
  - $i_{DC\ offset} = i_M - i\ \sin(\theta + 90°)$
    From the above, it can be seen that maximum value of DC offset would occur when $\theta = 180°$ with a magnitude of $(i_M + i)$.
  - The maximum value of steady state current occurs when $\theta = 0°$, at which point the steady state current has a magnitude of i, and the corresponding DC offset current has a minimum magnitude of $(i_M - i)$. This is shown in Figure 3.30(b).
  - At time $t = 0$, expression (3.24) can be restated as:

    $$i_T = i\sin(\theta + \Phi) + [i_M - i\sin(\theta + \Phi)].$$

    From the above, it can be deduced that i_T is equal in magnitude to i_M no matter the point on wave θ, of switching.
- From the above, it can be concluded that it is theoretically possible for the total current to exceed $2i$ (i.e. twice the steady state value of i). In practice, it would be usual for the circuit to contain some series inductance, the impact of which would be to add an oscillatory component to the DC offset.

3.9 HVDC transmission

3.9.1 HVDC transmission – introduction

HVDC transmission is a huge subject in its own right and therefore the intention of this text is to merely summarise the basics. The number of HVDC transmission systems is steadily increasing, not only in the United Kingdom but also worldwide. The criterion necessitating the installation of an HVDC system is as follows:

- Enables the connection of two power systems operating at different frequencies.
- Enables the connection of two substations where the alternative use of long cable is impracticable and an OHL is not an option. This is because the capacitive reactance of an AC cable circuit decreases with increasing length. This in turn results in increasing MVAr flow into the cable with resulting losses and voltage control problems (e.g. Ferranti effect – see Section 7.4.1.5). Maximum AC cable lengths are typically between 30 and 50 km.
- Enables the connection of two substations where the inability to obtain land rights of way for a new AC transmission circuit (either OHL or cable) necessitates the installation of an under-sea cable of such a length that only HVDC is a practical solution.

3.9.2 HVDC transmission – simplified description

Figure 3.31 illustrates the key requirements of an HVDC transmission system, which are summarised as follows:

- By appropriate firing of the thyristors at the rectifier end, the AC input is rectified in to a DC voltage, with a slight ripple.
- Similarly, by appropriate firing of the thyristor valves at the invertor end, the DC voltage is reconverted to an AC voltage, with a stepped waveform, but with a fundamental frequency of the AC system to which it is connected.
- The process results in harmonics in the AC networks which are removed by filter banks.
- The DC transmission line contains resistance and capacitance (*R* and *C*). The capacitance helps smooth the waveform, and in some instances, additional capacitors are installed for this very purpose.
- The power *P* transferred is given by:

  $$P = (V_A I - 2I^2 R) = V_B I.$$

- By selective control of the thyristor point on wave firing angle, the system can be made to generate VArs.
- Power can be made to flow in the reverse direction if the magnitude of V_B is arranged to exceed that of V_A. This can be achieved by tap changing, via the transformers.
- Both the rectifier and invertor ends are termed 'convertor' stations.

3.9.3 HVDC convertors – types

HVDC convertors may be categorised into two major types, which are as follows:

1. Current source
2. Voltage source

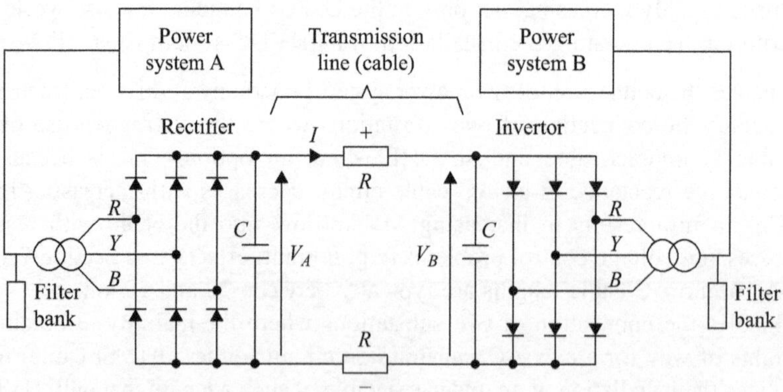

Figure 3.31 HVDC transmission system – simplified

1. **Current source HVDC convertor**
   Most HVDC convertors in operation today are of the 'current source' type. They are based upon thyristor technology, a feature of which is that they can only be turned 'on' and not 'off'. They rely on the AC power system direction of current (i.e. at a zero crossing) to affect the turn-off process. A consequence of this is that the DC current flows only one way around the DC system. Power flow is reversed by increasing the DC voltage at the other end, and reversing the polarity of the DC transmission line – but as such the current flow around the DC system remains in the same direction as before. A stable current flow on the DC side (i.e. minimisation of ripple) is usually aided by the installation of a large inductor placed in series with the DC transmission line, at either one or both of the convertor stations.
2. **Voltage source HVDC convertor**
   With this type of convertor, the rectifier and invertor process is not based upon thyristor technology but that of semi-conductor devices such as insulated-gate bipolar transistors. These have the advantage that they can be turned both on and off. In these type of convertors, the DC voltage polarity of the transmission line is usually constant (i.e. remains the same), and power flow reversal is achieved by reversing the direction of flow of the DC current. A salient advantage of this technology is that the convertor process no longer relies on an AC source of voltage connected at the invertor end, for its operation, and therefore can feed power into an AC network consisting only of passive loads (i.e. no generation). This cannot be achieved with current source convertors. This technology also reduces the requirement for harmonic filters, so resulting in physically smaller convertor stations. This is very beneficial where there are HVDC connections to off-shore wind generation platforms, where space is a premium.

Chapter 4

Power system fault analysis

4.1 Power system fault analysis – requirements

Power system design is dependent upon a complete understanding of power system behaviour and the ability to predict that behaviour across the range of power system conditions. Within this context, current flow through the power system is divisible into two main categories, namely three-phase balanced load current (when the power system is operating in a normal healthy state) and fault current (when the power system is subject to one of a wide range of fault conditions).

With power network construction schemes that involve the addition to, or reconfiguration of, an existing power network, there is invariably a requirement to calculate fault currents both at salient points on the new or reconfigured power network, and at other salient points on the existing power network. The reasons include the following:

- The determination of power system equipment short-circuit withstand capability, and circuit breaker fault making and breaking capacity
- Assessment of generator stability (where the construction work has an impact on generation)
- Protection systems application and the determination of protection settings
- The determination of substation rise in earth potential.

It is also worthy of note that fault situations occur when it is necessary to calculate not only the currents which arise during a fault but also the voltages at certain points on the power network (e.g. for protection application and performance assessment purposes). Furthermore, there is periodically a requirement to determine the currents and voltages associated with open-circuit conditions.

Power system fault calculations are invariably undertaken using the mathematical technique termed 'symmetrical components' analysis (alternatively termed 'phase-sequence components' analysis). A wealth of literature exists on this subject and therefore the purpose of this text is to provide an abbreviated, concise and relevant explanation focusing on fundamental concepts and practical requirements relevant to power network construction. There are numerous computer-based systems available for undertaking power system fault calculations, but instances arise in practice of where it is much quicker and convenient to undertake hand calculations (with the aid of a calculator), or even to carry out hand calculations as a rough check to provide assurance that a computer calculation is correct (since computer output depends on correct data input). Furthermore, the ability to undertake hand calculations most importantly requires a mastery of the principles and concepts involved. This in turn leads to an in-built understanding of comparative equipment impedances and the typical current flows that arise on the power system, i.e. an appreciation of numbers, scale and size. Suffice it to say that where an analysis of an extensive or complex part of the power system is concerned – then in those instances recourse to a computer-based solution is invariably essential.

This chapter will examine and summarise the following:

- Symmetrical components – fundamentals
- Equipment phase-sequence impedance derivation
- Generator performance under fault conditions
- Phase-sequence networks – for common fault conditions
- Maximum and minimum fault level studies
- Fault currents – methods and techniques
- Power system comparative impedances.

4.2 Symmetrical components fundamentals

4.2.1 Symmetrical components – basic concepts

The theory of symmetrical components is based upon the concept of analysing any power system fault, or condition, through the application and superposition of three individual and distinct three-phase balanced power systems. With reference to Figure 4.1, these comprise:

1. **Positive phase-sequence network (+ve seq or PPS)**
   Whose vectors (i.e. *R*, *Y*, *B*) are 120° displaced with a conventional (i.e. anticlockwise) direction of rotation, see Figure 4.1(a).
2. **Negative phase-sequence network (−ve seq or NPS)**
   Whose vectors (i.e. *R*, *Y*, *B*) are 120° displaced with a non-conventional (i.e. clockwise) direction of rotation, see Figure 4.1(b). NB: This is sometimes shown with a conventional direction of rotation with the position of vectors *Y* and *B* swopped over.

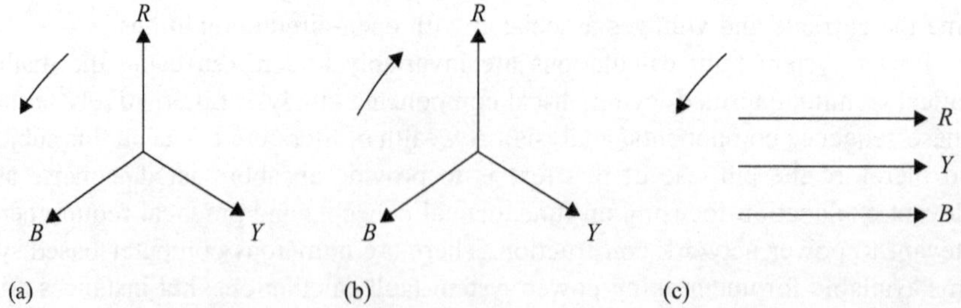

Figure 4.1 Symmetrical components – phasor relationships: (a) positive sequence (+ve), (b) negative sequence (−ve) and (c) zero sequence (zero)

3. **Zero phase-sequence network (zero seq or ZPS)**
   Whose vectors (i.e. *R*, *Y*, *B*) are all in phase with a conventional (i.e. clockwise) direction of rotation, see Figure 4.1(c).

The advantage of this form of analysis is that a three-phase balanced power system can be represented by a single-phase equivalent circuit. This greatly simplifies the analysis.

The expressions below define the relationships between power system phase quantities and their three symmetrical components – for both voltage and current. It is worthy of note that the subscripts *R*, *Y* and *B* denote the phase colours and the subscripts 1, 2 and 0 denote the positive, negative and zero-sequence components, respectively. Furthermore, by applying the *a* operator (see Section 3.2.3), all quantities can be expressed in terms of Red phase (usually taken as the reference phase):

Thus, $$V_R = V_{R1} + V_{R2} + V_{R0} \quad (4.1)$$

$$V_Y = V_{Y1} + V_{Y2} + V_{Y0} = a^2V_{R1} + aV_{R2} + V_{R0} \quad (4.2)$$

$$V_B = V_{B1} + V_{B2} + V_{B0} = aV_{R1} + a^2V_{R2} + V_{R0} \quad (4.3)$$

From the above-defined relationships, the sequence components of voltage can be derived from the phase quantities, which are as follows:

$$V_{R1} = 1/3\left(V_R + aV_Y + a^2V_B\right) \quad (4.4)$$

$$V_{R2} = 1/3\left(V_R + a^2V_Y + aV_B\right) \quad (4.5)$$

$$V_{R0} = 1/3(V_R + V_Y + V_B) \quad (4.6)$$

Similar expressions exist for current as follows:

$$I_R = I_{R1} + I_{R2} + I_{R0} \quad (4.7)$$

$$I_Y = I_{Y1} + I_{Y2} + I_{Y0} = a^2I_{R1} + aI_{R2} + I_{R0} \quad (4.8)$$

$$I_B = I_{B1} + I_{B2} + I_{B0} = aI_{R1} + a^2I_{R2} + I_{R0} \quad (4.9)$$

and $$I_{R1} = 1/3\left(I_R + aI_Y + a^2I_B\right) \quad (4.10)$$

$$I_{R2} = 1/3\left(I_R + a^2I_Y + aI_B\right) \quad (4.11)$$

$$I_{R0} = 1/3(I_R + I_Y + I_B) \quad (4.12)$$

As stated earlier, the Red phase is usually taken as the reference phase, and in recognition of this, and for reasons of simplification, in practice, the *R* subscript is often dropped, thus for example expressions (4.7) and (4.8) become:

$$I_R = I_1 + I_2 + I_0 \quad (4.13)$$

and $$I_Y = a^2I_1 + aI_2 + I_0 \quad (4.14)$$

4.2.2 Practical determination of phase-sequence impedances

All electrical equipment possess positive, negative and zero-sequence impedances termed Z_1, Z_2 and Z_0, respectively. The sequence impedances of any item of equipment can be determined by means of test by applying three-phase voltages of each phase-sequence in turn and measuring the resultant currents, and then applying Ohm's law to determine the impedance. From this, a single-phase equivalent circuit for each sequence impedance can be obtained. To be assured that the equivalent circuit is correct, the impedances seen by four test conditions need to be satisfied, which are as follows:

- Application of the test voltage to each end of the equipment in turn – with the other end short-circuited.
- Application of the test voltage to each end of the equipment in turn – with the other end open-circuited.

Not all four tests need to be carried out on each and every item of equipment – the number being dependent upon the complexity of the equipment.

4.2.3 Phase-sequence impedances of an OHL (or HV cable)

With reference to Figure 4.2(a), the positive-sequence impedance of an OHL (or an HV cable) is obtained by applying a PPS (i.e. balanced three-phase voltage with anticlockwise phase rotation) to the terminals at one end of the OHL, with the terminals at the other end subject to a three-phase short-circuit clear of earth. The resulting single-phase, positive-sequence equivalent circuit is as shown. In this instance, and by inspection, only a single test is required to determine the equivalent circuit, i.e. the other three test conditions as specified in Section 4.2.2 being unnecessary.

The negative-sequence impedance is obtained similarly, but this time applying NPS voltages. As a general rule, the positive- and negative-sequence impedances of static plant (not including generators) are identical – and as such, the derivation of the negative-sequence impedance is not shown in Figure 4.2 – nor in any future examples.

The zero-sequence impedance is similarly obtained by applying ZPS voltages (i.e. three in-phase voltages with anticlockwise phase rotation) to the terminals at one end of the OHL, with the terminals at the other end subject to a three-phase short-circuit to earth. Figure 4.2(b) shows the resulting zero-sequence, single-phase, equivalent circuit.

4.2.4 Phase-sequence impedances of a star-delta transformer

Transformers play a significant role in fault calculations, and their equivalent circuits are somewhat more complex to determine than that of an OHL.

Figure 4.3 illustrates how the sequence impedances are determined for a commonly utilised two winding transformer, the star-delta (similar to that found on the UK 132/33 kV network). As with the OHL, it is only necessary to carry out the test from one side of the transformer, although it is always wise to consider whether

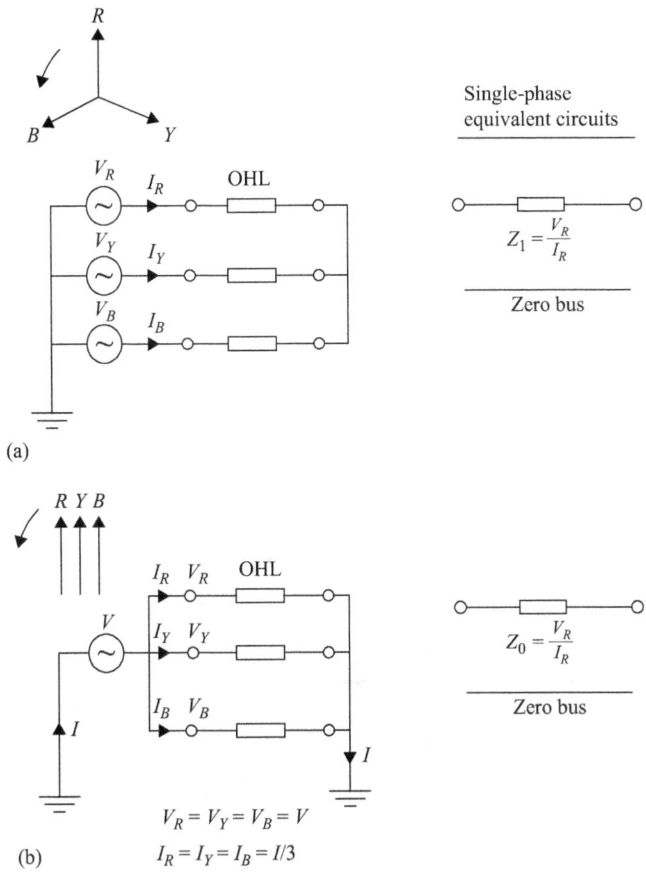

Figure 4.2 Sequence impedance measurement – OHL: (a) positive-sequence impedance and (b) zero-sequence impedance

the equivalent circuit satisfies all four test conditions as described in Section 4.2.2. It is worthy of note that the zero-sequence current circulates around the delta winding. It is also worthy of note that the impedances shown in the equivalent circuits are on a per-cent basis and converted from ohms (see Section 3.5), and as such the transformer windings do not need to be shown on the equivalent circuit, and this simplifies the analysis. In practice, it is usual to use per-cent impedance (on 100 MVA base) on all sequence network single-phase equivalent circuits, and this will be the case for the remainder of this chapter. Other two winding transformer impedances are obtained using the method described in Figure 4.3.

4.2.5 Phase-sequence impedance of an auto-transformer

Figure 4.4 shows the sequence impedance single-phase equivalent circuits for an autotransformer with a delta tertiary winding (which commonly exists in practice).

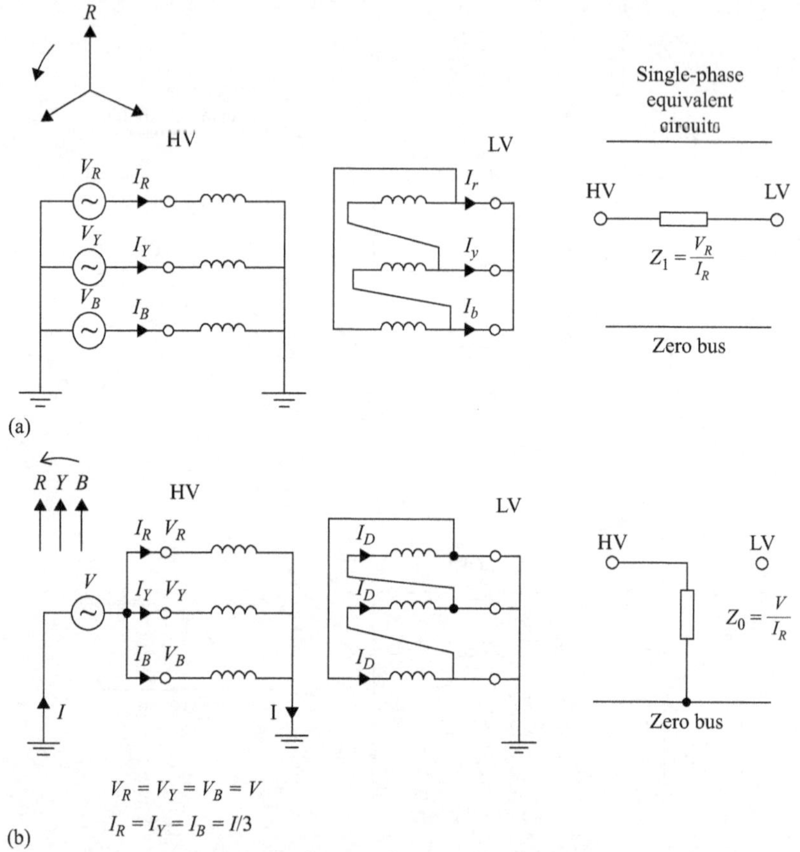

Figure 4.3 Sequence impedance measurement – two winding transformer: (a) positive-sequence impedance and (b) zero-sequence impedance

In this instance, the derivation of the equivalent circuit would require test voltages to be injected from the HV and LV sides of the transformer in turn, with each of the remote end terminals subject to short-circuit, in turn, respectively – and the resulting equivalent circuit impedances determined.

4.2.6 Phase-sequence impedances of an earthing transformer

Earthing transformers are commonly used on the UK 33 kV system – but are also found elsewhere on the power system. They are an important and frequently used item of equipment, but relatively complex to analyse – and for this reason, it is instructive to include it as part of this chapter.

Figure 4.5(a) shows a typical earthing transformer arrangement, as found in practice. The neutral of an interconnected-star (zig-zag) transformer is connected to earth via a neutral earthing resistance (NER). In this instance, a star-connected

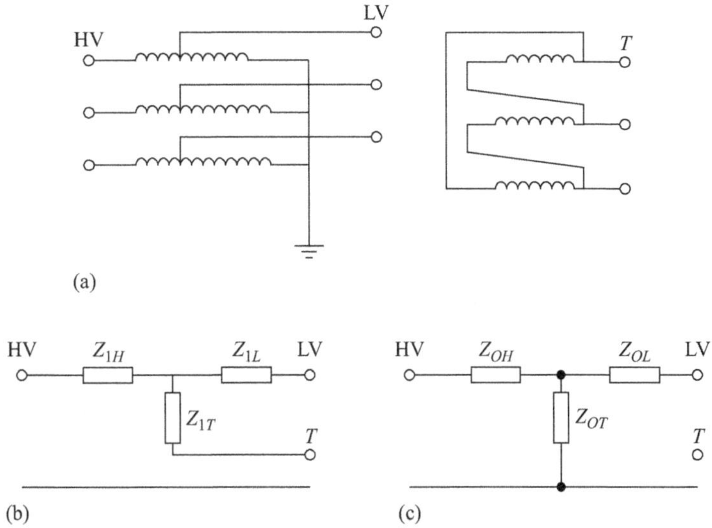

Figure 4.4 Sequence impedance equivalent circuit – auto-transformer: (a) autotransformer with a delta tertiary, (b) positive phase sequence and (c) zero phase sequence

secondary winding is employed to provide LV supplies. The star winding has no bearing on the performance of the earthing transformer, as will be explained below.

The requirements of an earthing transformer, as described below, are that it should provide a high impedance to earth for the flow of both positive-sequence and negative-sequence currents, but a low impedance to earth for the flow of zero-sequence currents.

With reference to Figure 4.5(b), the characteristic of the NER, when connected to the interconnected star neutral point, is that it appears as open-circuit when subject to either positive-sequence or negative-sequence voltages, but a resistance of $3R$ when subject to zero-sequence voltages. This is because a zero-sequence voltage of V, if applied between terminals N and E, would result in a current of $3I$ flowing through the NER (i.e. I from each phase making $3I$ in total).

So if $V = 3IR$ the apparent zero-sequence resistance per phase is $\frac{V}{I} = 3R$.

With reference to Figure 4.5(c) and considering the positive-sequence impedance equivalent circuit, the per-cent impedance Z_1 comprises the positive-sequence leakage reactance of both the HV (interconnected star) and LV (star) windings. Had the LV winding not been present, then the LV terminal would not be present and the HV terminal would be looking into an open-circuit. NB: In practice, the finite but very high impedance of the interconnected star magnetising impedance would be present, see Section 3.4.3. Thus, the interconnected star earthing transformer itself presents a high-impedance to positive-sequence (and negative-sequence) voltages.

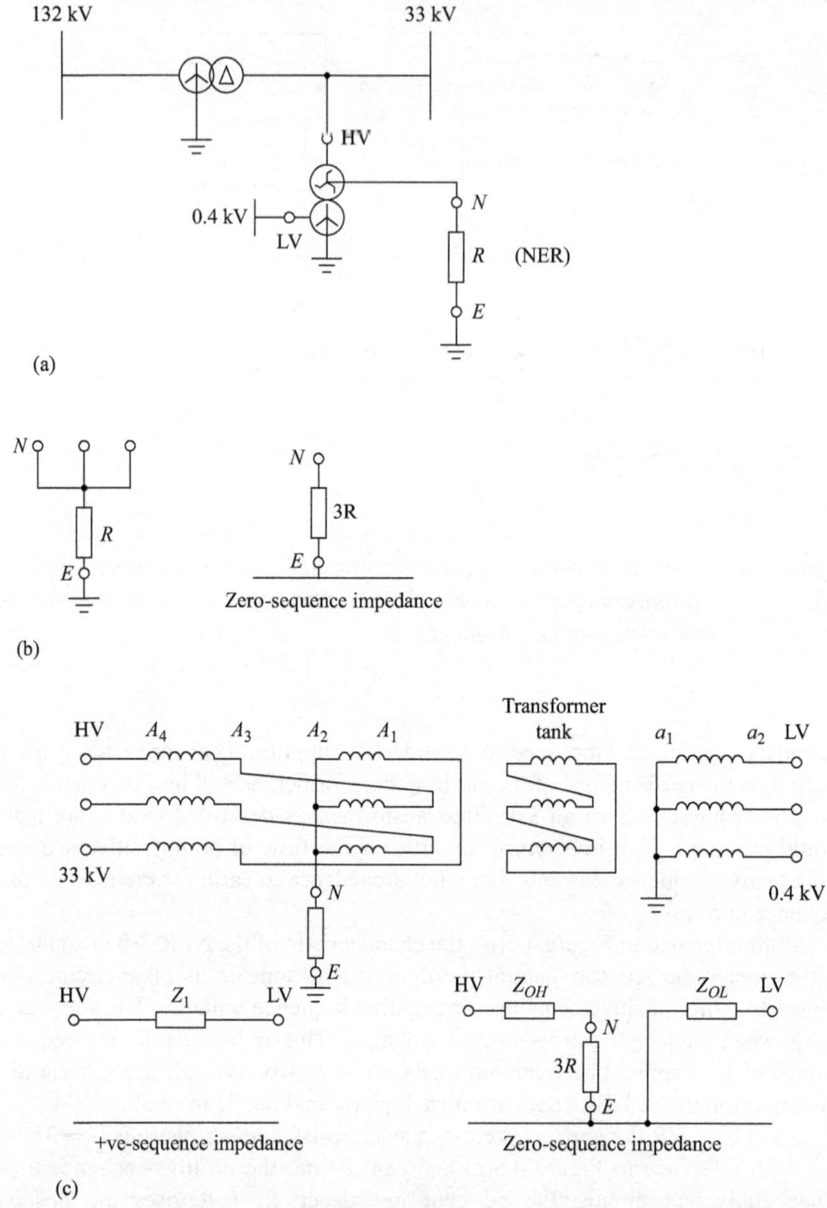

Figure 4.5 Sequence impedances equivalent circuits – earthing transformer: (a) single-line diagram, (b) NER sequence impedances and (c) zig-zag/star transformer sequence impedances

When zero-sequence voltages are applied to the transformer's HV terminals, with the LV terminals subject to an open-circuit, the resulting currents flowing into the interconnected star cause equal and opposite fluxes in the windings on the same limb (of the magnetic core), so creating a low-impedance circuit such that the current is only limited by the winding leakage reactance Z_{OH} (and the resistance of the NER). In this instance, no magnetic flux links the LV winding. Thus, the windings of the interconnected star present a low-impedance to zero-sequence voltages (and as such, network earth faults).

If the zero-sequence voltages are now applied to the LV terminals, with the HV terminals subject to a three-phase short-circuit to earth, equal and opposite induced voltages arise in each pair of connected windings (in the interconnected star). Thus, the interconnected star windings appear as an open circuit. However, if the transformer core is three limbed (the usual case), the three in-phase fluxes in the LV star winding circulate via the transformer tank, and in doing so induce currents into the transformer tank causing it to behave like a relatively high-impedance delta winding of magnitude Z_{OL}.

The resulting equivalent circuit that satisfies the above conditions is as shown in Figure 4.5. Thankfully, not all equivalent circuits are so complex to determine.

4.2.7 OHL single and double circuit impedances

Each phase of a single circuit HV OHL is subject to a self-impedance Z_P (essentially inductive) arising from its own current, and a mutual impedance Z_{PP} (again essentially inductive) arising from each of the currents in the other two phases of the same circuit. For the instance of a double circuit OHL, each phase of the first circuit is subject to a mutual impedance Z_M (essentially inductive) arising from the currents in the phases of the second circuit. With reference to Figure 4.6, a range of impedances can be derived, depending upon whether one or both circuits are in-service and whether the current flow of each circuit, when both circuits are in-service, is in the same or opposite direction. It is assumed that the current I is the same magnitude in both circuits. It is worthy of note that the impedances associated with zero-sequence currents are significantly greater than those associated with positive or negative sequence, as Figure 4.6 shows. This is because zero-sequence currents are in-phase and therefore the fluxes (and inductive reactances) are additive.

These impedances have great relevance to many fault calculations, and the determination of distance protection settings on OHL in particular.

4.2.8 Impedances database

Many power network companies hold a database of equipment sequence impedances for each item of equipment on the network. As stated earlier, at 132 kV and above all equipment impedances are essentially reactive (inductive). In addition, a catalogue of all generic OHL impedances for the various types of OHL configuration will usually be held. Generator impedances will need to be obtained from generation companies. The accuracy of all of this data is critically important for determining fault levels (and for power flow studies) and the database must therefore be totally accurate and constantly kept up to date.

	+ve sequence	Zero sequence
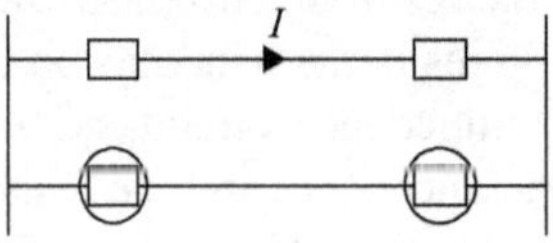 (a)	$Z_1 = Z_P - Z_{PP}$ $= 3.0 - 1.1$ $= 1.9\ \%$	$Z_o = Z_P + 2Z_{PP}$ $= 3.0 + 2.2$ $= 5.2\ \%$
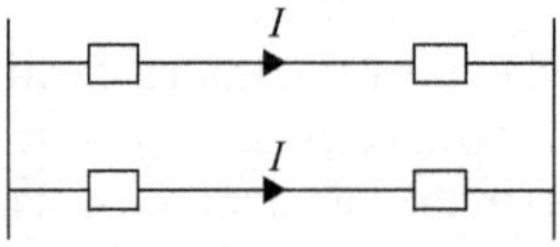 (b)	$Z_{1dc} = Z_1 + Z_{1m}$ $= 1.9 + (-0.1)$ $= 1.8\ \%$	$Z_{odc} = Z_o + Z_{om}$ $= 5.2 + 2.8$ $= 8.0\ \%$
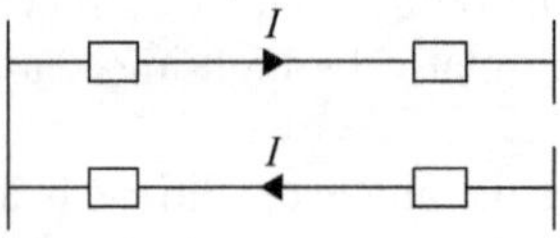 (c)	$Z_{1cd} = Z_1 - Z_{1m}$ $= 1.9 - (-0.1)$ $= 2.0\ \%$	$Z_{ocd} = Z_o - Z_{om}$ $= 5.2 - 2.8$ $= 2.4\ \%$

Z_P = Self-impedance of one phase

Z_{PP} = Mutual impedance between two phases on same circuit

Z_1 = Positive-sequence impedance of one circuit

Z_{1m} = Positive-sequence mutual impedance between adjacent circuits

Z_{1dc} = Positive-sequence impedance of double circuit OHL with currents in same direction

Z_o = Zero-sequence impedance of one circuit

Z_{om} = Zero-sequence mutual impedance between adjacent circuits

Z_{odc} = Zero-sequence impedance of double circuit OHL with currents flow in same direction

Z_{1cd} = Positive-sequence impedance of double circuit OHL with currents in opposite direction

Z_{ocd} = Zero-sequence impedance of double circuit OHL with current in opposite direction

NB = All impedances are percentage impedance of 100 km of typical 400 kV OHL on 100 MVA base

Figure 4.6 OHL impedances: (a) single circuit, (b) double circuit same current direction and (c) double circuit opposite current direction

4.3 Generator short-circuit performance

4.3.1 Generator short-circuit considerations

When undertaking fault calculations on the power network, it is necessary to possess an adequate understanding of the performance requirements of generators, given that they are the source of fault current infeed. The following sections will examine the behaviour of the synchronous generator, when subject to a fault on its terminals.

NB: It was pointed out in Chapter 3 that the fault current infeed from most wind generators (or any asynchronous generator) is usually very small in comparison to synchronous generators. The following will be examined:

- Unloaded generator subject to a three-phase fault
- Generator fault current – practical considerations
- Loaded generator subject to a three-phase fault
- Generator sequence impedances.

4.3.2 Rotor and stator windings

Chapter 3 examined the physical and electrical characteristics of the synchronous generator. Within this context, the rotor, in addition to carrying the rotor winding that provides the DC current (which in turn creates the rotor magnetic field), also carries an additional winding known as the 'damper winding'. This winding is short-circuited and its function is to dampen rotor oscillations during transient conditions. However, this winding also has a significant impact on the stator output current during short-circuit. Figure 4.7 illustrates the electrical arrangements.

4.3.3 Unloaded generator subject to three-phase short-circuit at generator terminals

With reference to Figure 4.7, consider a generator on no load, with an open-circuit voltage *E*, subject to a three-phase short-circuit at its terminals *T*. Fault current flows into the short-circuit, but as it does so it establishes a rotating stator (armature) flux that commences to rise from zero. Unlike the normal steady-state three-phase balanced load current condition, whose rotating flux induces no current into the rotor (see Section 3.6.1.1), the rising flux induces an emf into both the rotor field winding and damper windings (essentially via transformer action) – which causes circulating currents in these windings – which in turn produces a magnetic

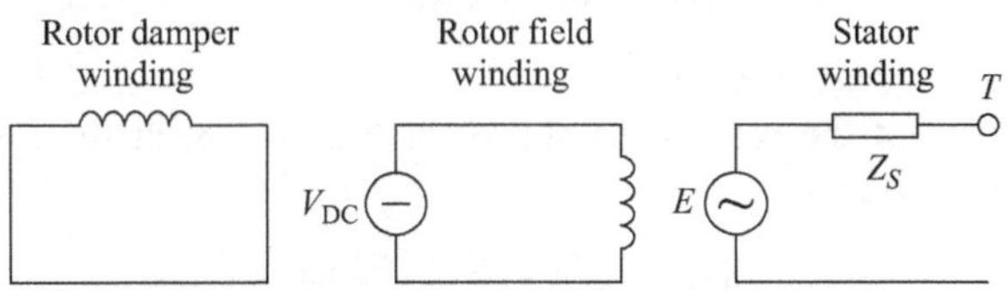

Figure 4.7 Rotor and stator windings

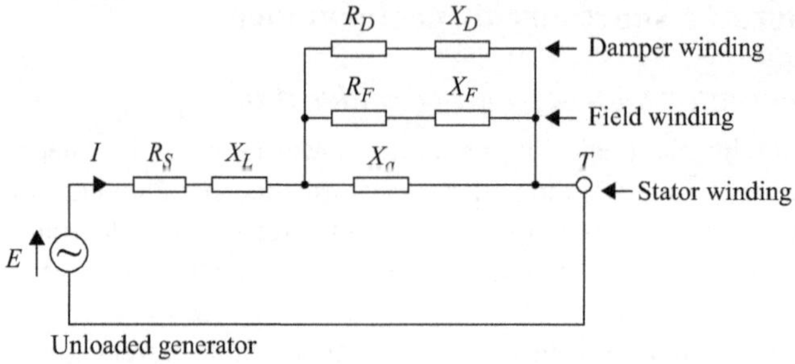

Figure 4.8 Unloaded generator equivalent circuit at instant of fault

flux which opposes and virtually cancels the stator flux (i.e. by Lenz's law). At this stage, the circuitry is equivalent to a three-winding transformer with the stator winding acting as the primary and both the field and damper windings acting as the secondary (both of which are in a short-circuit condition). The generator equivalent circuit at the instant of fault, with impedances shown as per-cent impedances is illustrated in Figure 4.8, in which:

$E =$ generator open-circuit voltage
$X_L =$ stator leakage reactance
$X_a =$ stator armature reaction reactance
$R_S =$ stator field winding resistance
$X_F =$ rotor field winding reactance
$R_F =$ rotor field winding resistance
$X_D =$ rotor damper winding reactance
$R_D =$ rotor damper winding resistance
NB: The resistances are usually relatively small compared to the reactances.

With reference to Figure 4.8, as the stator flux (and associated current) rises from zero to its final steady-state value, the damper and field winding reactances may be considered to transiently increase in value (in accordance with circuit time constants) until, when there is no further rate of increase of flux, they in effect become open-circuit (X_D achieves the open-circuit status in advance of X_F). At this point, the steady-state impedances of X_a and X_L remain, as does the stator steady-state rotating flux. Within this context, classical generator theory defines three reactances (ignoring the resistance) as follows:

1. **Sub-transient reactance X_{ST}**
   This consists of the reactances shown in Figure 4.8, at the instant of fault, defined as follows:

$$X_{ST} = X_L + \frac{1}{\frac{1}{X_a} + \frac{1}{X_F} + \frac{1}{X_D}} \tag{4.15}$$

2. **Transient reactance X_T**
   This comprises the reactances shown in Figure 4.8 minus the X_D, defined as follows:

$$X_T = X_L + \frac{1}{\frac{1}{X_a} + \frac{1}{X_F}} \tag{4.16}$$

3. **Synchronous reactance X_S**
   This is the steady-state reactance, i.e. the reactances shown in Figure 4.8 minus X_D and X_F, defined as follows:

$$X_S = X_L + X_a \tag{4.17}$$

The generator short-circuit current envelope is illustrated in Figure 4.9 (not to scale). It is subdivided into three periods as follows:

1. **Sub-transient period**
   This period commences at the instant of fault and continues until the damper winding transient period is ended (i.e. the damper winding effectively appears as an open-circuit). This period typically lasts 80–120 ms.
2. **Transient period**
   This period commences at the end of the sub-transient period and continues until the rotor winding transient period has ended (i.e. the rotor winding effectively appears as an opencircuit). This period typically lasts for 2 s.
3. **Steady state period**
   This period commences at the end of the transient period and is ongoing.

The expression for the fault current I flowing though the stator winding, and into the fault, is again taken from classical generator theory, and is as follows:

$$I = \left[\frac{E}{X_{ST}} - \frac{E}{X_T}\right] e^{-\left(\frac{t}{T_{ST}}\right)} + \left[\frac{E}{X_T} - \frac{E}{X_S}\right] e^{-\left(\frac{t}{T_T}\right)} + \frac{E}{X_S} \tag{4.18}$$

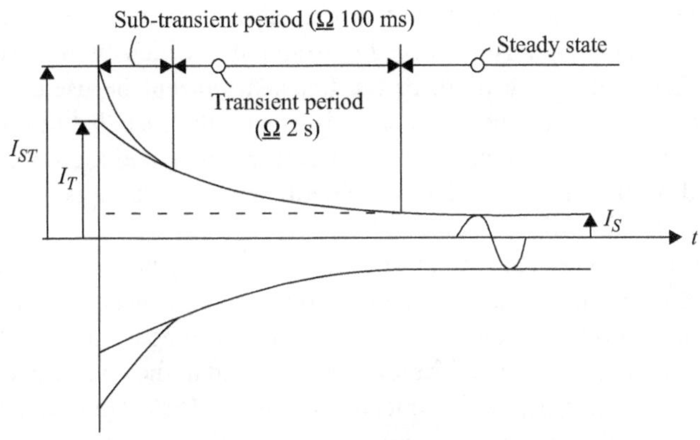

Figure 4.9 Generator short-circuit current envelope

where E is the stator open-circuit voltage, and the sub-transient and transient time constants, respectively are given by:

$$T_{ST} = \frac{X_D}{2\pi f R_D} \text{ and } T_T = \frac{X_F}{2\pi f R_F} \tag{4.19}$$

where f is the frequency (i.e. 50 Hz).

Expression (4.18) can also be directly expressed in terms of the currents in Figure 4.9:

$$I = [I_{ST} - I_T]e^{-\left(\frac{t}{T_{ST}}\right)} + [I_T - I_S]e^{-\left(\frac{t}{T_T}\right)} + I_S \tag{4.20}$$

where $I_{ST} = \frac{E}{X_{ST}}$, $I_T = \frac{E}{X_T}$, and $I_S = \frac{E}{X_S}$.

It is worthy of note that the ratio of $I_{ST}/I_T/I_S$ is typically 12/9/1 (for generators typically in the range 100–660 MW).

It was stated in Section 3.8 that inductive circuits, at the time of being energised (i.e. the instant of current flow) exhibit a DC offset transient. This applies equally to the generator subject to fault. The magnitude of the DC offset depends of course upon the point on wave of fault inception. It needs to be added to the fault current defined in expression (4.20) to obtain the total fault current and may (depending upon point on wave of fault inception) increase the magnitude of maximum fault current by up to a factor of two.

The theory of generator fault current determination is arguably one of the most difficult subjects in power engineering to grasp. However, in practice, it is only necessary to apply the equations – and not necessarily know how they were derived – but in the author's experience, it is very helpful in practice to understand the concept that underpin the equations.

4.3.4 Generator fault current – practical considerations

With reference to Figure 4.9, consideration needs to be given to which value on the current envelope (usually termed the 'current decrement' curve) is relevant to any practical situation under consideration. For example in which circumstances should the sub-transient current be used, or the transient current be used, or any other current in-between? Within this context, it is usually the case that the sub-transient current is used to evaluate main protection performance, and the transient current may be used as the current which the circuit breaker interrupts – but there are exceptions to this as will be discussed later.

It is worthy of note that the sub-transient current I_{ST} has a maximum value at the commencement of the sub-transient period (i.e. fault inception); however, the maximum value of the transient current is not at the commencement of the transient period, but again at the instant of fault inception – and at the commencement of the transient period I_T has decayed, typically, by about 15%. However, in practice, when considering the current magnitude at the start of the transient period, it is often usual to consider I_T, albeit that it is an overestimate of current.

4.3.5 Loaded generator subject to a three-phase short-circuit at the generator terminals

In practice, most generators subject to a short-circuit are likely to be supplying load prior to the short-circuit. As illustrated in Figure 4.10, the generated voltage, E, in such circumstances often has a magnitude which is significantly greater than system nominal voltage V, the latter being equal to the magnitude of E when the generator is unloaded. At first sight, it may therefore appear that the fault current associated with a loaded generator (for a terminal fault at T) would be significantly greater than that of the unloaded generator (due to the much greater value of E). However, this is not quite the case since a rotating magnetic stator flux already exists at the time of fault inception, as a result of the load current, and therefore at the instant of fault inception the stator flux is not rising from zero (as with the unloaded generator). As a consequence of an existing stator current and flux, classical texts on generator theory reason that the fault current I, which occurs with a loaded generator is given by the following expression:

$$I = \left[\frac{E_{ST}}{X_{ST}} - \frac{E_T}{X_T}\right] e^{-\left(\frac{t}{T_{ST}}\right)} + \left[\frac{E_T}{X_T} - \frac{E}{X_S}\right] e^{-\left(\frac{t}{T_T}\right)} + \frac{E}{X_S} \tag{4.21}$$

where $\frac{E_{ST}}{X_{ST}} = I_{ST}$, $\frac{E_T}{X_T} = I_T$ and $\frac{E}{X_S} = I_S$.

With reference to expression (4.21) and Figure 4.11, the quantity E_{ST} is termed the 'voltage behind the sub-transient reactance' and is given as follows:

$$E_{ST} = V + I_L X_{ST} \tag{4.22}$$

where I_L is the pre-fault load current.

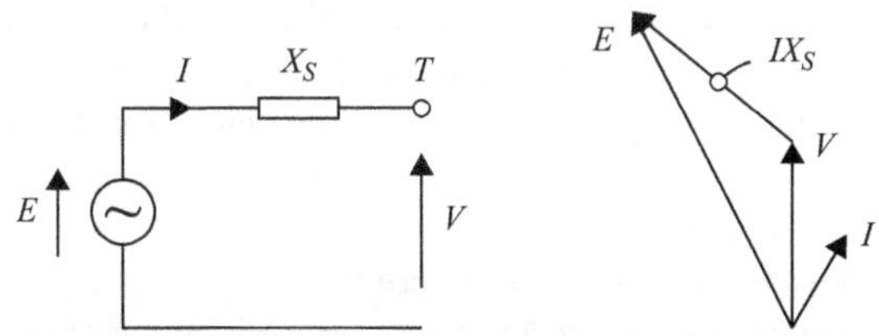

Figure 4.10 Generator loaded prior to fault

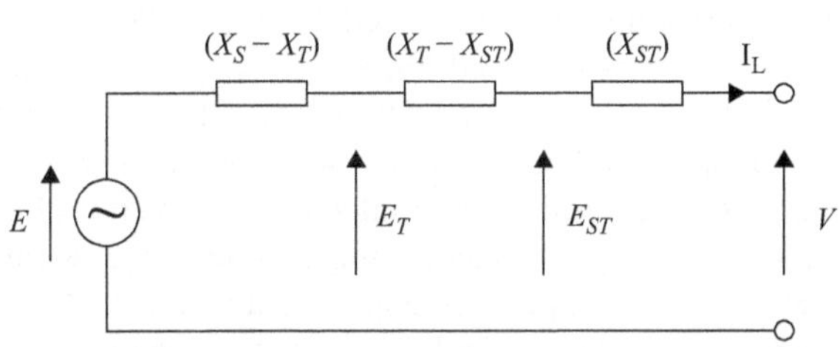

Figure 4.11 Voltage derivation for a generator on-load at the instant of fault

Similarly, E_T is termed the voltage behind the transient reactance and is given by:

$$E_T = V + I_L X_T \tag{4.23}$$

where, again, I_L is the pre-fault load current.

It is to be noted that for the instance of I being zero, i.e. an unloaded generator then expression (4.21) is one and the same as expression (4.18).

Calculations based upon expression (4.18), the unloaded generator, and expression (4.21) the loaded generator, show that in practice the fault current for a generator on full-load output is unlikely to exceed that of an unloaded generator by more than 10%. Thus, if the application under consideration can accept an error of 10%, then the simpler expression of (4.18) may be utilised (which is often the case).

4.3.6 Generator sequence impedances

It is worthy of note that the term generator impedance and generator reactance are reasonably interchangeable, since the resistive component of generator impedance is relatively small. Generator sequence impedances can be obtained as follows:

1. **Generator positive-sequence impedance**
   With reference to Section 4.3.3, there are three values of generator positive-sequence impedance, which are as follows:
   (i) Sub-transient
   (ii) Transient
   (iii) Synchronous
   The values can be obtained by short-circuit test, with the short placed at the generator terminals. An oscilloscope (or other similar device) is required to measure the current magnitudes. The pre-fault open-circuit voltage is usually the system nominal voltage. The transient impedance can be obtained by open-circuiting the damper winding. Generator models are also now available for impedance determination.
2. **Generator negative-sequence impedance**
   Negative-sequence impedance may be obtained by running the generator at synchronous speed with the rotor field winding un-exited and applying NPS voltage to the generator terminals.

   The rotating armature field produced by the NPS current in the stator winding rotates at the same speed but in the opposite direction to the rotor, thereby inducing voltages and circulating currents into both the rotor field winding, and damper winding – resulting in transformer action between stator and rotor. The resulting currents are limited by both the rotor field winding and damping winding leakage reactances, together with the stator leakage reactance. As such, the NPS impedance approximates to the PPS impedance sub-transient reactance X_{ST}. Again, the NPS impedance can often also be obtained by generator modelling.

3. **Generator zero-sequence impedance**
   Generator zero-sequence impedance is determined in a similar manner to that of the NPS impedance – but this time injecting ZPS voltage into the generator terminals. In this instance, the armature fluxes produced by the ZPS stator currents summate to zero in the rotor body and induce no voltage into either the rotor field winding or the damper winding. The zero-sequence impedance therefore approximates to that of the stator leakage reactance. Once again, this may also be obtained through generator modelling.

4.3.7 Generator subject to an open-circuit

An onerous condition for a generator is that of supplying maximum load current, and then instantly being subject to an open-circuit either at the generator terminals or on the network side of the generator transformer. This is termed a 'full-load rejection'. In this instance, the full open-circuit voltage E, as shown in Figure 4.10, is applied to the generator terminals and possibly the generator transformer (depending on the location of the open-circuit). Both the generator and generator transformer must be designed to withstand this possibility. Within this context, the generator AVR will rapidly come into play to reduce the voltage to the system nominal value.

4.4 Sequence networks for common fault conditions

4.4.1 Sequence networks

When faults arise on the power system, the currents that flow in the three phase-sequence networks can best be analysed by a diagram (unique to the type of fault) that interconnects the three phase-sequence networks (each shown as a single-phase equivalent circuit) – and which satisfies the symmetrical component mathematical solution for the fault type in question. This section will examine the sequence network arrangements for the following common fault conditions:

- Three-phase
- Single-phase-to-earth
- Phase-to-phase
- Open-circuit

When examining the faults, the following needs to be noted:

- Only the positive-sequence network contains a source voltage E since generators only produce PPS voltages. The NPS and ZPS voltage generators are therefore replaced by a short-circuit.
- As stated earlier, when deriving phase-sequence quantities, it is usual to express them in terms of the Red phase, i.e. the reference phase. The Yellow and Blue phase quantities are then obtained by applying the a operator to the Red-phase-sequence quantities, as indicated in expressions (4.2) and (4.3).
- In practice, it is often usual to use a short-hand by abbreviating I_{R1}, I_{R2} and I_{R0} to I_1, I_2, and I_0, respectively.

- All impedances are reactive and expressed as per-cent impedance on 100 MVA base.
- The driving voltage E is taken to be system nominal voltage.

4.4.2 Three-phase fault

Figure 4.12 shows the sequence network solution for a three-phase fault which is either clear of earth or connected to earth. It is a balanced three-phase condition and as such only PPS currents can flow.

The sequence diagram therefore only involves the +ve sequence network as shown. NB: It is assumed that the pre-fault currents are negligible:

$$\text{Therefore, } I_R = I_{R1}, I_{R2} = 0, I_{R0} = 0 \text{ and } I_Y = a^2 I_R, I_B = aI_R \tag{4.24}$$

4.4.3 Single-phase-to-earth fault

Figure 4.13 illustrates the single-phase-to-earth-fault network. It is reasoned (in more specialist texts) that the sequence currents arise from a series circuit connecting the three-sequence networks, as shown. Current relationships are as follows:

$$I_{R1} = I_{R2} = I_{R0} \tag{4.25}$$

$$\text{So } I_R = I_{R1} + I_{R2} + I_{R0} = 3I_{R1}$$

$$\text{So } I_Y = a^2 I_{R1} + aI_{R2} + I_{R0} = I_{R1}(a^2 + a + 1) = 0$$

$$\text{Similarly, } I_B = aI_{R1} + a^2 I_{R2} + I_{R0} = I_{R1}(a + a^2 + 1) = 0$$

At the point of fault, the Red-phase voltage falls to zero; however, individual sequence components of voltage exist as shown. To determine the voltage at the point

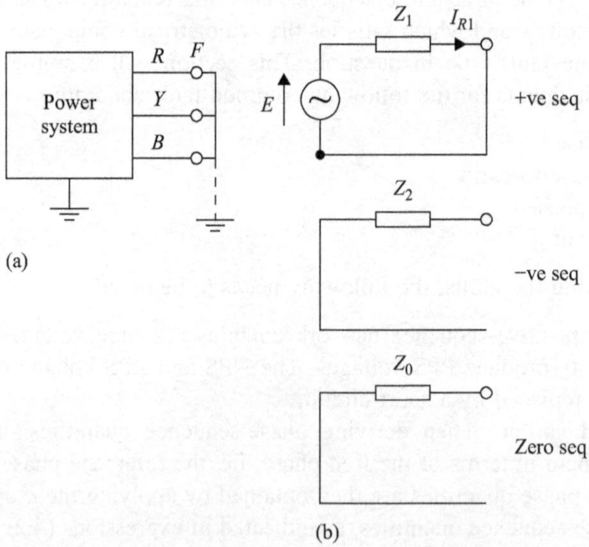

Figure 4.12 Sequence network for a three-phase fault: (a) power system subject to fault and (b) sequence network

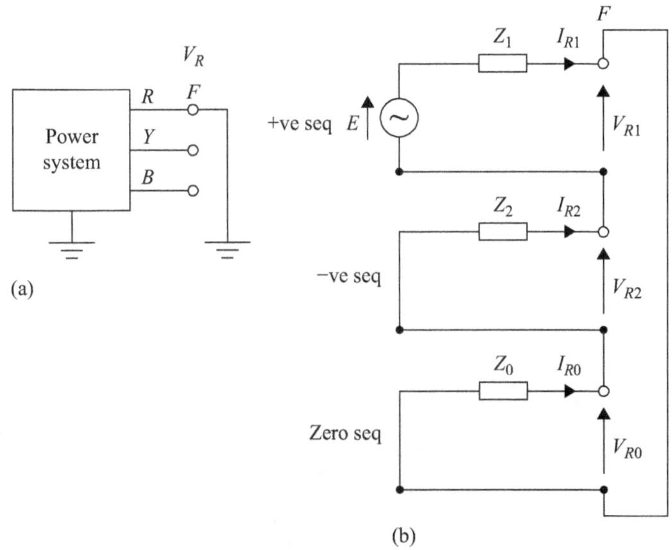

Figure 4.13 Sequence network for a single-phase-to-earth fault: (a) power system subject to fault and (b) sequence network

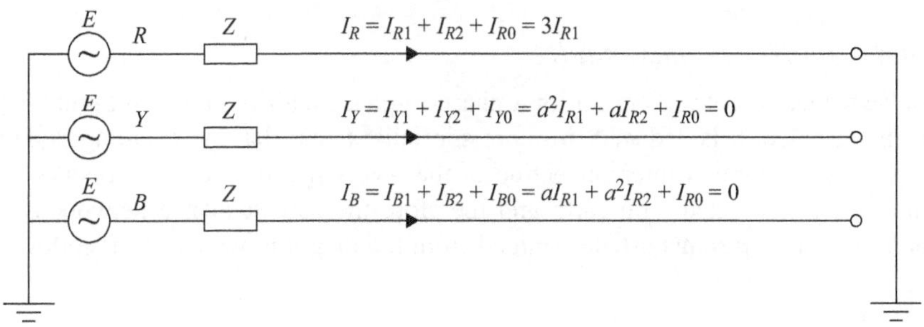

Figure 4.14 Superposition of three sequence-network currents to obtain currents in faulted network – single-phase-to-earth fault

of fault, the negative-sequence and zero-sequence networks are considered to contain a voltage generator of zero-voltage output. Thus, at the point of fault, it can be said:

$$V_R = V_{R1} + V_{R2} + V_{R0}$$

$$\text{Or } V_R = (E - I_{R1}Z_1) + (0 - I_{R2}Z_2) + (0 - I_{R0}Z_0)$$

$$\text{Or } V_R = E - I_{R1}Z_1 - I_{R2}Z_2 - I_{R0}Z_0$$

and applying Kirchhoff's second law around the closed network – see Section 3.3.1.1:

$$V_R = 0$$

For illustration purposes, Figure 4.14 shows the superposition of the three-sequence network currents onto the power system to derive the power system currents.

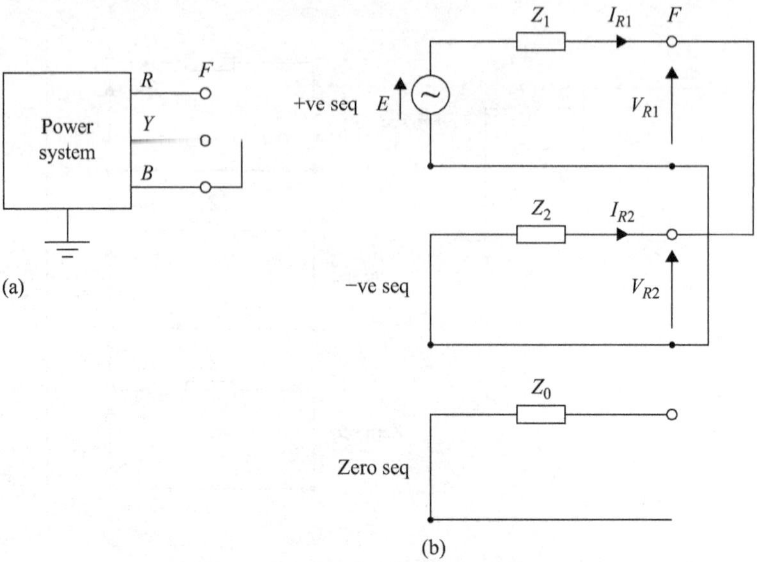

Figure 4.15 Sequence network for a phase-to-phase fault: (a) power system subject to fault and (b) sequence network

4.4.4 *Phase-to-phase fault*

The sequence network diagram for a phase-to-phase fault is shown in Figure 4.15. In this instance, it is deduced (in more specialist texts) that the resulting diagram comprises the parallel interconnection of the +ve seq and −ve seq networks. The fault is not associated with earth and therefore the zero-sequence network is not connected. The currents are determined from the diagram, which are as follows:

$$\begin{aligned} I_{R1} &= -I_{R2} \\ I_R &= I_{R1} + I_{R2} = 0 \\ I_Y &= a^2 I_{R1} + a I_{R2} = I_{R1}(a^2 - a) \\ I_B &= a I_{R1} + a^2 I_{R2} = I_{R1}(a - a^2) \end{aligned} \tag{4.26}$$

Therefore, $I_Y = -I_B$.

4.4.5 *Single-phase open-circuit*

An open-circuit on the power system can be almost as problematic as a short-circuit – and often difficult to locate. Figure 4.16 illustrates the sequence network arrangements for an open-circuit. As can be seen:

$$I_{R1} = -(I_{R2} + I_{R0}) \tag{4.27}$$

Thus, an open-circuit (which may arise on an OHL, circuit-breaker interrupters, disconnector blades not properly closed, etc.) also causes both negative and

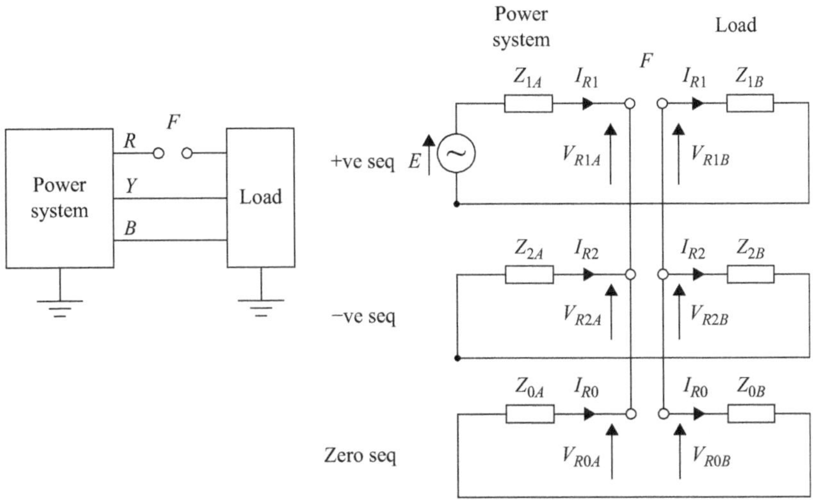

Figure 4.16 Sequence network for an open-circuit

zero-sequence currents to arise. This may cause the operation of protection systems which are responsive to such currents.

4.4.6 Phase-sequence current flow analysis

Figure 4.17 provides an analysis of phase-sequence current flow for a practical but relatively complex situation involving an earthing transformer – as an instructive and practical example of symmetrical component analysis. Salient points are as follows:

- Figure 4.17(a) shows a single-line diagram of a circuit subject to Red phase fault to earth. The position of the current transformers on the circuit is shown, and it is instructive to evaluate the flow of sequence currents through the current transformers – particularly when considering protection application and settings.
- Figure 4.17(b) shows the sequence impedance network diagram based upon the circuitry in Figures 4.3, 4.5 and 4.14. Within this context, impedance Z_{ET0} (in Figure 4.17(b)) is one and the same as Z_{OH} in Figure 4.5(c).
- Figure 4.17(c) shows the superposition of the three-sequence network currents – and the resulting fault current. As can be seen, the zero-sequence currents from each phase flow into the fault, and the current in each part of the actual network is shown. The following is worthy of note:
  - Zero-sequence current flows in CT C – but not CT's A and B.
  - With reference to Section 5.4.1.3, when PPS currents flow through a Yd1 transformer, they are phase shifted by −30° from HV to LV side. Alternatively, NPS currents are phase shifted by +30° when flowing through the same transformer.

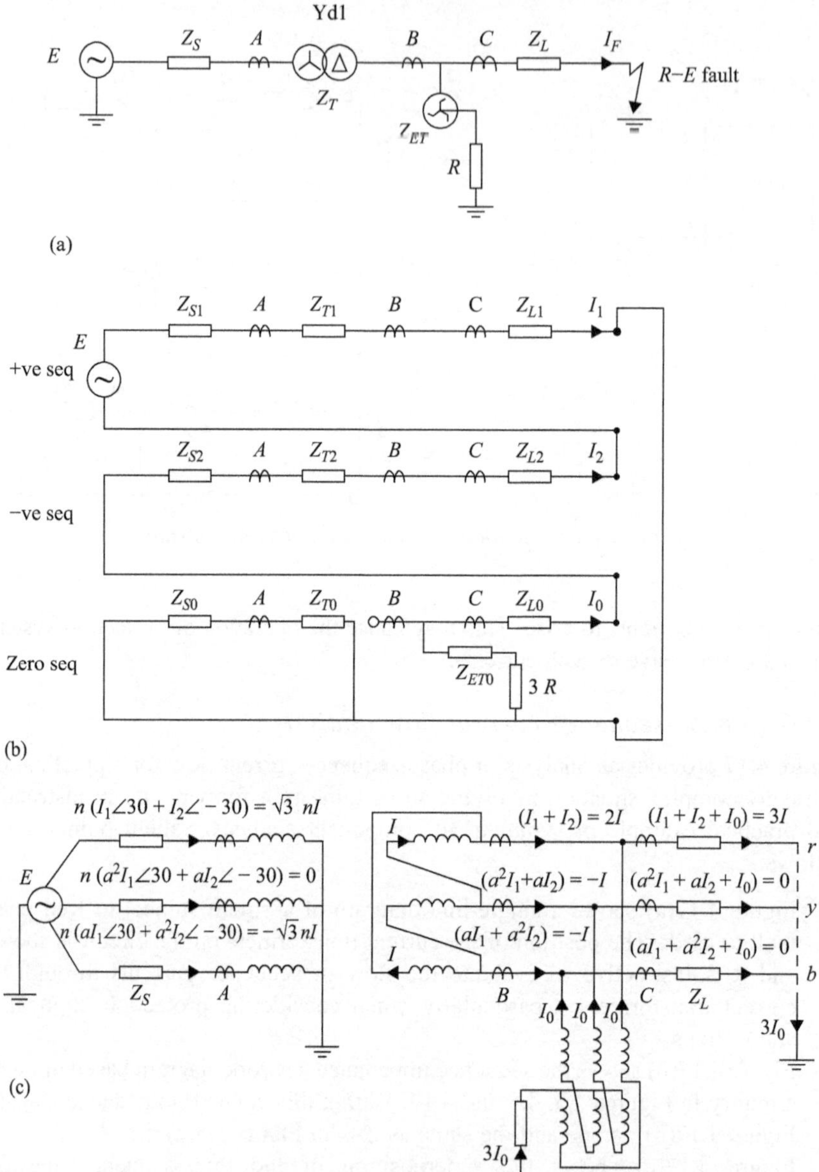

Figure 4.17 Phase-sequence current flow – analysis: (a) circuit subject to fault, (b) sequence-impedance network and (c) three-phase network

- As can be seen, although current flow can be mapped directly from the sequence network to the actual network, the phase-shifts through the transformer must be taken into account.
- On the diagram, *n* represents the HV/LV winding ratio of the transformer.

– The a operator and polar notation on the primary winding (middle phase) are manipulated as given in the following example (see Section 3.2.1):

$$a^2\angle 30^\circ + a\angle -30^\circ = 1\angle 270^\circ + 1\angle 90^\circ = 0$$

4.5 Maximum and minimum fault level studies

4.5.1 Maximum and minimum busbar fault levels

Each year, many power network companies undertake fault level studies for each and every substation across the network. The study places a fault on the substation busbars (usually with all circuits in service) to obtain the current flow in every feeder into the substation, and the resulting fault current on the busbars. The following studies are usually undertaken:

- Winter maximum study with maximum generation in service – with all circuits on the power system in service.
- Summer minimum study with summer minimum generation – with all circuits on the power system in service.
- Studies for both three-phase faults and single-phase-to-earth faults.
- Studies of both the sub-transient fault current at the instant of fault, and the transient fault current typically declared 120 ms after fault inception (i.e. at the end of the sub-transient period.

Figure 4.18 summarises a study for a typical UK 400-kV substation busbars to which generation is connected.

As a generalisation, the three-phase and single-phase fault currents for each of the conditions (e.g. winter max sub-transient) are usually within 25% of each other.

4.5.2 Maximum and minimum fault level studies – feeders

For each of the conditions shown in Figure 4.18, the individual feeder contributions to the busbar fault are also usually provided. A typical diagrammatic example is provided in Figure 4.19 (although they are usually presented in tabular format) which shows the Faraday 400-kV substation busbars subject to a winter max

Faraday 400 kV substation							
Winter max (kA)				**Summer min (kA)**			
ST		***T***		***ST***		***T***	
3 Ph	**1 Ph**	**3 Ph**	**1 Ph**	**3 Ph**	**1 Ph**	**3 Ph**	**1 Ph**
37	38	29	31	31	32	21	23

Figure 4.18 Typical substation max and min fault levels

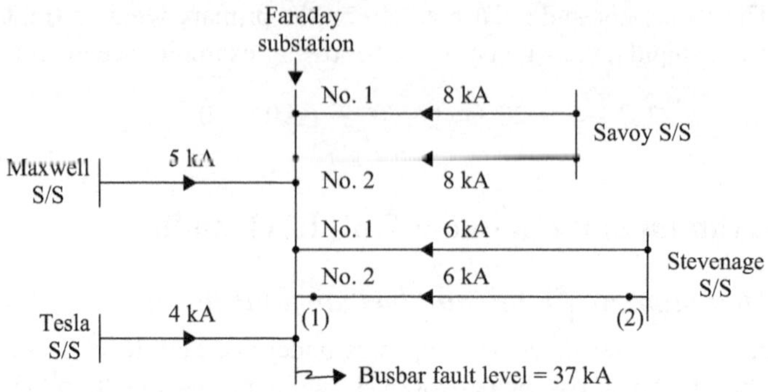

Figure 4.19 Example of busbar and feeder fault level study

sub-transient, three-phase fault, i.e. a busbar current of 37 kA as shown in Figure 4.18.

4.5.3 Use of maximum and minimum fault level studies

The maximum and minimum fault level studies as described above serve a number of purposes:

- To enable confirmation that HV equipment is not subject to fault level current beyond rated capacity.
- To provide maximum and minimum fault current data for the setting of new protection relays, during construction, and reassessing the settings of existing relays.
- To provide fault currents for assessing substation rise of earth potential.

4.6 Fault currents – methods and techniques

4.6.1 Fault calculations – overview

In many instances, the power system fault information required for a particular application is limited either to the point of fault, or a small part of the network containing the fault. Such instances require only a small amount of the network to be considered, which readily facilitates calculations by hand rather than resort to computer-based fault studies – which not only take time to set up and arrange but usually requires the resource of specialist engineers with limited available time. The following sections will examine fault calculation methods and approximation techniques – which assist with the task of fault calculations.

4.6.2 Generator driving voltage

Section 4.3.4 stated the generator open-circuit voltage *E*, associated with Figures 4.12 to 4.16 should be taken to be the system nominal voltage. This good approximation is acceptable practice but not fully accurate. Within this context,

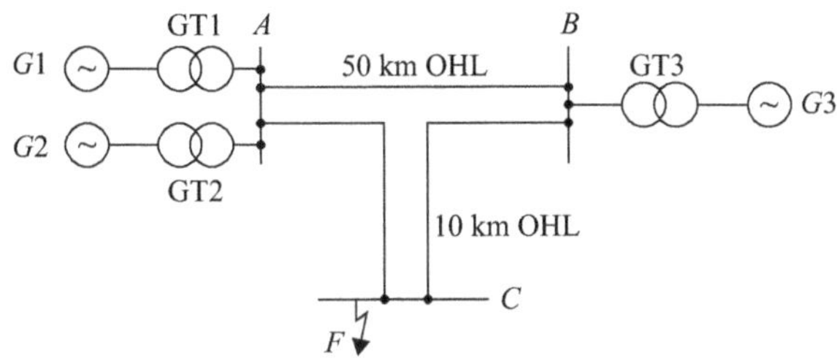

Figure 4.20 Power system under consideration

Section 4.3.5 pointed out that the generator driving voltage at the instant of fault is determined by the generator pre-fault load current, as defined in expression (4.21). However, the text in Section 4.3.5 states that the difference between the generator driving voltage when taking account of pre-fault load current, and the generator system nominal voltage (i.e. the open-circuit voltage on no load) is likely to create an error of 10% or less – which for many applications is acceptable. Thus, if absolute accuracy is required, expression (4.21) should be used, but if an error up to 10% can be tolerated, system nominal voltage and expression (4.18) can be utilised.

4.6.3 Methods and techniques

Figure 4.20 shows a simple power system subject to a fault at *F*. Depending upon the type of fault, e.g. three-phase, single-phase-to-earth fault, etc. the sequence networks in Figures 4.12–4.16 should be used to determine the fault current. Within this context, some of the common techniques available for calculating the current in the sequence networks, both in the fault and in the wider power system are as follows:

1. **Network equations**
   Network equations as described by both the Kirchhoff method or the Maxwell method, see Chapter 3, could be employed to solve for the fault current, and all other currents on the power system. However, unless the scale of the network is reasonably small, such a solution can be tedious and would generally not be used. In any event, most engineers would prefer not to solve simultaneous equations.
2. **Superposition**
   The superposition method, as described in Section 3.3.1.3, could be employed with the solution obtained by network reduction to determine the current in the fault from each generator source, and back substitution to obtain the currents in the power system and fault.
3. **Thevenin**
   If the current solely in the fault is required, the Thevenin method as described in Section 3.3.1.4 could be advantageously used, since only one source of generation, i.e. the pre-fault voltage at the point of fault is required – and as stated earlier, for most applications, the system nominal voltage would be used. If the pre-fault voltage had been different to nominal voltage, the error, in practice, is usually only a few per-cent.

4. **Extended superposition**
   In most instances, the fault current far exceeds the pre-fault load current, and therefore the load current can be ignored. However, instances may arise of high load currents in conjunction with a relatively low fault current, and it is necessary to know the total current in those parts of the network close to the fault (e.g. to assess biased differential feeder protection performance) – and for this the extended superposition can be used, see Section 3.3.1.5. It would of course be necessary to know the pre-fault load current in order to utilise this method.
5. **Single equivalent generator**
   The network shown in Figure 4.20 can be simplified by use of a single equivalent generator as shown in Figure 4.21. Single equivalent generators can theoretically be derived in terms of voltage magnitude, phase-angle and generator reactances for any number of connected generators – however, the mathematics becomes complex unless the voltage, phase-angle and reactances are similar. If this is the case (e.g. voltages within 10%, phase-angles within 30° and reactances within 20%) an equivalent generator with parameters equal to one of the generators can be employed to provide 'approximate' fault calculation figures. This greatly simplifies the analysis.

4.6.4 Use of circuit maximum and minimum fault levels

When evaluating protection application settings, there is a frequent requirement to evaluate the maximum and minimum fault currents to which a circuit may be subject, when looking from each end in turn. With reference to the data in Figure 4.18, consider this requirement with reference to Figure 4.19 and the protection at Faraday substation on the Stevenage No. 2 circuit.

The maximum fault current seen by the protection is that of a terminal fault [i.e. position (1) on Figure 4.19] when the current is that of the winter maximum sub-transient. However, the fault position is such that this is the same current as that for the busbar fault in Figure 4.19.

The minimum fault current experienced by the protection at Faraday is that due to a fault at the Stevenage No. 2 end of the circuit [i.e. position (2) on Figure 4.19]

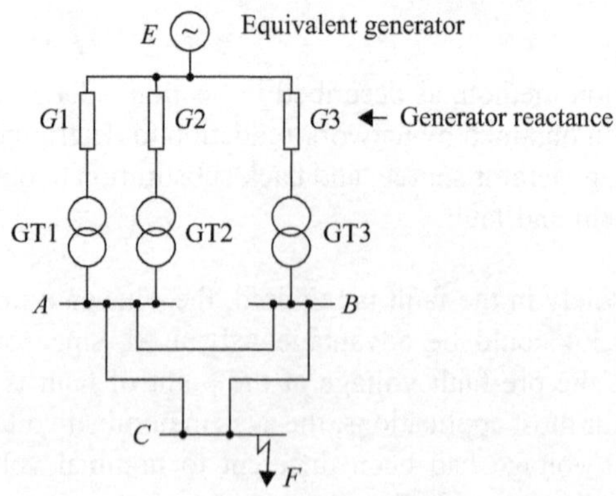

Figure 4.21 Equivalent generator concept

under summer minimum transient conditions. This is the same magnitude of fault current that would flow from Faraday to a busbar fault at Stevenage (i.e. as given in the max/min busbar fault current study for Stevenage substation).

There is also frequently a requirement to calculate the minimum fault current (at Faraday) for a fault at the far end of the (Stevenage No. 2) circuit when the circuit breaker at Stevenage is open, i.e. a single-end fed fault. An approximation technique for determining the fault current is as follows:

1. **Step 1**
   Populate the diagram shown in Figure 4.19 with the summer minimum transient fault currents – for a busbar fault at Faraday substation (consider initially – for simplicity – three-phase fault currents). NB: The transient currents are taken as 120 ms after fault inception.
2. **Step 2**
   On the diagram shown in Figure 4.19, open the circuit breaker at Stevenage on the Faraday No. 2 circuit. This removes the infeed from Stevenage into the busbar fault at Faraday.
3. **Step 3**
   Since summer minimum conditions are with all circuits in service (which is an unlikely running arrangement), remove the largest infeed into the busbar fault at Faraday substation – to represent practical summer conditions. This is likely to be one of the infeeds from Savoy substation. NB: Other infeeds may be withdrawn as local considerations dictate.
4. **Step 4**
   Remove the Step 2 and Step 3 current infeeds from the existing busbar fault current at Faraday to determine the new busbar fault current I_F.
5. **Step 5**
   Convert the new busbar fault current into a source impedance Z_S (see Figure 4.22), i.e.:

   $$Z_S = \frac{V_L}{\sqrt{3}I_F}\,\Omega$$

6. **Step 6**
   With reference to Figure 4.22, connect the source impedance in series with the Faraday – Stevenage circuit impedance Z_L, to determine the minimum fault current through the circuit.

This approximation technique is likely to result in a fault current which is lower than reality, since the removal of current infeeds is likely to increase the remaining

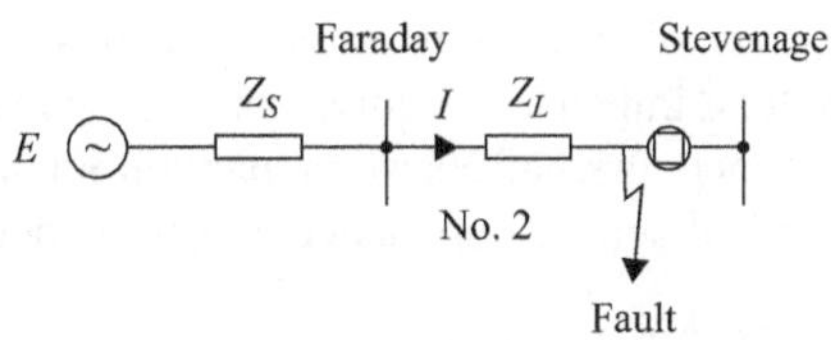

Figure 4.22 Minimum fault level circuit

fault current infeeds and so lower the source impedance. But if the protection can be satisfactorily set using the fault current determined (adding a factor of safety), then this is a useful, and relatively simple technique – and is often used in practice in preference to computer studies.

4.6.5 Fault current decrement and protection operation

Section 4.3.3 pointed out that a fault at the terminals of a generator results in a current decrement which results in a fault current reduction from the sub-transient value to the synchronous value, typically by a factor of 12, although the generator AVR, on sensing a reduction in terminal volts may decrease the factor to (typically) 3. The extent to which current decrement occurs is dependent upon the step magnitude of change to generator terminal voltage. Therefore, faults on the power system that are electrically remote from a generator will have little effect on generator terminal voltage and the size of the current decrement is accordingly reduced. Generally, on the lower voltage networks little account needs to be taken of the current decrement, and the fault current is a near constant magnitude (until cleared).

On the transmission networks, where the main protection operating time is typically between 15 and 40 ms, the current decrement exists, but rarely needs to be taken into account since the sub-transient current will not have decayed significantly in this time-scale. However, IDMT overcurrent and earth-fault relays, where the operating time may be in excess of 1 s, are subject to current decrement considerations – particularly near sources of generation. Generally, at network voltage of 132 kV and below the current decrement does not need to be considered with reference to protection application – unless significant generation is connected in the vicinity.

4.6.6 Transformer tap changers

Fault calculations are usually undertaken on the assumption that transformers are operating on nominal ratio tap position, the error introduced by this assumption is normally quite small when considering short-circuit currents. However, instances may arise where it is necessary to take the tap changer and winding ratio into account. Fault investigations in particular should consider whether the tap changer was operating towards the end of its range.

4.6.7 The infinite busbar

The term infinite busbar is a useful concept, frequently used to represent a fixed voltage source V, that can provide/absorb infinite power (MVA, MVAr, MW) through a source impedance of zero ohms. A more practical approach is to consider the infeed being through an impedance equivalent to system maximum fault level. For example, if a 400 kV network had a system maximum fault level of 63 kA, then the system maximum infeed source impedance Z_S can be determined as follows:

$$Z_S = \frac{V_L}{(\sqrt{3})I} = \frac{400\,\text{kV}}{(\sqrt{3})63\,\text{kA}} = 3.67\,\Omega$$

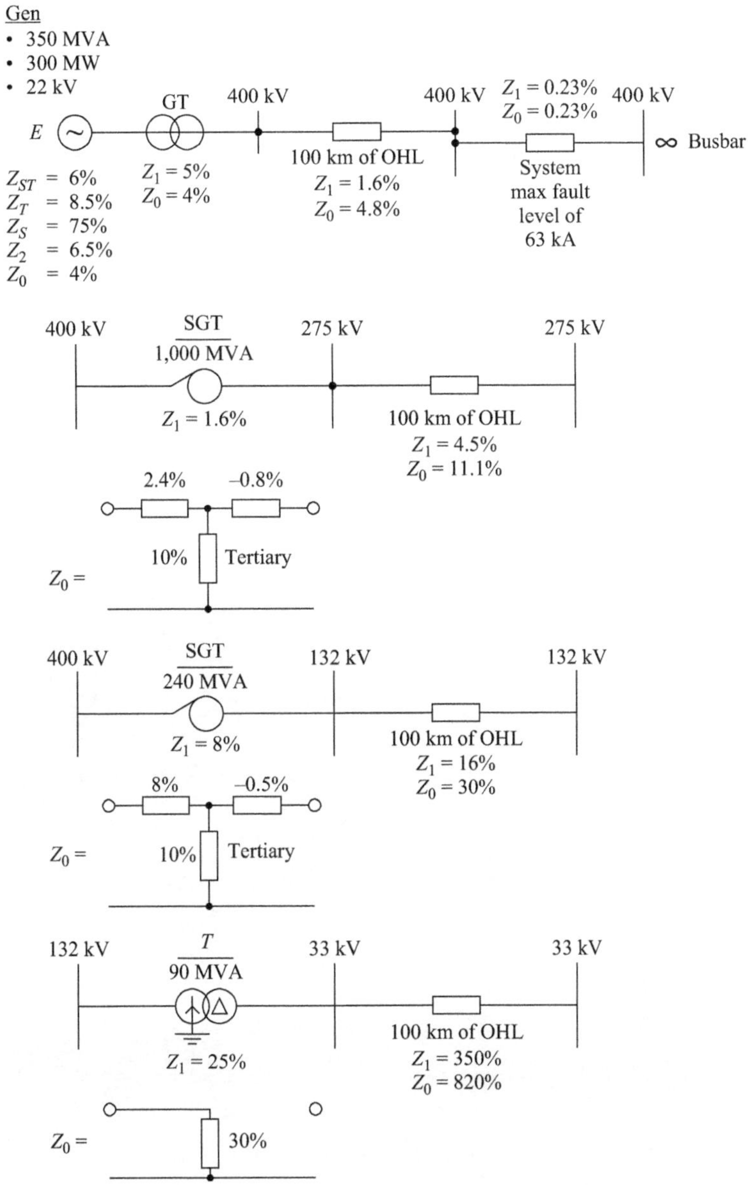

Figure 4.23 Comparative impedances on 100 MVA base

From expression (3.22) $\%Z = \frac{3.67\times100\times10^6}{(400\times10^3)^2} \times 100\% = 0.23\%$ on 100 MVA base.

This impedance limits the maximum infeed that this particular network can supply to a fault. This is illustrated in Figure 4.23.

4.6.8 Comparative impedances

Figure 4.23 illustrates comparative and typical impedances as found on the UK network (and typical of many in the world). In the author's experience, such a diagram provides finger tips, rough and ready indication of the orders of magnitude involved and helps in providing a greater understanding of, and perspective on, the wider power system.

Chapter 5
HV network design

5.1 Introduction

In Chapter 1, it was stated that a construction design specification essentially comprises two separate but interdependent categories of design specification: one that is project specific (i.e. design requirements that are unique to the project) and the other that is project generic (i.e. design requirements that are usually common to all projects).

The project-generic design specification (see Section 1.4.1) is usually derived from two types of documentation suites as follows:

1. **Technical policy**
   Technical policy usually relates to a documentation suite which is internal to a power network company. In the context of a construction scheme, the design requirements arising from technical policy are (usually) determined by the power network company prior to sanction. Such requirements more usually relate to network design standards (e.g. the criterion for levels of security of supply, etc.) as opposed to equipment design standards. The proposed network design would of course need to be conveyed to the contractor as part of the tender/contract documentation.
2. **Equipment technical specifications**
   Equipment technical specifications are essentially procurement specifications, usually based on equipment design standards (e.g. IEC, BS, etc.) and prepared by a power network company for a contractor to deliver, as part of the contract. Such specifications, of necessity, also stipulate some network design requirements (e.g. network over-voltages, maximum short-circuit currents, required fault clearance times, etc.).

This chapter will commence by examining the project-generic design requirements concerned with HV network design (with the requirements for HV equipment design covered in Chapters 6–8). The HV network design requirements examined may be categorised into:

- HV network planning standards.
- HV network design standards – requiring power system studies.
- HV network design standards – with fixed design parameters.

The chapter will then go on to examine project-specific design, at the HV network design level (and the interaction with project-generic design) and conclude with future challenges facing the design of power networks.

5.2 HV network planning standards

5.2.1 Network planning standards – purpose

Network planning standards are essentially concerned with security of electricity supply and the standard of service to the customer. In the United Kingdom, these standards are defined in, and imposed by, both the grid code and the distribution code, for the transmission and distribution systems, respectively, see Chapter 2.

NB: Similar types of documents exist in many other countries. These documents are known as the 'Licence Standards', and in general, they cover:

- Security of supply (i.e. criterion which limits the loss of supply)
- Quality of supply (covering, e.g., system nominal voltages and allowable deviations)
- System phasing standard (across all networks).

In practice, the criterion stated in the Licence Standards often leads to standard network designs (of which many designs are long-standing), e.g. the type of busbar system (i.e. single busbar, double busbar, etc.), the maximum number of supply transformers (and capacity) connected to a substation, the maximum number of in-feeding circuits into a substation, etc. Even so, it is perfectly possible for non-standard designs to be the optimum solution.

The task of network planning (i.e. the substation and circuit design proposal which satisfy both the need case and planning standards) is one of the first tasks to be undertaken in power network construction and forms part of the system design report (or other similar document), see Section 17.3.8.

In the United Kingdom, the two key network planning standard documents are:

- National electricity transmission system security and quality of supply standard (SQSS) – published by National Grid plc.
- Engineering recommendation P2/6: security of supply – published by ENA.

NB: The above documents will be paralleled by similar documents in many other countries.

Key requirements of these documents will be briefly examined below.

5.2.1.1 Security and quality of supply standard

The SQSS is applicable to the UK transmission system (i.e. 400, 275, and some 132 kV networks). It sets out a coordinated set of criteria and methodologies that applies for both planning and operational purposes – both onshore and offshore. The SQSS is a detailed and comprehensive document, and therefore the following text will provide only a summary of the key design considerations (as an indication of the document content and requirements).

The following SQSS requirements apply to the onshore transmission system. This is stated to comprise three interconnected components: generation points of interconnection (i.e. power station infeed); the main interconnected transmission system (MITS); grid supply points (GSPs) where demand is connected.

1. **Key planning factors**
   When planning a transmission substation, the following is required to be considered: security of supply; extendability; maintainability; operational flexibility; protection arrangements; short-circuit limitations; land area and cost.
2. **Substation types**
   (i) Both generation point of connection substations, and marshalling substations, shall be a double busbar design.
   (ii) GSP (transformer) connections may be either a double busbar substation design, four switch mesh, single transformer teed from the transmission system, or designs in between as circumstances dictate.
3. **Transmission design requirements – to limit generation loss**
   Generation connections to the transmission system should be planned such that commencing with an intact system the following is required:
   (i) Following a fault outage of any single transmission circuit, no loss of power infeed (generation) shall occur.
   (ii) Following the planned outage of any single section of busbar, or mesh corner, no loss of power infeed (generation) shall occur.
   (iii) Following a fault outage on any single generation circuit, or a section of busbar or a mesh corner the loss of power infeed (generation) shall not exceed 1,320 MW (1,800 MW in future), under which circumstances there shall be no frequency deviation on the power system outside specified limits.
4. **MITS transmission capacity**
   The minimum transmission capacity of the MITS shall be designed in accordance with the following requirements:
   (i) Prior to any fault arising, the transmission system shall not be subject to: equipment loadings exceeding the pre-fault rating limits; voltages outside the pre-fault planning limits; or system instability.
   (ii) For the situations specified below, there shall not be a loss of supply (with some exceptions); unacceptable overloading of transmission equipment; unacceptable voltage conditions; or system instability. The situations are:
      (a) Fault on a single transmission circuit
      (b) Fault on a double circuit OHL
      (c) Fault on a section of busbar or mesh corner
      (d) Fault on a transmission circuit, or generation unit, coincident with a prior outage of another transmission circuit.
5. **Demand (load) connection capacity**
   The SQSS also contains detailed requirements of the HV network design associated with a 'demand group' (i.e. a substation or number of substations

that collectively take power from the transmission system). Again this requires that there shall be no loss of supply (greater than 1 MW), unacceptable equipment overloading, no voltages outside limits and no system instability for a range of circumstances covering:

(i) The in-tact system
(ii) Planned outages
(iii) Fault outages.

6. **Circuit complexity**
Design restrictions on circuit complexity for the purpose of minimising the time to isolate and earth circuits for maintenance work, and to minimise the potential for human error are as follows:
(i) Facilities for the isolation and earthing of transmission circuits shall not be located at more than three sites.
(ii) Isolation and earthing of transmission circuits shall not require the operation of more than five circuit breakers.
(iii) No more than three transformers shall be connected together and controlled by the same circuit breaker.

5.2.1.2 Engineering recommendation P2/6: security of supply

P2/6, although referencing the transmission system, is the principle document for influencing network design on the distribution system, i.e. essentially 132 kV and below. The document defines the required levels of security of supply in terms of the timescales to restore electricity supplies to customers following any interruption. Less time is allowed for larger groups of load, as may be expected. The document also takes account of the impact of generation directly into the distribution system (which is incrementally increasing). A summary of the defined security of supply levels is given below, with the aim of providing an insight into the structure of the document. The reader is directed to the document itself should a more comprehensive understanding be required. It is worthy of note that different arrangements apply to parts of the distribution network in London.

P2/6 is structured around the following criterion:

1. **Class of supply**
The network supplies (essentially substations) are classed as follows:
——— Group Demand ———
(i) Class A = Up to 1 MW
(ii) Class B = Over 1 MW and up to 12 MW
(iii) Class C = Over 12 MW and up to 60 MW
(iv) Class D = Over 60 MW and up to 300 MW
(v) Class E = Over 300 MW and up to 1,500 MW
(vi) Class F = Over 1,500 MW refer to SQSS.
NB: Group demand is effectively the total maximum demand on the substation.

2. **Criterion**
   The criterion is associated with the minimum demand to be met (i.e. supplied) after
   (i) A first outage, defined as (n – 1), where n is the in-tact system, and a second outage of (n – 2).
3. **Examples of criterion applied to class of supply**
   (i) Class A
      (a) For (n – 1), the time requirement is the restoration of the group demand in the actual repair time
      (b) For (n – 2), there are no stated time requirements.
   (ii) Class E
      (a) For (n – 1), the time requirement is group demand within 60 s
      (b) For (n – 2), the time requirement is two-third of group demand within 60 s (assumes that this is a summer outage and that all customers will be averaging two thirds of group demand) – followed by the full group demand in the time to restore the arranged outage.

Applying the P2/6 requirements has influenced the development of numerous standard substation designs and associated automatic control systems (e.g. auto-reclose).

5.3 HV network design standards – power system studies

5.3.1 System studies – requirements

Power system studies are required to be undertaken to ensure that the proposed construction design will not subject either the power system, or connected equipment, to conditions which exceed pre-defined limits. Such limits may be defined in the grid code, or the distribution code, or power network company technical policy or associated operational standards.

The studies are invariably carried out using tried and tested computer-based solutions. They are usually undertaken by specialist engineers who are competent in the field of HV power system design. Such studies usually include some or all of the following:

1. Load flow
2. Power system stability
3. Short-circuit current levels
4. Harmonic distortion levels
5. Power system unbalance levels
6. Power system X/R ratios
7. Ferroresonance.

5.3.1.1 Load flow

Load flow studies comprise an evaluation of a network under simulated maximum steady state load-current conditions, including short-term over-loads as described in Chapter 20. Two factors are usually evaluated: network (equipment) thermal capability and network voltage limits – which are described as follows:

1. **Network thermal capability limits**
   The level of power that can flow through an electrical network is limited by the thermal capability (i.e. the current rating) of the equipment. Proposed current magnitudes are evaluated under a variety of outage conditions (as defined by the network planning standards) to confirm that the equipment (circuit breakers, OHL, cables, etc.) is not subject to over-load – otherwise, the proposed network would need to be re-designed.

   The term 'firm thermal capability', i.e. the capability of a network after the loss of one or two circuits, as specified in the appropriate network planning standard, is often used to define the power flow limit across a defined network boundary. In addition the term 'firm supply point' in a network (usually a substation) is used to describe an arrangement where a fault or maintenance outage on one point (e.g. substation) on the system will not prevent a supply being available at that point. Duplication of circuits/equipment usually satisfies this requirement.

   Figure 5.1 summarises maximum continuous load currents (i.e. post-fault continuous) to which switchgear may be subject in the United Kingdom. Self-evidently the maximum loading to which OHL and HV cables will also be subject will not be expected to exceed these values. NB: In some special circumstances a lower level may be specified.

   The studies are required to confirm that both the proposed section of network to be constructed and the adjoining existing network satisfies the thermal capability rating limits (i.e. in Figure 5.1) – or otherwise that accepted restrictions apply.

   NB: It is also worthy of note that the allowable load currents must take account of statutory OHL clearance (sag) limits.
2. **Network voltage limits**
   The voltage profile of both the proposed and existing networks must be evaluated across the range of maximum real and reactive power flows, as determined by the thermal capacity limits. Again, the voltage limits should not exceed stipulated (statutory) requirements, when outages in accordance with

System nominal voltage (kV)	Maximum continuous current rating
400	4 kA (5 kA in future)
275	3.15 kA
132	3.15 kA (other ratings are also used)
33	3.15 kA (other ratings are also used)

Figure 5.1 Switchgear continuous current ratings – commonly used in the United Kingdom

Network nominal voltage (kV)	Voltage range limits
400	Continuous = −10% to +5% (360 to 420 kV) 15 min = +5% to +10% (420 to 440 kV)
275	−10% to +10% (247 to 303 kV)
132	−10% to +10% (119 to 145 kV)
33	−6% to +6% (31 to 35 kV)

Figure 5.2 Statutory voltage limits – UK networks

network planning standards are taken into account. The main causes of voltage step changes on the power system are as follows:

(i) Switching of circuits
(ii) Switching of major loads
(iii) Switching of reactive compensation
(iv) System disturbances (usually faults).

Figure 5.2 specifies the statutory voltage limits on the UK networks (relevant to this text).

5.3.1.2 Power system stability

Power system stability is concerned with the ability of each synchronous generator to remain in synchronism with the rest of the power system as a result of changing conditions on the power system. To do otherwise would result in generator pole-slipping as described in Section 3.6.1.5. Power system stability studies may be categorised as follows:

- Steady-state stability
- Transient stability
- Dynamic stability.

These will be briefly examined below.

- **Steady-state stability**
  This is the ability of a generator or group of generators to remain in synchronism during gradual increases in load (i.e. MW). This is governed by the generator power angle relationship as described in Section 3.2.5, and the generator performance chart as described in Section 3.6.1.4. The study would evaluate the maximum load to which a generator would be subject to assess the margin of stability.
- **Transient stability**
  This is concerned with the impact of a sudden change in generator loading conditions, typically arising from a system fault, or the sudden loss of either other generators, or a part of the network. In such instances, the imbalance of generator mechanical input power and electrical output restraining power results in relatively rapid movement of the rotor towards the generator stability limit with potential loss of synchronism – again governed by the power angle

relationship as described in Section 3.2.5. Generator AVRs (i.e. automatic voltage regulators) and network auto-reclose systems are helpful in maintaining stability during transient disturbances. The studies would evaluate transient stability against a range of power system operational conditions.

- **Dynamic stability**
  Dynamic stability is concerned with the behaviour of the power system in the interval between the transient event arising and steady state conditions (hopefully) being re-established. The studies look at the behaviour and interaction, during this period, of generator fuel flow and boilers, generator governors and automatic load-shedding systems.

NB: Power system stability studies would normally only be undertaken when power network construction proposals could potentially have an impact on generator performance – or when new generation requires reinforcement of the existing power network.

5.3.1.3 Short-circuit level limits

The maximum allowable short-circuit current to which a power network may be subject is limited by the short-circuit rating of the equipment – particularly circuit breakers which are required to interrupt the short-circuit current. Power system studies are therefore undertaken to be assured that the short-circuit currents, arising as a result of the proposed construction design (usually considering three-phase, and single-phase-to-earth, faults) do not exceed the designed maximum short-circuit current limit for the network – as determined by equipment short-circuit rating limits. To avoid excessive heating of the equipment (and prolonged mechanical forces), networks and their equipment are also specified in terms of the maximum duration of short-circuit current. Figure 5.3 shows the maximum short-circuit ratings in general use in the United Kingdom.

5.3.1.4 Harmonic distortion

Harmonic distortion of the voltage waveform due to non-sinusoidal currents generally arises from the following sources:

- Consumer demand connected at the lower voltages – particularly devices involving half-wave rectifiers.
- Transformer magnetising inrush current – recognising that this is a transient current.

Network nominal voltage (kV)	Maximum short-circuit current rating (3 phase and 1 phase) (kA)	Maximum duration of short-circuit (s)
400	63	1
275	40	1
132	40	3
33	17.5	3

Figure 5.3 Short-circuit ratings in general use in the United Kingdom

- Reactive compensation plant, especially SVCs (i.e. static VAr compensators).
- HV DC convertors.
- Single-phase or two-phase loads, e.g. traction systems.

The harmonic distortion arising from the above can adversely affect certain types of equipment and may lead to network resonance situations. It can also affect the performance of certain types of protection equipment and CT and VT performance.

Harmonics observed on the UK power networks generally range up to the 11th harmonic with the 2nd, 3rd, 5th and 7th harmonics being more prevalent Both the Grid Code and Distribution Code place limits on the levels of harmonics. Within thus context, total harmonic distortion is typically required to be less than a maximum of 2% nominal voltage (depending upon the network in question). System studies on harmonic distortion are undertaken when new connections are likely to result in equipment causing harmonics. In some instances, harmonic filters will be required to reduce the harmonic distortion to the maximum limits allowed.

5.3.1.5 Power system unbalance limits

The term 'unbalance' refers to the existence of either negative-sequence or zero-sequence currents and voltages (see Chapter 4) during normal steady state loading of the power networks. Unbalance usually arises from the following sources:

- Unbalanced load current, in particular traction supply points (since traction supplies are usually two phases)
- Mutual coupling effects between conductors – usually on OHL.

Fault currents of course also create unbalance but only for the duration of the fault clearance time. Equally, unbalance is caused by open-circuit fault conditions and remains until the open-circuit position is located and the condition remedied.

Unbalanced currents can impact on generator and motor performance and cause overheating. They can also cause unwanted operation of some protection systems. Again both the Grid Code and the Distribution Code stipulate maximum levels of NPS and ZPS currents. For example on the 400 kV network the maximum continuous, NPS and ZPS currents are generally are 470 and 165 A, i.e. approximately 12% and 4%, respectively, of network maximum continuous loading, of 4,000 A.

Power system studies to evaluate the level of unbalance will only be undertaken when the construction work is likely to have unbalance implications. In the instance of unbalanced load currents, the level of unbalance may have to be reduced by the installation of a load balancer. OHL unbalance arising from unequal mutual coupling of the three phases is usually partly corrected on the transmission system by endeavouring to create a rough balance across the whole system through the use of standard phasing arrangements on double circuit OHLs. These arrangements are generally as shown in Figure 5.4. The UK transmission networks are therefore partly transposed networks.

5.3.1.6 Power system X/R ratio

Section 3.8.1 explained that when switching an inductive circuit (such as the instance of a power system fault) a transient DC offset current arises which is

	Arrangement 1		Arrangement 2		Arrangement 3	
	Cct 1	**Cct 2**	**Cct 1**	**Cct 2**	**Cct 1**	**Cct 2**
Top phase	R	B	Y	R	B	Y
Middle phase	Y	Y	B	B	R	R
Bottom phase	B	R	R	Y	Y	B

NB: Cct, circuit.

Figure 5.4 Transmission network double circuit OHL phasing arrangements

super-imposed upon the steady state AC fault current. The DC offset causes two main concerns:

- Increased peak fault current arising from the summation of the DC and AC components of fault current – to which circuit breaker interrupters are subject when clearing the fault.
- The possibility of CT saturation arising from the unidirectional DC current (and associated magnetic flux) – which in turn may affect protection operation.

As a result, power network equipment is usually designed to accommodate stipulated maximum X/R ratios, which determines the time constant (and hence the duration) of the DC offset. Where construction involves equipment which is highly inductive an evaluation must be made to ensure the X/R ratio does not exceed the stipulated design maximum value.

NB: Circuit breaker specifications stipulate the maximum X/R ratio to which the circuit breaker may be subject.

5.3.1.7 Ferroresonance

Figure 5.5(a) shows a double circuit OHL with one of the circuits terminating in a transformer feeder arrangement (although with reference to the following text both circuits could be transformer feeders). At the instant of de-energising circuit 2 and depending upon the magnitudes of capacitances C_1 and C_2 and the winding inductance of transformer *T1*, it is possible to achieve a sustained electrical oscillation between the capacitances and the inductance, resulting in a significant voltage being applied to *T1* – with the energy supplied from the live circuit. This oscillation is termed 'ferroresonance'.

Ferroresonance voltage waveforms usually exhibit a flat peak as a result of the transformer magnetic core being driven into saturation, as illustrated in Figure 5.5(b). The frequency of oscillation may be 50 Hz with an amplitude of 1 per-unit on one phase and 1/2 per-unit on the other two phases. Another common oscillation is a one-third subharmonic of 16 2/3 Hz with an amplitude of 1/3 per-unit on one phase and 1/6 per-unit on the other two phases.

Sustained ferroresonance can result in excessive transformer core mechanical vibration and overheating – resulting in damage. Ferroresonance is usually quenched (and damage prevented) through the automatic application (following

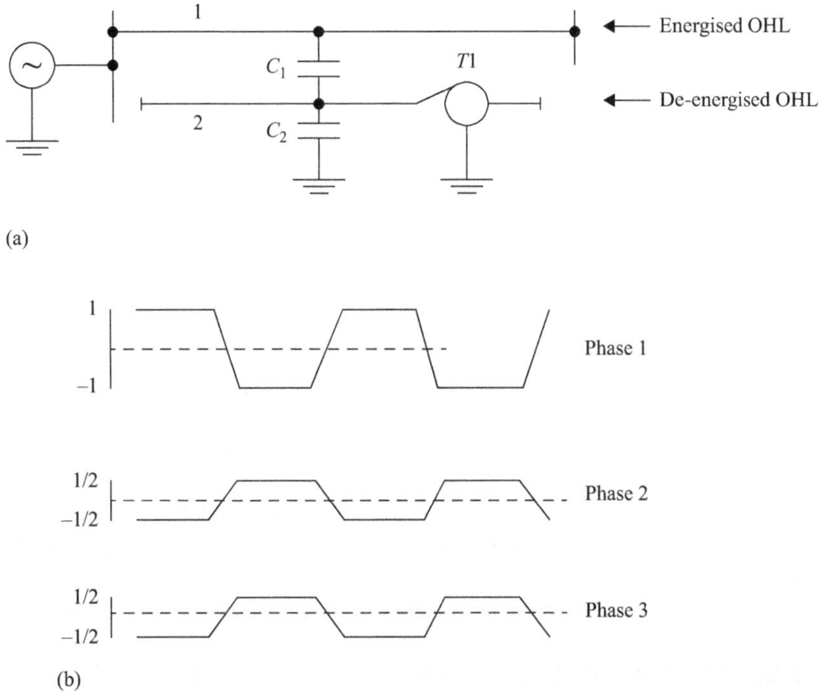

Figure 5.5 Ferroresonance conditions: (a) system under consideration and (b) voltage waveforms

operation of ferroresonance detection equipment) of high speed earth switches designed for this duty. Older ferroresonance quenching systems employed the opening of the OHL disconnector (again designed for this duty) adjacent to the transformer. Ferroresonance is more likely to arise on the transmission networks where the levels of stored energy in the capacitances (which instigates the ferroresonance process) is greater than with the distribution networks.

Experience gained over the years has resulted in an ability to identify which network configurations will be subject to ferroresonance – however, some configurations must be subject to studies to determine the likelihood of ferroresonance.

5.4 HV network design – fixed design parameters

5.4.1 Fixed design standards

Certain standards pertaining to HV network design may be considered as fixed, and do not require power system studies to confirm that the proposed design is within defined limits. These standards are either in the Grid or Distribution Codes, or power network company technical policy and will be quoted in power network company technical specifications. In summary, these include the following:

- Power system frequency
- Insulation coordination
- Transformer neutral earthing
- Transformer winding configurations – and network phasing
- Power system fault clearance times
- Environmental service conditions design

5.4.1.1 Power system frequency

Worldwide two dominant power system frequencies exist: 50 Hz as used in (and originated from) Europe including the United Kingdom; and 60 Hz as used in (and originated from) North America. Both frequencies have equal technical merit and the adoption of each is essentially due to historical influences.

In the United Kingdom, the frequency range is mandated via the ESQC Regulations, as discussed in Chapter 2, and subsequently mandated in the Grid and Distribution Codes. The stipulated frequency range limits are summarised in Figure 5.6.

NB: In the United Kingdom, National Grid plc has responsibility for frequency control.

NB: Equipment installed on the UK power system must be capable of continuous operation in the range 47.5–52.0 Hz with excursions of between 47.0 and 47.5 Hz for 20 s.

5.4.1.2 Insulation coordination

Insulation coordination is the method that ensures that the electrical insulation of the equipment, that comprises the electrical network, is of sufficient strength to prevent insulation breakdown, when subject to the full range of network operational conditions – and expected range of voltages. In general, the purpose of insulation coordination is to ensure the following:

- General public and network workforce safety – by preventing insulation failure and possible explosion, and scattering of fragments, etc.
- Avoidance of damage to HV equipment
- Minimise loss of supply to electricity consumers.

• Nominal frequency	= 50.0 Hz
• Target operational frequency	= 49.8–50.2 Hz
• Allowable 'normal' condition frequency range	= ±1% (49.5–50.5 Hz)
• Allowable 'exceptional' condition frequency range	= 47.5–52.0 Hz 47.0–47.5 Hz for 20 s

Figure 5.6 Frequency range limits

The insulation coordination process generally comprises three stages, which are as follows:

- Identification of the range of network conditions and calculation of the associated over-voltages
- Choice of a suitable level of equipment insulation
- Application of protective devices to prevent insulation damage from over-voltages.

Many networks have standard levels of equipment insulation and therefore during power network construction no insulation coordination studies are required – unless, of course, non-standard network arrangements or unusually complex circuit arrangements are proposed.

The over-voltages to which equipment may be subject may generally be categorised as follows:

- Temporary over-voltages – arising from operation of the network
- Switching over-voltages
- Lightning over-voltages.

These will be briefly described as follows:

- **Temporary over-voltages**
  Temporary over-voltages arise from the following sources:
  - Phase-to-earth faults resulting in a rise of healthy phase-to-earth voltages
  - Sudden load rejection
  - Ferroresonance
  - Ferranti effect

  The above typically result in over-voltages of up to 1.7 per-unit and last from typically 20 ms to a few tens of seconds.
- **Switching over-voltages**
  Depending upon the equipment concerned certain switching operations can result in over-voltages. Examples include:
  - Energisation of an OHL
  - Energisation of a transformer feeder with the transformer LV open
  - Energisation of a line or cable through equipment possessing series inductance
  - Interruption of transformer, or reactor, in-rush currents
  - Re-energisation of a cable possessing trapped charge
  - Current chopping, see Section 9.3.1.4.

  The above typically result in over-voltages up to a maximum of 3 per-unit but with a time duration ranging from about 80 μs to 10 ms. These voltages typically comprise a high frequency oscillation superimposed on the power frequency voltage. Switching over-voltages tend to be more onerous on networks above 132 kV.
- **Lightening over-voltages**
  Lightning mostly impacts on OHL. Steel towers are usually fitted with a continuous aerial earth wire which acts as a shield to lightning hitting the phase conductors. Should a lightning stroke hit the earth wire, the current will flow

along the earth wire and down through the towers and discharge to earth. The combined impedance of the tower inductance and earth resistance will result in the voltage on the tower rising so producing a high-voltage difference between the tower and phase conductors – which may lead to a 'back flashover' between tower and phase conductor. Current flow in the earth wires will also result in induced voltages into the phase conductors.

A direct stroke to the phase wire may also occur if the phase wire is not well shielded by the earth wire – or alternatively that no earth wire exists (as with most 33-kV wood pole OHL). Lightening over-voltages are unidirectional (i.e. DC) and again superimpose themselves on the power frequency voltage. Typical lighting surge currents may range from a few kA to up to 200 kA – but with an average of about 20 kA. Typical voltage magnitudes may be of the order of four or five per unit (depending on the network nominal voltage) but with durations of between 0.5 and 80 μs. Lightening over-voltages are proportionately larger (in comparison to the nominal network voltage) and therefore more problematic on lower voltage networks.

Equipment insulation must be designed to withstand the above over-voltages. Such design is not an insignificant task since insulation breakdown is a function of both the magnitude of the applied voltage, and the duration of that voltage – therefore, the insulation withstand voltage needs to be time graded. IEC 60071 – Insulation Coordination – addresses the subject of insulation coordination and specifies recommended withstand voltages. However, power network companies in many countries, including the United Kingdom, developed their own policies prior to the issue of the IEC, and these often remain in place to the present day (although the values in most instances will be close to those in the IEC).

Figure 5.7 provides an indicative example of the levels of insulation required. The IEC also stipulates voltage wave shapes and durations for each type of

Nominal voltage (kV rms)	Rated voltage (kV rms)	Rated short duration power frequency withstand voltage (kV rms)	Rated switching impulse withstand voltage (kV pk)		Rated lightning impulse withstand voltage* (kV pk)
			Ph–E	Ph–Ph	
33	36	70	N/A	N/A	145
132	145	275	N/A	N/A	650
275	300	380	850	1,275	1,050
400	420	520	1,050	1,575	1,425

NB: (1) * = Applies to both Ph–E and Ph–Ph.

(2) In the United Kingdom it is also usual to specify withstand voltages across open switching/isolating devices.

Figure 5.7 Insulation withstand requirements – typical

over-voltage. Within this context, equipment is usually subject to the equipment insulation withstand voltages during type testing.

NB: The term 'basic insulation level' (BIL) is commonly used to describe the insulation withstand voltages – particularly the lightning impulse voltage.

With reference to insulation levels two associated design factors are worthy of note: insulation protection and phase clearances. These will be briefly examined.

Insulation protection

Even though equipment insulation is designed to withstand the voltage levels described in Section 5.4.1.2 it is usual to add devices to further protect the insulation from damage. The simplest and cheapest form of protection is the rod (or spark) gap, also called arcing horn. They are usually installed in parallel with insulators, connected between the equipment live terminal and earth. The gap distance setting is arranged such that flashover occurs at voltages well below the insulation breakdown level of the equipment being protected. It is usual to reduce the gap setting across OHL insulators (i.e. tower OHL) as the OHL approaches a substation to afford increased protection to the substation insulation.

The disadvantage of rod gaps is that once they have broken down (to dissipate an over-voltage to earth) they effectively create a fault on the power network, with subsequent removal of the circuit in question from service via protection operation. In addition, rod gaps have a relatively slow response time, and as a result the front of any over-voltage wave could have progressed down the circuit (and say into a substation) before the rod gap has operated. It is worthy of note that at the higher transmission voltages of 400 and 275 kV, rod gaps tend not to be used because of the levels of corona discharge (which arises from the high electric field intensity caused by sharp/pointed objects such as the rod) and subsequent radio interference – but rather a variation comprising a metallic loop, with a radius, so reducing the electric field intensity.

For critical or expensive items of equipment, additional protective measures, in the form of surge arrestors, may be installed. They alternatively may be used on equipment where the insulation withstand is designed to a lesser voltage than the predicted over-voltages. Surge arrestors have the advantage of resealing following breakdown and discharge of the over-voltage (i.e. a subsequent power network fault does not arise, as with a rod gap) thereby retaining the circuit in service. Examples of where surge arrestors may be applied include the following:

- Across transformer/reactor/quadrature booster bushings
- Across cable sealing ends
- In parallel with circuit entries into GIS substations (this is because, unlike AIS substations GIS, once broken down, requires an inspection to ensure no physical damage has occurred – which if it did would result in long and expensive repair times).

Some countries install surge arrestors across all incoming circuits to a substation, and this may allow reduced equipment insulation levels within the substation. This is not usual UK practice.

Phase clearances

The required clearances between equipment phase conductors and earth, and phase conductor to phase conductor are self-evidently related to equipment withstand voltages otherwise flash-over would occur, both for AIS and GIS substations (phase-segregated GIS involve only phase-earth clearances). In AIS substations, this forms the basis of the safety distances, as examined in Chapters 6 and 18.

5.4.1.3 Transformer winding configuration and network phasing

By changing the windings configuration of two winding transformers, it is possible to obtain different phase shifts between the HV and LV windings. In most countries, including the United Kingdom, both the phases shift across the transformers, and the relative phase position of each network are standardised. Therefore, when connecting new transformers to a particular network, it is essential that the winding configuration is suitable for the network phasing in question. Within this context, this section will briefly examine:

- Transformer vector group
- Phase shifts arising from rotation of phase connections
- Phase shifts arising from negative-phase sequence connections
- Standard network phasing.

Transformer vector group

With reference to Figure 5.8(a), standard practice is to designate transformer HV terminals (to which the windings are connected) with capital letters A, B, C and the LV terminals with lower case letters a, b, c. Each winding has two ends designated by subscripts 1 and 2. If the induced emf in the HV phase is in the direction A_1 to A_2 at any instant, then the induced emf in the corresponding LV phase at the same instant will be a_1 to a_2, and the HV and LV voltages are designated as being 'in-phase'.

NB: The HV winding induced emf is the same magnitude and in-phase with the applied network voltage.

Transformer are normally categorised in terms of their 'vector group' which defines both the HV and LV winding configuration, and the transformer phase shift from HV to corresponding LV terminals – expressed in terms of clock-face hour and minute-hand positions. The HV vector (phase) position is taken as being 12 o'clock, i.e. the reference position and the position of the hour hand, and the corresponding LV vector position is represented by the minute hand. Within this context, the transformer shown in Figure 5.8(a) is designated as having a vector group of Yd11. That is, the HV winding (designated by a capital letter) is star configuration, the LV winding (designated by a lower case letter) is delta configuration, and the 11 is the clock-face position of the LV vector – which is leading the HV vector (positioned at 12 O'clock) by 30°. The vector diagram in Figure 5.8(a) shows how the HV and LV in-phase vectors are aligned to deduce the Yd11 arrangement.

In practice, transformer windings fall into three main types: star connected windings, delta connected windings, and interconnected star (zig-zag) connected

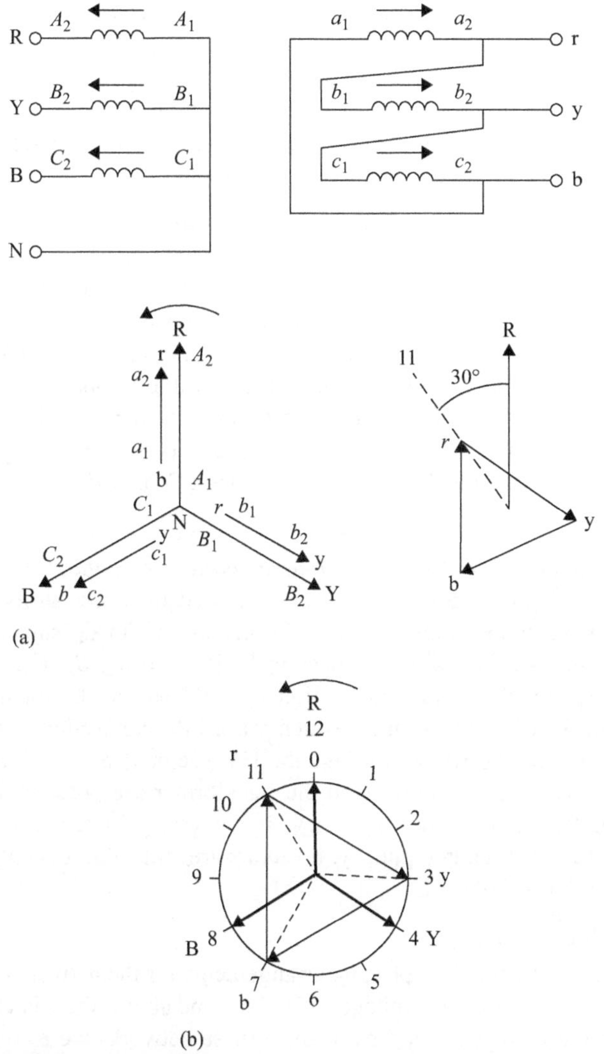

Figure 5.8 Two winding transformer vector group: (a) Yd11 and (b) clock face vector alignment

windings. From these three types, phase shifts of 0°, 180°, +30° and −30° can be arranged. They are generally categorised into four groups, as follows:

- Group 1: 0° phase shifts, e.g. Yy0, etc.
- Group 2: 180° phase shifts, e.g. Yy6, etc.
- Group 3: −30° phase shifts, e.g. Dy1, Yd1, etc.
- Group 4: +30° phase shifts, e.g. Dy11, Yd11, etc.

Most power transformers have accessible connections termed 'vector group links' which allow the windings to be configured either as Group 1 or 2, or Group 3 or 4.

A relatively simple and visual method for analysing/deducing winding phase shifts is illustrated in Figure 5.8(b) where both windings are positioned on the clock face with the HV and LV in-phase windings aligned (i.e. in parallel) with each other. The example shows the Yd11 vector positions.

Phase-shifts arising from rotation of phase connections
The conventional connections arrangement is to connect the R, Y, B phases to A_2, B_2, C_2 (i.e. HV connections) and r, y, b to a_2, b_2, c_2 (i.e. LV connections), respectively. This is sometimes referred to as a positive-phase sequence (PPS) connection. However by rotating the phase connections on the LV side of the transformer, with respect to the HV side, it is possible to obtain additional phase shifts of multiples of 120°. For example in Figure 5.8(b) by rotating the LV phase connections such that r is now at clock position 3, y is at clock position 7, and b at clock position 11, the transformer now becomes a Yd3. NB: This is still a PPS connection.

Phase shifts arising from negative-phase sequence connections
With reference to the PPS connection as discussed in 'phase-shifts arising from rotation of phase connections', the transformer can be made subject to an NPS connection. This is achieved by connecting R, B, Y to A_2, B_2, C_2, and r, b, y to a_2, b_2, c_2, respectively. If this connection is analysed via the method shown in Figure 5.8(b), then the transformer (which via a PPS connection is a Yd11) now becomes a Yd1, i.e. the LV vector lags the HV vector R by +30°. Similarly, the current vectors when flowing through the transformer are phase shifted by −30° when flowing from transformer HV to LV. It is worthy of note that in this instance both the input R, Y, B and output r, y, b vectors are still PPS, it is only the method of connection that is NPS.

Standard network phasing
Figure 5.9 shows the standard phasing arrangements for the networks in the United Kingdom. At the transmission voltages of 132 kV and above, there is virtually 100% adherence to the arrangements shown, since these networks were constructed to a single common design. However, the networks at 33 kV and below, although greatly standardised today, were constructed in the very early days of the power system and to diverse designs – and therefore, historical variations continue to exist.

NB: The 400 V network is usually solidly earthed with a clock reference position of 11 o'clock (using DY11 transformers) – although there are numerous exceptions to this.

5.4.1.4 Neutral earthing

The standard earthing arrangements on the UK power system are illustrated in Figure 5.9. The method of earthing influences the level of insulation, particularly on transformers. At 132 kV and above (where each networks is interconnected), it is economically beneficial to solidly earth all networks, via transformer neutrals.

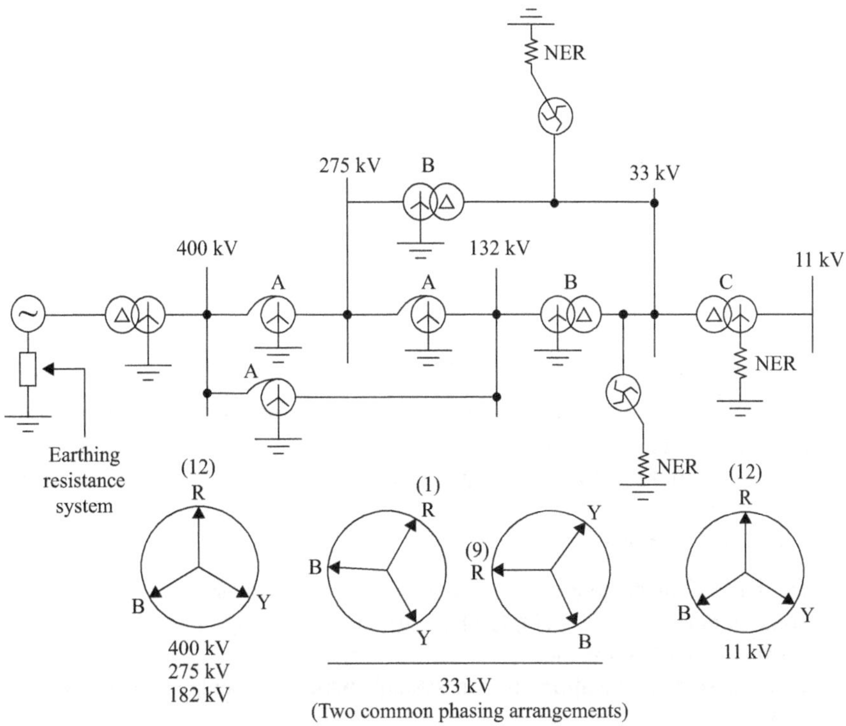

Figure 5.9 UK transformer winding configurations – and network phasing

At 132 kV and below, impedance earthing (usually NER) is widely used without any economic penalty (excepting that the 400 V network is solidly earthed). Impedance earthing has the advantage of limiting the fault current, and as such equipment damage, at the point of fault. The magnitude of fault current with impedance earthing is usually limited to the network full load current – i.e. sufficiently high to facilitate protection operation.

When an earth fault takes place, the voltage between the healthy phase and earth rises, depending upon the impedance of the earth current return path from the point of fault to the source of supply of fault current – the latter being the transformer neutral when the source is a transformer star winding – or to the phase conductors connected to an earthing transformer for the instance of a transformer secondary being a delta winding. The term 'earth fault factor' describes the ratio of the rms healthy phase to earth voltage (at the point of fault), to the rms phase-to-earth voltage without a fault. This factor determines the magnitude of voltage rise on the healthy phases during an earth fault. Networks which are solidly earthed are those which have an earth fault factor less than 1.4. Impedance earthing systems have an earth fault factor up to 1.8. These factors are usually specified in equipment technical specifications.

5.4.1.5 Fault clearance times

Fault clearance time is the summated time of protection operation and circuit breaker opening time, when clearing a fault on the power system. The rapid and effective clearance of a fault is critical for the following reasons:

- Retention of power system stability.
- Minimisation of damage to both the equipment subject to the fault, and to healthy equipment which carries the fault current.
- Minimisation of disruption of supply to electrical consumers.

Fault clearance times are referenced by both the Grid and Distribution Codes. Figure 5.10 specifies target fault clearance times, noting that at 132 kV and below, there may be slight variations in accordance with power network company's technical policy.

The following notes apply to Figure 5.10:

- The faults considered are three-phase – with a three-phase fault clearance. NB: Some countries employ single-phase tripping with high speed auto-reclose – but this is not UK practice.
- Worst case fault location from a system disruption perspective is usually a busbar fault.
- The back-up clearance of 500 ms generally refers to zone 2 of distance protection, and the 1,000 ms clearance to both zone 3 distance protection and residual earth fault protection. NB: Different power network companies use slightly different values.
- The 400 and 275 kV transmission networks are also fitted with circuit breaker fail protection with a target maximum fault clearance time of 300 ms.
- At 400 kV, the 80-ms clearance time is critical for the preservation of system stability for the instance of a three-phase fault on certain high infeed busbars in the network. However, a single-phase to earth fault at the same location could allow a fault clearance time of up to one second (typically) before loss of stability occurs.

Nominal voltage (kV)	Target maximum fault clearance time (ms)	Target back-up clearance time (ms)
400	80	500/1,000
275	100	500/1,000
132	(120–200)	1,500
33	(200–500)	Various

Figure 5.10 Target fault clearance times – United Kingdom

5.4.1.6 Environmental service conditions design

Environmental service conditions design generally covers two requirements:

- The temperature range to which the equipment will be subject.
- Levels of pollution to which outdoor equipment will be subject.

These will be briefly examined.

- **Temperature range**
  Temperature is usually categorised as relating to indoor or outdoor equipment. Options are defined in IEC 61936 – Power Installations Exceeding 1 kV AC. Indoor equipment is generally required to operate down to temperatures of −5°C and outdoor equipment down to −25°C. A range of maximum design temperatures are also defined in the IEC.
- **Pollution control**
  Pollution considerations are generally applicable to insulators. The level of pollution is dependent upon the insulator location, with the highest levels being experienced in coastal (saline pollution) or industrial locations. IEC 60815 – Guide for the Selection of Insulators in Respect of Polluted Conditions, specifies a range of design options in accordance with the severity of the pollution conditions and whether the insulators are mounted vertically or horizontally. Pollution withstand is obtained by increasing the creepage distance of the insulation, specified in mm/kV.

5.5 Project-specific design

5.5.1 Project-specific design considerations

With reference to HV network design, the project-specific design specification is first proposed and summarised in a system design report and later finalised in the scheme design specification (or other similar titles – see Chapter 17). In effect, the project-specific design specifies a bespoke design solution which both satisfies the need case and the boundary requirements specified by power network company project-generic design specification(s). The project-specific design specification is generally required to balance and optimise three prime requirements:

- Security of supply (in accordance with Project Generic Design limit requirements)
- Cost
- Public acceptance (particularly relating to OHL routes, substation siting and the extent of disruption caused by the construction site installation stage).

Many project-specific designs employ previously used design solutions, since this usually removes error, minimises engineering resource, minimises completion timescales, and usually reduces cost. However, instances frequently arise of where a unique design is used, perhaps involving new technology, since this provides the optimum solution. Suffice it to say that any deviation from project-generic design specifications requires great care and attention to detail with an exploration of all potential operational conditions to which the proposed equipment may be subject.

5.6 Future power network challenges

5.6.1 Power flows

With increasing connection of generation on the distribution networks, a number of technical challenges arise as follows:

- Two-way flow of power across the distribution network and reverse flow of power into the transmission network with concomitant problems for:
  - Equipment load current ratings
  - Voltage profiles.
- Increase in short-circuit current levels with the potential to over-stress equipment.
- Network control considerations, e.g. loss of an HV infeed to a network-containing generation and subsequent reconnection process.

5.6.2 Smart grid

With ever-increasing levels of intermittent none fossil-fuel generation being connected to both the transmission and distribution networks, coupled with both the society based pressures to minimise the building of new OHL (for facilitating increasing power flows), and ever-increasing pinch points of network congestion – there is a momentum in both the power industry, and government, to manage these challenges through the development of a 'smart grid'. Contemporary literature has a variety of definitions of a smart grid – and it would seem to be a term still in the making – but generally, it covers the following initiatives:

- Employment of computer-based systems in protection, control and telecommunications for the management of frequency/voltage/power flow.
- Installation of intelligent metering in domestic/commercial premises to facilitate the management of electricity consumption. The objective is reduced cost for consumers, and reduced demand for additional generation and power network reinforcement.
- Demand response control through contracts with generators for frequency control and operational reserve – and contracts with major consumers for demand control. In addition, it is envisaged that in future, domestic and commercial consumers will be offered instantly variable tariffs, via intelligent metering, to level out the demand profile.
- Equipment thermal rating monitoring – through the use of embedded temperature sensors (e.g. in transformers and cables) with communications to a control centre to enable control engineers to exploit maximum power flows through equipment.
- Installation of GPS synchronised phasor measurement devices at selected locations in the transmission system for evaluation of voltages magnitudes and comparative phase angles – to exploit maximum power transfers and evaluation of the risk of system instability.
- Measures to minimise new build (i.e. optimise the existing power system).

Chapter 6
Overhead line design

6.1 Introduction

Overhead lines (OHLs) comprise one of the major power engineering technologies, and although the basic principles of OHL design were mostly established at the dawn of electrical power systems, it is a technology that is subject to ongoing development and change. Within this context, the construction of a new OHL remains both complex and challenging. This chapter will examine and summarise salient requirements of OHL design covering the following:

- OHL design over-view
- Tower, pole and foundation design
- Conductor system design
- Insulators and OHL fittings design
- Tower earth-wires and earth resistance
- Conductor tension and sag
- Tower height and spacing
- OHL design philosophy
- Routing and siting design
- OHL asset replacement.

6.2 OHL design overview

6.2.1 OHL design – key requirements

Figure 6.1 summarises the key requirements for the design of an OHL under the three main categories of:

- Geographical and geological – the determination of the physical route and eventual position of the OHL supports
- Electrical – essentially concerned with the power transfer capability of the OHL and electrical safety considerations
- Civil and structural – the design of both the OHL support and associated foundations, and conductor tension and sag considerations.

Iteration may be required between the three categories as the design progresses.

NB: The term OHL 'support' covers both OHL towers and poles.

Geographical and geological	Electrical	Civil and structural
• Strategic routing options • Initial route selection • Environmental impact assessment • Final and detailed route selection • Ground-line survey and profile plotting • OHL support plotting • OHL support pegging	• OHL power transfer capability • Conductor selection and arrangement • Insulator selection • Shielding angle considerations • Electrical safety access evaluation • Electromagnetic field proximity evaluation • Tower footing resistance evaluation	• Conductor tension limits and sag • OHL support selection/design • Soil survey • Foundation design

Figure 6.1 OHL design – key requirements

6.3 Tower, pole and foundation design

6.3.1 Tower design

OHL towers in the United Kingdom are generally employed at network voltages of 132 kV and above. They are usually selected from a suite of towers types, mostly designed many years ago. That being said, instances do arise of a requirement for new tower designs – either to address a specific situation or to carry a new type of conductor.

Towers in the United Kingdom have traditionally been of a lattice steel, self-supporting design – usually double circuit with vertically arranged phases. This is mainly because such towers can be erected in virtually all terrains, have a relatively small footprint and are mostly accepted (or tolerated) from an aesthetic perspective by the general public. However, a double circuit OHL has the disadvantage that a double circuit fault could occur, arising from, say, a lightning strike, severe weather conditions or accidental damage due to collisions with vehicles/aircraft – but it is none the less more cost effective than two single circuit OHLs.

Figure 6.2 illustrates typical towers installed on the UK networks. Horizontal or triangular tower designs result in a lower OHL profile, but because of required clearances between phases necessitate a wider strip of ground. As a result, the double circuit, vertical arrangement predominates.

Tower design and selection is usually based upon the following key requirements:

- The optimum tower profile to satisfy the constraints (usually aesthetic/public acceptability) of the route in question.
- Tower strength – to withstand and support the loads to which the tower is subject.
- Electrical clearances – i.e. conductor phase-to-phase and phase-to-earth clearances for both insulation coordination and statutory (i.e. sag limits) purposes.

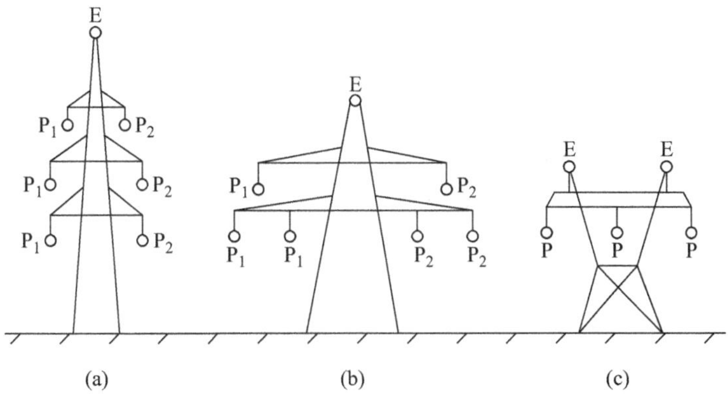

Figure 6.2 Tower designs – typical: (a) vertical, (b) triangular and (c) horizontal single circuit

The above will be briefly examined.

- **Optimum tower profile**
  With reference to Figure 6.2, the optimum tower profile (shape) will be chosen after considering the OHL route profile and features of the project in question. In the United Kingdom, the default tower is usually the vertical arrangement in Figure 6.2(a) – for the reasons outlined above – but instances arise of where the height profile of the vertical arrangement is not aesthetically acceptable and the triangular or horizontal is preferred.
- **Tower strength**
  Towers are categorised as one of two types: tension towers which are used for either angles in an OHL, terminating a section of OHL, or as a terminal tower; or suspension towers which support an OHL in between tension towers. Both categories of tower must be designed to withstand the various loads to which the tower will be subject. These are defined in terms of three axes *X*, *Y* and *Z* known as transverse, longitudinal and vertical, respectively. Figure 6.3 illustrates the transverse and longitudinal axis – with the vertical axis being that of the height (vertical) elevation of the tower.
  - **Transverse loads**
    The transverse loads on the tower comprise:
    - Wind loads on both the conductors and the insulators. The term 'design wind span' as used in the calculation of wind load is defined as half the sum of the two spans on either side of the tower in question.
    - Wind loads on the tower itself.
    - The component of conductor tension, which arises on terminal towers resolved along the '*X*' axis.

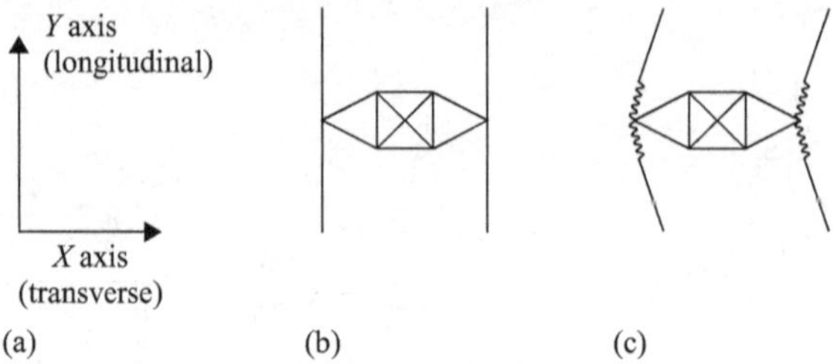

Figure 6.3 Transverse and longitudinal axis loads: (a) axis, (b) suspension tower plan view and (c) tension tower plan view

- **Longitudinal loads**
  The longitudinal loads on the tower comprise:
  - Tension tower conductor loads resolved along the '*Y*' axis. Suspension towers are not subject to conductor tension loads since they merely suspend the conductor.
  - Broken conductor load on one side of a tension tower. The most onerous broken conductor design condition is all three phases on one side of a tension tower. Again, suspension towers are much less impacted since the insulators swing and reduce the tension.
  - Wind loads on the tower.
- **Vertical loads**
  The vertical loads on the tower comprise:
  - The dead weight of the conductors, insulators and fittings. The term 'design weight span', used in the calculation of conductor weight, is defined as the horizontal distance between the lowest point of the conductors of the two spans either side of the tower.
  - Tower self-weight, including a stipulated covering of ice to which the tower members, including the four supporting legs, are subject.

    In addition, two further loads need to be considered, which are as described in the following points, and resolved into three load axis as defined above.
- **Short-circuit loads**
  Under power system fault conditions, short-circuit currents cause the conductors to either mechanically attract or repel. This causes the conductors to physically move, and often bounce when the short-circuit is removed. This is a particularly onerous consideration for terminal towers with reference to the short-circuit forces on the down leads to the substation.
- **Maintenance loads**
  Consideration must also be given to the loads imposed on a tower during maintenance activities.

Forces on towers, tower members and foundations may be either tension or compression. Towers may also be subject to uplift forces particularly when the tower route crosses highly undulating ground. Traditionally, towers have been designed by static methods employing force diagrams and hand calculation with proof of adequate design being demonstrated by means of type test. More recently, computer and finite element analysis, using elastic techniques, have been utilised.

- **Electrical clearances**
  Electrical clearances are the prime determinant of the tower physical dimensions, especially tower height. The main clearances are (see also Section 6.8.1):
  - Statutory ground clearance – minimum allowed distance between phase conductors and normal flat ground.
  - Conductor sag – which occurs under maximum conductor temperature and still air.
  - Insulator set length – i.e. between cross-arm and conductor (applies only to suspension towers).
  - Vertical/horizontal distance between phase conductors – usually determined by safety distances and maintenance access requirements.
  - Minimum height of the earth-wire above the highest phase conductor.
  - The minimum acceptable earth-wire shield angle between the earth-wire and highest phase conductor.

Historically, tower design in the United Kingdom has comprised families of towers. Each family contains both tension towers (angle, section and terminal) and suspension towers – designed both to carry defined loads and types of conductor systems. Typical family designations and associated conductors are as follows:

- L4 family = 132 kV capable of carrying 1 × 175 mm^2 ACSR (Lynx)
- L7 family = 132 kV capable of carrying 2 × 175 mm^2 ACSR (Lynx)
- L2 family = 275 kV capable of carrying 2 × 400 mm^2 ACSR (Zebra)
- L6/2 family = 275 kV capable of carrying 2 × 700 mm^2 AAAC (Araucaria)
- L8/4 family = 400 kV capable of carrying 2 × 570 mm^2 AAAC (Sorbus)
- L12 family = 400 kV capable of carrying 2 × 700 mm^2 AAAC (Araucaria)

NB: Towers capable of carrying higher voltages are also capable of carrying lower voltage conductors. Variations exist on the same family e.g. L6, L6/1, L6/2.

The design of each family usually includes extensions (usually in increments of 3.0 or 3.5 m) to increase the height of the tower. Each family also usually includes a range of angle towers designated as follows:

- D = straight line
- D10 = angle deviation 0°–10°
- D30 = angle deviation 10°–30°
- D60 = angle deviation 30°–60°
- D90 = angle deviation 90°
- D7 = terminal tower, angle deviation = 60°–90°.

Special towers, e.g. for river crossings may have to be specially designed. It is worthy of note that angle towers usually have unequal length cross-arms both to provide more clearance (for access) on the outside of the angle and to provide a tipping moment to the outside which helps balance the inward pull of the conductor.

The majority of OHL towers both in the United Kingdom and in many parts of the world have been (and still are) constructed from galvanised, hot rolled mild or high tensile angle steel. Generally, high tensile steel requires a lower tonnage for the same design of tower.

6.3.2 Tower aesthetic design

Around 2010, in the United Kingdom, it was recognised that although lattice steel tower designs had served the networks well for many years, there was increasing public pressure to design and install towers that had a more acceptable aesthetic impact. To this end the National Grid Company established a competition for a new design of tower which was not only fit for purpose but aesthetically more pleasing. The winner was chosen by representatives of the wider community. The winning tower was the 'T' tower as illustrated in Figure 6.4. The height of this tower is lower than the lattice steel equivalent but overall is wider. This design of tower is currently being trialled.

6.3.3 Pole design

In the United Kingdom, as in many parts of the world, wood pole design predominates up to network voltages of 33 kV, although there are some at 132 kV. Wood as an OHL support material has the advantage of being readily available and mechanically strong, yet hard wearing and very adaptable in the way it can be used. The woods commonly in use are primarily softwoods such as Scots Pine – referred

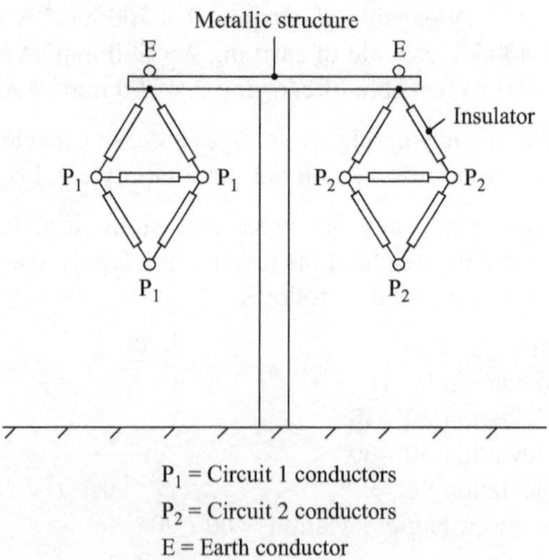

Figure 6.4 'T' tower design – key features

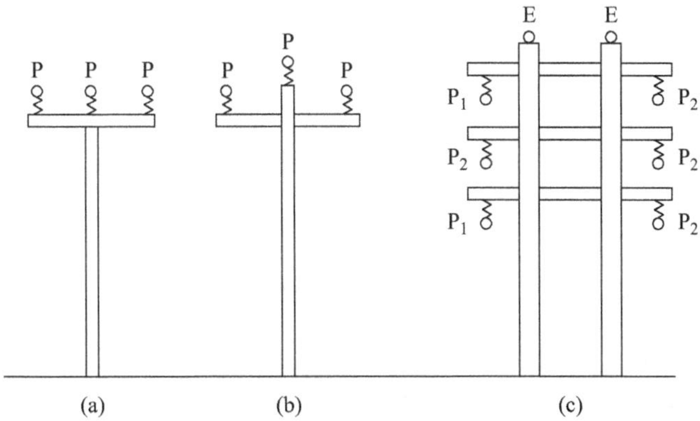

Figure 6.5 Pole designs – typical: (a) horizontal, (b) delta and (c) double circuit

to as Redwood. Such wood is readily found in much of Northern Europe, particularly Finland. It is usually impregnated with creosote or other chemicals to counteract decay. Instances also arise of thin-wall steel poles (at network voltages up to 132 kV) – but these are not common. Figure 6.5 illustrates the profile of typical pole designs. Most wooden poles have no earth-wire nor any earth connection to the cross-arm metalwork and many are single circuit.

Poles are subject to similar mechanical loading as those on towers, namely conductor loads, wind and ice loads, insulator and fittings loads, pole self-weight and short-circuit loads. Again, as with towers, poles are classified as either tension or suspension, with stays being used to counteract transverse loads on angle poles.

6.3.4 Foundations

The foundation of an OHL tower or pole is the means by which the towers/poles transfer the loads to which they are subject into the ground. Foundations are a critical element in the overall design of an OHL. A specific requirement of a foundation is that it must not be subject to settlement, nor lose strength with age.

Foundation design is dependent upon the properties of the soil in which it is placed – in particular the properties of compressibility, plasticity, moisture content, soil particle friction, and shear strength. At an early stage in the tower design, ground investigations are required to be undertaken by geotechnical experts (a specialised subject) using bore holes and trial pits. The number of soil investigations carried out is dependent upon an assessment of the range of soils in existence across the OHL route. Soil analysis may take place on site, but more usually in a laboratory where the soil attributes described above can be more accurately assessed. The results from the soil investigations allow the OHL civil engineer to decide which type of foundation will most economically support the tower.

The foundation design solution must also take into account the tower site access requirements for piling rigs and materials transportation, etc.

Foundations may be subject to a number of forces: compressive, uplift, overturning moments or shear forces. Usually, the most onerous and design limiting force is that of uplift. Resistance to uplift is usually achieved by burying a pyramid shaped concrete block at such a depth that restraint is achieved through soil weight and soil cohesion. Within this context, tower foundations are generally divided into a number of different types. Figure 6.6 illustrates two common types: the 'pyramid and chimney' and the 'pad and chimney'. The usual construction method is to excavate the soil, install a 'stub leg', which projects above the ground, for later connecting to the main tower leg members. The stub leg is cast into the reinforced concrete blocks, the shape of which is achieved by pouring concrete into 'former boxes' which are constructed for the particular design. With very poor soils, piled foundations may additionally be required. Clearly to ensure the angle and height dimensions of the resulting tower are within design limits the positioning of the foundation and stubs must be undertaken with significant accuracy.

Poles differ from towers in as much as they pivot about some point below ground. Both compressive and uplift forces are countered by the pole to soil interface resistance force (i.e. friction). Instances may arise of the use of a wooden baulk (like a railway sleeper) being used to assist the pole in resisting both overturning and compressive (downwards) forces. Baulks are typically mounted horizontally at a depth of 50-cm-below ground level. In addition, the pole may also be supported by stays especially where the ground is relatively soft (to prevent the pole leaning at an angle). Stays of course are also used to stabilise angle (in the OHL) poles. Instances also arise of 'bog shoes' being connected to poles to provide stability in boggy ground. These usually comprise a wooden framework of considerable width connected to the pole below ground.

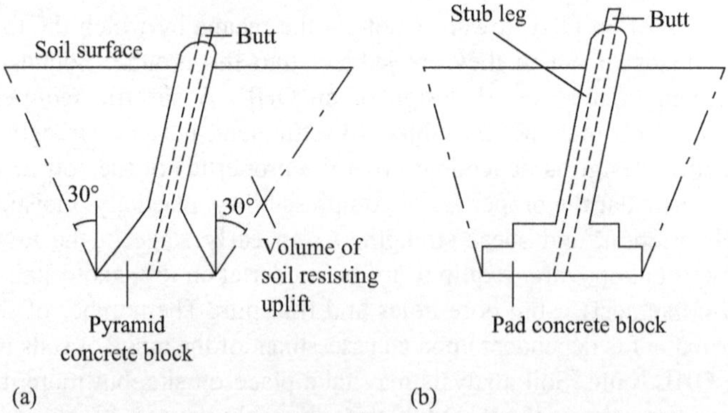

Figure 6.6 Common types of foundations: (a) pyramid and chimney and (b) pad and chimney

6.4 Conductor system design

6.4.1 Conductor requirements

OHL conductors are required to possess a number of important characteristics which are as follows:

1. **Electrical properties**
   The first requirement of OHL design is the specification of the power transfer capability (current rating) of the conductor. Increased current ratings (both load and fault current) are achieved through the property of high electrical conductivity. This property minimises conductor resistive heating and power loss. It is also minimises voltage drop – although, generally, voltage drop is more associated with the inductive reactance of the conductor system.
2. **Mechanical properties**
   The conductor must also possess high tensile strength to enable operation under tension, good flexibility to avoid vibrational fatigue and relatively low weight to minimise conductor selfload when strung.
3. **Thermal properties**
   The two main thermal properties are low coefficient of thermal expansion to minimise conductor sag and ability to withstand high and low operating temperatures without conductor degradation.
4. **Conductor shape and size**
   The design of the conductor from a shape/size perspective must take account of the following:
   (i) Must limit corona discharge and associated radio interference. NB: Corona effect diminishes with increased conductor radius.
   (ii) Must minimise wind noise and vibration – this usually requires a relatively smooth surface.
5. **Cost**
   The conductor must be economic in terms of the manufactured cost, the cost of installation and the ongoing operation and maintenance cost.

6.4.2 Conductors – materials and types

The earliest conductor materials were copper and cadmium copper – but these are no longer used due to cost – although some remain in service. Copper has now largely been replaced by aluminium-based conductors. All aluminium conductors can be used on the lower voltage networks with lower span lengths – where the strength to weight ratio is acceptable.

On the transmission networks, 'aluminium conductor steel reinforced' conductor (ACSR) was the dominant conductor type for many years – and much still remains in service. This type of conductor comprises a central core of galvanised steel (to provide mechanical strength) surrounded by aluminium conductors (which effectively carry the current) – see Figure 6.8(a). ACSR required the stranded conductor to be subject to a grease coating, both to alleviate the formation of aluminium oxide and to prevent rubbing and grazing of the conductor individual wires.

The melting point of the grease limited the temperature at which the conductor could be operated, and hence the current rating of the cable.

Dating from *ca.* 1990, ACSR has been progressively replaced by 'all aluminium alloy conductor' (AAAC). AAAC has a greater strength to weight ratio than ACSR and does not necessitate the steel inner core required by ACSR. Modern conductors also include 'aluminium conductor alloy reinforced' and 'aluminium alloy conductor steel reinforced'.

At network voltages of 400 and 275 kV (and some 132 kV), the OHL conductors are usually erected in bundles of twin, triple or quad. Bundle conductors possess the following salient characteristics:

1. **Erection**
   Each conductor in a bundle is easier to handle and erect than a single large conductor.
2. **Skin effect**
   Skin effect (see Section 7.4.1.2) limits penetration of current into the conductor. At 50 Hz, most of the current is contained within 1 mm of the conductor surface. Thus, a bundle of conductors has greater current carrying capacity than a single large conductor.
3. **Corona loss**
   Corona loss results from the breakdown of air at a critical level of electrical field intensity of approximately 30 kV/cm. It additionally causes a hissing sound, radio interference and a small power loss. The level of corona decreases as the conductor diameter increases. Bundle conductors have a greater effective radius than a single conductor serving the same purpose and therefore reduce the level of corona. The term 'geometric mean radius' defines the equivalent radius of a bundle of conductors (or the equivalent of the stranded make up of a single conductor).
4. **Inductive reactance**
   It can be shown that the inductive reactance of a bundle of conductors is less than that of a single equivalent conductor.

Specific code names are allocated to conductors by national standards organisations. The codes are categorised into specific families – for example in the United Kingdom, ACSR conductors are named after mammals and AAAC conductor are named after trees. Figure 6.7 provides typical examples.

Code name	**ACSR (mm^2)**	**AAAC (mm^2)**
• Horse	70	N/A
• Dog	100	N/A
• Lynx	175	N/A
• Zebra	400	N/A
• Upas	N/A	300
• Rubus	N/A	500
• Araucaria	N/A	700
• Redwood	N/A	850

Figure 6.7 Common conductor code names

6.4.3 Gap conductor

With the ever increasing requirement for conductors to carry higher load currents, a new design of conductor termed 'Gap' was developed (mostly in Japan) and first trialled in the United Kingdom in the early years of the twenty-first century. This conductor uses heat resistant zirconium–aluminium alloy over a steel core. A small annular gap, in which high thermal resistant grease is inserted, is maintained between a high-strength steel core and the first layer of aluminium alloy strands, see Figure 6.8(b).

The principle of Gap conductor is that it can be terminated on the steel core alone. The salient advantages of Gap are:

- A high current capacity (up to 1.6 times that of equivalent conventional conductors).
- Relatively low sag at high temperatures.
- Continuous operation at temperatures up to 200°C – without adverse effects on mechanical or electrical properties.

The code name for Gap conductor is 'Matthew', and its abbreviation is GZTACSR.

6.4.4 Conductor selection

When selecting a conductor for a scheme/project, the following requirements need to be considered:

- Maximum power to be carried by the OHL, and the current carrying capacity of the conductor.
- Tower design to support the conductor. In some instances, lighter duty towers may be more acceptable to the general public but must be able to carry the conductor system proposed.
- Corona interference and noise characteristics of the conductor, particularly if passing through a sensitive (either urban or rural) area. Some conductors are 'noisier' than others.
- Impedance characteristics, if magnitude of impedance is significant to the power transfer requirement.

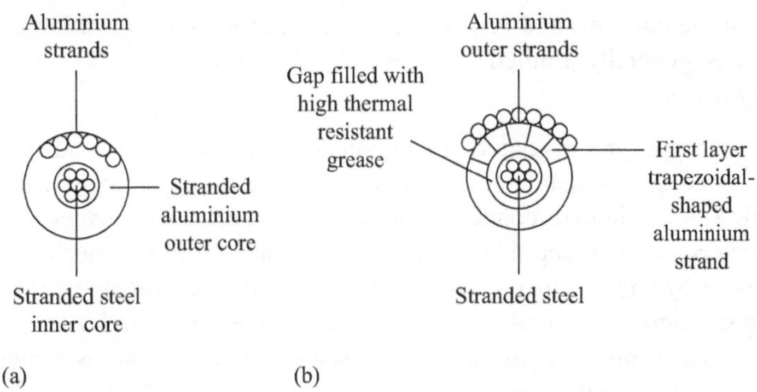

Figure 6.8 Conductor types: (a) ACSR conductor and (b) gap (GZTACSR) conductor

6.5 Insulators and OHL fittings design

6.5.1 Insulator requirements

Insulators are essential to the successful operation of OHL. In general, insulators must possess three prime quantities:

1. **Mechanical strength**
   Mechanical strength is required to withstand the various loads imposed upon them by the conductor system.
2. **Electrical strength**
   Electrical strength is required both to withstand electrical (dielectric) breakdown and to prevent flashover over the range of environmental conditions experienced. The latter is assisted by shaping and profiling the insulator sheds.
3. **Cost effectiveness**
   Given the huge number of insulators in operations across all the networks, they must be cost effective in terms of manufacturing cost, reliability and operational lifetime.

There are three basic types of insulator design, which are as follows:

- Pin
- Cap and pin
- Post

These will be briefly examined.

6.5.1.1 Pin-type insulators

As the name suggests, the pin-type insulator is attached to a steel bolt or pin which is vertically mounted on, and secured to, the cross-arm of a pole (or in some instances a tower). Figure 6.9(a) illustrates the profile and features of a typical pin-type insulator. The core of the insulator is fitted with a screw thread ferrule into which the pin is screwed. The insulator sheds are shaped to conform, as far as possible, to electric field equipotential zones, with the leakage resistance, and capacitance, of the sheds approximately equal, all of which assist in both minimising leakage currents, and insulator electrical stress. The application of pin-type insulators is generally limited to network voltages of up to 33 kV – for the following reasons:

- With the pin-type insulator, the conductor is physically bound (via binding wire) onto the insulator top – so making the resulting arrangement relatively rigid. The rigidity (in contrast to a flexible arrangement) makes the pin-type insulators more susceptible to fatigue when subject to the mechanical forces imposed by the conductor system and therefore only suitable for the lighter weight conductor systems experienced at voltages up to 33 kV.
- The cost of pin-type insulators increases rapidly as the working voltage increases making them uneconomic at voltages above 33 kV.

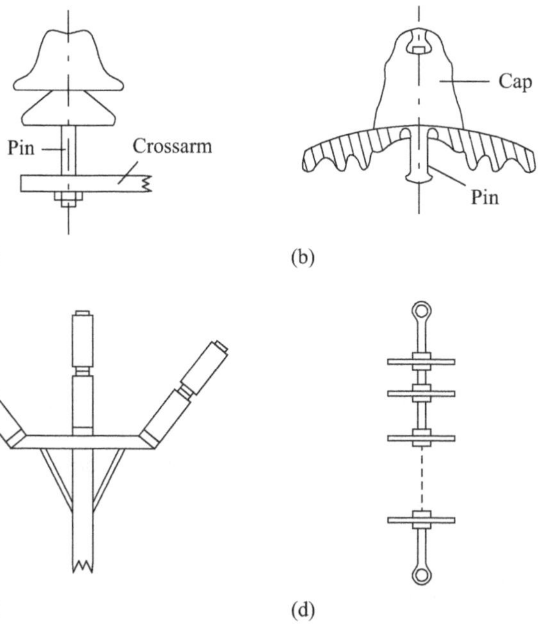

Figure 6.9 Types of insulators: (a) pin-type insulator, (b) cap and pin type, (c) post-type insulator arrangement and (d) composite insulator typical profile

6.5.1.2 Cap and pin insulators

Cap and pin insulators (alternatively termed ball and socket) are usually employed at network voltages above 33 kV. A number is usually connected in series to form an insulator string, as such they form a flexible arrangement more capable of accommodating the conductor system mechanical forces (than the pin-type insulator, described above). Each cap and pin insulator is designed to withstand a comparatively low working voltage with the total insulation strength obtained through the string assembly. Because of the relatively heavy weight, the insulator string is suspended from a metallic tower. This insulator arrangement has the advantage that should a single insulator in the string fail, sufficient insulation strength remains to retain the OHL in service – and in addition, only one insulator and not the whole string requires replacing. This design also has the advantage that with an increase in the OHL operating voltage, the new insulation requirement can be met by adding an appropriate number of units to the string – instead of replacing all units as would be necessary with the pin-type insulator. A disadvantage of this arrangement is because the insulators are suspended, (i.e. hang down) the height of the OHL support is increased.

Figure 6.8(b) illustrates a typical cap and pin insulator. The pin of each insulator slots into the cap of the insulator below and is locked into position by a security clip ('w' pin) which is inserted into the slot of the cap. Fog is a particular problem for insulators since it may deposit droplets of water at any point of the whole insulator surface – which can lead to intermittent arcing and tracking damage. Polluted air also causes risk of flashover. As a result, in locations susceptible to fog or pollution insulator designs known as 'anti-fog' with long creepage lengths on the underside of the insulator may be installed.

6.5.1.3 Post-type insulators

In recent years, wood poles for voltages up to 132 kV have been developed using post-type insulators, similar to those found in substations, and arranged in trident formation, see Figure 6.9(c). Post-type insulators are more or less similar to pin-type insulators but have a greater number of sheds (petticoats) and greater height. They are manufactured in multiple shed units with the ability to bolt the units together (as illustrated) for higher voltage applications. The conductor is usually fixed to the top insulator by means of a connector clamp. The advantage of this type of OHL construction is that it is relatively lightweight and as such easy to transport and erect.

6.5.2 *Insulator materials*

The three main materials used in insulator design are:

- Porcelain
- Glass
- Composite

These will be briefly examined.

- **Porcelain**
  Porcelain has long been the main material used in the design and manufacture of insulators. It has the advantage of being easy to manufacture and has both good mechanical loading and electrical with-stand properties. Importantly, it is relatively cheap.
- **Glass**
  Toughened glass is a ready alternative to porcelain, and there is little difference in either technical performance or cost. Glass has the advantage that broken insulators tend to shatter completely (e.g. when subject to impact) and are therefore more easy to spot than porcelain during OHL maintenance and inspection. Conversely, a porcelain insulator will not necessarily shatter when damaged and as such continues to provide better electrical insulation than glass. For this reason, porcelain is used almost exclusively in substations (where inspections are easier to undertake) while both porcelain and glass are used on OHL. It is also argued by some that porcelain is aesthetically (visually) more acceptable than glass.
- **Composite**
  Composite insulators have been under development for many years and as such are only in service in relatively small numbers – often for trial

purposes – although numbers are increasing. Composites are manufactured from polymeric materials and supported by a resin bonded glass fibre core; a typical example is illustrated in Figure 6.9(d). To protect the core and provide an extended creepage path, the core is encompassed by weather sheds manufactured from a range of polymers (e.g. polysulphide rubber). The advantage of composite insulators is that they are relatively light weight, visually less obtrusive and possess greater resistance to physical damage (particularly vandal attacks). The main disadvantage is cost (although this is reducing with time), and long-term reliability in service is yet to be concluded.

6.5.3 OHL fittings

OHL technology involves a number of fittings, which include the following:

- Arcing horns
- Spacers and dampers
- Sag adjusters
- Tower clamps
- Conductor joints

These will be briefly examined.

6.5.3.1 Arcing horns

Arcing horns, alternatively termed rod or spark gaps, are connected across insulator strings, between the cross-arm of the tower and the (live) conductor, as described in section 'Insulation protection' in Chapter 5. They are designed to provide overvoltage protection to protect the insulators from both lightning strikes (to the line), and at the higher network voltages, from switching over-voltages. The live conductor end arcing horn is often ring shaped to assist with improved voltage grading along the insulator string (i.e. an evening out of the insulator string capacitance) and is constructed from steel tubing to reduce corona loss.

Most wood pole circuits run with the cross-arms unearthed (the advantages of being unearthed outweighing the disadvantages) with the pole acting in effect as an insulator. However, with a lightning strike to one of the phase conductors, there is effectively the equivalent of a duplex gap between adjacent phases via the metallic cross-arm and flashover may occur.

6.5.3.2 Spacers and dampers

Spacers are required to maintain sub-conductor separation of bundle conductors, thereby avoiding conductor clashing. They are usually fitted at intervals of 60 m or less. Some spacers are semi-flexible through the use of rubber bushes (enclosing the conductors) – although these are not usually applied to jumpers.

Vibration dampers are used to dampen out conductor vibration and oscillations. The most commonly used is the 'Stocksbridge' damper comprising two weights separated by a connecting rod, the centre of which is clamped onto the conductor. Spacers and dampers may also be combined to become a spacer/damper.

The most common types of conductor oscillation/vibration to be counteracted are as follows:

1. **Aeolian vibration**
   Conductors subject to low speed laminar wind flow may experience relatively high frequency vibrations – between 5 and 30 Hz – termed Aeolian vibration. This may cause damage to conductor strands, particularly at points where the conductor is clamped. The dampers, which are positioned at the end of the conductor spans, are not always fully successful in dampening out Aeolian vibration – especially on those parts of the conductor remote from the dampers.
2. **Conductor galloping**
   Conductor galloping is a low frequency (0.1 to 1.0 Hz), high amplitude (up to 4 m and mostly vertical) phenomenon which can effect both single and bundle conductors. It is usually associated with a build-up of ice and snow on the conductors which are then subject to wind – particularly in exposed areas. In addition to possible fatigue damage to conductors and fittings, galloping may result in conductor clashing – i.e. the touching of different phases resulting in a system fault condition. Design measures to counteract conductor galloping include the selective removal of spacers, although this increases the risk of conductor clashing – and may necessitate larger spacers, or the application of dampers with either a pendulum property or an aerodynamic shape – both of which dampen the impact of wind. Ensuring that the circuit is subject to a continuous high load so heating the conductor and preventing ice formation can also alleviate galloping.
3. **Sub-conductor oscillation**
   Sub-conductor oscillation occurs on bundle conductors i.e. twin, triple or quad which have pairs of sub-conductors lying in the same horizontal plane. The oscillations are usually in the frequency range 0.5 to 3.0 Hz and the cause is wind turbulence. Solutions comprise:
   (i) Installation of rubber inserts into the spacers.
   (ii) The vertical arrangement of conductors – with no spacers – although this risks conductor clashing.
   (iii) The angular rotation of the bundle such that each conductor is out of the wind wake of the other.
   (iv) Increase conductor separation.

6.5.3.3 Sag adjusters

Sag adjusters comprise a plate located between the tension tower cross-arm and the insulator string, with adjustment holes to allow both flexibility and ease in the initial sagging of the conductor and (most importantly) enables the sag of a conductor to be subsequently varied e.g. to compensate for conductor expansion following erection.

6.5.3.4 Tower clamps and joints

Suspension clamps support the conductor at the bottom of the suspension insulator set. They grip the conductor very lightly such that the conductor can slip through the clamp and not act as an anchor if the conductor should break on one side of the tower.

When a conductor needs to be anchored to a tension insulator set, it is done via a joint arrangement. Occasions also arise of a joint being required within a span. In both instances, the joint must be equal or better than the conductor both electrically and mechanically. For most aluminium-based conductors, mechanical compression joints with a hexagonal design have proved to be the most superior.

Joints and anchor clamps for tower earth-wires are identical to those for line conductors.

6.6 Tower earth-wires and earth resistance

6.6.1 OHL earth-wires

Most OHLs erected on towers are fitted with one (or more) earth-wires, usually bonded to the top of the tower. The earth-wire serves a number of purposes.

- As a lightning protection system
- As a means of returning fault current to source
- The OHL tower and earth-wire is usually connected to the substation earth. In so doing, it effectively serves to extend the substation earth system beyond the substation, so lowering the overall resistance to earth and the associated substation rise in earth potential.

With reference to lightning protection, and Figure 6.10, the 'shielding angle' is a measure of the degree of protection of the phase conductor by the earth-wire from a direct hit from lightning. It can be shown that the lower the angle, the greater the shielding. Practical experience has shown that a shield angle of 45° intercepts most lightning stokes – although on the UK 400-kV network a shield angle of 35° has been used.

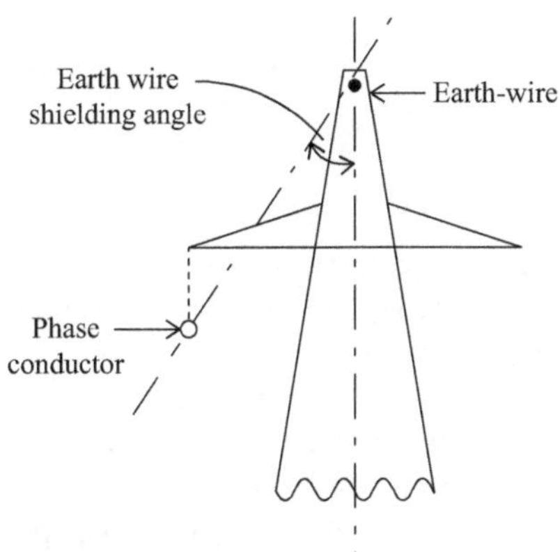

Figure 6.10 Earth-wire shield angle

6.6.2 Tower earth resistance

Tower steelwork is naturally earthed through the foundation arrangement, and earthing is generally required for the following reasons:

- Maintaining the earth-wire at or close to earth potential – so facilitating the shielding effect – and providing an effective discharge route to earth for a lightning strike to the earth-wire.
- Providing a connection to the earth-wire for returning earth fault current to source.

As explained in Section 5.1.4.2, the magnitude of tower earth resistance (i.e. tower footing resistance) influences the rise in voltage of the tower above true earth during the flow of lightning discharge current through the tower. This may lead to a 'back flashover' between the tower and the tower HV line conductors. Generally, at 132 kV, a tower earthing resistance of 30 Ω or less proves satisfactory, and at 275 and 400 kV 10 Ω or less. However, precise limits are difficult to specify. With very high values of tower earth resistance, the advantages of earth-wire shielding may be completely lost, and it is necessary to reduce the earth resistance by installing extra earthing electrodes usually in the form of driven rods. Instances may arise of a few towers only with high earth resistance, and this may be considered a statistically low risk. Suffice it to say that with electrically exposed towers such as those at the top of a hill or with tall river crossing towers, then measures must be taken to keep the earth resistance within the values specified above.

6.7 Conductor tension and sag

6.7.1 Tension and sag – considerations

In order to design suitable tower heights and dimensions for an OHL, it is necessary to calculate the conductor sag (the minimum height of which is governed by the ESQC regulations, see Section 2.4.2) and associated conductor tension. The following sections will therefore consider the fundamental (mathematical) relationships governing sag and tension.

6.7.2 Sag and tension – basic theory

The theory underpinning conductor sag and tension usually commences by considering a conductor hung freely between two supports, under which circumstances it will assume a 'catenary' shape. For the range of spans experienced in practice, the catenary shape is very close to that of another mathematical curve, namely that of the 'parabola'. The formula obtained from treating the conductor curve as a parabola is much simpler (when preparing sag curves) than that of a catenary – and from an accuracy perspective acceptable so long as the span to sag ratio is greater than 20:1, which is the case for virtually all OHL in practice. Within this context, the parabolic formula is accurate to approximately 0.5%.

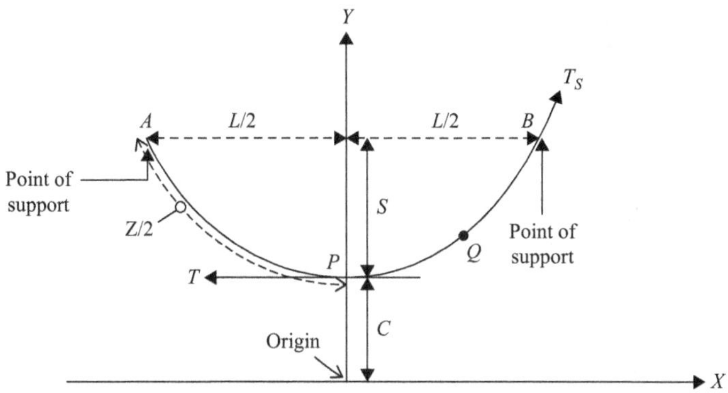

Figure 6.11 OHL sag vs tension curve

6.7.3 Determination of sag

With reference to Figure 6.11, a conductor A–B is supported by rigid supports on the same horizontal plane at the positions A and B. The following parameters are applied:

Let

- Z = true conductor length
- L = span length
- S = sag
- T = horizontal tension at the lowest point, P, on the conductor
- T_S = tension at the support positions A and B
- W = weight per unit length of the conductor
- P = lowest point of the conductor
- C = distance between the origin and the point P.

1. **Catenary expression**
   The expression of a catenary from points P to B is given by:

$$y = C \cosh \frac{X}{C} \tag{6.1}$$

   NB: The origin is usually chosen for reasons of mathematical convenience so that:

$$C = \frac{T}{W} \tag{6.2}$$

2. **Parabola expression**
   The equivalent expression of a parabola from points P to B is given by:

$$y = \frac{X^2}{2C} + C \tag{6.3}$$

   This expression is easier to plot and use than that of a catenary.

3. **Conductor sag**
   Using the parabolic expression, it can be deduced that:

$$\text{Sag } S = \frac{WL^2}{8T} \tag{6.4}$$

   Further, it can be deduced that the total conductor length is given by:

$$Z = L + \frac{W^2L^3}{24T^2} \text{ or } L + \frac{8S^2}{3L} \tag{6.5}$$

   It can also be shown that the tension at the supports is given by:

$$T_s = T + WS$$

   where the sag is small, then WS in the above expression may be ignored, and the tension is then constant throughout the whole conductor span (and the whole section of OHL).

6.7.4 Conductor tension factors

The determination of conductor tension (stress) must factor in two physical characteristics of a conductor as follows:

1. **Thermal expansion/contraction**
   Any conductor subject to temperature change will have an associated increase/decrease in length depending upon whether the temperature rises/falls, the change in length being dependent upon the 'coefficient of temperature linear expansion' of the material in question.
2. **Elastic expansion/contraction**
   In addition, any conductor subject to tension will suffer an elastic elongation in length, the length of elongation being a dependent on 'Young's modulus of elasticity' of the material in question.

6.7.5 Temperature/ice/wind considerations

The maximum conductor tension usually occurs at minimum temperature (i.e. minimum thermal expansion) which is usually taken as being at −5.6°C and when the conductor is subject to maximum ice loading, usually taken as 125-sq-mm radial thickness, i.e. 1/2 in. – and with wind loading applied (currently determined by probabilistic type criterion, see Sections 6.9.2 and 6.9.3).

The maximum conductor sag usually occurs at the maximum conductor temperature (i.e. at maximum thermal expansion). This is usually taken as being 75°C, but as was pointed out in Section 6.4.3, Gap conductor can operate at much higher temperatures. It is worthy of note that in some countries, maximum sag may not occur at maximum temperature but at a lower temperature but with maximum wind/ice loading.

6.7.6 Change of state equation

Using expressions (6.3)–(6.5) and factoring in the thermal and elastic considerations, the well-known 'change of state equation' can be determined, which is as follows:

$$EA \propto (t_2 - t_1) + \frac{(W_1)^2 L^2 EA}{24(T_1)^2} - T_1 = \frac{(W_2 L)^2 EA}{24(T_2)^2} - T_2 \tag{6.6}$$

where E is the Young's modulus of elasticity, A is the conductor cross-sectional area, $\propto$ is the coefficient of temperature linear expansion, t_1 is the initial temperature, t_2 is the final temperature, T_1 is the initial tension, T_2 is the final tension, W_1 is the initial conductor weight (including effect of ice/wind), W_2 is the final conductor weight (including effect of ice/wind) and L is the span length.

The above somewhat formidable expression, which can be expressed in a variety of ways, can be used to determine conductor tension for certain pre-defined conditions from which, using expression (6.4), the associated sag can be determined (and subsequently tower/pole height). For a specific type of conductor and span length, all of the above parameters except, tension, temperature and weight are fixed, and the variable parameters can be used in the expression to determine and ensure that the key limits for conductor erection are not violated. The limits in question comprise

1. **Maximum working tension (MWT)**
   The maximum working tension is usually required to be 50% of the breaking tension (i.e. load) of the conductor (this value may be different if a probabilistic type of design is undertaken – see Sections 6.9.2 and 6.9.3). At this tension, the conductor is usually taken as operating at a temperature of −5.6°C with wind and ice loading. Historically, wind and ice loadings were taken as 12.5-mm radial thickness of ice subject to a wind pressure of 380 Newtons per square metre – however, probabilistic type of design values may now be employed, again see Sections 6.9.2 and 6.9.3.
2. **Everyday working tension (EDT)**
   The everyday working tension is that which exists over the largest portion of the life of the conductor and therefore at a mean temperature with little or no loads imposed. Experience has demonstrated that if conductor tension does not exceed about 20% of the breaking tension under EDT conditions, then excessive fatigue from Aeolian vibration is avoided with the use of normal Stocksbridge dampers. EDT is therefore usually calculated at 20% breaking capacity, with no wind/ice considerations and a mean temperature, which, depending upon the standards adopted, ranges from 5°C (see ENA TS 43-40) up to 16°C.
3. **Maximum erection tension (MET)**
   This is the erection tension taking into account the effect, over time, of conductor creep. The latter is an inelastic (i.e. permanent) elongation of the conductor when subject to tension. Aluminium is particularly susceptible to conductor creep. Conductor creep is usually compensated for by sagging the conductor to a higher tension, which is equated to a (equivalent) temperature

(on sag charts) lower than the ambient temperature at the time sagging is undertaken. NB: The effect of lower temperature equates to higher tension.

4. **Sag at maximum temperature**
   The conductor size and rating in the first instance is chosen to match the maximum power flows through the OHL, and the maximum sag of any specific conductor takes place at maximum conductor temperature. Within this context, the design must ensure that the maximum sag for any span in any line section does not exceed the statutory limits defined in the ESQCC regulations.

With reference to the above, a calculation must be carried out to determine which of the criterion of the MWT or the EDT is the dominant criterion (since they may be in conflict). If in expression (6.6), the MWT values of tension (i.e. 50% breaking tension), temperature and weight and the EDT values of temperature and weight are inserted, the resulting EDT value of tension can be calculated. If this value exceeds 20% of the breaking tension, then this is not acceptable since 20% is a limiting value. The 20% figure is then inserted into the expression and the new MWT value calculated (which will now be less than 50% and with a greater sag). Using these values, the magnitudes of the following can be calculated again using expression (6.6).

- The MET can be calculated for a range of erection temperatures. A check should be made to ensure this tension (including the compensation for creep) does not cause the MWT to be exceeded. Should this be the case, the MWT (and EDT) would need to be reduced again using expression (6.6) until no limit is exceeded.
  NB: In practice, it is usual to provide erection tensions/sags for each individual span over a range of ambient temperatures, the temperature being corrected to reflect creep compensation, and the tensions and sags calculated accordingly.
- The tension at maximum temperature can be calculated, and then using expression (6.4), the maximum sag. Once the maximum sag is determined, the required tower/pole height can be determined (to ensure statutory clearances are not infringed).

Today much of the above is accomplished by the use of software packages, but as always, it is important to understand the basis upon which the software operates.

6.7.7 *Equivalent span*

In practice of course, each section of OHL comprises numerous spans, which in theory necessitates a tension/sag calculation for each span which is tedious. However, providing all span lengths are approximately the same, single representative or 'basic' span can be used to represent all spans. However, if the span lengths differ greatly, the concept of the 'equivalent span' (or ruling span) can be used. This comprises a weighted average concept and is derived as follows:

$$\text{Equivalent span } L_E = \sqrt{\left\{\frac{(L_1^3 + L_2^3 + L_3^3 + L_n^3)}{L_1 + L_2 + L_3 + L_n}\right\}} \tag{6.7}$$

where L_1, L_2, L_3, L_n, etc., are the span lengths.

6.7.8 Tension and sag charts

To simplify the procedure for deciding the correct tension and sag during the conductor erection and sagging process, tension and sag charts for each conductor type have been traditionally used. This comprised a graphical plot of tension against equivalent spans over a wide range of temperatures from which the correct sag can be determined. This was very useful, particularly given that it was a pictorial presentation and therefore easy to read. Today, software based solutions can provide the same data.

6.7.9 Conductor supports not on same level

Invariably, the land profile on OHL routes is uneven which necessitates the conductor supports (towers/poles) being on different levels, e.g. supports located at positions A and Q as shown in Figure 6.11. In this instance, the conductor profile is still that of a catenary/parabola and expressions for defining the sag are readily available (in more specialist texts). An important consideration from a tower design perspective is the inclination of the conductor at the lower level support. With reference to Figure 6.11, as position Q moves clockwise along the conductor profile, the tension becomes fully horizontal when point Q coincides with point P, and then on moving beyond point P the tension on point Q contains a vertically upwards component – which presents suspension insulators with difficulties and is to be avoided.

6.8 Tower height and spacing

6.8.1 Tower height and conductor clearances

With reference to Figure 6.12(a), tower height is usually determined as follows:

- Dimension H_1 = Minimum (statutory) permissible conductor clearance
- Dimension H_2 = Maximum conductor sag
- Dimension H_3 = Vertical spacing between bottom and top phase conductors as determined by insulation withstand clearances and insulator length
- Dimension H_4 = Vertical spacing between top phase conductor and tower earth conductor – as determined by the insulator length and the shielding angle.

The minimum statutory permissible conductor clearances in the United Kingdom over roads, fields and other situations are stipulated in the ESQC Regulations, see Section 2.4.2. Additional clearances (not covered by the ESQC Regulations), e.g. to trees and lampstands are covered in ENATS 43-8. Clearances must also be maintained to suspension tower insulator strings when inclined at an angle up to 45°, e.g. arising from wind pressure, see Figure 6.12(b).

6.8.2 Tower spacing

For any type of OHL construction, it is necessary to determine the economic span length for the positioning of towers (and equally so for poles). This must take into account the required clearances; the tower height (given that the general public do

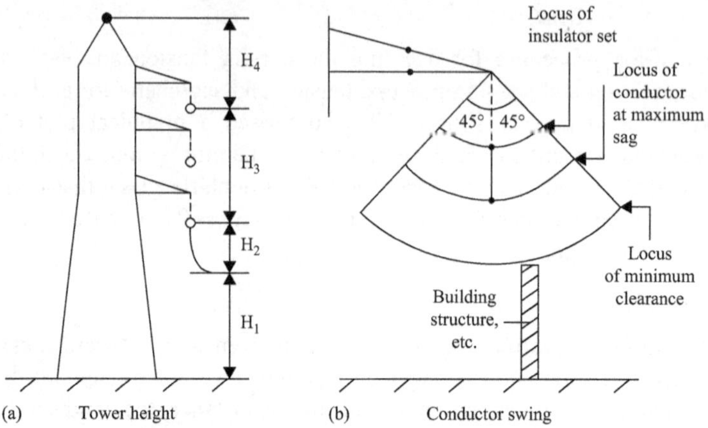

Figure 6.12 Tower height and conductor clearances

not favour high towers); the strength (and size of the tower members); the conductor characteristics; the insulators and fittings requirements; and of course the cost. OHL with short spans require lighter towers but more of them. For each family of towers and conductors, an optimised span length (termed the basic span) for flat ground can be deduced. However, instances arise when the design may have to be tailored to meet individual route circumstances.

6.9 OHL design philosophy

6.9.1 Deterministic design

Historically, OHL design had been based upon a 'deterministic' methodology. In this approach, working loads are based upon the defined breaking limit of a component divided by a factor of safety (FOS) (e.g. a FOS of two or more). The same design criterion was traditionally applied across the whole network, and this also applied to climatic conditions of wind, ice and temperature. Over the years, it was concluded that this type of design performed well in some parts of the country but not others – and the cost of increasing the strength of the various components and continue with a standard of design common to the whole network was uneconomic.

6.9.2 Probabilistic design

In reality, neither the load applied to an OHL component nor the load-bearing capability (resistance) of that component is absolute but a variable quantity following a probabilistic curve, see Figure 6.13. Correspondingly, there is always a risk (magnitude of probability) that design loads exceed, or component reliability is less than, that specified – and therefore complete reliability cannot be ensured.

As a result, and dating from *ca.* the late 1980s, the design of OHL, world-wide, has turned increasingly to a 'probabilistic' design. For this type of design to be successful, there must be sufficient data available to determine the statistical likelihood of a situation arising. Through use of this technique, a desired level of reliability can be achieved in terms of designing for an extreme event. These are

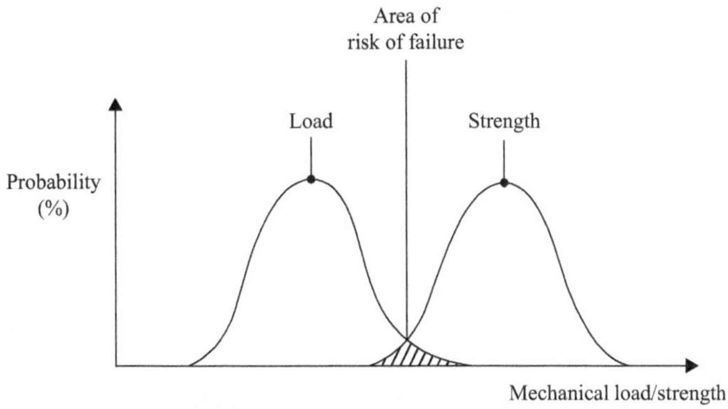

Figure 6.13 Diagrammatic illustration of the probabilistic method

usually taken as one in 50, 100 or 500 years (although other durations can be determined). In general, probabilistic design can be applied to the following:

- Tower design
- Foundation design
- Conductor design
- Climatic conditions, i.e. wind and ice
- Insulators
- Construction and maintenance loads (e.g. the probability of simultaneous broken phase conductors).

UK OHL design (using towers as supports) is generally a mix of deterministic and probabilistic – with climatic condition weather maps for the country having the biggest impact. Authoritative documents on probabilistic design include:

- IEC 60826 (2003): Design Criterion for Overhead Transmission Lines
- ASCE Manual No74 (2010) Guidelines for Electrical Transmission Line Structural Loading
- BS8100 (1999) Lattice Towers and Masts
- BSEN 1993 (2005) Design of Steel Structures.

6.9.3 Load factored design

The probabilistic design method, however, was not considered appropriate for wood pole designs, because of the large range of probabilities that applied to wood in both tension and compression. As a result, a new method of design termed 'load factored design', which is part probabilistic, has been introduced for wood poles. In summary, load factored design considers the following:

- Maps of severe weather areas to enable the development of location-specific designs
- Modelling of weather conditions during wet snow storms to develop OHL withstand capability
- Modelling the impacts of loadings on lines
- Modelling how conductors clash in the wind to enable design preventative measures.

The objective is to design OHLs which are both more capable of withstanding severe weather conditions (in locations subject to severe weather), and more economic for areas where performance has been historically good. An authoritative document on this type of design is:

- ENATS 43-40 (2004) Specification for Single Circuit OHL on Wood Poles up to and including 33 kV.

6.10 Routing and siting

6.10.1 Routing and siting – stages

There are a number of stages to be undertaken to establish both the OHL route, and tower/pole positions and heights. Generally, the stages comprise the following:

- Initial route and siting selection
- Inspection of the initial route
- Final route and siting selection
- Support spotting
- Support pegging.

Section 17.4.3 points out that OHL design in the United Kingdom should accord with the principles and guidelines laid down in the Holford Rules. These consider the impact of an OHL on communities, landscape, ecology, etc. In addition, it was pointed out in Chapter 1 that power network construction in general must strive to achieve public acceptance both during site installation and following completion, and none more-so than OHL construction. Within this context, an international survey conducted by CIGRE concluded that the concerns of the general public with reference to OHLs were as follows (and ranked in order):

- Visual impact
- Electromagnetic field effects
- Land value depreciation
- Ground occupation
- Impact on the natural environment
- Impact on agriculture
- Impact on cattle
- Audible noise
- Radio interference.

The various stages of routing and siting will be briefly examined.

6.10.2 Initial route and siting selection

This is usually a desk top study using maps typically ranging from 1:10,000 to 1:50,000 to propose the shortest possible route. The route proposed should consider the CIGRE criterion and also take into account:

- Terrain
- Urban areas (avoid if possible)

- Buildings
- Topological features
- Major roads/railways
- Other utilities infrastructure
- Access routes for construction
- Soil types
- Cost.

From this study, a proposed route corridor can be developed.

6.10.3 Inspection of the initial route

This comprises a physical inspection on the ground to assess obstructions and constraints, some of which may not have been apparent from the initial route selection. Sections of OHL that may be subject to construction difficulties, including access routes, should be noted. A register should be complied of all risks – and method of mitigation.

6.10.4 Final route and siting selection

This generally comprises the following:

- Ground-line survey of the proposed OHL centre route, measuring vertical and horizontal coordinates (clearances) to obstructions (e.g. buildings/roads/other power lines, etc.)
- Proposed tower/pole positions
- Confirmation of access routes (and any related temporary works design)
- Environmental impact assessment, see Section 17.4.2.

6.10.5 Support plotting

Support plotting comprises the following:

- Preparation of a ground-line profile
- Preparation of conductor sag templates
- Calculation of conductor loads
- Calculation of conductor maximum sags
- Determination of tower/pole heights
- Preparation of tower/pole position and conductor profile (sag) drawings
- Preparation of drawings showing conductor clearances to obstacles and adjacent infrastructure.

6.10.6 Support pegging

This comprises establishing the position of the tower/pole centres on the ground, with centre-line pegs. Should any final adjustments be required due to unforeseen obstacles, or last moment wayleaves problems etc., then the support plotting stage would have to be revisited.

It is worthy of note that in undertaking the above, the agreement of local authorities, and the obtaining of wayleaves, easements and consents must be achieved before applying (in the United Kingdom) to the Secretary of State to obtain Section 37 consent to erect a new OHL, see Section 2.2.2. In addition, major infrastructure OHL work should accord with the requirements of the Planning Act, see Section 17.4.2.

6.10.7 Aerial surveys and software solutions

OHL routing and siting is greatly assisted today by the use of aerial surveys and software solutions. With aerial surveys (using helicopters or planes) and using light detection and ranging techniques in conjunction with GPS, drawings of the full route plan and profile, typically covering as much as 300 km, can be achieved in a single day. Specialist contractors are usually employed for this purpose. This facility has the advantage of not raising undue concerns with either land-owners or the general public. In addition, computer-aided solutions are now widely available to determine maximum tensions and sags (using the more accurate catenary equation) and clearance distances and OHL profiles.

6.11 OHL asset replacement

6.11.1 Asset replacement

OHL asset replacement schemes/projects are, in most developed parts of the world, more common than new build. Such projects may comprise all or some of the following:

- Conductor replacement
- OHL tower bracing replacement (part)
- OHL tower/pole replacement (part or whole circuit)
- Foundation repair/replacement
- Insulator and fittings replacement.

Often these schemes/projects require a significant volume of small scale work to be undertaken to evaluate the condition of the OHL prior to concluding the scale of the work. Ongoing condition monitoring of the OHL greatly assets in determining the need for asset replacement. It is frequently the case that site surveys, inspections and intrusive examinations are required to assess the scale and scope of the work, requiring a preliminary works scheme to be raised to cover the cost of this work. The main works itself often require cost flexibility since additional replacement may become apparent as the actual work progresses.

6.12 Review

The previous text provides a brief summary of both OHL design requirements and the associated stages of routing and siting. In addition to the technological challenges, OHL construction, perhaps to a greater extent than any other category of power network construction, demands extensive interaction, and agreement, with landowners, the general public, local government, planning authorities and national government.

Chapter 7
High voltage and auxiliary cable design

7.1 Introduction

This chapter will examine characteristics of three types of electrical interconnection mediums that fall under the umbrella of cable design, which are as follows:

1. **HV cable design**
   This comprises the main interconnecting AC cables for power transmission and distribution, broadly covering the network voltages of 33 up to 400 kV – although 11 kV cables will be discussed where relevant.
2. **Gas insulated transmission lines (GILs)**
   Gas insulated transmission line (GIL) is included as part of this chapter since when installed, it generally serves as a replacement for HV cables.
3. **Auxiliary cables**
   Auxiliary cables refer to LV (i.e. low voltage) cables for interconnecting protection, control and telecommunications equipment and will include the following:
   (i) Multi-core cables
   (ii) Multi-pair cables
   (iii) Fibre optic cables.

7.2 HV cable design

7.2.1 HV cables – considerations

HV cables, like OHL, provide the means by which substations are interconnected, thereby facilitating both the transmission and distribution of electrical power. Generally, OHLs are preferred for economic reasons, as the cost is significantly less than their HV cable equivalent, typically by a factor of up to 12 times at 400 kV. As such, cables are usually only installed when it proves impracticable, or it is visually/aesthetically unacceptable to install an OHL. It is unusual to find cable installations in excess of 30 km in length – usually for cost/practical reasons – although at the higher voltages the high cable capacitance also becomes a limiting factor. Within this context, HV cables are generally installed in the following circumstances:

- To interconnect substations within urban/metropolitan areas, where the installation of an OHL is neither practical nor environmentally acceptable.
- To interconnect equipment within a substation, where the use of busbars or OHL connections would either be impractical or uneconomic.

- To provide a connection between an OHL terminal tower to an item of equipment within a substation – usually because the substation and surrounding land are highly congested for OHL access – and an HV cable provides a practical alternative.
- To form part of a circuit, where an OHL is either environmentally unacceptable or technically difficult to install.
- Where third party infrastructure requires the replacement of an OHL by an HV cable, e.g. airport runway or a major industrial, commercial or domestic complex.

The following sections will examine the design requirements for HV cables comprising:

- Historical development and types
- Cable technical characteristics
- Cable ratings
- Cable laying mediums
- Cable laying formations
- Impressed voltages
- Cable terminations and joints
- Cable-type tests
- Cable specification, design and routing.

7.3 Cable historical development and types

7.3.1 HV cable historical development

HV power cable development essentially covers four main types of cable dating from the very earliest networks to the present day. These comprise the following:

- Belted cables
- H and HSL solid cables
- Assisted cables
- Cross-linked polyethylene (XLPE).

The above-mentioned points will be briefly examined.

7.3.1.1 Belted cables

The earliest practical HV cables were developed in the early part of the twentieth century and were three phase, oil-impregnated paper insulation, lead sheathed and steel (tape) armoured. The extensive use of oil-impregnated paper is due to the fact the component materials and substances that comprise the insulation are stable both chemically and electrically and can be operated continuously at a high temperature (up to 85°C).

NB: It is worthy of note that early American power cables, unlike Europe, tended to favour rubber insulation (facilitated by the large American chemical industry), rather than paper insulation.

The paper insulated cable described above has evolved over time, and the design variant that is still in service today is termed the 'belted cable'. Figure 7.1(a) shows a modern paper insulated, corrugated aluminium sheath type of belted cable termed 'PICAS'. With this type of construction, part of the paper, insulation is applied around each phase conductor, and part is in the form of a 'belt' of insulation around the three-phase conductors, after they have been twisted together.

Modern paper insulation is 'mass impregnated', however, when such cables are laid on a gradient the impregnation fluid tends to flow to the lower level, therefore reducing the content, with a tendency to form voids at the higher level. Although this

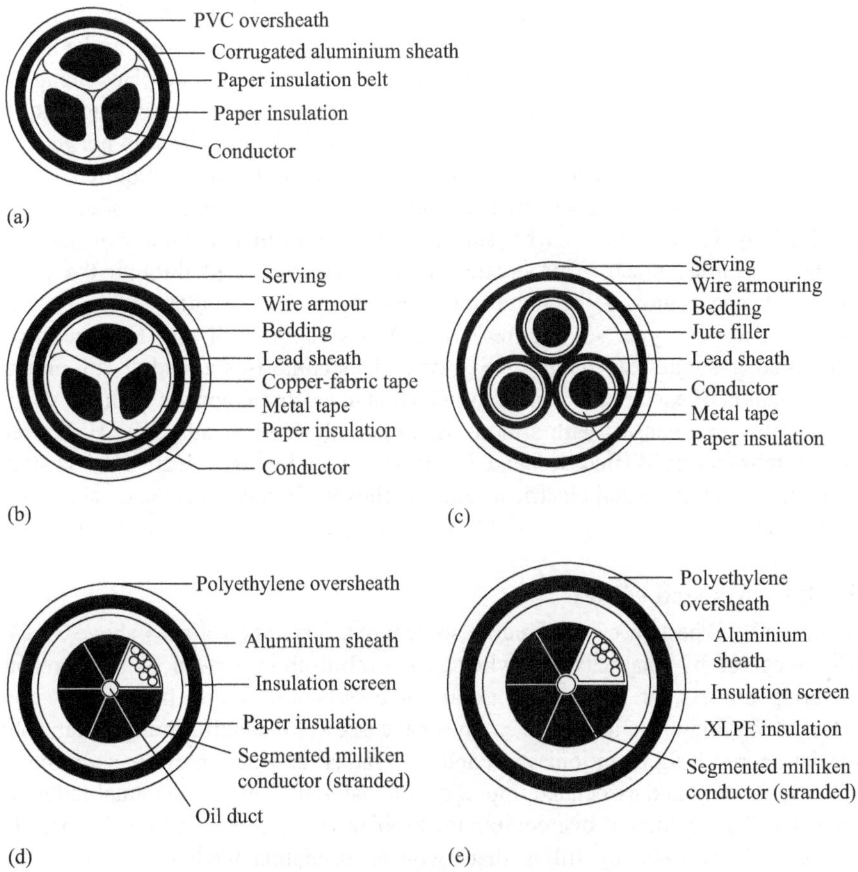

Figure 7.1 Cables – example types: (a) belted cable, (b) 'H'-type cable, (c) 'HSL'-type cable, (d) oil-assisted cable (single core) and (e) XLPE cable (single core)

is not a problem for cable voltages up to 11 kV, it is at higher voltages and this can lead to electrical breakdown and cable failure. This was overcome by the use of impregnation compounds that result in a waxy solid which does not flow and drain. The resulting cable is described as 'mass impregnated non-draining' or MIND for short, and this design has extended the use of belted cables up to network voltages as high as 33 kV. Belted cables, although rarely specified for construction projects today, have been widely installed in the past and many are still in service.

7.3.1.2 'H' and 'HSL' type solid cable

A plot of electrical field distribution in a three core cable with applied three-phase voltage reveals that there is a considerable component of electric field intensity directed along the paper laminations, especially near the centre of the cable. If the insulation (i.e. dielectric) is homogeneous, the direction of the electric field would be of little importance, but paper insulation is laminar and the electrical strength over the surface is only about 6 per cent of that through it. As a result, at the higher network voltages, both spark discharges and charring tend to take place along the paper laminations (i.e. tangential) with ultimate cable failure. For voltages up to about 33 kV, the electrical field strength is sufficiently low for this not to be a problem – but at higher voltages, an alternative design of cable is required.

Circa 1916, the weakness of HV cables to withstand tangential stresses was resolved by Hochstadter – who surrounded each insulated conductor with an earthed metallic screen. This created a radial electric field through the paper insulation and removed the tangential component. The resulting cable became known as an 'H' type, see Figure 7.1(b). A development of the 'H' type is the 'Hochstadter single lead or the 'HSL' type. This comprises a lead sheath around the insulation of each phase conductor which is in electrical contact with the sheath of the other two phases, with armouring around all three sheaths. The HSL overcomes terminating difficulties and is illustrated in Figure 7.1(c). The 'H'-type concept of creating radial electrical fields applies to all paper-insulated cables (and XLPE), at network voltages of 33 kV and above, either three or single core.

7.3.1.3 Assisted cables

A further weakness in cable design also occurs at network voltages above 33 kV. This is caused by small cavities, which arise both from imperfections in the manufacturing process, or from cable expansion/contraction arising from the heating effect of varying cable load. The cavities have a lower dielectric strength than solid insulation, resulting in an ionisation and the gradual destruction of the surrounding paper insulation, and associated impregnation material – causing eventual failure of the cable. The method of overcoming the problem of cavities was first proposed by Emmanueli *ca.* 1925, by filling them with an insulating medium of appropriate dielectric strength. The main techniques comprise the following:

- Filling the cable with an inert high pressure gas (such as commercial nitrogen). Such cables have been successfully used up to 132 kV – but with most installations at 33 kV.

- Maintaining the impregnation insulating oil in the cable under pressure by means of oil reservoirs (into which the oil can expand/contract). Such cables have been extensively used at all voltages in the United Kingdom up to 400 kV. A typical oil-assisted cable is illustrated in Figure 7.1(d).

7.3.1.4 XLPE cables

XLPE has been an emerging cable technology since the 1950s and is now the dominant cable technology. XLPE takes its name from the type of chemical bonding employed, termed cross-linking (of the polythene). The advantage of this type of cable is that it does not require the additional complexities of assisted cables, arising from the gas or oil. Furthermore, XLPE cables can operate at higher temperatures than paper insulation, up to as high as 90°C (with paper typically being between 65 and 85°C). XLPE cables do however require greater insulation thickness than their paper equivalent and as such are more difficult to physically handle.

XLPE-like paper insulation is subject to cavities, but the effects of these can be eliminated by use of a semi-conducting layer around both the conductor and the insulation, which has the effect of reducing the electrical field strength.

At voltages up to 33 kV, XLPE installation dates from circa 1985. However, installations at the higher network voltages were tempered with caution until satisfactory operational service was obtained – with installations at 400 kV (in the United Kingdom) dating from *ca.* 2005. Figure 7.1(e) illustrates a single-core XLPE cable.

7.3.2 Single-core vs three-core cable

At 132 kV and above, it is a common practice to use single-core cables. This is largely due to the fact that as the network voltage increases, the cable insulation thickness increases, with the following implications:

- Heat dissipation from the cable to the surrounding medium becomes more difficult, necessitating a reduced current rating. Within this context, the heat dissipation from three single-core cables is more effective than a single three-core cable, and therefore, higher cable ratings are achieved
- Single-core cables are easier to manually handle than an equivalent three-core cable
- Single-core cables enable longer cable drum lengths than a three-core cable, so requiring fewer joints.

It is worthy of note that one of the advantages of a three core cable is that the magnetic field arising from load current is contained almost entirely within the cable, with little induced voltage into the cable sheath. This is not the case with single-core cables – see Section 11.8.9.

7.3.3 Summary of cable types and installations

Figure 7.2 provides a summary of cable types installed both today and in the recent past.

Cable type	Voltage range and number of cores	Status
Paper insulated, lead covered, steel wire armoured (PILCSWA) (Belted)	11 kV (3 core – mainly)	Installed up to *ca.* 1990
Paper insulated, corrugated aluminium sheath (PICAS) (Belted)	11 kV (3 core – mainly)	Installed *ca.* 1985–2005
Paper insulated, lead covered, steel wire armoured (PILCSWA) (H and HSL solid)	33 kV (3 core or 3 single cores)	Installed up to *ca.* 1995
Gas filled, aluminium sheath (assisted)	33 kV (3 core or 3 single cores) 132 kV (3 single cores)	Installed up to *ca.* 1975
Oil filled, aluminium sheath (assisted)	33 kV (3 core or 3 single cores) 400/275/132 kV (single core)	33 kV installed up to *ca.* 1985 132 kV installed up to *ca.* 1995 400/275 kV installed up to *ca.* 2005
XLPE	11 kV (3 core – mainly) 33 kV (3 cores or 3 single cores) 400/275/132 kV (3 single cores)	Currently installed at all network voltages

Figure 7.2 Summary of cable types and installation voltages

7.4 Cable technical characteristics

7.4.1 Cable technical characteristics examined

The following salient technical characteristics of HV cables will be examined:

- Conductor resistance and material
- Skin effect
- Proximity effect
- Cable inductance
- Cable capacitance
- Dielectric loss angle
- Cable electromechanical forces
- Cable thermal expansion
- Cable bending radius
- Cable tensile strength
- Cable laying (section) length
- Cable laying depth

Characteristic	Ratio Copper: Aluminium	Comment
• Resistance	1.0:1.5	Copper has greater conductivity
• Weight	1.0:0.7	Aluminium is lighter
• Tensile strength	1.0:0.5	Copper is physically stronger
• Coefficient of thermal expansion	1.0:1.5	Aluminium expands more with heat
• Market price	1.0:0.3	Copper is more expensive

Figure 7.3 Comparison of copper and aluminium characteristics

7.4.1.1 Conductor resistance and material

The resistance of a conductor is a key consideration in the design of an HV cable, not only because resistance creates unwanted power losses (proportional to the square of the current) but also because the power losses create heat within the cable – which becomes a limiting factor in the cable current rating. The key factors defining cable resistance are the resistivity of the conductor (resistance is proportional to resistivity) and the cross-section area (resistance is inversely proportional to the cross-section area). From a cable conductor perspective, the two main materials in common use are copper and aluminium. Unlike OHL, copper is still an economic and a practical option for cables. Figure 7.3 shows a comparison of characteristics of copper and aluminium for the same cross-section area of conductor.

Although the market price of copper, as shown above, far exceeds that of aluminium, it does vary over time with the state of the market. In addition, the apparent price disadvantage of copper is reduced once the cost of manufacturing the cable is added, especially when it is considered that copper, being the better conductor, enables a smaller conductor cross-section area for the same current carrying capacity (as aluminium), and consequently lesser quantities of insulating and protection materials are required to finish the cable. As a result, copper cables tend to be physically smaller than the equivalent aluminium and therefore easier to handle. Thus, although price advantage favours aluminium, and it tends to be the more widely installed conductor, copper cables are still installed at virtually all network voltages, and many are in operational service.

7.4.1.2 Skin effect

When a conductor is carrying a steady direct current, the current will distribute itself uniformly over the whole conductor cross-section. However, with an alternating current, a magnetic field arises within the conductor, so causing a conductor self-inductance. The self-inductance is greater towards the centre of the conductor (because of greater flux linkages). As such, the electrical current is forced, to a greater extent, to flow around the periphery of the conductor. In doing so, the effective conductor resistance and associated conductor losses increase. This phenomenon is termed 'skin effect' and becomes more pronounced as the cross-section of the conductor increases. Skin effect can be minimised by the use of compact stranded conductor or by the 'Milliken' (segmented) design of conductor. Milliken conductor comprises individual conductor segments with a thin layer of insulating material in between the segments, as shown in Figure 7.1(d) and (e). The requirement for Milliken conductor is generally associated with conductors with larger cross-section areas as found on the transmission networks.

7.4.1.3 Proximity effect

The 'proximity effect' is similar in nature to the skin effect but in this instant arises from the mutual induction from the currents in adjacent conductors. Proximity effect is more significant in cables than, say, OHL or busbars, due to the relative closeness of the conductors. The magnitude of mutual inductance varies with the physical separation of the conductors concerned, and depending upon the relative direction of current flow in the conductors (at any given moment) causes different proportions of current to flow in the two halves of a conductor. This again has the effect of increasing the effective resistance of the conductor. Proximity effect tends to be smaller than skin effect and is negligible if the cable spacing exceeds eight times the cable outside diameter.

7.4.1.4 Cable inductance

Generally, the (series) inductance of a cable is not a prime consideration when designing a cable. This is because compared to an OHL, HV cable's lengths are relatively small, and therefore, the effect of cable inductance on either voltage drop or power transfer is usually not significant.

However, there are significant differences worthy of note between the ZPS reactance/impedance (see Chapter 4) of a cable compared to an OHL. Cables generally exhibit a proportionately larger ZPS resistive component and lower ZPS reactive component than an equivalent OHL, with the magnitude of the resistive component often exceeding the reactive component. This arises because zero sequence currents flowing in the conductor and sheath (which are in close proximity) cause the magnetic fields to cancel, so reducing the inductance. This used to cause a problem (prior to numerical relays) when applying distance protection to cables as the residual compensation setting was not always available to match the ZPS/PPS ratio of the cable.

7.4.1.5 Cable capacitance

As previously stated, the shunt capacitive reactance of a cable is significantly lower than that of an OHL, resulting in a relatively large flow of capacitive VArs into the cable. Figure 7.4 provides a comparison for the instance of each circuit being energised from one end only, thereby resulting in the flow of circuit capacitance current only – for typical OHL and HV cables.

Figure 7.4 clearly illustrates the much larger capacitive current flowing into cables. As a result, the following is worthy of note:

- The power loss, arising from the flow of charging current in the cable conductor resistance, is considerably greater in HV cables (per km of route length) than OHLs.
- Cables are a major source of the 'Ferranti Effect'. That is the flow of capacitive VArs (into a circuit shunt capacitance) through the circuit's series inductive reactance – resulting in a rise in circuit voltage. This is illustrated with reference to Figure 7.5 in which a composite circuit comprising an OHL and cable

Circuit voltage (kV)	OHL charging current		HV cable charging current	
	(Amps/km)	(MVAr/km)	(Amps/km)	(MVAr/km)
33	Neg	Neg	2.5	0.14
132	0.2	0.05	7.0	1.6
275	0.6	0.3	12.0	5.7
400	1.0	0.7	15.0	10.4

Figure 7.4 OHL and cable comparative charging currents

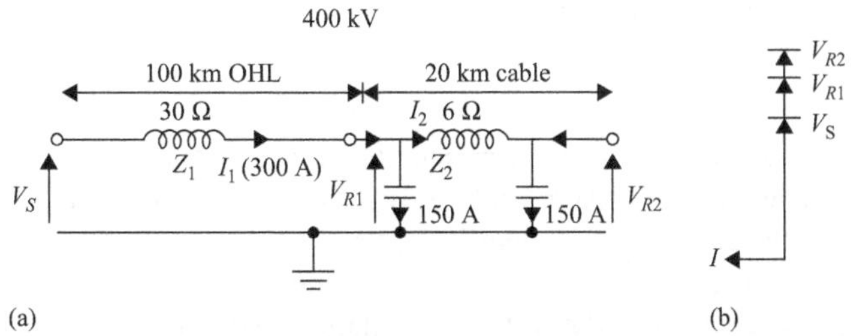

$$V_{R1} = V_S + I_1 Z_1 = V_S + 9 \text{ kV}$$
$$V_{R2} = V_S + I_1 Z_1 + I_2 Z_2 = V_S + 9.9 \text{ kV}$$

Figure 7.5 Ferranti effect – simplified analysis: (a) OHL + cable circuit and (b) vector diagram

is subject to little or no load through the circuit. As a result, the flow of VArs into the cable shunt capacitance causes an additive voltage drop across both the OHL and cable series inductive reactances. This causes the receive end voltages of V_{R1} and V_{R2} to exceed the send end voltage V_S. The Ferranti effect can be particularly problematic on transmission systems at times of light loading, e.g. in the early morning and during the summer. This may necessitate either the temporary switching out of cable circuits from service – or the switching in of shunt reactors (the effect of which is to cause voltage reduction).

7.4.1.6 Cable dielectric loss and loss angle

If the insulation of a cable is free from defects, the cable acts as a perfect capacitor, and hence, the insulation is often referred to as the dielectric, in which the leakage current through the insulation leads the conductor voltage by 90°. However, if there are impurities in the insulation, the resistance of the insulation decreases, resulting in a resistive component of current through the insulation which in turn causes the generation of heat and cable losses. If the resulting heat is excessive, this may cause further deterioration of the cable, with further decrease in insulation resistance.

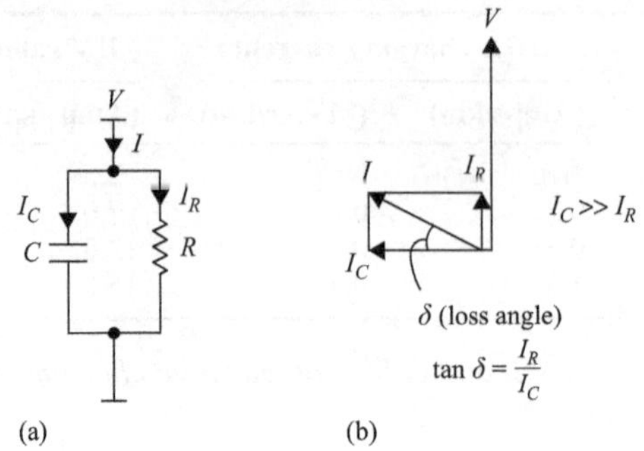

Figure 7.6 Dielectric loss angle: (a) cable equivalent circuit and (b) vector diagram

Figure 7.6 shows the cable insulation equivalent circuit and corresponding vector diagram. It is usual to measure and express the insulation quality through the tangent of the 'loss angle', as shown in Figure 7.6(b). In a perfect cable, the loss angle would be close to 0°. An increasing loss angle over time is indicative of the increasing resistive current through the insulation and a measure of the level of degradation of the cable. During cable manufacture, it is usual to measure the loss angle – which can be used as a comparison with later in-service test values as an assessment of cable condition.

The component of current through the dielectric, which is in phase with the conductor voltage results in power loss in the dielectric. With reference to Figure 7.6, the power loss (dielectric loss) is given by the following expression:

$$\text{Dielectric loss} = \frac{V^2}{X_C} \tan\delta \text{ W/m}$$

where X_C is the reactance of capacitance C as shown in Figure 7.6(a), per metre of cable length.

7.4.1.7 Cable electromechanical forces

Under conditions of heavy phase-to-phase or three-phase short-circuits, severe mechanical forces arise in a cable as a result of the intensive magnetic fields caused by the short-circuit currents. With short-circuit currents of the opposite direction, the forces are repulsive, with the result that they tend to endeavour to burst three core cables, or force apart single-core cables. The forces can also be attractive for the instance of two adjacent cables carrying single-phase fault current in the same direction. It is worthy of note that balanced three-phase currents create both repulsive and attractive forces. With cables laid directly in the ground, there is considerable mechanical support to stabilise cable bursting or movement. However, where cables are supported by cleats, the fixings must be strong enough to withstand these forces. Cable design must take this into account.

7.4.1.8 Cable coefficient of thermal expansion

Both cable short-circuit currents and load currents give rise to cable temperature rise and subsequent expansion of the cable conductors (and the cable metallic sheath). Figure 7.2 shows that aluminium will expand more than copper for the same temperature rise (and same cross-section area of conductor). The extent of expansion is not insubstantial – e.g. a temperature rise of 150°C in a copper conductor would cause an expansion of the order of 2.25 m/km of cable. Expansion/elongation of a cable can result in 'snaking' or elongation at bends in the cable – to relieve the resulting mechanical stress. The effect of expansion may be more severe at joint positions, particularly with buried cables where the surrounding ground holds the cable more rigid. Cable terminations may also be more at risk, since if the expansion is not alleviated elsewhere, it will extend into the terminations. Where cables are held by clamps, the clamping arrangement must be strong enough to withstand the forces arising from expansion. Cable thermal expansion must be considered as an integral part of cable design.

7.4.1.9 Cable bending radius

Cables are designed to accommodate a minimum bending radius, which if exceeded may lead to disturbance and failure of the insulation. Bending radius must be taken into account when planning a cable route. It must also be taken into account when planning the entry of cables into cable sealing ends and switchgear, etc. Clearly, the greater the diameter of the cable, the greater is the bending radius. Bending radii generally range from 12 times cable outside diameter for typical 11 kV cables to 35 times outside diameter for 400 kV cables.

7.4.1.10 Cable tensile strength

Cables are designed with a defined tensile strength, and if a mechanical pulling force exceeds the tensile strength elastic limit, the cable will fracture. The force applied when pulling a cable into position must therefore be compatible with the cable tensile limit – and considered as part of the cable design.

7.4.1.11 Cable laying lengths

In general, it is usual to lay the longest length of cable possible to minimise the number of joints. However, a number of factors may limit the length of laid cable (usually termed a section of cable), as follows:

- The maximum continuous length that can be manufactured
- The cable drum size
- Drum transportation limits
- Induced voltage limits into the cable sheath
- Limitations on the length of continuous trench that can be opened
- Physical limitations on the positions of the joint bays
- Balancing of cable lengths in a cross-bonded cable installation.

All of the above needs are to be considered, initially in outline at the scheme development stage, and then finally at the detail design stage – to optimise the lengths of cable to be manufactured.

Voltage (kV)	Depth in roads (m)	Depth in good agricultural land (m)	Depth in open countryside (m)
11	0.75	0.91–1.2	0.70–1.05
33	0.75–0.9	0.91–1.2	0.75–1.05
132	0.75–1.0	0.91–1.2	0.75–1.2
275 and 400	0.75	0.90	1.00

Figure 7.7 Typical range of cable laying depths

7.4.1.12 Cable laying depth

In the United Kingdom, there is no specific legal requirement which defines the depth of cable laying. The depth is governed primarily by that which is considered good practice to avoid cable damage. Suffice it to say that the depth of a buried cable usually does not have an effect on thermal resistance – except at transmission voltages where the overall cable and installation design tends to be more bespoke. Guidance on cable laying depth is provided in ENA TS 09-02. This guidance is largely consistent with similar practice world-wide. There is a range of recommended depths depending upon ground circumstances. Figure 7.7 gives typical values used by UK power network companies.

NB: The marginal inconsistencies represent different company practices.

7.4.2 Cable temperature monitoring

It is worthy of note that many modern XLPE cables, particularly those suitable for transmission network voltages, frequently contain an integral fibre optic sensor along the cable length, to enable cable distributed temperature monitoring (i.e. at all positions along the cable route), in conjunction with a distributed temperature sensor system located at end of the cable.

7.5 Cable ratings

7.5.1 Cable continuous current rating

The limiting factor in the continuous current rating of a cable is the maximum permissible temperature at which its component parts may be operated with a reasonable factor of safety. The limiting component is usually the insulating material (i.e. the dielectric) – which is subject to ageing with increased temperature. To ensure a reasonable cable life (typically 40 years minimum), cable design must ensure that the temperature is kept within limits.

The current carried by a cable conductor raises the temperature of the conductor as a result of the conductor resistance. With an increase in current, the temperature of the conductor commences to rise until equilibrium is established and the heat generated is equal to the heat dissipated (via thermal conduction) through the cable insulation, metallic sheath, the (polythene) over-sheath and finally the surrounding earth to air.

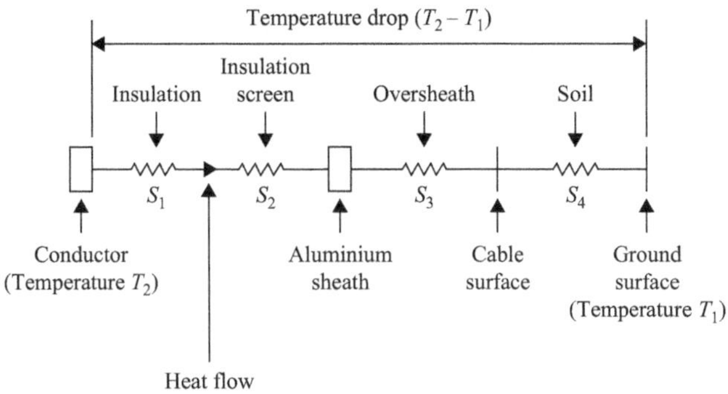

Figure 7.8 Cable thermal circuit – example

Figure 7.8 illustrates the (simplified) mechanism of heat flow for a single-core XLPE cable. It can be seen that the heat flow arrangement is similar to that of an electrical circuit. This analogy is used in practice where the temperature drop is analogous to voltage, and the power flow in watts is analogous to current. Within this context, the thermal form of ohms law defines a 'thermal ohm', 'S', as being the temperature difference in degrees Celsius between opposite faces of a 1 metre cube of material by the flow of 1 W of heat. It is expressed in units of °C/W/m. The following is worthy of note:

- Some texts use the absolute temperature measure in degree Kelvin. However, since it is temperature difference that is measured, then either degree Celsius or Kelvin may be used since 1°C is equal to 1°K.
- All materials have a thermal resistivity 'g' which is similar to electrical resistivity, and the following similar expression holds true:

$$S = \frac{gl}{a}\ {}^{\circ}\mathrm{C/Watt/m} \tag{7.1}$$

  where 'l' is the length of the material in metres, and 'a' is the cross-sectional area of the material in square metres.
- In an electrical circuit, Ohms law for current flow is expressed as:

$$I = \frac{V}{R}$$

  Similarly, the thermal equivalent in terms of heat flow is expressed as:

$$I^2R = \frac{T_2 - T_1}{\sum S}\ \mathrm{Watt/m} \tag{7.2}$$

  where I^2R is the power in watts, I is the current flow through the cable, R is the conductor resistance/metre, $T_2 - T_1$ is the temperature difference between conductor and ground or air respectively and $\sum S$ is the summation of all the thermal resistances (i.e. thermal ohms) as shown in Figure 7.8, i.e. S_1, etc.
- Metals are assumed to have zero thermal resistance, i.e. heat transfer is immediate.

If with reference to expression (7.2), all the parameters are known (i.e. specified) except the current '*I*', i.e. the cable continuous current carrying capacity, then this quantity can be calculated. The above is a simplified analysis and in practice account must be taken of the following:

- Whether the cable is a single core or three core
- If single core whether it laid flat or in trefoil,
- Whether it is laid in soil or air, or other medium
- The heat impact of any adjacent cables.

IEC 6087 – Electric Cables – Calculation of Current Ratings provides authoritative guidance. Specified values for XLPE cable are as follows:

- Maximum permissible continuous core conductor temperature (i.e. with continuous current) = 90°C
- Ambient air temperature = 25°C
- Ground temperature = 15°C.

7.5.2 Cyclic rating

In comparison to OHL, the heating and cooling times of cables, particularly when buried, are several times greater. This is a result of the long thermal time constant of cables and cable installations. A representative heating curve is shown in Figure 7.9. However, in practice, the loading of a cable is usually not continuous but follows a daily loading cycle. This cycling allows cables to operate above their maximum continuous current ratings for short period of time, on the basis that the load current subsequently decreases – and the loading cycle does not result in temperature excursions exceeding the maximum allowable conductor temperature (and therefore no degradation of the cable insulation). A typical cyclic rating comprises 6 h at 125 per cent maximum continuous loading followed by 18 h at

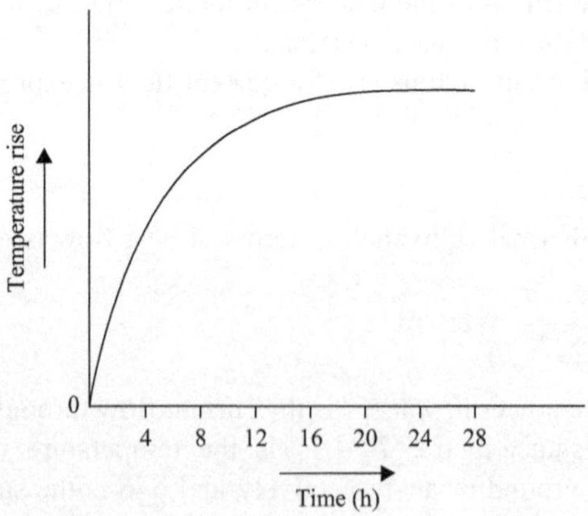

Figure 7.9 Rate of conductor temperature rise – when continuously loaded

60 per cent maximum continuous loading, see also Section 20.4.4, on thermal rating schedules.

7.5.3 Cable short-circuit current rating

Each conductor in a three-phase circuit must be capable of carrying, without damage or unacceptable deterioration, the maximum short-circuit current to which it will be subject. Within this context, the short-circuit may occur when the cable is carrying its maximum permissible load current and operating at its maximum permissible continuous temperature. The additional short-circuit current, which may only flow for a short time, may be many times the magnitude of the load current and therefore must not cause an additional temperature rise which exceeds the maximum short time temperature of the cable. Ratings are usually taken over a 1-s fault duration period with a conductor temperature specified as not exceeding 160°C for paper insulated cables and 250°C for XLPE insulated cables. The formula for calculating the allowable short-circuit current I_S is given by:

$$I_S = \frac{KA}{\sqrt{t}} \text{ Amps} \tag{7.3}$$

where K is the a constant, depending upon the conductor material, and the initial and final temperatures associated with the short-circuit, e.g. for XLPE the initial temperature is usually taken as the maximum continuous loading temperature of 90°C, and the final temperature as the maximum permissible short-circuit current, temperature of between 210 and 250°C, A is the conductor cross-section area in mm and t is the duration of the short-circuit in seconds.

The above expression is valid for short-circuit durations in the range 0.2 to 5.0 s. A reference value of I_S for a 1-s duration of short-circuit current may be determined, termed I_{ONE} from which the following relationship and other values of I_S may be obtained:

$$I_S = \frac{I_{\text{ONE}}}{\sqrt{t}} \text{ Amps} \tag{7.4}$$

It is worthy of note that the short-circuit current level, rather than the maximum load current, may dictate the size (i.e. cross-section area) of the conductor to be used.

7.6 Cable laying mediums, formations and impressed voltages

7.6.1 Types of cable laying mediums

Cables are generally installed in the following mediums:

- Air
- Ground (soil)
- Ducts
- Tunnels

These will be briefly examined:

- **Air**
  Cables laid in air may be laid horizontally or vertically and supported by cleats. The cleat system may be rigid that allows no movement of the cable (e.g. from either thermal expansion or short-circuit forces), similar to being buried in the ground, or flexible that allows movement. Specialist calculations must be undertaken to assess the forces involved. Air has a higher thermal resistance than soil, so the conductor will heat up much more quickly in air under both steady state and short-circuit conditions.
- **Ground**
  Most cables are laid in the ground. They may be either direct buried (problem to maintain) or installed in a performed trench (easier to maintain). It is usually the cheapest method of installation. The cable excavation trenches themselves may only be a few metres wide but may require a land width of 20 m or more (with the larger, high voltage cables) to accommodate the excavation and vehicle movement requirements. At the higher voltages where the heat flow calculations are cable specific, it is often usual to surround the cable with thermally advantageous backfill (such as cement-based sand) which must be carefully compacted to remove all air pockets.
- **Ducts**
  Where there is a requirement for a cable to be installed in urban areas, involving major roads and requiring traffic management schemes, an alternative but more expensive option to that of direct burial is the installation of ducts buried in the ground. Generally, only one power cable per duct is installed. Ducts have the salient advantage that they can be installed in advance of cable delivery and at a time which minimises disruption to the general public. Cable ratings in ducts are less than those of direct ground burial.
- **Tunnels**
  Tunnels are generally found in urban and metropolitan areas where direct ground burial (even in ducts) would cause unacceptable interruption and disruption. They are mostly used at transmission voltages where the construction cost per MW of power flow is economically viable. Most tunnels are lined with bolted segments and sealed. They are typically at a depth of 25 m or greater with a diameter of about 4 m. They could run for many kilometres. Instances also arise of the tunnel being shared with other utilities. Access to the tunnel is usually via a 'head house' building. A purpose built tunnel would usually be sized for the installation of two cables (i.e. circuits) one each in either side of the tunnel and mounted vertically. The phase position of the conductors within the tunnel must be arranged to minimise impressed voltages (IVs) into the cable sheaths. Fans are generally installed to provide forced air cooling to increase the cable rating. Consideration needs to be given, when designing the tunnel, to access arrangements not only for initial installation but also for maintenance, inspection and repair – this may necessitate a rail mounted access vehicle. Consideration must also be given to the removal, and disposal, of very

considerable volumes of soil when excavating the tunnel. In the United Kingdom, there are numerous cable tunnels, particularly in the London area. In addition, one is installed under the river Mersey (shared with a road) and one tunnel crossing the Pennines in the north of England utilising a disused railway tunnel.

7.6.2 Single-core cable formations

With reference to Figure 7.10, single-core cables are laid and arranged in two main formations:

- Trefoil phase formation
- Flat phase formation.

These will be briefly reviewed.

- **Trefoil phase formation**
  Trefoil arranges the three cables in a triangular format with the phase centres the same distance apart. Trefoil has two main advantages:
  - A reduction in installation space due to the close proximity of the cables
  - Minimises induced voltages into the cable sheaths, arising from three-phase load currents. This is due to the close proximity and symmetry of the cables, resulting in significant reduction in the resultant magnetic field.

  The main disadvantage of trefoil is that the close proximity of the cables reduces heat dissipation, which may lead to a reduction in the current rating of the cable. Trefoil is mostly used at voltages up to 132 kV and on shorter lengths of cable with few, or no, joints.
- **Flat phase formation**
  In flat formation, the three cables are laid horizontally and therefore not symmetrical in arrangement. Flat formation reverses both the advantages and disadvantages of trefoil, namely:
  - Due to the separation of phases, it improves heat dissipation and cable rating
  - It increases the magnitude of induced voltages (arising from currents in the cable conductor) in the cable sheath – with a larger magnitude in the central cable sheath than the two outer cables.
  - It increases the spacing between the cables – which is advantageous in both cable laying and cable jointing.

  Generally, at network voltages above 132 kV, both economic and engineering considerations favour flat over trefoil.

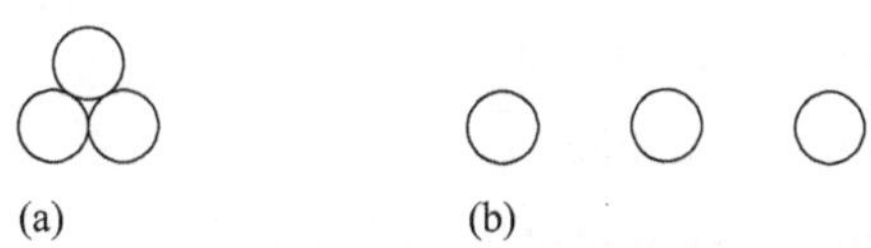

Figure 7.10 Cable laying formations: (a) trefoil and (b) flat

7.6.3 Impressed voltage

HV cables are subject to IV into the cable sheath arising from inductive coupling from the cable core conductors – both from balanced load current and fault current conditions. The magnitude of IV and potential circulating currents in the sheaths is potentially the greatest in single-core cables. This fact greatly influences the design of these cable installations in minimising the impact of IV – a necessary requirement both for ensuring personnel safety – and maximising cable ratings. Chapter 11 covers the subject of IV in some depth, with Section 11.8.9 specifically addressing HV cables.

7.7 Cable terminations and joints

7.7.1 Terminations and joints – considerations

In its entirety, a cable circuit comprises not only the HV cable itself but also the cable terminations at each end for connecting to either switchgear or OHL, and the joints for connecting sections of cable. Both terminations and joints are integral to the overall cable installation design, requiring specialist consideration. The salient requirements are summarised below. Only XLPE will be considered since it is now by far the dominant cable technology for new installations. It is worthy of note that most cable faults are associated with cable terminations and joints, and therefore, their correct design and on-site installation are of critical importance.

7.7.2 Cable terminations

The major concern associated with cable terminations relates to the dielectric screen and the concentration of the electric field that occurs at the point at which the screen ends. Without special measures, the higher voltages associated with the concentrated field could lead to tracking and breakdown. The traditional means employed to relieve electrical stress is the application of a stress cone. This effectively controls the capacitance in the area of the screen termination by use of a high permittivity (lower capacitive reactance) material, thereby reducing the stress in the dielectric to an acceptable level.

Cable terminations broadly fall into the following categories:

- Cable connections to distribution voltage metal-clad switchgear
- Cable sealing ends for connecting to OHL down leads or down droppers, or to AIS switchgear
- Cable sealing end connections directly into GIS switchgear
- Cable connections directly into transformers.

The above will be briefly examined.

- **Distribution voltage metal-clad switchgear**
  This type of switchgear is usually installed indoor, and at voltages up to 33 kV. In this design, the cable terminates in a cable termination box integral to the

switchgear. The stress control mechanism may be achieved by pre-moulded rubber-based sleeves which are attached to the cable via heat or cold shrinkable techniques, which provide a water/moisture tight seal. Similar techniques can be applied to 33 kV cables connecting to an OHL.

- **Cable sealing end connections to OHL and AIS switchgear**
  Cable sealing ends, as stated earlier, provide outdoor cable terminations for connecting to either an OHL or to AIS switchgear connections (e.g. busbars). A simplified illustration is provided in Figure 7.11 for an XLPE cable. The core screen is usually terminated by a slip-on factory moulded rubber-based stress cone. The insulator provides protection from the external environment and is filled with a viscous silicone oil – which has a high impulse voltage withstand performance and minimises risks of leaks from the seals. At 132 kV, cable sealing ends may be mounted on platforms which are connected to a tower, but at 400 and 275 kV, they are usually ground mounted and, if external to the substation, located in fenced off compounds. In some instances, the compounds may also house associated protection and control equipment.
- **GIS terminations**
  Designs also exist for the termination of cables directly into GIS substations i.e. SF6 metal-enclosed equipment.
- **Transformer terminations**
  As with GIS, designs exist for a cable sealing end to connect to and terminate directly in a transformer.

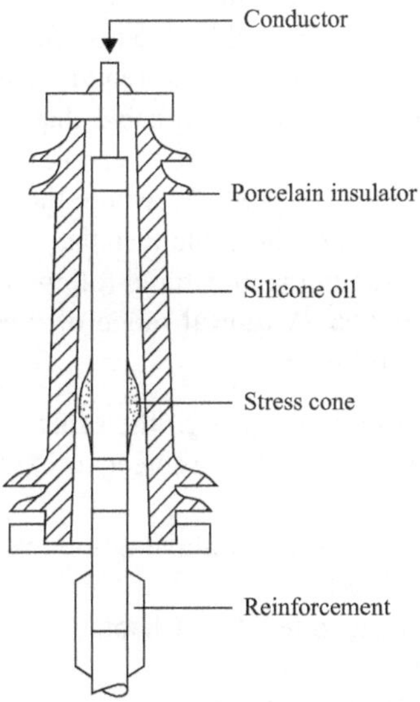

Figure 7.11 XLPE cable sealing end

7.7.3 Cable joints

HV cable joints usually comprise the following types:

- Straight joints – to join sections of a cable to form the complete cable circuit.
- Transition joints – to join cables of different technology types, e.g. XLPE to oil filled.
- Stop joints – to provide a hydraulic barrier in an oil-filled cable.
- Branch joints – to provide a 'T' connection. Mostly used at lower voltages.

At the higher voltages, jointing techniques must again consider appropriate measures to manage the impact of high electrical field stress. The location of both the joint, and joint bay during construction, requires advance planning to minimise disruption to third parties and to maximise construction access. At transmission voltages, the joint bays are usually relatively big and require a clean, humidity controlled environment – this may necessitate the installation of concrete floors and side retaining walls during construction.

7.8 HV-cable type tests

7.8.1 Cable type tests – requirements

Chapter 21 describes two factory-based testing regimes: type tests and routine tests. The former are usually carried out on the very first item of manufactured equipment of its type and are usually very comprehensive to be fully assured that the equipment meets its specification. The latter is carried out on subsequent items of manufactured equipment (of the same type) and is less comprehensive. At distribution voltages, e.g. 11 kV, where cables are installed in relatively large volumes (and usually held in stock), and the technical requirements are less numerous, then most power network projects would install cable that was subject to routine tests (the type test having been carried out on the very first production run). However, at the higher network voltages e.g. 400 kV, where the number of installations are relatively small, and the design of the cable installation more bespoke and tailored to the project, the cable is usually manufactured following placement of order – and type tests would be undertaken. A typical range of type tests carried out on HV cables would comprise the following:

- Bending test to confirm the integrity of the cable sheath
- Load cycle test where the cable is subject to loadings more severe than in normal service
- Dielectric loss angle test (usually following the load cycle test)
- Over-voltage withstand test
- Lightning impulse withstand test (see Chapter 4)
- Thermal stability test
- Partial discharge test (see Chapter 15)
- Tests on the cable protective finishes.

7.9 Cable specification, design and routing

7.9.1 Cable specification – considerations

A specification for an HV cable should consider the following:

- Maximum operating voltage
- Over-voltage withstand levels
- Maximum continuous current rating
- Short-duration current rating
- Short-circuit current rating
- Cyclic (current) rating profile
- Insulation type (e.g. XLPE)
- Cable length
- Three core or single core
- If single core, the number of cables per phase
- If single core, flat or trefoil – if flat the cable spacing
- Sheath bonding arrangements
- Joint positions
- Cable laying medium (i.e. air/ground/duct/tunnel)
- Chemical composition of soil (if laid in soil)
- Thermal resistivity of soil (if laid in soil)
- Termination types
- Cable rout and obstacles
- Cable laying depth
- Trench width
- Trench depth
- Back-fill material
- Cable protective covers
- Cable markers (e.g. 'Electric cable – danger') and location
- Access requirements.

7.9.2 Cable overall design and routing

The following provides typical requirements for the overall design and routing of an HV cable construction project.

1. **Electrical specification**
   Specification of the substation locations, or end points, to be connected together with the cable power system requirements (voltage, currents, etc.), i.e. as specified in Section 7.9.1.
2. **Route proposal**
   A desk top study to determine a proposed route, using maps of suitable scale, and taking into account obstacles – i.e. roads, railways, other services, etc. – and land ownership.
3. **Route evaluation**
   A walk of the proposed route resulting in a refinement of (2).

4. **Soil analysis**
   The taking of soil samples for thermal and mechanical considerations. An assessment will need to be undertaken on the type of soils that comprise the cable route – to determine the number of samples required.
5. **Consultation**
   Consultation with local authorities and land owners to agree the route and access arrangements.
6. **Final route design**
   Final proposal with reference to
   (i) Cable sections and joint locations
   (ii) Link pillar arrangements
   (iii) Termination arrangements
   (iv) Access and special arrangements for installation
7. **Cable specification**
   As specified in Section 7.9.1
8. **Temporary works design**
   The design of the temporary works including
   (i) Excavation requirements
   (ii) Trench shuttering
   (iii) Access route arrangements
   (iv) Joint bay preparation.
9. **Drawings preparation**
   The following typical drawings are required to be prepared:
   (i) Route plan (large scale overview)
   (ii) Route record (small scale, e.g. 1:200)
   (iii) Cross-section drawings (roads, bridges, tunnel, railway crossings, etc.)
   (iv) Phasing, cross-bonding and earthing electrical drawing (where required)
   (v) Link pillar physical arrangements
   (vi) Schedule of cable warning marker posts and plates locations (e.g. in proximity to railway)
   (vii) Fully dimensioned drawings of
      (a) Cable
      (b) Cable joints
      (c) Cable sealing ends.

7.10 Gas insulated transmission line

7.10.1 GIL background

From circa the year 2000, developments in sulphur hexafluoride (SF6)-based technology have led to ever increasing numbers of GILs. Although over long distances OHL remains the most economic and practical method of high-power electricity transmission, situations arise when it is necessary to install an HV cable (such as in both urban and visually sensitive environments). GIL now provides a viable alternative to HV cables. Although GIL possesses many advantages over HV cables (shortly to be examined), it does possess one very significant disadvantage, namely that it contains significant volumes of SF6, a highly polluting green-house gas, exacerbating global warming. Although SF6 leakage from GIL is (and is expected to be) relatively low, it is not insignificant. Consideration of the use of GIL in power network construction should therefore ideally follow a risk assessment.

7.10.2 GIL construction

With reference to Figure 7.12, GIL consists of two concentric aluminium tubes: an inner tube, the conductor, which is hollow (to minimise skin effect) with a typical diameter of 180 mm, and an outer tube which provides protection and support, with a typical diameter of 500 mm. The tube sections are generally up to 20 m in length and are welded on site to make the arrangement gas tight. At the boundary between the sections, the conductors connect via a sliding contact system. This compensates for the differences in thermal expansion between the conductor and outer protective tube. The conductor support insulators are placed at intervals of 10–12 m approximately. They usually comprise pairs, made out of epoxy cast resin or similar material, and retain the conductor at the centre of the housing – they are usually fixed to the conductor but slide on the enclosure housing (outer tube) to compensate for thermal expansion differences between conductor and housing. The insulating gas is typically 80 per cent nitrogen and 20 per cent SF6 and operates at about seven bar. GILs are usually divided by barriers into gas zones which are separately alarmed to indicate high/low pressure.

7.10.3 GIL installations

GIL installations in the United Kingdom have typically only been a few kilometres in length, but world-wide there are installations in excess of 30 km (which is typical of maximum cable length) – however, they are capable of much longer installations. The bending radius of GIL is much larger than that of a cable being typically of the order of 400 m – but if sharper deviations are required, elbow modules can be fitted. GIL can be laid above ground and mounted on structures (usually at a low level), or directly buried in the ground (similar to a gas pipe), or installed in tunnels. GIL may also be used in substations as a substitute for an inter-circuit cable or to provide high current connections between busbars. They can be designed for direct entry into a GIS bay.

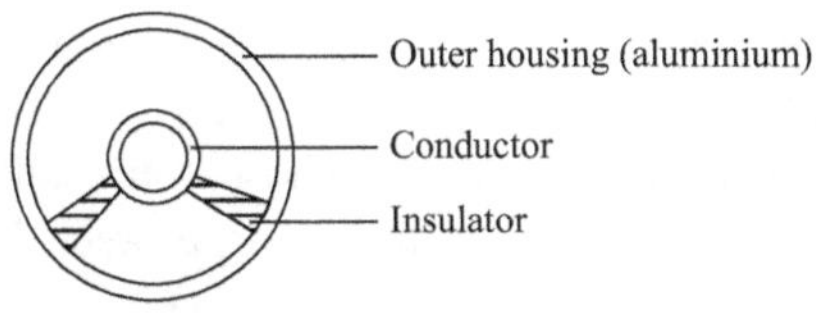

Figure 7.12 GIL cross-section

7.10.4 Advantages of GIL

GIL possesses the following advantages:

- Higher current ratings than cables are achievable – although when a relatively low current rating is required, cables may be economically advantageous.
- No fire risk, unlike cables.
- Much reduced magnetic field compared to either a cable or an OHL. This is because voltages induced into the outer housing (i.e. tube) by currents flowing in the inner conductor cause circulating currents in the outer housing which may be up to 95 per cent of the inner conductor current. This is due to the relatively low impedance of the outer housing. In accordance with Lenz's law, the housing current and associated magnetic field are oppositely directed to that of the inner conductor, resulting in a total magnetic field external to the GIL housing of virtually zero. To achieve this, the outer housing is periodically connected to earth.
- GIL current rating is relatively unaffected by high ambient temperatures.
- GIL unlike cables do not exhibit high capacitance, and therefore do not consume high levels of capacitive VArs, nor exhibit the Ferranti effect to the same extent as cables. As such, GIL is suitable for being installed in greater lengths than cables. Furthermore, GIL does not require shunt reactor compensation to be installed.

7.11 Auxiliary cables

7.11.1 Auxiliary cable types

Auxiliary cables are used to interconnect protection, control and telecommunications equipment and HV equipment secondary wiring – usually within a substation but also between substations. They generally fall into three types:

- Multi-core cables
- Multi-pair cables (PVC insulated)
- Multi-pair cables (polythene insulated).

These will be briefly examined below.

7.11.1.1 Multi-core cables

Multi-core cables have traditionally been used to interconnect both protection and control systems, and switchgear, including that of circuit breaker trip/close circuitry. They typically comprise the following specification:

- 2.5 square millimetre copper conductor (larger conductors also exist but are used by exception).
- PVC insulated and sheathed. Cables whose route takes them outside of a building are usually additionally armoured to protect the cable from mechanical damage.
- Maximum working voltage is usually 600 V AC rms, and 1,000 V DC.

- Induced voltage withstand is typically 5 kV rms for 1 min between core and earth.
- Standard cables sizes are usually 2, 3, 4, 7, 12, 19 and 37 cores with identification numbering on each core.

Typical use of multi-core cables is:

- Connecting CTs, VTs, circuit breakers, disconnectors, etc. to a relay/control room
- Interconnection of protection and control facilities within a relay room
- Connections from the battery distribution board to protection equipment and circuit breakers, disconnectors, etc.
- Site common wiring such as that for interlocking, synchronising, older types of busbar protection and circuit breaker fail systems, etc.
- Connection of AC supplies from a distribution board to items such as heating, lighting, power outlets, motors, pumps, etc.

Multi-core cables are normally terminated with the use of glands connected to the protective sheath/armouring, from which the individual cable cores emanate and connect to terminal blocks in the relay panel, cubicle, etc. It is usual to earth the armouring at the gland position at both ends of the cable – this provides screening of the cores from IVs which are ever present in a substation environment.

During the construction process, the design of the multi-core cable arrangements must take account of a number of considerations as follows:

- Careful evaluation of the required interconnection points, the routing of the cables, the number of cables required and the number of cores required in each cable – to optimise the number of cables and associated cost. Consideration should be given for leaving a number of cores unused (spare) in each multi-core for potential future use.
- Most multi-core cables are laid in trenches, which usually come in a number of standard sizes. It is important to ensure that no trench exceeds about 70 per cent capacity (i.e. volume) to allow for future project cable requirements. Suffice it to say that if during future projects a cable has to be withdrawn from an over capacity trench to make room for new cables – then, this can be physically difficult – and the alternative may be to install new trenches (both costly and difficult). Careful design of multi-core trench sizing is therefore essential.
- That as far as possible, different functions are allocated to different cables, e.g. segregate 230 V AC from 110 V DC wires; segregate CT and VT wires; keep all interconnecting buswires (between two locations) in the same cable.
- Some multi-core cables may be as long as 500 m. This may result in unacceptable voltage drop – requiring doubling of cores or larger conductors. Voltages at circuit breaker trip coils when subject to trip coil current must be kept within defined limits, see Section 10.25.1.4.

Multi-core cables have historically comprised crimped terminations, connected to the terminal blocks by nuts and washers – but these have now largely given way to bare wire ends being held by clamp and screw terminal blocks. Although for new installations, multi-core cables are being replaced by fibre optic cables, they still remain a major interconnection medium within a substation.

7.11.1.2 Multi-pair cable – PVC insulated

These cables have traditionally been used to interconnect equipment associated with telecommunication and Scada systems and typically comprise the following specification:

- 0.5 square millimetre copper conductors
- PVC insulated and sheathed, and armoured if laid outside of a building
- Maximum working voltage is usually 110 V AC and 150 V DC
- Induced voltage withstand is typically 5 kV AC rms for 1 min between core and earth
- Each pair of conductors is usually given a unique colour combination
- Standard sizes are usually 2, 5, 10, 40, 60 and 100 pair.

These cables are mostly associated with equipment supplied from a 48-V nominal battery. However, with an increasing tendency to standardise on 110-V nominal battery systems, their use has declined. Again, it is usual to earth the armouring at both ends of the cable to provide screening from the impact of IVs. It is worthy of note that because the cable pairs are twisted together, this type of cable is less susceptible to induced voltages than multi-core cables (which have a helical wormed lay).

7.11.1.3 Multi-pair cable – polythene insulated

These cables have traditionally been employed for the connection of protection systems, including inter-tripping, etc. between substations and comprise pilot cables laid alongside HV power cables. They have been historically used as the communications medium for pilot-wire protection, and although pilot wire protection is in relative decline many systems remain in service. A typical specification comprises:

- 0.5-square-millimetre copper conductor
- Polythene insulated and sheathed, and steel wire armoured
- A range of maximum working voltages is available
- Induced voltage withstand can be either 5 or 15 kV AC rms for 1 min between core and earth
- Each pair of conductors is usually given a unique colour combination.
- Standard sizes are usually 4, 7, 10, 19, 37 and 61 pair.

Polythene cables have an inter-core capacitance that is typically one-third that of the equivalent PVC cable (i.e. higher capacitive reactance) making them superior for long distance communications. Again, it is usual to earth the cable armouring at both ends to provide screening. Maximum circuit lengths are typically up to 30 km requiring intermediate straight joints.

7.11.2 Fibre optic cables

Fibre optic cables are increasingly being used in preference to metallic core cables both within and between substations. When used between substations the fibre cable (network) is usually owned, installed and maintained by a separate telecommunications company – and many are incorporated into OHL earthwires. Fibre optic communications generally offer the following advantages over metallic cables:

- Cost effectiveness, since fewer fibre cables than metallic cables are required for the same volume of communications
- Immune to IV interference.

Fibre cables are also subject to the following disadvantages compared to metallic cables:

- Usually more costly to effect changes
- Requires a power supply in order to function
- Usually not practical (or economic) when the volume of communications is small.

Generally fibre optic cables are employed as follows:

- Interconnecting protection and control equipment within a substation.
- Connecting protection equipment located in different substations.
- Connecting protection and control equipment (usually in a relay/control room) to communications equipment, e.g. multiplexor (usually in a telecommunications room).
- Part of a substation local area network connecting a substation control system to each and every bay for the purposes of control, alarms, indications and metering.

It is worthy of note that a new substation may contain communications mediums which are mostly fibre. However, with older substations that have predominately metallic core cables already installed, the installation of fibre will take place as new construction work arises, i.e. a 'work towards'. Inter-substation fibre communications is highly dependent upon the availability of a fibre network in the area (as provided by a telecommunications company).

Chapter 8
Substation design

8.1 Substation design considerations

Substation design is one of the major tasks in power network construction requiring a solution which balances operational performance against cost – taking into account land space, location and aesthetic considerations, etc. From a technical perspective, substations may be categorised in a variety of ways but generally fall under the following type of headings:

1. Distribution voltage metalclad
2. Air-insulated switchgear (AIS)
3. Gas-insulated switchgear (GIS).

The essential characteristics and spatial design considerations relating to each will be briefly overviewed, as follows:

1. **Distribution voltage metalclad substation**
   With reference to the circuit breaker arrangement illustrated in Figure 9.3(a), this type of substation is a dead-tank design where the live conductors and insulation medium are contained within an earthed metallic enclosure, with the circuits separated via metallic partitions. This type of substation has a relatively small footprint, typically occupying between 300–800 m^2 at 33 kV (i.e. the space within the perimeter fence), and the modular, physical arrangement of the substation is such that there is virtually no design required in terms of spatial layout considerations. The main consideration is the circuit designations of the modular bays (i.e. which circuit is adjacent to which). These substations are usually housed in a building (which usually also accommodates switchgear control cubicles, relay cubicles and general substation facilities), although they may be designed to be outdoors in a weatherproof cubicle. A key design consideration is therefore the positioning and orientation of the building(s). This type of substation is generally installed at network voltages of 33 kV and below.
2. **AIS substation**
   AIS substations contain HV equipment with live open terminals and exposed busbars (see Figure 9.3(b)–(d) for typical circuit breaker designs). Within this context, the (atmospheric) air acts as an insulating medium (i.e. air is positioned between the live exposed conductor and ground). Careful consideration needs to

be given to the spatial positioning of equipment and layout of the substation. Within this context, the exposed HV conductors require electrical clearances both for the satisfactory performance of the equipment and for maintenance and general-substation-access purposes – which influence the layout and physical size of the substation. AIS substations vary significantly in the land space required but typically may range from 800 m^2 for a 33-kV substation (i.e. land space within the perimeter fence) to 8,000 m^2 for a large 400-kV substation. AIS substations may be located outdoors, or indoors in purpose designed buildings.

3. **GIS substation**
   GIS substations, as outlined in Figure 9.3(e), are also of the dead tank design to which the comments in (1) above apply. Again, they may be located indoors (usually in a building) or outdoors (either in a weatherproof cubicle or with a weatherproof design). They are usually installed at network voltages of 132 kV and above (although some exist at 33 kV) and again occupy a relatively small area, typically 500 m^2 for a 132-kV substation to 1500 m^2 for a 400-kV substation (i.e. the land space within the perimeter fence). Again, relatively little layout design is required.

The remainder of this chapter will examine the following substation design considerations:

(i) Busbar systems
(ii) Merits of AIS vs GIS substations
(iii) AIS substation clearances
(iv) Busbar design and forces
(v) Merits of indoor vs outdoor substations
(vi) Substation design principles and specification.

8.2 Busbar systems

8.2.1 Busbar systems – design considerations

The term 'busbar system' describes the configuration of the busbars and the point of connection of circuits, circuit breakers, disconnectors, etc. They are invariably shown on a single line diagram. Busbar systems apply equally to distribution metalclad, AIS or GIS substations. In selecting a busbar system a number of considerations need to be taken into account as follows:

1. **Security of supply**
   This concerns the retention of transformers and generators in service, when the busbar system is subject to either faults (both circuit and busbar faults) or maintenance (i.e. requiring outages).
2. **Continuity of power flow**
   Many substations are as much concerned with power flow through the substation, as the retention in service of transformers and generators. Again, both fault situations and maintenance requirements must be considered.

3. **Substation extendability**
   The design should ideally allow for future extendability. In some instances, skeleton bays may be provided to facilitate this purpose. Within this context, the potential future location of OHL or HV cable circuits should be considered.
4. **Maintainability**
   The design must also take account of how equipment will be maintained, i.e. required outages, required access equipment, required clearances.
5. **Operational flexibility**
   Operational flexibility relates to the ability to reposition circuits to facilitate required power flows, or remove circuits from service without loss of supply. It also relates to the number of circuit breakers that are tripped when a fault occurs, and the implications for security of supply and power flow.
6. **Short-circuit fault levels**
   Consideration needs to be given to the substation short-circuit fault level, for each busbar when running arrangements are evaluated – and the contribution to fault level provided by each infeeding circuit.
7. **Land availability**
   The available land space may dictate the configuration of the busbar system and the location of incoming circuits.
8. **Operating skill**
   This relates to the level of skill and experience required by the operating personnel for a particular busbar system.
9. **Strategic importance**
   The strategic importance of the connecting circuits (e.g. major generators) and the position of the substation in the network is a key consideration in selecting a busbar system.
10. **Cost**
    All of the above need to balance against the cost of the substation with reference to the busbar system – such that an optimised technical and economic design is selected.

Over the years, attempts have been made to derive mathematical or logical methods for optimising busbar system design, but the value of such techniques is limited by the unpredictable nature of much of the data (e.g. likelihood of a particular fault occurring coincident with certain running conditions) – and the difficulty in logically linking the criteria to the cost. As a result, a number of well proven busbar systems have evolved over the years and in effect become 'standard' designs. Some of the most common, with worldwide application, are summarised below and will be briefly reviewed.

- Single busbar
- Double busbar
- Mesh (limited application worldwide)
- One and a half circuit breaker
- Transformer feeder configurations.

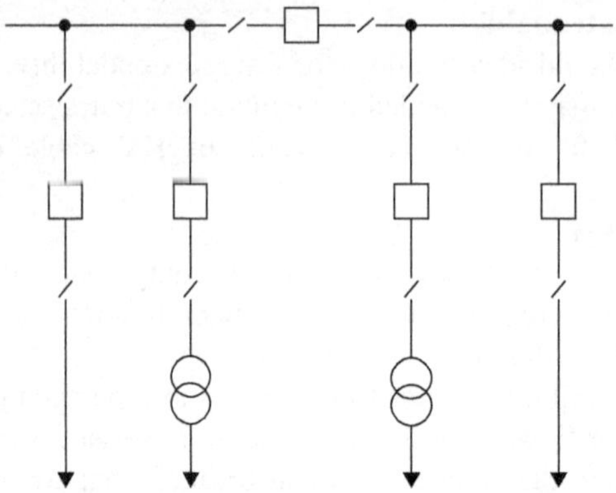

Figure 8.1 Single busbar substation

8.2.1.1 Single busbar substation

Figure 8.1 illustrates the configuration of a single busbar substation. This is one of the simplest busbar systems. Characteristics are as follows:

- Each circuit has its own circuit breaker, allowing each circuit to be removed from service (as a result of a maintenance or a fault outage), without involving other circuits.
- An outage on a busbar due to either a fault or maintenance results in the loss of half the substation.
- A bus section circuit breaker fault results in the loss of the whole substation.
- No flexibility to route a circuit to another busbar.
- The number of circuit breakers per connected circuit (i.e. feeder, transformer, etc.) is given by $(N+1)/N$ where 'N' is the number of circuits. So, if $N=8$, then $(N+1)/N$ is approximately 1.12.

This design of substation is generally found on the distribution networks of 33 kV and below, where operational flexibility is sacrificed for a relatively low-cost design.

8.2.1.2 Double busbar substation

Figure 8.2 illustrates a double busbar substation(s) (alternatively termed a duplicate busbar). It is a busbar system that is widely used throughout the world. Characteristics are as follows:

- Each circuit has its own circuit breaker, allowing each circuit to be removed from service (as a result of a maintenance or fault outage), without involving other circuits.
- An outage on a busbar due to maintenance allows circuits to be pre-selected, on load, to the alternative busbar, without loss of any connected circuits.
- A fault outage on a busbar results in loss of approximately one quarter of the substation's circuits.

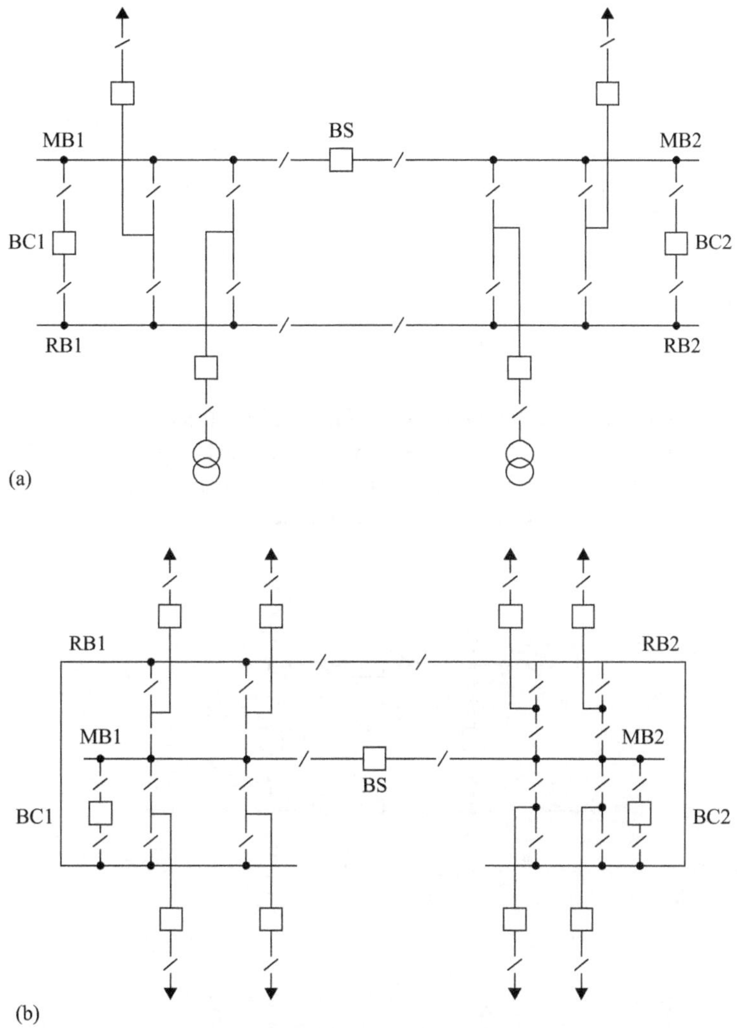

Figure 8.2 Double busbar substations: (a) straight type and (b) wrap around or 'U' type

- A bus section or bus coupler fault results in the loss of half the substation.
- The existence of main (MB1 and MB2) and reserve (RB1 and RB2) busbars together with the bus section and bus coupler circuit breakers allows the busbar selection disconnectors to be moved from one busbar to the other, whilst on load (as a result of a parallel power flow path) – providing significant operational flexibility.
- The 'straight' busbar design (see Figure 8.2(a)) is best suited to sites that are long and narrow, while the 'wrap around' busbar design (see Figure 8.2(b)), of which there are numerous variants, is suited to more compact sites.

- The number of circuit breakers per circuit is given by $(N + 3)/N$ where N is the number of connected circuits. Therefore for $N = 8$, then $(N + 3)/N$ is equal to 1.38.
- This design is significantly more costly than the single busbar.

The double busbar design is most suitable for highly connected power network, in which the ability to group circuits together in a variety of combinations, changeable at will, is of major importance. These substations are generally found at all network voltages of 132 kV and above. In the United Kingdom virtually all major generating stations are connected to a double busbar substation.

8.2.1.3 Mesh substation

With reference to Figure 8.3(a), the characteristics of mesh substations are as follows:

- A maintenance or fault outage on a circuit requires three-circuit breakers to be opened (including the transformer LV circuit breaker) together with the de-energisation of the other circuit connected to the mesh corner. In doing so, it also changes the power flow routes across the mesh.

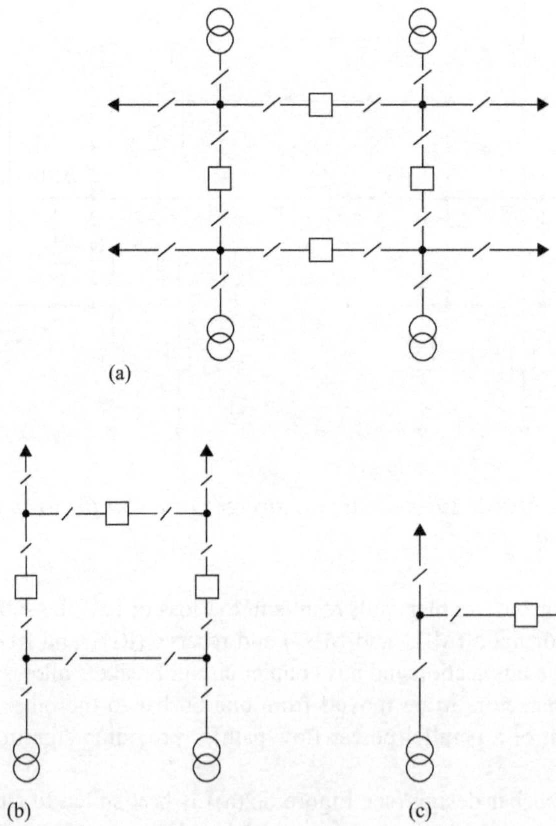

Figure 8.3 Mesh substations: (a) four-switch mesh, (b) three-switch mesh and (c) single-switch mesh

- Maintenance can be carried out on a mesh circuit breaker without the requirement for any circuit to be switched out of service.
- A circuit breaker outage (e.g. maintenance) followed by a fault on the circuits remote from the circuit breaker in question results in the substation being split into two.
- The number of circuit breakers per circuit is 0.5, substantially lower than that for any other substation.

Mesh substations in the UK tend to be installed as part of the 275- and 132-kV networks where operational flexibility is reduced in favour of substantial cost advantage and reduced land space. They are a relatively complex design from both an HV equipment and protection and control perspective. Figure 8.3(b) and (c) illustrates variants in mesh busbar systems where, again, operational flexibility and security are sacrificed for cost advantage.

8.2.1.4 One-and-a-half circuit-breaker substation

The one-and-a-half circuit-breaker substation combines the flexibility of the double busbar substation, with the ability to remove circuit breakers from service without the requirement to switch out another circuit as with the mesh substation. This busbar system is rarely used in the United Kingdom but is common in much of the rest of the world. Its characteristics comprise the following:

- A maintenance or fault outage on a busbar has no impact on any circuit connected to the substation.
- Removal of a circuit from service as a result of a maintenance or fault outage requires the opening of two-circuit breakers but does not affect any other circuit.
- Significant flexibility in power flow options across the substations, which can be reconfigured at will.
- One-and-a-half switch substations require substantially more land than double busbar substation for the same number of circuits.
- The number of circuit breakers per circuit is 1.5 irrespective of the number of circuits. This is greater than that required for a double busbar substation.
- This is the mostly costly substation design of all of those examined.

The one-and-a-half circuit breaker design contains high security against loss of supply, and significant operational flexibility – but at a relatively high cost. For this reason, it is often associated with large and important power stations or locations in the network where large quantities of power flow over individual circuits, and across the substation (Figure 8.4).

8.2.1.5 Transformer feeder configurations

Transformer supply points are not always associated with substations on both the HV and LV sides of the transformers. Figure 8.5 provides two examples of what is commonly termed 'transformer feeders'. Figure 8.5(a) illustrates the instance of a direct transformer feeder, and Figure 8.5(b) illustrates a 'T' off arrangement.

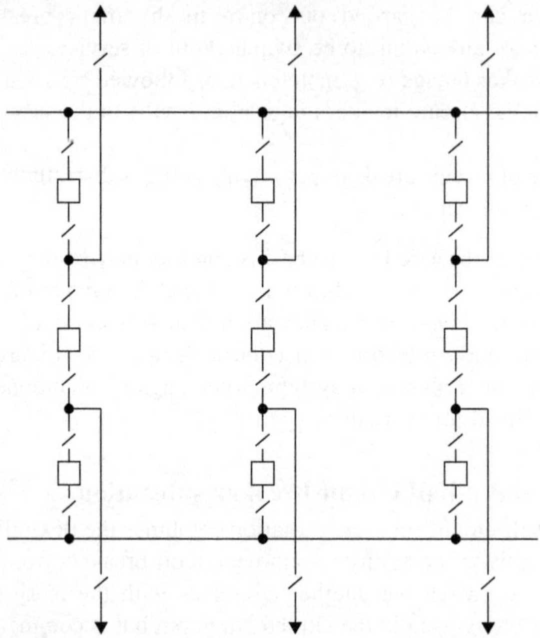

Figure 8.4 One-and-a-half circuit breaker substation

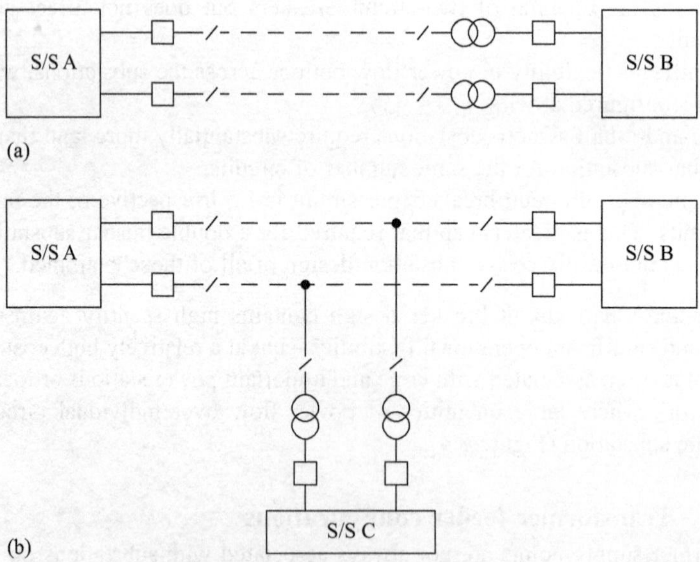

Figure 8.5 Transformer supply arrangements: (a) transformer feeder and (b) 'T' off

8.3 Merits of AIS vs GIS substations

8.3.1 AIS vs GIS considerations

Economic considerations usually dictate that GIS substations are only installed at network voltages of 132 kV and above (although some installations do exist at 33 kV). A decision therefore needs to be taken, when planning new, replacement or extended substations for these networks, on whether to install AIS or GIS. Generally, AIS is the preferred (and default) design both on cost grounds, and for the reason that SF6 (should it leak) is a heavily polluting greenhouse gas, and therefore not preferred from an environmental perspective. However, situations do arise that make the case for the installation of GIS substations, as summarised below.

- Levels of pollution at the substation site (largely arising from industrial and coastal sources) renders an AIS substation unsuitable – taking into account the lifetime costs for managing the pollution.
- AIS has an unacceptable impact on visual amenity.
- Land space constraints make a new or extended AIS impractical.
- When replacing an existing substation, outage constraints require a new substation to be built on adjacent land, rather than replace in situ – with insufficient land space for a new AIS substation.
- Proximity of other infrastructure causes significant risks (particularly to health and safety) during both construction and maintenance, if AIS is the chosen design.

When GIS is being considered a risk assessment should be undertaken to substantiate the resulting decision.

8.4 AIS substation clearances

8.4.1 AIS substation – required clearances

Whereas the design of metalclad and GIS substations is effectively frozen at time of factory manufacture, the design of AIS substations is more bespoke and needs to consider a range of clearances. The clearances are associated with both the satisfactory electrical performance of equipment, and the enabling of safe working in the substation, particularly while construction and maintenance work is being undertaken. It is worthy of note that the clearances stipulated in power network company construction specifications do not, for historical or other pertinent reasons, always accord with national/international specifications – but the differences are usually marginal. The clearances stipulated in the following sections are representative of those used in the United Kingdom (see also BS 7354 and BSEN 61936).

- Electrical clearances
- Insulation height
- Safety distances (clearances)
- Design clearances
- Boundary clearances

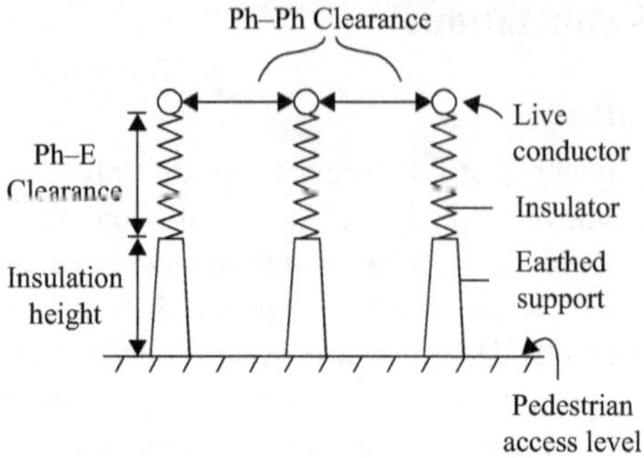

Clearances (m)	Ph–Ph	Ph–E
33 kV	= 0.5	0.43
132 kV	= 1.4	1.1
275 kV	= 2.4	2.1
400 kV	= 3.6	2.8

Figure 8.6 Electrical clearances and typical values

8.4.1.1 Electrical clearances

Section 5.4.1.2 examined the range of over-voltages to which power networks may be subject – and must withstand. Such over-voltages define the minimum electrical clearances between phase conductors, and a phase conductor and earth. Figure 8.6 illustrates the clearances and commonly used values.

8.4.1.2 Insulation height

'Insulation height' (alternatively termed 'ground clearance') is defined to be the lowest point of a support insulator to a point of pedestrian access (i.e. to where a person's feet stand). It is shown in Figure 8.6 and a commonly used dimension, covering all voltages (up to 400 kV), is 2.4 m.

NB: 2.4 m is taken to be the vertical reach of a person with an upstretched hand, as shown in Figure 8.7.

8.4.1.3 Safety distance (clearance)

The 'safety distance' is concerned with clearances under the power network company safety rules, and with safe access. It is defined as that distance from an exposed HV conductor which must be maintained (i.e. not infringed) to avoid danger (to people). Safety distance is discussed in more detail in Section 18.6.3.

8.4.1.4 Design clearances

The 'design clearance' (or safe working clearance) is the minimum distance from a live conductor to a point where pedestrian access is permitted. The design clearance defines the boundary of where work (e.g. maintenance) may be undertaken. The term 'section clearance' was previously used to define the limits of a maintenance zone. With reference to Figure 8.7, design clearance is specified both vertically and horizontally as follows:

1. **Design clearance (vertical)**
   This is equal to:
   Safety distance + 2.4 m (where 2.4 m = vertical reach of a person with upstretched hand).

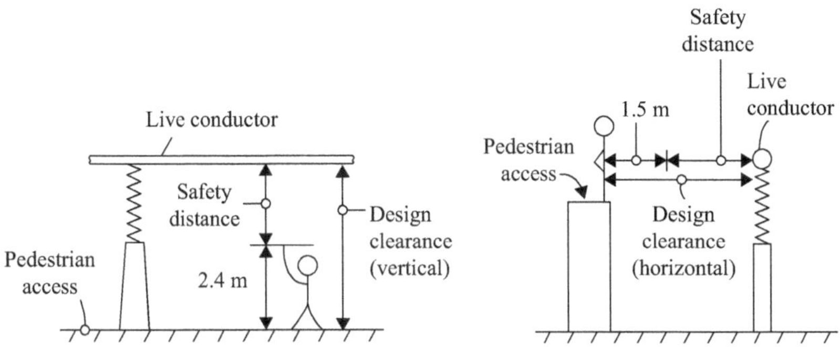

Design clearance (vertical)
(m)
33 kV = 3.2
132 kV = 3.8
275 kV = 4.8
400 kV = 5.5

Design clearance (horizontal)
(m)
33 kV = 2.3
132 kV = 2.9
275 kV = 3.9
400 kV = 4.6

Figure 8.7 Design clearances and typical values

2. **Design clearance (horizontal)**
   This is equal to:
   Safety distance + 1.5 m (where 1.5 m = horizontal reach of a person with outstretched hand).

Ideally, the design clearance (vertical) should also be applied in the horizontal direction unless space limitations do not allow. Design clearances have a significant impact on the physical dimensions of an AIS substation.

8.4.1.5 Boundary clearances

A number of other clearances are also usually defined which generally comes under the title 'boundary clearances'. These are defined in the ESQCC Regulations (see Chapter 2) and ENATS 43-8, OHL clearances. It is worthy of note that some power network companies may adopt greater clearances than those defined in these documents. With reference to Figure 8.8, boundary clearances include:

1. **Clearances to substation roadways**
   With reference to Figure 8.8(a), this is the vertical distance from an exposed conductor to the roadway. It generally comprises the distance from the roadway to the top of the vehicle (usually taken as being a maximum height of 5 m) plus the safety distance (of the voltage in question) plus 0.5 m.
2. **Clearances to substation boundary fences**
   With reference to Figure 8.8(b), this is typically taken as being the same as the height above ground as allowed for an OHL – as given in Figure 2.2 (e.g. 7.3 m at 400 kV).

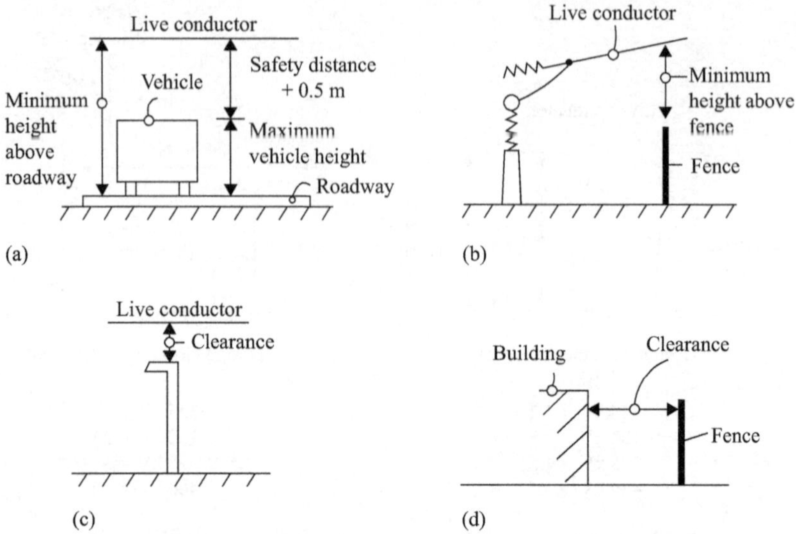

Figure 8.8 Boundary clearances: (a) roadway clearance, (b) fence clearance, (c) lighting columns clearance and (d) building to fence clearance

3. **Clearances to lighting columns**
   As defined in ENA 43-8, this is 4 m at 400 kV, down to 1.7 m at 33 kV, see Figure 8.8(c).
4. **Building clearances to fences**
   This is usually taken as being a horizontal distance of 2 m, for all voltages – this distance being taken as the length of a person's outstretched arms, see Figure 8.8(d) and also Chapter 10 on substation fence earthing.

All of the above must be taken into account in substation design.

8.4.2 Bay centres and bay separation

The term bay generally relates to those items of equipment which form part of a circuit that are within the substation boundary. It will typically include circuit breakers, disconnectors, CTs, VTs and connecting bars. When designing an AIS substation, it is convenient to determine a standard distance in-between bays (based upon clearances criterion). Historically, the distance termed 'bay centres' was used, but this has largely been replaced by 'bay separation' – since the latter does not need to account for inter-phase separation, see Figure 8.9.

The bay separation distance is influenced by the maximum safe working zone – usually involving one or more of the following considerations.

- MEWP access
- Crane access

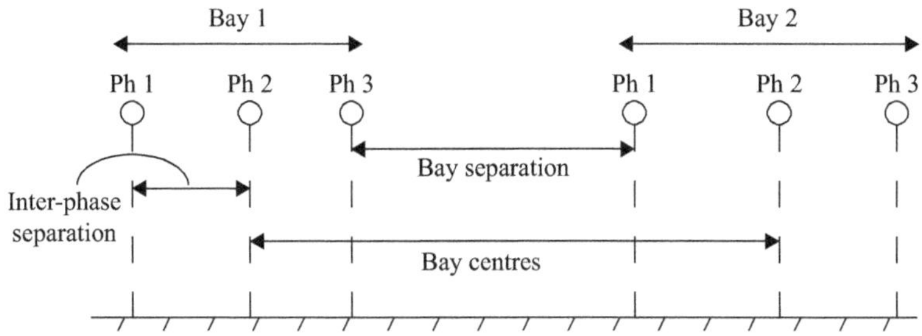

Figure 8.9 Bay centres and bay separation

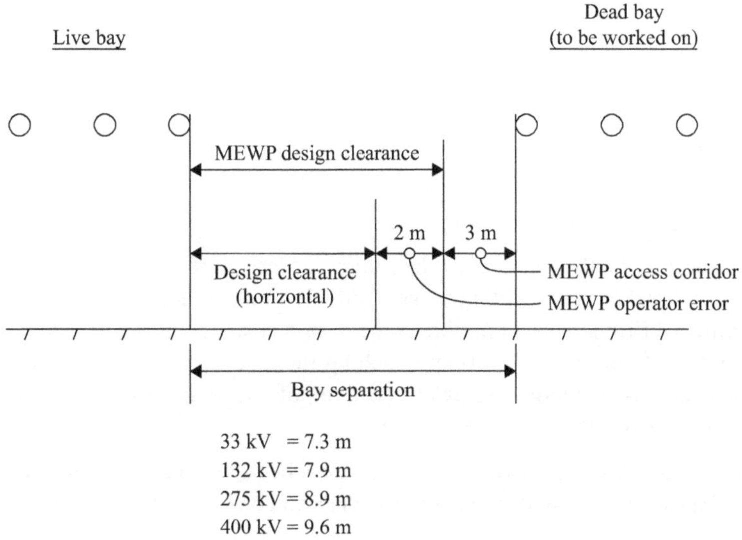

Figure 8.10 Bay separation considering MEWP access

- **MEWP access**
  The term MEWP stands for 'mobile elevated working platform', i.e. a mobile platform whose vertical, and often horizontal, position, can be controlled by operating facilities on the platform. It is a flexible device for accessing points of work – without the need for scaffolding. Figure 8.10 illustrates a possible MEWP horizontal design clearance strategy for the purpose of ensuring adequate clearance from live HV conductors when undertaking work from a MEWP. NB: larger MEWPs would require greater MEWP access clearances.

  It is to be noted that substation design must also take account of the vertical distance between a MEWP (at the end of its vertical range) and an over-sailing live HV conductor. This is typically taken to be:

  MEWP (vertical clearance) = design clearance (vertical) + 2 m (operator error)

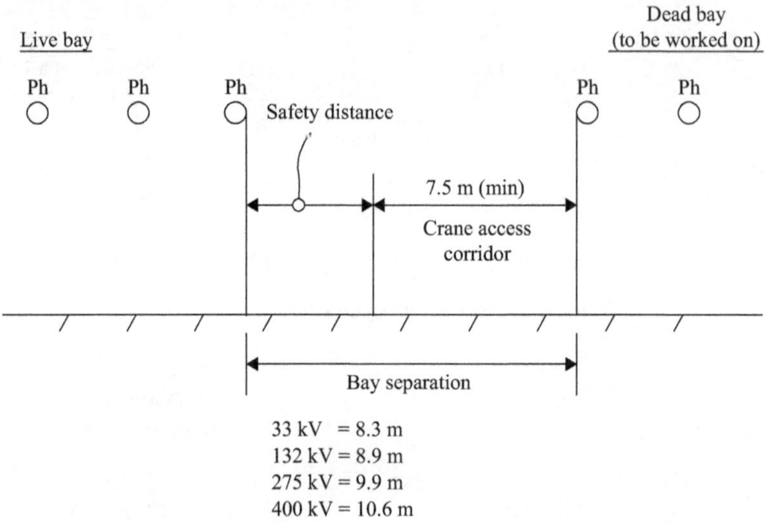

Figure 8.11 Bay separation considering crane access

- **Crane access**
  Instances will also occur of when crane access is required, e.g. replacing a circuit breaker, or building a new circuit breaker adjacent to a live bay, Figure 8.11 illustrates a possible crane design clearance strategy. Here, the risk is mostly from the crane jib approaching the live conductors. In addition, the crane access corridor also takes account of the manoeuvring of the crane before it is in the final work position.

The above shows the considerations that must be taken into account when considering bay separation with respect to MEWPs and cranes. Salient observations are as follows:

- If crane access is required then the crane clearance determines the bay centres – however, not all bays necessitate crane access, and in such instances, MEWP clearance will determine the bay centre.
- At 33 and 132 kV, the bay separation stipulated in Figures 8.10 and 8.11 may become disproportionate to the switchgear and site dimensions. As such, alternative means of access may be derived to reduce bay separation, i.e. crane access could be behind the circuit and not by alongside. Alternatively, it may be possible to switch out the adjacent circuit – or the case may be made for installing a GIS substation.
- Individual power network companies and contractors need to define their own criteria to conclude standard bay separation distances for each network voltage.

8.4.3 Over-sailing conductors

Over-sailing conductors are exposed HV live conductors which are above an area of work, and because they are outside design clearance, may remain energised whilst the work is being undertaken. Some power network companies (included the National Grid company) are requiring new substation design to eliminate over-sailing conductors – due to the latent safety risk – particularly when MEWPs are involved with the work.

8.5 Busbar design and forces

8.5.1 Busbars – overview

Busbars provide the electrical node for the connecting of incoming circuits to the substation. In doing so, they comprise long runs of conductor along the length and breadth of the substation. Figure 20.18 shows a typical single-line diagram busbar arrangement including connecting circuit breakers, disconnectors and earth switches.

Busbar current ratings usually accord with network maximum current ratings, since they are likely to carry the highest loadings on the network. As with most large current carrying conductors, busbars are subject to skin and proximity affects (see Sections 7.4.1.2 and 7.4.1.3) – and to counteract these effects, busbars are usually provided as a solid tubular conductor (either copper or aluminium), however at lower voltages some comprise stranded flexible conductors.

High-load and short-circuit currents cause busbar heating and associated expansion. To accommodate the expansion and to alleviate mechanical forces the joints connecting the busbars usually incorporate a flexible expansion arrangement.

8.5.2 Busbar short-circuit forces

Short-circuit currents flowing through conductors of various items of power system equipment create electromagnetic forces and associated mechanical stresses. If the current is flowing in two conductors in the same direction, the force is attractive, and if in the opposite direction, the force is repulsive.

Busbars which comprise long lengths of conductor are particularly subject to short-circuit forces. Within this context, the electromagnetic force F acting on three ridged, parallel and round conductors located in the same plane (such as an in a three-phase busbar arrangement), which arises from the instance of three-phase fault current, results in maximum force in the middle conductor which is given by the following equation:

$$F = \frac{\sqrt{3}\mu_0}{4\pi} \times \frac{l}{d} \times i^2 \tag{8.1}$$

where μ_0 is the permeability of free space, l is the conductor (busbar) length, d is the distance from central conductor to each of the other two conductors and i is the peak current.

NB: Peak current will usually include a DC offset – see Section 3.8.2.

It can also be shown that for the instance of a phase-to-phase fault the maximum repulsive force is given by the following equation:

$$F = \frac{\mu_0}{2\pi} \times \frac{l}{d} \times i^2 \tag{8.2}$$

Given that a phase-to-phase fault current is $\sqrt{3}/2$ of the three-phase current for a fault at the same location, expression (8.1) results in a marginally greater force than expression (8.2). From the above two expressions, it can be seen that the force is decreased with increasing phase separation, and the greater the number of busbar supports (i.e. reducing the distance l between each support) the less the force applied to each support. IEC 60865, short-circuit currents – calculation of effects, is an authoritative standard on busbar forces.

8.5.3 Busbar withstand forces

The total forces to which an AIS substation busbar is subject comprises the following:

- Short-circuit force
- Wind force
- Ice loading
- Dead (self) weight.

These are shown collectively in Figure 8.12. In addition, operational loads, such as the operation of connected disconnectors or earth switches, create mechanical loadings on both busbars and busbar support structures.

8.5.4 Busbar forces – design considerations

Busbar supports, both above and below ground, must be designed to withstand the various forces described in Section 8.5.3. The forces on the support structures may be categorised as

- Bending moment
- Shear force
- Torsion

which may exist separately or collectively.

It is worthy of note that the wind force may result in busbar resonance and Aeolian vibration (see Chapter 6), requiring the busbar to be damped. This may be

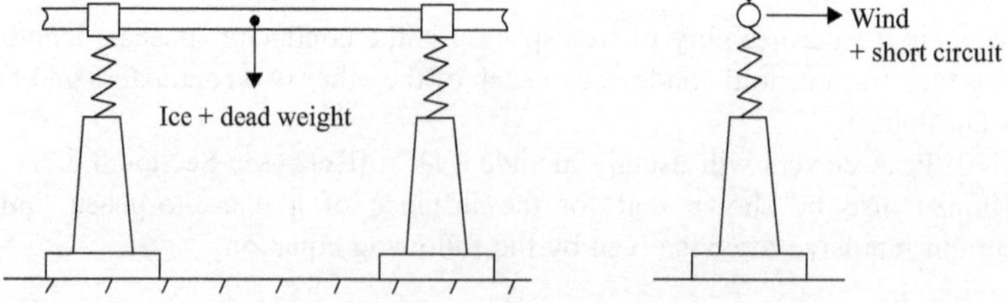

Figure 8.12 Busbar forces

achieved by positioning a conductor of the same material as the tubular busbar within the busbar – fixed at one end.

Where flexible busbars are concerned, they may additionally be subject to:

1. Tensile swing out
2. Tensile drop
3. Sub-conductor tensile pinch

All of the above forces must be factored into busbar support and foundation design.

8.5.5 Busbar insulators

Integral to busbar design is the insulators that support the busbars. As such, insulators must be capable of mechanically supporting the busbars across the range of loading and short-circuit conditions, whilst preventing flashover under worst weather and pollution conditions.

Insulator design is covered in some detail in Chapter 6, when considering OHL. Within this context, the most commonly used type of insulator for supporting busbars is the 'pedestal post'. These are available in single units, or depending upon the insulation requirements of the voltage, alternatively available as insulator stacks.

8.6 Design merits of indoor vs outdoor substations

8.6.1 Indoor vs outdoor

The decision to place an AIS substation inside is usually influenced by a number of factors, as follows:

1. **Atmospheric pollution**
   Atmospheric pollution lowers insulation security and causes corrosion and deterioration of equipment, which is prevented through use of an indoor substation. As an alternative to an indoor substation the insulator creepage distance may be increased, or live insulator washing installed.
2. **Small site area**
   Although not always the case a more compact AIS design can usually be achieved through use of an indoor substation.
3. **Aesthetics**
   Some communities, particularly in urban areas, would much prefer to look on to a building (i.e. indoor substation) than an outdoor AIS substation.

Indoor substations generally fall into two types. The first is the open hall type which is effectively an enclosed outdoor substation. The second is cellular which reduces the physical size by the use of solid barriers between circuits – through which the busbars must be designed to pass.

Generally, indoor substations are more costly than outdoor, and if an outdoor AIS substation is not an acceptable design solution, then a GIS substation may have greater merit than AIS indoor.

8.7 Substation design principles and specification

8.7.1 Substation design principles

Substation design must consider the impact of maintenance and construction work on the operational running of a substation. Within this context, it is helpful to define a generic set of design criterion and principles (although some requirements are invariably substation specific). Such criterion may include:

- When undertaking maintenance and construction work, the number of circuits and/or busbars that must be removed from service, either individually or simultaneously should be considered – either for undertaking circuit switching or for undertaking the work.
- Consideration should be given to whether MEWP or crane access can be achieved without the adjacent circuit (or busbar) being removed from service (i.e. bay separation considerations).
- Definition of the access and egress routes for the largest items of equipment in the substation (i.e. consideration of safety distances), should be evaluated.
- Whether, if large items of equipment are brought on to site, it is permissible to switch some circuits out of service to avoid infringing safety distance.
- Whether it is permissible from an operational perspective for circuits comprising a double feeder circuit to be selected to the same busbar(s).
- The criterion specified in Section 8.2.1 should additionally be considered.

8.7.2 Phasing diagram

It is self-evident and critically important that circuits and busbars are connected to the correct phase (and not crossed), both within the substation and the circuit connections to the substation. To facilitate this requirement, a 'phasing diagram' should be prepared and integrated into the design. This shows the three-phase physical arrangements and phasing of the substation. For purposes of illustration, Figure 8.13 shows a simple phasing diagram relating to an OHL connection to a single busbar substation. Phasing diagrams are particularly important where two winding transformers are concerned to ensure vector group consistency. In some instances, a three-phase physical arrangement plan drawing of the substation may suffice as a phasing diagram – although the detail contained in such a drawing may obscure clear observation of the phasing. The phasing diagram also facilitates correct phasing connections of the secondary equipment.

8.7.3 Substation design specification

The following summarises the considerations for a substation specification:

1. **Physical**
   - (i) Size and shape of the site
   - (ii) Flat or sloping ground
   - (iii) Site access arrangements
   - (iv) MEWP and crane access

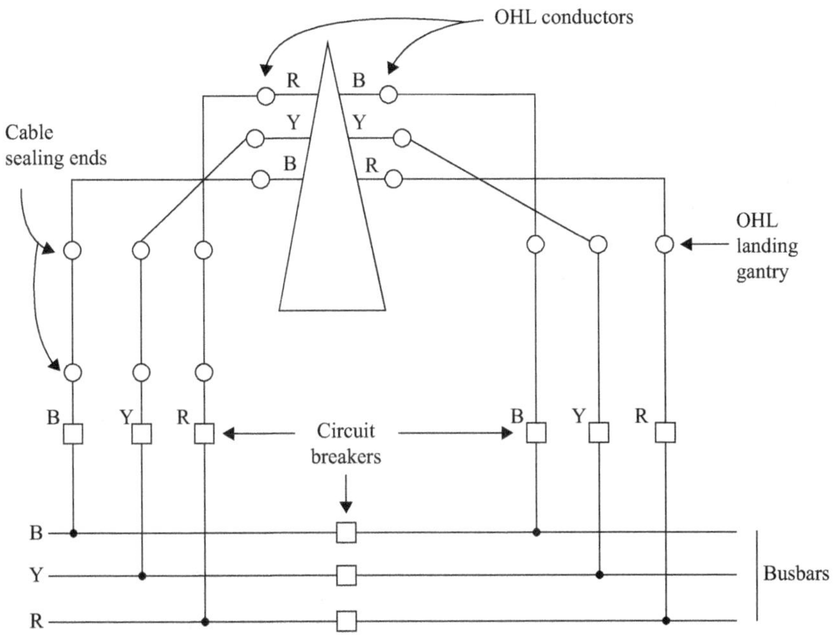

Figure 8.13 Phasing diagram – simple illustration

(v) Substation road arrangements
(vi) Type and location of circuit entry (i.e. OHL, cables)
(vii) Aesthetic considerations (screens, trees, etc.)
(viii) Ground composition (dictating foundation design/piling)
(ix) Indoor or outdoor substation
(x) Measures to combat atmospheric pollution
(xi) Planning authority requirements.

2. **Technical**
   (i) Nominal voltage and basic installation level
   (ii) Busbar system
      (a) Number of circuits
      (b) Type of circuits (e.g. feeder/transformer, etc.)
      (c) Future extendibility
      (d) Future potential conversion (mesh to double busbar, etc.)
   (iii) Type of substation (metalclad, AIS, GIS)
   (iv) Continuous and short term current ratings of busbars and circuits
   (v) Equipment and busbar short-circuit withstand levels
   (vi) Methods of equipment earthing
   (vii) Method of lightening protection
   (viii) VT and CT requirements (protection, control and metering, etc.)
   (ix) Bypass facilities
   (x) Substation rise of earth potential

3. **Operations and maintenance**
   (i) Safe system of work and safety rules (which organisation's documents are to be used during site installation and commissioning)
   (ii) Bay separation (MEWP and crane access)
   (iii) Electrical design clearances to adjacent live equipment
   (iv) Boundary responsibilities (with other power network companies)
   (v) Maintenance earthing arrangements
   (vi) Provision of ladders, screens, walkways, etc.
   (vii) Location of access roads and pathways within a substation.

Chapter 9
Substation HV equipment design

9.1 Introduction

Substation HV equipment comprises a wide variety of power network assets that are connected together, usually by busbars, to form the types of substation layouts described in Chapter 8. This chapter will examine underpinning design principles and performance characteristics of the main items of substation equipment, which are as follows:

- Circuit breakers (CBs)
- Earth switches
- Disconnectors
- Interlocking
- Power transformers
- Reactors
- Quadrature boosters (QBs)
- Manually switched capacitors (MSCs)
- Static VAr compensators (SVCs)
- Voltage transformers (VTs)
- Current transformers (CTs).

9.2 Circuit breakers

9.2.1 Circuit breaker – duty

CBs are probably the leading item of equipment that falls under the generic heading of 'switchgear' – a term that also includes switches, disconnectors, earth switches, HRC fuses, CTs, VTs and in some definitions protection and control equipment. CBs are one of the key items of equipment in the design of the power system, and although many items of equipment are essentially static in performance, the CB's prime task is associated with movement – namely the movement of electrical contacts to either interrupt or connect (make) current flow in a circuit. The duties of CBs are stipulated in many technical documents and are summarised as follows:

- To make, break and carry normal-rated (load) current
- To make and break fault current (short-circuit current)

- To carry fault current for a defined period
- To make, break and carry circuit charging current.

With reference to the above, a CB differs from an electrical switch in as much as the latter is not capable of breaking fault current.

The following salient aspects of CBs will be examined:

- Arc interruption mechanism
- Switching duties
- Arc interruption mediums and methods
- CB type classification
- CB tripping and close times
- CB specification.

9.2.2 Arc interruption mechanism

When two current carrying contacts are closed, actual contact is made at a number of points on the two surfaces, the number of points being a function of the pressure holding the surfaces together. When the contacts begin to part, as within a CB, the contact resistance commences to increase with an associated I^2R temperature rise (where I is the current through the contacts and R is the resistance of the contacts. The temperature rise causes thermionic emissions (i.e. emission of electrons) from the contact surface, thereby maintaining the current flow via an arc. As the contacts begin to part, the arc increases in resistance, and as result, the voltage across the contact gap increases. Provided that the temperature rise and potential gradient in the region of the arc are high enough, the arc will be intensified by further electrons obtained from the surrounding medium, and through ionisation by collision. However, still further increasing the contact gap will commence to reduce the voltage gradient and increase the arc resistance, which in turn will lower the current and temperature rise, and eventually (as the contact gap continues to increase) the arc will extinguish. The length of contact separation in a CB is however limited and therefore other means of extinguishing the arc must be considered.

In CBs, an arc interruption medium (e.g. SF6, vacuum, oil, air, etc.) is introduced into the contact separation gap, the purpose of which is to remove energy from the arc path at a greater rate than it is produced in the arc. In AC systems, the situation is greatly helped by virtue that the current reduces naturally to zero once every half cycle, at which point the arc is normally extinguished. Successful arc extinction requires the rate of increase of dielectric strength in the arc gap (i.e. in the interrupting medium) both during and immediately after current zero, to exceed the ability of the voltage across the gap to break down the dielectric (and any arc products) to re-establish the arc. Suffice it to say that CB opening is invariably a more onerous duty than CB closing.

9.3 Circuit breaker switching duties

9.3.1 Circuit breaker interrupter duties – summary

CB contacts are usually referred to as interrupters, and the chamber within which arc extinction takes place is termed the interrupter chamber. Within this context, the current interruption duties placed upon CBs may be summarised as follows:

- Short-circuit fault current interruption – fault at CB terminals
- Short-circuit fault current interruption – fault a short way down an OHL (short line fault)
- Load current interruption – resistive load
- Load current interruption – inductive load
- Load current interruption – capacitive load
- CB opening – asynchronous switching
- Current interruption – modern CBs

These duties will be briefly reviewed.

9.3.1.1 Short-circuit fault current interruption – fault at circuit breaker terminals

Figure 9.1(a) shows the instance of a CB subject to a fault at its terminals. The following points are relevant:

- The impedance to fault will be essentially inductive (via L_1). Assuming the CB opens and successfully interrupts the fault at a current zero (on the AC wave), the voltage E will be at the peak of the AC wave.
- L_1 and C_1 depict the equivalent circuit of the distributed inductance and capacitance which comprises any power network. At the instant of current zero (i.e. fault current interruption), the situation may be likened to voltage E suddenly being applied to the L_1 C_1 circuit. This will result in an oscillatory transient voltage being imposed upon C_1 and hence the interrupters. The duration of the transient decay will depend upon the value of circuit series resistance (not shown), which will dampen the oscillation. With reference to Figure 9.1(b), the voltage V across the CB interrupters will therefore comprise the oscillatory voltage (as just described) superimposed upon the ongoing steady-state voltage arising from the voltage source E. As a result, the voltage across the interrupters is described in two stages which are as follows:
  - A transient oscillatory voltage termed the 'restrike voltage' whose peak magnitude could rise to twice that of E. For successful arc interruption, the build-up of dielectric strength in the interrupter gap must exceed the voltage imposed by the restrike voltage.
  - An ongoing steady-state normal frequency (50 Hz) voltage termed the recovery voltage.

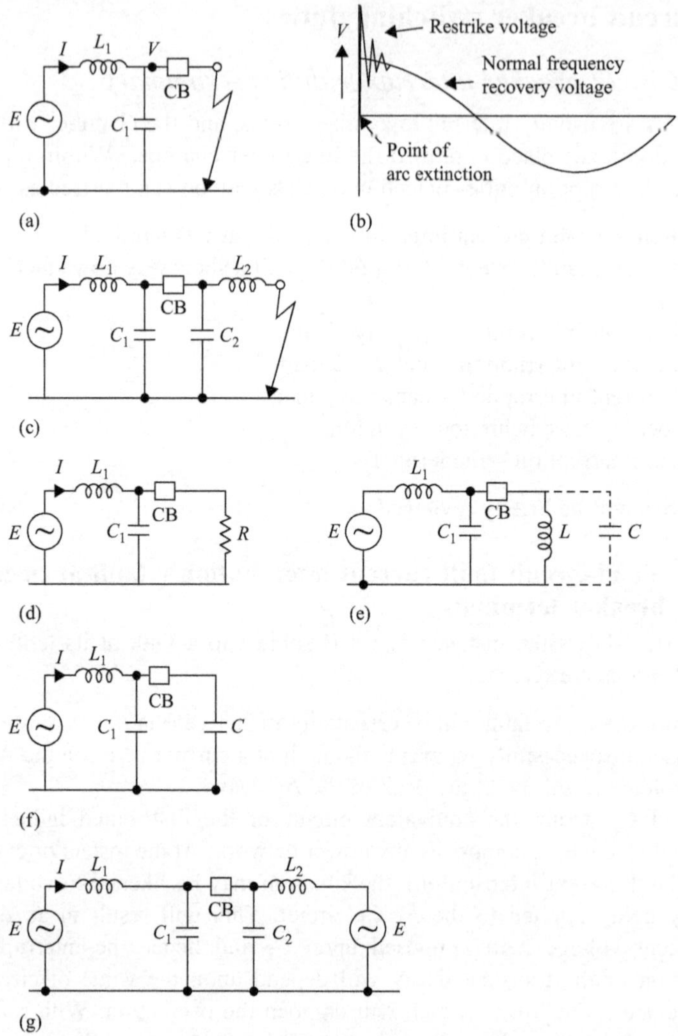

Figure 9.1 Circuit breaker switching duties: (a) circuit breaker terminal fault, (b) restrike and recovery voltages, (c) short line fault, (d) resistive load, (e) inductive load, (f) capacitive load and (g) asynchronous switching

- The oscillatory voltage frequency f is given by:

$$f = 1/\left(2\pi\sqrt{L_1C_1}\right) \tag{9.1}$$

- It was pointed out in Section 3.8.2 that an inductive circuit subject to a short-circuit fault would contain a DC transient component of current – whose magnitude is a function of point on wave of the fault, and whose duration is a function of the resistance-to-inductance ratio of the impedance to fault. The

DC transient current when added to the steady-state AC fault current can have the effect of delaying current zero (i.e. in the total fault current) – see Figure 3.29. This is an important consideration in CB interrupter design – and power network companies usually declare in their construction specification, the maximum value of DC transient decay time. This is usually expressed as the time constant of the series inductance/resistance circuit, as determined by the ratio of inductance to resistance (L/R) of the circuit. A common time constant for the 400-kV system is 45 ms.

It is also worthy of note that CBs that either switch generators, or are close to generators, are subject to both the generator fault current decrement and the DC offset, as described in Chapter 3. This may result in a current zero not arising for a number of cycles – which may delay CB arc interruption. Some generator CBs are installed with a delayed trip arrangement to be sure that interrupter opening is close to current zero.

- It is also worthy of note that at the instant a CB is required to open, each of the three phase voltages, and currents, will be at a different point in the 50-Hz cycle – therefore, the level of DC offset and arcing time will differ from pole to pole.

9.3.1.2 Short-circuit fault current interruption – short line fault

Figure 9.1(c) shows the instance of an OHL fault, a short distance from the CB terminals. The fault current I will be less than that for a fault at the CB terminals, and therefore the heating effect sustaining the CB arc will be less. At first sight, it may be thought that this is a less onerous fault to interrupt. However, the distributed inductance and capacitance, and associated stored energy, on the OHL side of the CB results in another oscillatory voltage transient – which is added to the voltage wave shown in Figure 9.1(b) to obtain the total voltage across the (open) interrupters. The resulting voltage peak may exceed the voltage shown in Figure 9.1(b), thus creating a greater probability of interrupter restrike. This condition is most onerous when the fault is a few kilometres from the CB. As a result, it is known as either the 'short line fault' or the 'kilometric fault'.

CBs must be designed to withstand this fault condition. It may be of interest to note that the oscillatory transients arising from L_1 C_1 are typically of the order of 2 kHz, and from L_2 C_2 of the order of 10–50 kHz.

9.3.1.3 Load current interruption – resistive loads

The switching of resistive loads (or loads with a high power factor) is the most common use of CBs. With reference to Figure 9.1(d) in this instance the voltage and current are in phase, and at time of the interrupter opening at current zero the voltage will be at a minimum. As a result, the restrike voltage magnitude will be much reduced. This is one of the least onerous conditions for CBs.

9.3.1.4 Load current interruption – inductive loads

CBs are occasionally required to interrupt small inductive loads, see Figure 9.1(e), which may arise from shunt reactors or transformer magnetising inrush currents (i.e. fault on energisation). When interrupting current magnitudes below the CB current rating, 'current chopping' may occur, i.e. the current is interrupted before current zero. This leaves stored energy in the inductances which causes oscillatory transients on either, or both, sides of the CB. On the inductive load side, the oscillation may be in conjunction with leakage capacitance *C*. This may result in a restrike voltage across the interrupters of up to three times the peak phase to neutral voltage. With some types of CBs, multiple restrikes of the arc may occur until the (reducing) current becomes low enough to allow successful interruption. Interrupter design should take account of the risk of current chopping.

9.3.1.5 Load current interruption – capacitive loads

With reference to Figure 9.1(f), switching (i.e. breaking) a capacitive load such as a capacitor bank or unloaded HV cable may also result in a particularly onerous duty for CB interrupters. In this instance when the current is interrupted at current zero, the capacitor remains charged at peak network voltage. After half a cycle (10 ms), the source voltage reaches peak voltage of opposite polarity, and twice the peak network voltage appears across the CB. Should the contact gap break down and restrike occur, and the arc extinguish at the next current zero, the capacitor now becomes charged to twice peak network voltage. Subsequent restrike and arc extinction causes increased high voltages on the capacitor and network. Interrupters, for CBs, required to break capacitive loads must be designed to avoid restrike. Similarly, as with reactors, capacitive loads may also be subject to current chopping.

9.3.1.6 Circuit breaker opening – asynchronous switching

Under conditions of severe network disturbance, such as pole-slipping (see Section 3.6.1.5), the sources of generation may be 180° out of phase when a CB is required to open and split the system, see Figure 9.1(g). In this case, the voltage across the CB interrupters (after opening) would be twice system phase to neutral voltage. CBs located in parts of the networks that may be subject to asynchronous switching should be rated accordingly.

9.3.1.7 Current interruption modern circuit breakers

Modern CBs are designed with all of the above scenarios in mind, with the aim to interrupt the current ideally at the first current zero, but if not the first, certainly the second current zero. Older CBs may be subject to more restrikes, and therefore, the current interruption process would take longer, and not as efficient.

9.4 Arc interruption mediums and methods

9.4.1 Arc interruption mediums and methods – categories

Arc interruption mediums and methods may generally be categorised as follows:

- Oil (older designs)
- Vacuum (current design)

- Air blast (older designs)
- SF6 (current design).

These common designs will be briefly reviewed.

9.4.1.1 Oil circuit breakers

Oil circuit breakers (OCBs,) containing bulk oil, provided the back bone of power networks for almost a century but have now been superseded by other types (mostly vacuum and SF6). However, many still remain in service at all voltages up to and including 275 kV (with some exceeding 50 years of operational service) – they were never installed at 400 kV. The principle of operation is illustrated in Figure 9.2(a). The contacts are housed in an arc control chamber as shown, which is immersed in a tank of oil (hence the term bulk oil). Oil is used because of its excellent insulating and arc interruption properties.

When the interrupter contacts separate, an arc is created, the heat from which decomposes the surrounding oil into a number of gasses, some 80% of which is hydrogen – which has the property of efficiently conducting the heat of the arc into the cooler surrounding oil. The limited space in the arc control chamber causes a build-up of pressure by the expanding gasses, which drives the oil within the chamber through the arc splitter plates, so fragmenting and interrupting the arc, and out of the splitter vents. As the moving contact travels downwards, the arc is eventually extinguished at current zero. The arc products are expelled from the interrupter chamber and replaced by cool oil. The oil therefore serves three purposes:

- As an insulating medium
- As a source of hydrogen for assisting extinction of the arc
- To displace the arc products following extinction.

Oil CBs have provided good service over very many years but have now become obsolete (for new installations) for the following reasons:

- Oil maintenance cost – of maintaining the oil in good condition
- General maintenance cost – particularly associated with the interrupters
- Fire risk
- Environmental pollution risk arising from holding and handling bulk oil.

9.4.1.2 Vacuum circuit breakers

Vacuum CBs have become increasingly popular at distribution voltages, particularly 11 and 33 kV. They are rarely installed above 33 kV due to design limitations (although some have been designed for 132 kV). They do not suffer many of the disadvantages of oil CBs and therefore are progressively replacing oil. With reference to Figure 9.2(b), vacuum CBs comprise interrupter contacts within a sealed chamber (typically one microbar). Opening of the contacts causes an arc to arise which vaporises the metal of the contacts. At current zero, the vapour rapidly diffuses away from the contact gap to be absorbed by 'sputter shields' inside the vacuum tube walls. In doing so, the shields are instrumental in restoring the

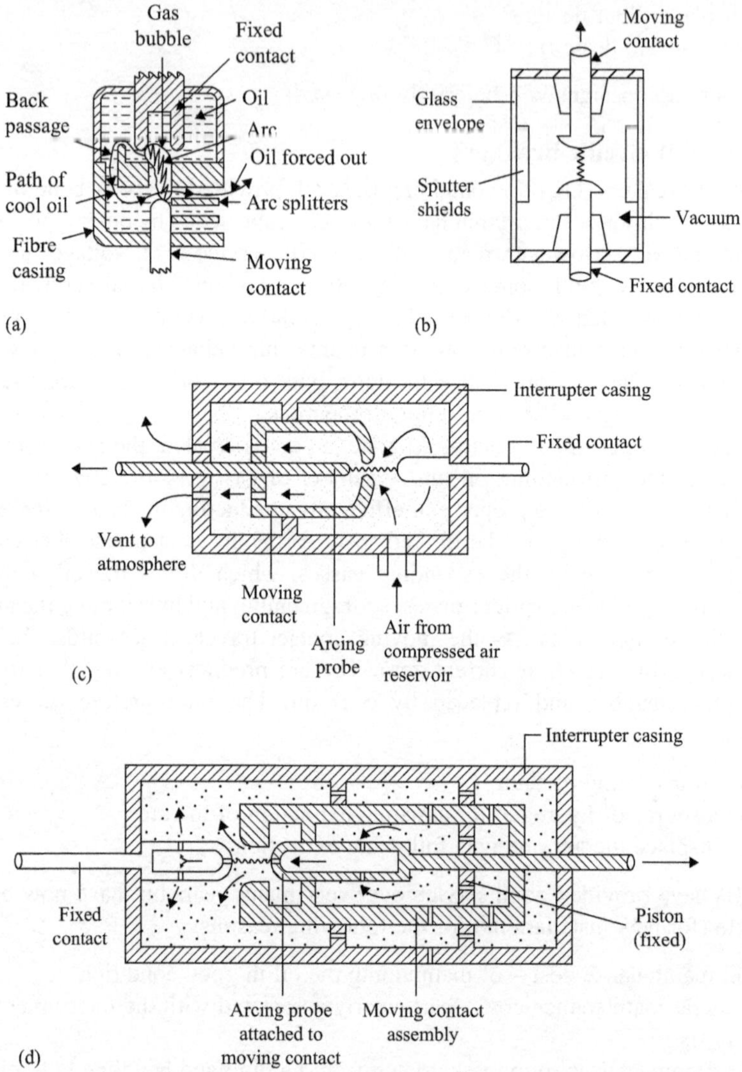

Figure 9.2 Arc interruption methods – simplified: (a) oil circuit breaker, (b) vacuum circuit breaker, (c) air blast circuit breaker and (d) SF6 circuit breaker – puffer type

strength of the vacuum – which acts as both an insulating and arc extinction medium. Designs are available for both indoor and outdoor installations.

Advantages of vacuum CBs are as follows:

- Low maintenance requirements. Vacuum CBs are usually capable of 20,000 switching operations and 100 short-circuit fault clearances typically.
- Virtually no fire risk.

- Compact and relatively small.
- Arc interruption is generally within half of a 50-Hz cycle, i.e. 10 ms.

9.4.1.3 Air blast circuit breaker

Air blast circuit breakers (ABCBs) were the main stay of the 400- and 275-kV transmission networks (and some 132 kV) for almost 50 years, dating from *ca.* 1950. They have largely been replaced by SF6, but a sizable number still remain in operational service. The principle of arc extinction is illustrated with reference to Figure 9.2(c). Separation of the interrupter contacts results in a blast of high pressure (compressed) air through a nozzle of the moving contact. As it does so, the arc is extended and transferred to an arcing probe. The air conducts heat very rapidly and efficiently from the arc to achieve extinction at current zero.

ABCB installation was largely discontinued for the following reasons:

- Relatively high maintenance costs
- Dependent upon an ongoing supply of compressed air, requiring a compressor and storage vessel installation on site – adding additional capital and maintenance costs
- Very noisy in operation (as the air expands on CB tripping), and unpopular in built up areas
- Physically large and relatively complex CB.

9.4.1.4 Sulphur hexafluoride (SF6) circuit breaker

The installation of SF6 CBs in significant numbers dates back to *ca.* 1990. This technology now dominates installations at 132 kV and above (although with some installations at 33 kV). SF6 is an odourless, colourless, non-toxic, non-inflammable gas that is five times heavier than air with approximately 2.5 times the dielectric strength. The molecules of the SF6 gas rapidly absorb the free electrons found in a CB arc. SF6 also has good thermal conductivity for removing heat from the arc – it is therefore a good arc quenching medium. In addition, it provides a good insulating medium.

The principles of one type of interrupter design of an SF6 CB are illustrated in Figure 9.2(d) – i.e. the 'puffer design'. As the contacts separate, the moving contact assembly compresses the SF6 gas (by use of a piston type mechanism), which causes a high pressure flow of gas through the contacts resulting in a build-up of high dielectric strength and arc extinction at current zero. The arrangement shows a sealed interrupter chamber assembly which is a common arrangement – but alternative designs involve a tank of SF6 (both AIS and GIS – see Figure 9.3). Advantages of SF6 CBs are

- Low maintenance with low maintenance costs
- Absence of fire risk
- Relatively simple operating mechanism (often spring loaded) and physically more compact than ABCB
- Relatively silent in operation

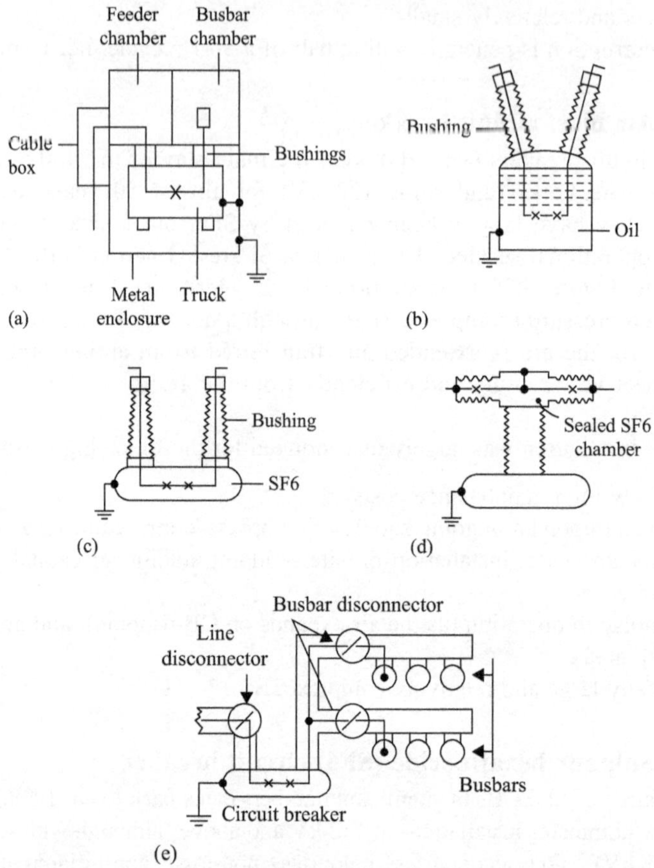

Figure 9.3 Circuit breaker – types: (a) metalclad (distribution voltage), (b) dead tank – open terminal (bulk oil), (c) dead tank – open terminal (SF6), (d) live tank – open terminal (air, SF6) and (e) metalclad (GIS)

Disadvantages of SF6 are:

- SF6 is a green-house gas – global warming potential is some 24,000 times greater than carbon dioxide
- Will not support life if inhaled (causes suffocation)
- Requires special handling equipment on site.

9.5 Circuit-breaker-type classification

9.5.1 Circuit breaker type – characteristics

CB types are generally categorised according to the following physical characteristics:

1. **Dead tank**
   In this design, the interrupters are housed in a metallic tank which is not bonded to the interrupters, but is bonded to the earth (i.e. dead). The dead tank

Figure	Arrangements			Typical voltages (kV)	Interrupting medium
	Tank	**Terminals**	**Pressurised**		
9.3(a)	Dead	Enclosed	No	33, 11	Oil
9.3(a)	Dead	Enclosed	Vacuum	33, 11	Vacuum
9.3(b)	Dead	Open	No	275, 132, 33, 11	Oil
9.3(c)	Dead	Open	Yes	400, 275, 132, 33	SF6
9.3(d)	Live	Open	Yes	400, 275, 132	SF6/ABCB
9.3(e)	Dead	Enclosed	Yes	400, 275, 132, 33	SF6

Figure 9.4 Circuit-breaker-type classification

design usually houses a large volume of the insulating, and arc quenching medium, see Figure 9.3(a–c) and (e).

2. **Live tank**
   In this design, the interrupters are usually housed in a chamber which is bonded to the interrupters (i.e. live) and therefore stands at the voltage of the network, see Figure 9.3(d).
3. **Open terminal**
   In the open design, the interrupter terminals are open to the air (i.e. AIS), see Figures 9.3(b),(c) and (d).
4. **Enclosed terminal**
   In the enclosed terminal design, the interrupters are enclosed and not open to the air (i.e. metalclad and GIS), see Figure 9.3(a) and (e).
5. **Pressurised**
   In the pressurised design, the interrupter chamber is pressurised, see Figure 9.3(c),(d) and (e).
6. **Non-pressurised**
   In this design, the interrupter chamber is not pressurised, see Figure 9.3(a) and (b).

9.5.2 Circuit-breaker-type and voltage-range classification

The classification of the various types of CBs is summarised in Figure 9.4.

With reference to Figure 9.3(a) and (e), both of these CBs are dead tank, enclosed terminal arrangements; however, it is usual in practice to refer to refer to Figure 9.4(a) as 'distribution voltage metalclad', and Figure 9.3(e) as GIS. Metal-clad switchgear is housed indoors, but GIS although usually housed indoors may be designed for outdoor installations.

9.6 Circuit breaker tripping and close times

9.6.1 Tripping and closing considerations

The process of CB tripping and closing and the associated operating times are key considerations in CB design. The following will be considered:

- CB tripping time
- CB closing considerations

- Resistance switching and capacitance grading
- Trip and close operating mechanism
- Point on wave switching
- Trip and close coil circuitry.

9.6.1.1 Circuit breaker tripping time

CB tripping times are generally required to be as follows (see IEC 62271-100):

- CB contact opening time = 22–35 ms
- Arcing time = 10–23 ms
- Total tripping time = 32–58 ms.

If the protection operating time is taken as being 10 ms (as allowed for in the IEC), then the total fault clearance time would be 42–68 ms (or 32–45 ms with the absence of an arc).

A consideration for CB opening time is interrupter opening when considerable DC offset is still present (see Section 9.3.1.1), i.e. if the interrupters open too early when the DC offset remains high, the interruption process is more onerous.

9.6.1.2 Circuit breaker closing considerations

When a CB is closed onto a short circuit, the DC offset discussed in Section 9.3.1.1 will be present with the possibility of almost doubling the resulting current magnitude. Within this context, the CB latching arrangements must be robust enough to ensure successful closing in the presence of the magnetic forces associated with the peak value of current – followed by a controlled open.

CBs are usually also specified in terms of close–open performance, i.e. a trip on close. IEC 62271-100 defines close–open performance in the following (example) way:

O–0.3 s–CO–3 min–CO

The above is interpreted as follows:

1. O = open
2. 0.3 s = 300 ms delay
3. CO = close–open (i.e. trip on close)
4. 3 min = 3 min delay
5. CO = a further close–open (i.e. trip on close).

Suffice it to say that the above specifies a design capability limit of the CB – which should not be exceeded. In the above, item 3 would typically be a trip resulting from a fault following a CB auto-reclose.

9.6.1.3 Resistance switching and capacitance grading

The magnitude of the restrike voltage as examined in Section 9.3.1 may be limited by use of a shunt resistance connected across the CBs main interrupter(s) as shown in Figure 9.5(a). The impact of the resistance is to dampen and diminish the transient

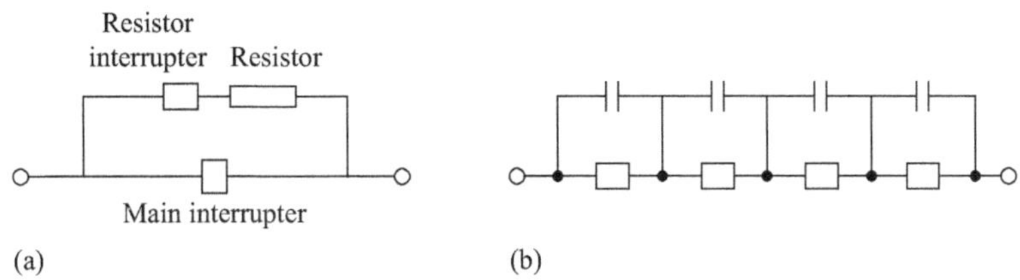

Figure 9.5 Resistance switching and grading capacitors: (a) resistance switching and (b) grading capacitors

oscillatory component of the restrike voltage. With reference to Figure 9.5(a), the time sequence for insertion of the resistor is as follows:

- CB trip coil energised (requiring CB to trip)
- Resistor interrupter closes
- Main interrupter opens
- Resistor interrupter opens.

With designs of CBs that have multiple interrupters per phase, grading capacitors are frequently placed across the interrupters to ensure voltage sharing when the CB is open. See Figure 9.5(b).

9.6.1.4 Trip and close operating mechanism

A CB operating mechanism is an electromechanical arrangement that receives an open or close signal to instruct the mechanism to open or close the CB interrupters as appropriate. The signals are usually received by trip or close coils which operate a latch mechanism to set the rest of the mechanism in motion. The mechanism needs to be fast acting and reliable. Operating mechanisms tend to be powered by either hydraulic, pneumatic or spring (mechanical) sources of energy, or a combination. All need to possess stored energy to ensure a fast acting mechanism. The UK power system operates on the basis of a three-pole trip and close mechanism (with some exceptions). This contrasts much of the rest of the world that utilises single-pole trip and close mechanisms (usually to accommodate a single pole trip and auto-reclose scheme). In such instances, a mechanism per phase is required.

Many (but not all) CBs are designed with a 'trip-free' mechanism. This is a feature such that when a CB is closed coincident with a trip relay being operated (i.e. in the tripped position), the mechanism is immediately collapsed into the open state the instance the trip coil becomes energised. The latter is accomplished via the change of position of the CB auxiliary contacts, which are in series with the trip coil and driven by the CB mechanism.

9.6.1.5 Point on wave switching

Closing a CB to energise capacitor banks (e.g. MSCs) may (depending upon point of wave of closing) result in closing transient harmonics (see Section 3.8.3), which can cause unacceptable voltage fluctuations on the power system. To greatly diminish, or eliminate, the transient, CBs may be equipped with 'point on wave closing' delays to ensure that the closing of each phase coincides with a system voltage zero (i.e. minimum transient, see Section 3.8.3). This requires CBs with a separate mechanism per pole, with the second and third pole arranged to close at 3.35 ms intervals, respectively, after the first (i.e. increments of 60°, to ensure coincidence with voltage zero).

Similarly, shunt reactors, depending upon point on wave on opening, may produce overvoltages at the reactor terminals which may cause CB interrupter reignition (see also Section 9.3.1.4). To combat this, 'point on wave opening' may be added to the CB mechanisms. This usually requires opening at voltage maximum, i.e. current minimum to eliminate the over-voltages (see Section 9.3). Again, pole switching stagger, as described for capacitance switching, above, would be required.

9.6.1.6 Trip and close-coil circuitry

Opening (tripping) and closing of a CB is usually via the energisation of an open or close coil, respectively, (as stated above) which in turn operates a latch arrangement (one each for trip and close) to initiate the operating mechanism. The coil supply voltage is usually 110 V nominal DC. At transmission voltages of 400 and 275 kV, it is usual practice to employ two trip coils, fed from the first and second main supply systems, respectively, see Figure 10.41. Both trip coils operate the same mechanism latch. A manual opening of the CB would usually be via one trip coil only, whereas a protection operation would involve both trip coils.

The trip and close coil circuitry is usually interlocked with series contacts which provide safety features to prevent the CB operating until all relevant conditions are satisfactory. Such conditions include:

- SF6 gas pressure within limits
- All associated disconnectors in either the fully open or fully closed positions
- If the CB open circuit is initiated, the close circuit cannot be simultaneously initiated.

9.7 Circuit-breaker specification

9.7.1 Electrical design specification

A CB electrical design specification typically comprises the following:

1. **Rated RMS voltage**
   This refers to the maximum continuous voltage, which are as follows:
   (i) 420 kV for the 400 kV network
   (ii) 300 kV for the 275 kV network
   (iii) 145 kV for the 132 kV network
   (iv) 36 kV for the 33 kV network
   (v) 12 kV for the 11 kV network.
2. **Rated RMS short-circuit breaking current**
   This refers to the symmetrical three-phase and single-phase (symmetrical) breaking current.

3. **Peak making current**
   This refers to the peak current that the CB may experience when closing.
4. **Rated RMS continuous and short-term currents**
   The continuous current refers to the post-fault continuous current as described in Section 20.4.2, and the short-term currents are those described in Section 20.4.3, e.g. 20, 10, 5 min, etc.
5. **Rated duration of short-circuit current with the circuit breaker closed**
   This refers to the duration that the maximum current (usually equal to the short-circuit breaking current) can be carried without exceeding the thermal rating of the equipment. This is usually 1 s at 400 and 275 kV, and 3 s for lower voltages.
6. **Maximum tripping time**
   This refers to trip coil initiation to arc extinction. Typically, 32–58 ms as given in Section 9.6.1.1.
7. **Close–open time**
   This is the maximum time to open if the CB is closed on to an operated trip relay.
8. **Rated line and cable charging breaking current**
   This is the ability to interrupt capacitance current, see Section 9.3.1.5.
9. **Rated small inductive break**
   This relates to the ability to interrupt inductive currents as described in Section 9.3.1.4.
10. **Power System DC Time Constant**
    This relates to the maximum inductive reactance to resistance ratio of the circuit, See section 5.3.1.6.

9.7.2 Circuit-breaker specification standards

The following standard documents define CB specifications requirements relevant to the above sections.

- IEC 62271: HV switchgear and controlgear
  - Part 1: Common specifications
  - Part 100: Alternating current CBs
  - Part 110: Inductive current switching
  - Part 302: AC CBs with intentionally non-simultaneous pole operation.

9.8 Earthing devices

9.8.1 Earthing switches – design characteristics

An earthing switch is a mechanical device which, when closed, connects an item of equipment (via the earth switch) to the substation earth mat, usually for the purposes of enabling safe working on the item of equipment. Salient characteristics are as follows:

1. **AIS substation physical arrangement**
   The earth switch is usually three-phase mechanically coupled so that the three phases make and break simultaneously. They are usually hand operated by hand wheel, or handle, adjacent to the mechanism; however, some, particularly

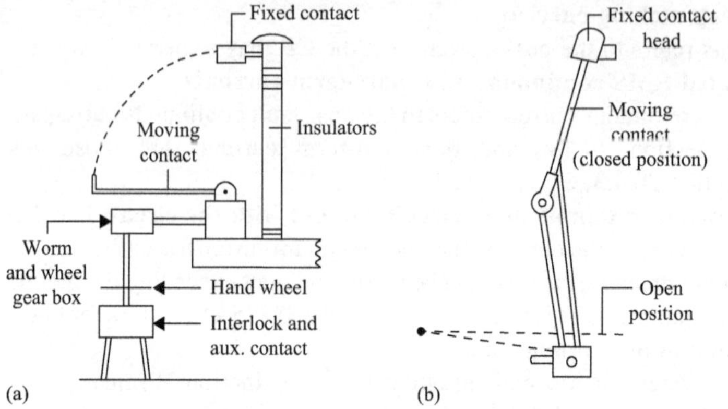

Figure 9.6 Earth switches: (a) swinging arm design and (b) pantograph design

at the higher voltages, are electrically powered so allowing remote control. The earthing devices often comprise a straight swinging (single blade) arm which moves in an arc from the fully open horizontal position to the fully closed vertical position. When fully closed, the moving contact pushes into a fixed contact housing, see Figure 9.6(a).

At higher network voltages, particularly at 400 kV, the design is often a pantographic arrangement, in which the two arms of the moving contact are folded and are almost horizontal when open, and almost vertical when closed, see Figure 9.6(b). This arrangement economises on space when in the open position, since the single blade design would need increased space both when open and when moving to the closed position. The earth switch is often integral to an associated disconnector. Figures 20.15 and 20.16 show typical earth switch points for a mesh and double busbar substation, respectively.

2. **GIS substation physical arrangement**
   With GIS substations, the earth switch is totally enclosed within the GIS housing. It comprises a physical gap (when open) of sufficient dielectric strength to withstand steady state and transient overvoltages. See Figure 20.18 for typical arrangement.

9.8.2 Metalclad switchgear

Figure 9.3(a) illustrates the typical physical arrangement for a metalclad CB (i.e. switchgear). In this arrangement, the CB truck can be either lowered (from the in service position) and withdrawn, or lowered and repositioned within the truck chamber to effect circuit (cable) or busbar earthing – this is termed 'transfer earthing'. Typical arrangements for transfer earthing are illustrated in Figure 9.7. Mechanical interlocks can be provided to lock the truck in position when the appropriate earth position has been obtained.

9.8.3 Portable earths

To provide flexibility for the point of earthing in AIS substations, portable earths are employed. These usually comprise a flexible conductor covered by an

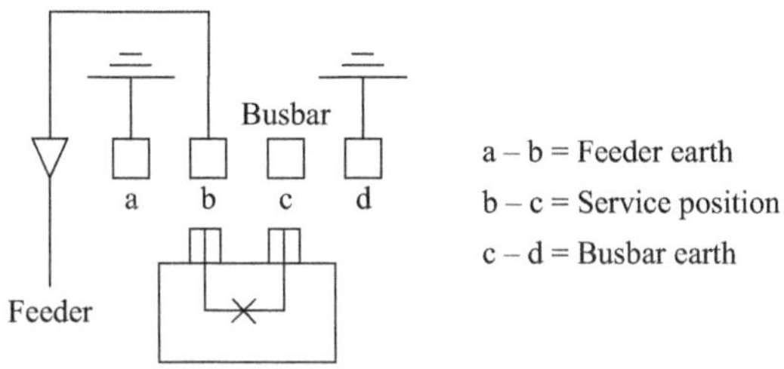

Figure 9.7 Transfer earthing

insulating material to which a conductor clamp is connected at one end, and an earth bar clamp at the other. Both clamps can be screw adjusted. The conductor end clamp screw is designed to be operated by a long insulated earthing pole. Spigots are usually incorporated into the busbar design for ease of connection of portable earths. When erecting a portable earth, the earth end clamp connection should be made first, for reasons of safety. Portable earths are usually categorised into two types:

1. **Circuit earths**
   These are designed to be capable of carrying short-circuit current (should a circuit be inadvertently energised). At the higher voltages, with higher short-circuit currents, the earth may comprise two flexible conductors within the same earth. As such, they can be difficult to handle and erect.
2. **Impressed voltage earths**
   These are designed to be capable of carrying currents associated with impressed voltages (IVs), mostly capacitive and inductive coupling.

9.8.4 Interlocking and indications

With reference to Figure 9.6, the earth switch arrangements usually include mechanical interlocks (e.g. castell key) for electrical safety management. When the earth switch is closed, the mechanical interlock can be removed and locked in a PTW key box to prevent interference with the earth switch when work is being undertaken. The interlock cannot usually be removed when the earth switch is in the open position.

The earth switch also usually includes electrical auxiliary switches (within a kiosk integral to the earth switch) driven by the earth switch mechanism, which must be synchronised to the open and closed positions and used for earth switch position indications.

9.8.5 Earthing device ratings

Earthing devices are usually rated as follows:

- To carry system maximum fault current (e.g. 63 kA for the 400 kV network) for a designed period (normally 1 s at 400 and 275 kV and 3 s at lower voltages)

- To make, carry and break the maximum circulating current as described in Section 11.8.8 on IVs
- To make onto and supress ferroresonance current, typically, 350 A peak at 400 kV, see Section 5.3.1.7 (contacts must be rated for this purpose).

9.9 Disconnectors

9.9.1 Disconnectors – design characteristics

A disconnector (previously termed an isolator) is a mechanical switching device, which when open provides physical separation and electrical isolation for an item of equipment(s), and when closed provides circuit continuity. They are used in the following circumstances:

- The provision of electrical isolation of an item of equipment to enable work to be carried out on that item. This is usually the major use.
- To carry out busbar disconnector changeover duties (contacts must be rated for this purpose).
- To make and break busbar charging current (contacts must be rated for this purpose).

Salient characteristics are as follows:

1. **AIS substations – physical arrangements**
   The most common type of disconnector is the centre rotating post, in which each phase has three insulator posts. The outer posts carry the fixed contacts (and busbar connections). The centre post carries the moving contact arm – which is arranged to rotate through 90° on its own axis (i.e. horizontal movement of the disconnector arm), see Figure 9.8. On later designs, the angle of rotation was reduced to 70°, to reduce the physical space occupied by the disconnector. With later 400 and 275 kV disconnector designs, once the rotating arm has reached the fixed contact, the arm is arranged to rotate on its

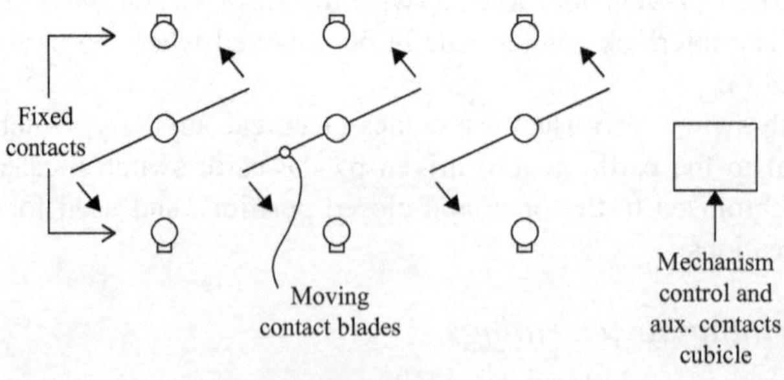

Figure 9.8 Centre rotating post disconnector – plan view

own axis until the full diameter of the arm is presented to the fixed contact metallic fingers – so obtaining optimum contact pressure without the need for a more powerful operating mechanism.

The disconnector is driven from a mechanism box near to ground level and by means of linkages drives all three phases simultaneously. All 400, 275 and some 132 kV disconnectors are usually operated by a gear motor mechanism, whilst those at lower voltages (which are physically smaller) are usually manually operated (typically by handle). The operating time (for both the open and close movements) for the motor-operated disconnectors is of the order of 15–20 s.

Disconnectors are also designed and manufactured in pantograph format, using a linkage arrangement similar to that illustrated in Figure 9.6(b). Again, the three phases are driven simultaneously – and the moving contacts connect with a spigot arrangement connected to the busbars. This design, again, reduces the physical area of the substation, but with possible increase in height.

2. **GIS substation – physical**
   With GIS substations, disconnectors are totally enclosed within the GIS trunking. When open, the disconnectors comprise a physical gap of sufficient dielectric strength to withstand maximum steady state and transient overvoltages. In some designs, the open and closed indication is backed up by a viewing window for assurance that the disconnector is open. See Figure 20.18 for an example arrangement.

9.9.2 Metalclad switchgear

Distribution voltage metalclad CBs, as illustrated in Figure 9.3(a), do not require a separate disconnector. Isolation is simply achieved by lowering the truck from the in-service position, and then locking the feeder or busbar spout shutters, as appropriate.

9.9.3 Disconnector interlocking and indications

With reference to Figure 9.8, the mechanism cubicle usually contains mechanical interlocks (e.g. castell keys or similar) which can be withdrawn when the disconnector is in the fully open or fully closed position – and once withdrawn, it locks the disconnector in this position. The keys are used for safety-management purposes (see later sections on interlocking).

The disconnector drive mechanism is also linked to a number of auxiliary contacts which reflect either the disconnector open or closed position – and can be used for control indications, and other protection and control interlocking purposes. Many disconnectors are fitted with a 'magnetic bolt arrangement' which is energised when the disconnector is moving to either the open or closed position. Failure to fully complete the open/close cycle results in the magnetic bolt causing the auxiliary contacts to be physically left in an indeterminate position which, with reference to control indications, is denoted by a condition termed DBI (don't believe it).

9.9.4 Sequential disconnectors

Some CBs are accompanied by a sequential disconnector, which opens some seconds after the CB main interrupters open. This provides electrical isolation of the CB usually for two reasons:

- First, to prevent lightning strikes to a connected OHL from impinging on the open interrupters
- Second, to ensure electrical isolation of a CB fitted with a force close mechanism, which causes the CB to close on loss of the interrupting medium.

9.9.5 Disconnector ratings

With reference to Chapter 20, disconnector ratings include the following:

- Rated RMS voltage (maximum continuous voltage)
- Rated RMS continuous current (usually the post-fault continuous current)
- Rated RMS short-term current (i.e. 5, 10 min rating, etc.)
- Rated magnitude and duration of short-circuit current
- Busbar charging make/break current
- Busbar transfer (changeover current) – usually associated with double busbar substations.

9.10 Interlocking

9.10.1 Interlocking – purpose

Although interlocking is self-evidently not an item of HV equipment, it is intrinsically connected with the operation of CBs, disconnectors and earth switches, and for this reason included in this chapter. The purpose of interlocking is to ensure the correct sequencing, and/or immobilisation (in the open or closed position), of the previously mentioned equipment, to protect the equipment from the dangers of incorrect operation. Interlocking may be categorised as follows:

- Operational interlocking (sometimes termed electrical interlocking) – concerned with the operational sequencing of disconnectors
- Safety interlocking (sometimes termed mechanical interlocking) concerned with repositioning of keys (or software equivalent) to facilitate safe working of switchgear
- Metalclad switchgear interlocking concerning with CB operation and truck repositioning and involving a combination of electrical and mechanical interlocking.

The above will be briefly reviewed.

9.10.1.1 Operational interlocking

Operational interlocking is concerned with the operational sequencing of disconnectors (both on and off load) when network connections are being reconfigured. It essentially concerns the interlocking, and enabling, of disconnector open

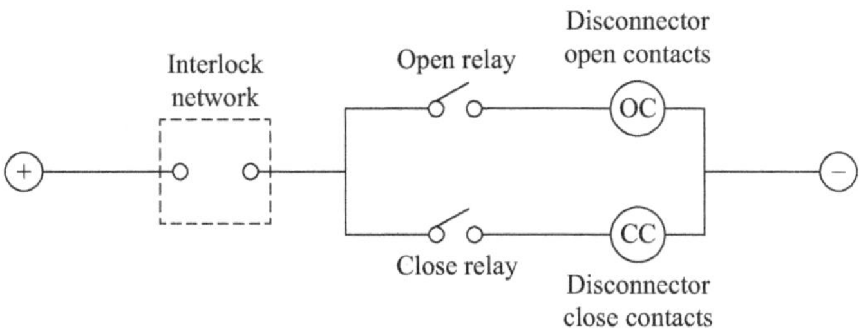

Figure 9.9 Operational (electrical) interlocking circuitry

and close coils, as shown in the simplified arrangement in Figure 9.9. The interlock network comprises disconnector and CB auxiliary contacts which are connected to replicate the open or closed status of equipment in either the whole or part of the substation, as required to safely operate the disconnector in question. The interlock network is frequently arranged to interlock the magnetic bolt (as described in Section 9.9.3), which in turn, when operated, closes a contact to short circuit the contact arrangement enclosed by the dotted box in Figure 9.9 (and thus enable the open or close operation).

As an example, and with reference to Figure 20.16, and considering the disconnectors associated with Line 1, disconnector X104 can only be operated (i.e. open or closed) under the following conditions:

1. **No power path**
   When X105 and X106 are both open – under which circumstances X104 can be either open or closed (and permitted by the interlock contacts) without damage – i.e. neither making nor breaking a power path.
2. **Parallel path**
   When there is a parallel path across X104, then X104 may be either opened or closed, and as such it neither makes nor breaks a power path. Examples of the parallel (interlocking chain) path are
   (i) X106, X136, X130 and X134 all closed
   (ii) X106, X166, X169, X236, X230, X234, X128, X120 and X124 all closed.

This enables a busbar on-load changeover.

9.10.1.2 Safety interlocking

Safety interlocking is concerned with the sequential operation of items of switchgear, usually to enable the earthing of equipment for the purposes of providing safe access, particularly for maintenance, but also for construction when the work involves operational assets. It is usually achieved by an interlocking system using mechanical keys, e.g. castell type of key. The keys are trapped or released depending upon whether the equipment is in the open or closed position.

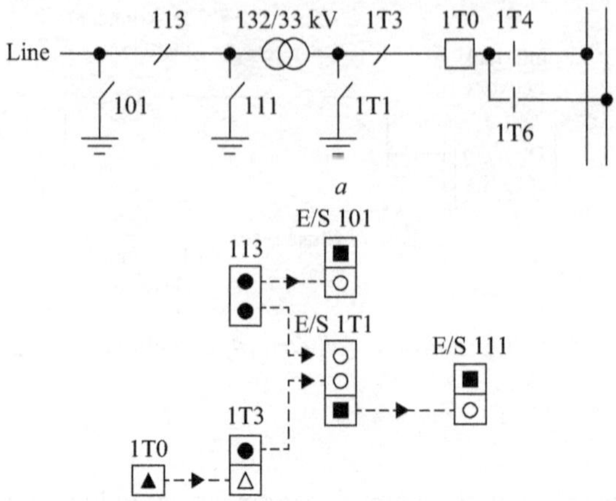

▲ = Key released when 1T0 is in the open position

△ = Key slot which releases key when 1T3 is in either open or closed position

● = Key released when disconnector is in open position

○ = Key slot which releases key when in open or closed position

■ = Key which is released when earth switch is in closed position

Figure 9.10 Safety (mechanical) interlocking – example

Figure 9.10 provides a typical example. To close earth switch 111, the following sequence would need to be undertaken:

1. De-energise the circuit by opening both the CB at the remote end and CB 1T0.
2. Remove the key (black triangle) from 1T0 and place in the disconnector 1T3 exchange box (white triangle). 1T3 can now be opened. On opening, the 1T3 (black circle) and 1T0 (white triangle) keys can now be released.
3. Transfer the released key (black circle) from 1T3 to the earth switch 1T1 exchange box (white circle).
4. Open disconnector 113 and transfer the 113 key (black circle) to the earth switch 1T1 exchange box (white circle). Earth switch 1T1 can now be closed, releasing a 1T1 key (black square).
5. Transfer the 1T1 key (black square) to the earth switch 111 exchange box (white circle). Earth switch 111 can now be closed.
6. On closing earth switch 111, a 111 key (black square) is now released which can be placed in a PTW safety lock out box to enable work on the transformer which is now earthed on both sides to commence.

A mechanical interlocking scheme for the closing of earth switches in a substation similar to that illustrated in Figure 20.16 is self-evidently much more complex, involving a significant number of key exchange boxes.

9.10.1.3 Metalclad switchgear

Interlocking arrangements associated with distribution voltage metalclad switchgear are usually a combination of electrical and mechanical and are concerned with the correct positioning of the CB. Salient arrangements are as follows:

1. **Electrical interlocking**
   (i) Ensures that the CB truck cannot be electrically lowered or raised unless the interrupter contacts are in the open position.
   (ii) Ensures that the CB interrupters can be closed only when the truck is in the fully raised or fully lowered position.
2. **Mechanical interlocks**
   (i) Ensures that the feeder and busbar spouts are automatically covered by shutters as the CB is lowered to the isolated position. This is achieved by a mechanically operated lever system.
   (ii) Ensure that the CB can only be raised and closed in the feeder or busbar earth position when the truck is physically located in the appropriate position. This is accomplished via positioning latches and mechanical levers (to open the feeder or busbar spout shutters).

9.10.2 Computer-based interlock systems

Many modern interlocking systems, both electrical and mechanical, are located within computer-based systems, particularly the interlocking for GIS substations, but with the operating logic based upon the arrangements described above. However, these systems are not the easiest to commission and can be demanding to extend (e.g. add new circuits). In addition, there is often a greater confidence by operational engineers in key-based systems for safety-management arrangements. As a result, many power network companies still prefer to specify the older electro/mechanical-based interlocking systems.

9.11 Power transformers

9.11.1 Power transformers – introduction

The power transformer is a key item of power system equipment enabling the transmission and distribution of electrical power to be undertaken at a variety of network voltages. It is one of the most costly items of equipment to be installed in a substation. Transformer theory and principles of operation of both two winding transformers and autotransformers are covered in Chapter 3, and transformer phase shifts and phasing are covered in Chapter 5. A wealth of literature exists already on power transformers, and therefore, this text will narrowly focus on design considerations relevant to power network construction, which are as follows:

- Transformer design – key considerations
- Tap changer
- Transformer protective devices

- Transformer ratings
- Transformer overfluxing
- Transformer noise
- Physical dimensions and weight
- Tertiary windings
- Transformers operating in parallel
- Capitalisation of losses
- Transformer specification.

9.11.1.1 Transformer design – key considerations

Figure 9.11 illustrates the key components of a power transformer and its cooling system, of which the following will be examined:

1. Core and windings
2. Oil
3. Conservator, cooler fans and pumps
4. Selector and diverter switches

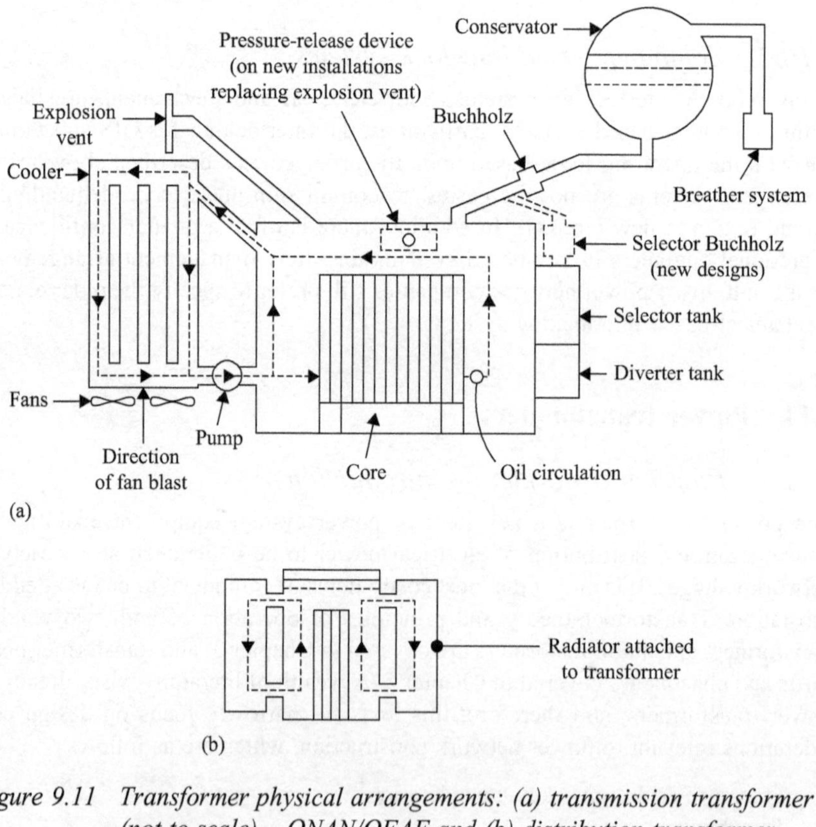

Figure 9.11 Transformer physical arrangements: (a) transmission transformer (not to scale) – ONAN/OFAF and (b) distribution transformer – ONAN

1. **Core and windings**
   Three-phase power transformers, as installed on the power networks, usually comprise a laminated three-limbed alloy steel core (or similar material). Generator transformers may be five-limbed cores (to lower the height) – or with larger generators three separate two limbed cores (one per phase). The primary and secondary windings of all transformers are usually paper insulated, coiled around each limb and connected to the HV and LV terminals. At the higher voltages, the HV and LV terminals are usually connected to the windings via bushings, but at distribution network voltages, this may be via cable boxes.
   The magnetic core and core support structure of a transformer is frequently designed with a connection that is brought out (through an oil tight seal) to a terminal (usually at the side and near the bottom of the transformer tank) for the purposes of connecting to earth (i.e. earthing the core). This enables the insulation resistance of the core to be measured without the need for oil handling.
2. **Oil**
   The transformer tank is filled with oil which serves two purposes:
   (i) As an insulation medium (in conjunction with the paper-covered windings)
   (ii) As a cooling medium capable of extracting heat from the transformer windings.
   The oil is usually a mineral oil with additives, possessing the properties of being non-sludging and non-oxidising.
3. **Conservator**
   The conservator acts as both an oil reservoir and an oil expansion vessel. It exhausts to atmosphere via a breather system whose properties restrict moisture intake during transformer cooling, and oil contraction.
4. **Coolers, fans and pumps**
   With reference to Figure 9.11, transformer cooling is assisted by the installation of coolers, which may stand apart from the transformer (transmission voltages) or attached to the transformer (distribution voltages). The traditional types of cooling are classified in Figure 9.12.
5. **Selector and diverter switches**
   The selector and diverter switches comprise part of the tap changer scheme. The selector tank is usually connected to the main tank (i.e. connecting oil) and contains clean oil.

Classification	Interpretation	Meaning
ONAN	Oil normal, air normal	Oil immersed, with natural air cooling
ONAF	Oil normal, air forced	Oil immersed, with fan-assisted cooling
OFAN	Oil forced, air normal	Oil immersed, with pumped oil and natural air cooling
OFAF	Oil forced, air forced	Oil immersed with pumped oil and fan assisted cooling

Figure 9.12 Transformer cooling classification

The diverter tank is usually self-contained and bolted to the main tank. It contains arc products arising from the tap change process, and the oil must be periodically exchanged for new clean oil.

9.11.1.2 Tap changer

Almost all power transformers installed on the electricity networks are fitted with tap changers. An exception on the UK system may be some 400/275 kV transmission transformers rated between 500 and 1,000 MVA. Most tap changers are designed to be 'on load' – an exception to this are 11/0.4 kV distribution transformers which are usually equipped with 'off load' tap changers. Generally, tap changers are located on the HV side of two winding transformers (for both economic and technical reasons) and usually located at the neutral end (where the HV winding contains a neutral) to minimise insulation requirements. They are also usually located towards the neutral end of autotransformers (although with some exceptions).

The purpose of a tap changer is to vary the tap position and therefore the turns ratio of a transformer, to maintain a target voltage on the secondary of the transformer under varying load conditions. With reference to Figures 3.14 and 4.3(a), the term 'transformer voltage regulation' is frequently used as a measure of the change in transformer secondary voltage when the load varies from no load to full load, with the transformer remaining on nominal tap. It is expressed as follows:

$$\text{Voltage regulation} = \frac{V_{\text{NO Load}} - V_{\text{Full load}}}{V_{\text{Full load}}} \times 100\% \tag{9.2}$$

where in the above all the voltages are transformer LV voltages.

Figure 9.13 illustrates a typical on load tap change scheme, which functions as follows:

Consider the tap changer to be selected initially to *tap 2*. The current flow is then down the winding to *tap 2*, then via switches *S2* and *D1* to the neutral terminal, as shown in the figure. Consider now the tap changer moving from *tap 2* to *tap 3*, the steps are as follows:

1. Stage 1: The diverter switch rotates to bridge contacts *D1* and *D2* and then rotates further such that it only makes contact with contact *D2*. At this stage, the load current flow is via resistor *R2*.
2. Stage 2: The diverter switch continues its travel bridging contacts *D2* and *D3*. At this stage, the load current flow divides via resistors *R1* and *R2*, with a circulating current, arising from the voltage across *taps 2* and *3*, circulating around the *tap2 – S2 – R2* – diverter switch – *R1 – S1 – tap 3* circuit. The resistors limit the magnitude of the circulating current.
3. Stage 3: As the diverter switch continues to rotate, it breaks contact *D2* (so breaking the circulating current and creating arc products) and connects only with contact *D3*. At this stage, the load current flow is via *tap 3 – S1 – R1 – D3* to the neutral.
4. Stage 4: The diverter switch rotates to bridge contacts *D3* and *D4* and rotates further such that it makes contact with only contact *D4*. Load current now flows via *tap 3*, *S1* and contact *D4* to neutral. The transition to *tap 3* is now complete.

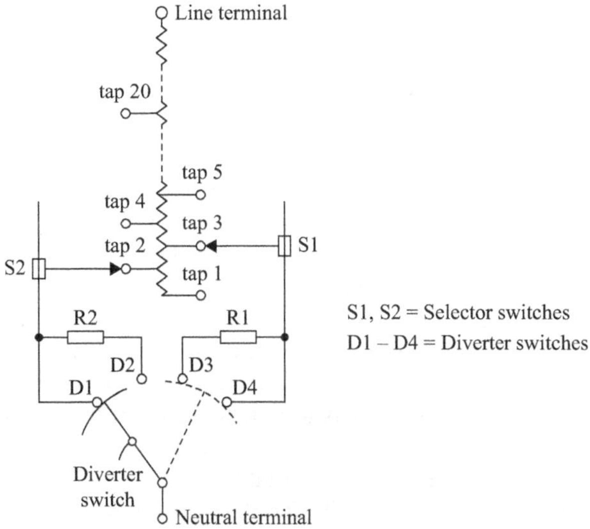

Figure 9.13 Tap changer – diagram of connections

If the next tap change is from *tap 3* to *tap 4*, the cycle commences with selector switch *S2* moving from the *tap 2* to *tap 4* position. In doing so, it neither breaks nor makes current. The diverter switches then enact the above stages in reverse, rotating from contact *D4* back to contact *D1*.

The resistors are designed to withstand the voltage imposed across them and the rated current flow through them. The tap change mechanism is fast acting, termed a 'high speed resistor' type, so that the resistances are only in service for tens of milliseconds. It is worthy of note that reactors have occasionally been used in preference to resistors, particularly in North America. The diverter switch contacts when transitioning to and from contacts *D2* and *D3*, and are subject to arc erosion as a result of breaking and making current, causing gradual deterioration of the condition of the diverter tank oil. The number of taps on the tap changer is usually standard for a particular voltage and typically may be between 15 and 20 in number with a typical voltage adjustment ranging between ±15% (depending upon the networks). Most tap changers are fitted with automatic tap change control as described in Chapter 10, in addition to local manual control.

9.11.1.3 Transformer protective devices

Transformer protection is examined more generally in Chapter 10. Within this context, HV power transformers at 33 kV and above are almost universally fitted with three particular devices, which are complementary to relays operated directly by fault current, as follows:

1. Buchholz relay
2. Winding temperature relay
3. Pressure relief device.

These will be briefly examined below.

1. **Buchholz relay**
   The Buchholz relay is positioned in the oil pipework connecting the conservator to the main tank, see Figure 9.11. It comprises a chamber filled with oil with two types of operating devices, which are as follows:
   (i) The first comprises an open-top container located at the top of the chamber, which, when the chamber is filled with oil (normal operating condition), is in the horizontal or balanced position. Should an electrical fault occur in the transformer tank, the fault arc causes the liberation of gasses which rise towards the conservator. In doing so, they become trapped in the top of the Buchholz chamber and displace the oil such that it causes the container (which retains its oil) to rotate in position and close two electrical contacts to initiate an alarm (usually termed 'Buchholz gas'). This alarm may also be initiated by slowly liberated gasses (e.g. a slowly developing arcing fault).
   (ii) The second comprises a deflector plate positioned lower in the Buchholz chamber and in the main oil/gas flow to the conservator. Should a major (arcing) fault occur in the transformer tank there will be violent liberation of gasses, which rapidly flow to the conservator and in doing so rotate the deflector plate, which in turn causes electrical contacts to close to trip the transformer HV and LV CBs – to remove the transformer from service. Operation of this device also initiates an alarm usually termed 'Buchholz trip'. The speed of operation of the Buchholz is highly dependent upon the volume of gasses generated by the fault. The angle of inclination of the pipework is important in optimising performance of the Buchholz relay.
2. **Winding temperature relay**
   The winding temperature relay is designed to measure the temperature at the transformer 'hot spot'. This is the location of the hottest point within the transformer core and surrounding windings, and as such the point of maximum transformer ageing and deterioration. The relay usually comprises a pocket of oil located at the top of the transformer tank, which will be at the same temperature as the top of tank oil. The pocket is usually heated by a heater fed from a CT (which carries a current proportional to the transformer load). The combined heating effect is designed to bring the pocket of oil to a temperature that reflects the hot spot temperature. The pocket of oil is traditionally connected to a capillary tube and bellows mechanism to drive the temperature indicating part of the relay (calibrated in degrees centigrade). With more modern relays, a temperature sensing device in the oil feeds directly into a microprocessor-based relay. When the temperature reaches the temperature setting(s) of the relay, contacts close to undertake the following functions:
   (i) Start and stop of transformer fans and pumps on rising and falling oil temperature, respectively, as the load varies, thereby ensuring the hotspot maximum allowable temperature is not reached.

(ii) If however the fans and pumps do not adequately control the transformer temperature (e.g. because they have failed or due to unacceptable loading conditions, etc.), and it continues to rise, then a 'winding temperature' alarm is given. If on further increases in temperature the hot-spot setting of the relay is reached, the relay will trip the LV CB, so removing the load and source of heating – an alarm 'winding temperature trip' is also usually given.

Transmission transformers operating at 400, 275 and 132 kV network voltages usually have a winding temperature relay associated with each winding.

3. **Pressure release device**

   With reference to Figure 9.11(a), the 'pressure release device' provides protection from high-pressure gas build up arising from an (arcing) fault internal to the transformer tank – which could be so great as to split the tank – with hazardous expulsion of hot oil. The pressure relief device must therefore be fast acting. It is usually spring operated and provides a controlled release of internal pressure.

   Older transformers, still in service, were fitted with an 'expansion vent', as shown in Figure 9.11(a). At the interface with the external air, the vent comprises a thin and relatively weak diaphragm, which under normal circumstances, seals the transformer from the atmosphere. Under transformer internal fault conditions, the excess pressure in the tank bursts the diaphragm to release the pressure – but in a less controlled way than the pressure release device.

9.11.1.4 Transformer ratings

Transformer ratings typically fall into the following categories:

- Rated power
- Emergency continuous rating
- Cyclic rating.

- **Rated power**

  This is the information normally provided on the transformer nameplate (fixed to the transformer) and is sometimes termed the nameplate rating. It specifies the continuous loading both with and without fans and pumps. Where fans and pumps are installed, the continuous rating in the ONAN operating state (i.e. without fans and pumps) is typically 50% of the OFAF operating state (i.e. with fans and pumps). OFAF is therefore the maximum-rated power.
- **Emergency continuous rating**

  Some transformers are given and designed with an emergency, and infrequent, continuous rating, typically 25% above maximum rated power. It is usually accepted that the transformer will suffer some increased ageing of the insulation under this condition, and to minimise the deterioration, this rating is usually linked to a low ambient temperature, e.g. 5°C.
- **Cyclic rating**

  Transformers are loaded cyclically to reflect daily load profiles, e.g. 8 h at 130% rated power followed by 16 h at 65% rated power. Cyclic loading is

Voltage ratio (kV)	Typical ratings (MVA)
400/275	1,000, 750, 500
400/132	240
275/132	180, 120
132/33	90, 60, 45
33/11	30, 18, 12

Figure 9.14 Typical UK transformer ratings

permissible without increased ageing of the transformer due to the long thermal time constants (i.e. slow heating) of the insulating medium. See also Section 20.4.5.

Typical UK transformer (maximum) ratings are given in Figure 9.14.

9.11.1.5 Transformer overfluxing

With larger power transformers, the design is such that a condition known as 'overfluxing' can arise whereby the transformer core is driven into saturation. This may cause a large increase in magnetising current resulting in increased transformer noise, vibration and temperature rise. Overfluxing arises when there is a high level of magnetic flux density in the transformer core. The level of flux density can be expressed as:

$$B = K\frac{E}{f} \tag{9.3}$$

where B is the flux density, K is the constant, E is the applied voltage and f is the frequency.

The worst combination for overfluxing is the highest possible applied voltage coincident with the lowest possible frequency. To minimise the risk of overfluxing, transformer cores must be designed around power system voltage and frequency limits.

9.11.1.6 Transformer noise

Transformer noise or 'hum' is caused by a phenomenon termed 'magnetostriction' which results from the extension or contraction of the transformer core when magnetised by an alternating flux. The vibration takes place at twice per cycle (moving from a positive to negative flux), i.e. for a 50 Hz power system frequency the vibration is 100 Hz.

Transformer noise can be a significant nuisance, particularly in urban areas, and where necessary transformers are housed in noise enclosures.

9.11.1.7 Transformer physical size and weight

Transformer physical size and weight are important when considering transportation to site. Within this context, it is usual to transport a transformer to site devoid of oil (with the tank pressurised with dry air or nitrogen), and then filled with oil on site. Once on site, physical size and weight are important considerations in relation to location, positioning, ground bearing capability and plinth size. Within this context, large transmission transformers may typically weigh as much as 220 t (when full of oil) with a typical footprint of 3.5 by 5.5 m and a height of 5 m.

9.11.1.8 Transformer tertiary windings

Many transmission transformers (e.g. 400/132, 275/132 kV) are fitted with a delta tertiary winding. This is usually rated at 13 kV and 60 MVA. The tertiary enables the connection of shunt reactors, MSCs and SVCs for the purpose of voltage control. The tertiary winding has the effect of reducing the zero sequence impedance of the transformer, see Section 4.2.5, which in turn increases power system fault level. When not in use, the tertiary is usually connected to the main transformer winding neutral/earth, see Figure 10.15.

9.11.1.9 Operation of transformers in parallel

It is often the case that transformers have to be operated in parallel, i.e. connected to the same HV and LV busbars. Considerations relating to parallel operation of transformers are as follows:

- The nominal voltage ratios of both transformers must be identical.
- The vector group of both transformers (when HV and LV phase connection arrangements are taken into account) must be identical, to ensure identical phasing – see Section 5.4.1.3.
- The MVA rating must be identical (or at least close) to maximise load sharing and utilisation (otherwise one will have to run under loaded).
- The per cent impedances should be identical (or at least close) to ensure equal sharing of the load. The per cent impedance also determines the fault level on the LV busbars and therefore the number of transformers that can be operated in parallel.
- Transformer tap changers must be compatible in terms of the voltage steps and the range of the tap changer – both to minimise transformer circulating currents and to ensure the network is retained within its design voltage limits.

It is worthy of note that when transformers operating in parallel are on different taps, a reactive circulating current (VAr flow) circulates around the transformer HV and LV connections, driven by the voltage difference between the taps. This is illustrated in Figure 9.15 which shows the instance of transformer *T1* operating at a higher tap (i.e. higher LV open-circuit voltage) than transformer *T2*. The circulating current is typically 4% of full load current for each tap the two transformers are apart.

9.11.1.10 Capitalisation of transformer losses

Transformers are subject to losses comprising both iron losses (no load losses), see Section 3.4.3, and copper losses (load-related losses) arising from the flow of load through the windings. During the lifetime of a transformer, the losses can be significant and are frequently factored into an evaluation of the overall cost of a transformer at time of contract tender evaluation.

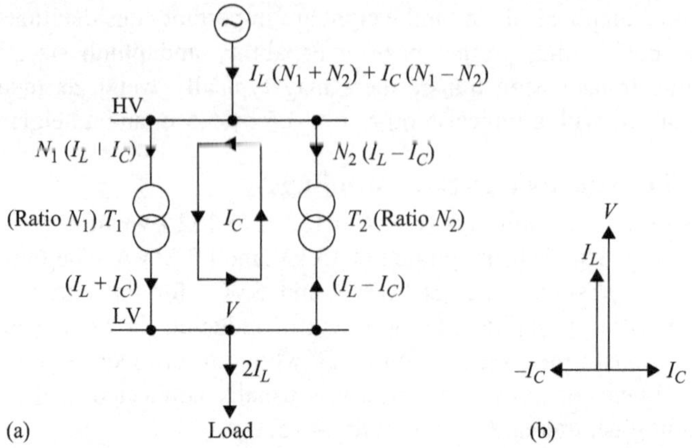

Figure 9.15 Transformer circulating current: (a) T_1 on higher tap than T_2 and (b) vector diagram

9.11.1.11 Transformer specification

A typical transformer specification would comprise the following:

- Number of phases
- Winding arrangements (e.g. autotransformer)
- Vector group
- Rated HV and LV voltages
- Rated power (MVA)
- Rated HV and LV current
- Overvoltage withstand capability
- Short-circuit per cent impedances
- Frequency
- Neutral earthing type
- Type of cooling (e.g. OFAF)
- Termination type (air, SF6)
- Noise containment requirements
- Tap changer (type, number of taps, tap increments, tap range, i.e. plus/minus voltages)
- Buchholz
- Winding temperature
- Pressure release device
- Capitalisation of losses.

The following standard documents are relevant to transformer specifications:

- IEC 60076 – Power transformers
- IEC 60214 – On load tap changers
- IEC 60354 – Loading guide for oil immersed transformers.

9.12 Reactors

9.12.1 Reactor types

Reactor types fall into two categories, which are as follows:

1. Shunt-connected reactors
2. Series-connected reactors

1. **Shunt-connected reactors**
   Shunt-connected reactors are used for power system voltage control by reducing system voltage, usually at times of light load to counteract the Ferranti effect on feeders (see Section 7.4.1.5). With reference to Figure 9.16, shunt reactors are generally connected as follows:
   (i) Directly connected to the network busbars, as shown in Figure 9.16(a). These reactors are typically up to 150 MVAr rating, and when switched into service have a direct bearing on the voltage of the network to which they are connected.
   (ii) Connected via a transformer tertiary, as shown in Figure 9.16(b). These reactors are usually connected at 13 kV and are often connected as a pair, each having a typical rating of 30 MVAr. The positioning of these reactors has a proportionately bigger impact on the transformer LV voltage, by virtue of the lagging VAr flow through the transformer, but also influences the HV voltage.
2. **Series-connected reactors**
   Series-connected reactors are usually installed to limit fault levels (usually on a busbar), but in some instances, their installation is to limit load flow (or redirect load flow). Typical ratings range from 1,320 MVA at 400 kV to 50 MVA at 33 kV. They are usually connected with a bypass CB, see Figure 9.16(c).

Reactor construction comprises two types: air insulated and oil immersed. Air insulated tends to be associated with both series reactors, and shunt reactors up to about 33 kV. Air-insulated reactors often do not have an iron core and frequently

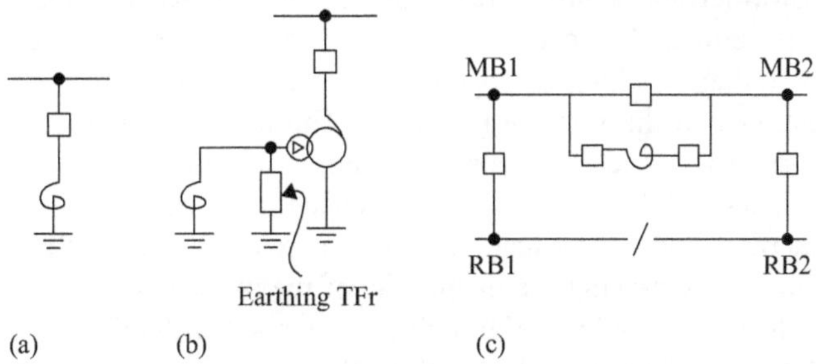

Figure 9.16 Reactor types and connections: (a) directly connected reactor, (b) tertiary-connected reactor and (c) series reactor

comprise embedded circular coils of stranded copper conductor located in a series of concrete pillars arranged in a circular formation. The conductor may be bare (i.e. not insulated), and if so, the reactor must be placed in a non-magnetic enclosed cell. Some designs are surrounded by a magnetic shield to retain the magnetic flux within immediate vicinity of the reactor.

Oil-immersed reactors employ an insulation medium and cooling similar to that of a transformer. Shunt reactors connected at 132 kV and above are generally of this design but differ from a transformer in as much as transformers have a lower level of magnetising current, whereas reactors are designed with a high level of magnetising current. To prevent saturation, the magnetic core usually contains an air gap.

A consideration for all reactor design is the minimisation of stray magnetic flux – which risks causing unwanted induced voltages into equipment and conductors with metallic loops – which in turn causes heating.

9.13 Quadrature boosters

9.13.1 Quadrature boosters – purpose

QBs are used for controlling both the magnitude and direction of active power flow (i.e. MW), usually over parallel connected circuits in order to optimise overall network power transfer capability. In doing so by increasing/decreasing power in one circuit (i.e. the circuit in which the QB is installed), the power in the parallel connected circuits decreases/increases, respectively. This ensures that an optimised power flow across all the circuits can be achieved, thereby avoiding circuit overloading on a more highly loaded circuit(s). As a result, the requirement to construct additional circuits (both at considerable cost, and often with adverse visual/aesthetic impact) can be avoided. QBs are usually installed on transmission networks, which have numerous parallel connected circuits, but they have been installed on network voltages as low as 33 kV.

9.13.2 Quadrature booster – winding arrangements

Figure 9.17(a) shows a single line diagram of a QB. It comprises two transformers: a shunt transformer and a series transformer. The primary winding of the shunt transformer is termed the 'exciting winding' and the secondary is termed the 'regulating winding' – which is tapped (not shown) to enable a 'buck' or 'boost' voltage injection into the series transformer. The delta winding (primary) of the series transformer is called the 'booster winding', and the star connected (secondary) winding is termed the 'series winding', since it is in series with the feeder. The control of the magnitude and direction of the active (MW) power flow through the feeder is achieved by varying the magnitude of the phase shifted voltage which is injected into the series winding. Within this context and with reference to the vector diagram in Figure 9.17(b), it can be seen that the injected voltage V_{QB} is arranged to be at right angles (i.e. quadrature) to each respective phase of the feeder voltage (e.g. V_R as shown).

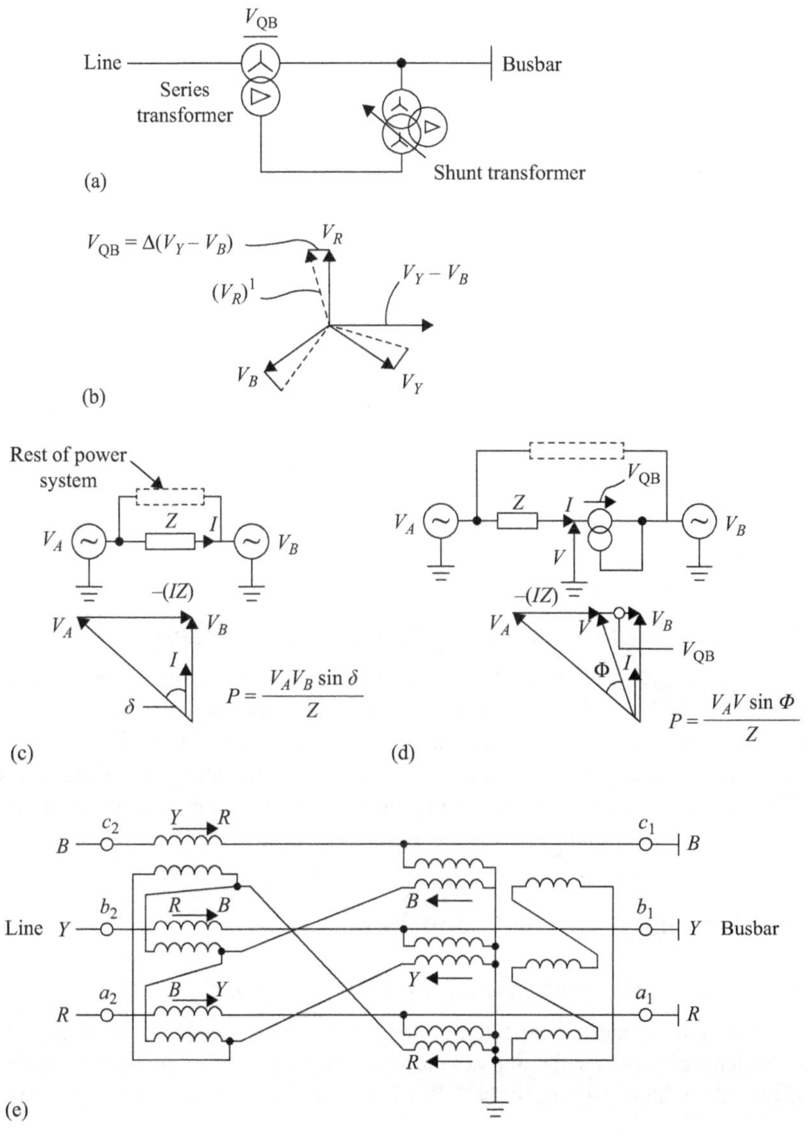

Figure 9.17 Quadrature booster: (a) single-line diagram, (b) vector diagram, (c) without QB, (d) with QB and (e) winding arrangement

9.13.3 Quadrature booster theory

Prior to the QB being installed, the power flow P along the feeder in question is as stated in the Figure 9.17(c) (see also Section 3.2.5). Consider Figure 9.17(d) showing the QB now installed. It is assumed that the magnitude and phase angle of V_A and V_B are fixed by the rest of the power system (i.e. stationary) as a result of a power system with numerous parallel circuits. As a result, the impact of the injected voltage V_{QB} is to cause the voltage at the QB end of impedance Z to move from V_B to V. As a consequence, the angle across Z is now reduced, and the power flow along the feeder is reduced accordingly. Within this context, the change in voltage magnitude at the QB end of Z (i.e. from V_B to V) is relatively small, and the main determinant of power flow change is determined by the change in power transfer angle (i.e. from δ to Φ).

The above explanation relates to the instance of the QB causing a reduction in power transfer (i.e. power buck); however, the tap changer can be arranged to inject V_{QB} in the opposite direction to that shown (i.e. power boost) to increase the power flow along the feeder.

Figure 9.17(e) illustrates a QB three-phase winding arrangement. In practice, the winding arrangement also modifies the MVAr flow along the circuit, but this is relatively small compared to the impact on MWs.

It is worthy of note that an alternative way of examining the performance of a QB is to consider the impact of the injected voltage on the current flow around the network concerned. This analysis is analogous to the superposition method as described in Section 3.3.1.3.

9.13.4 Quadrature booster – physical arrangement

The rating of QBs may typically range from 2,750 MVA for a 400-kV unit to 250 MVA at 132 kV. They usually comprise (i.e. manufactured as) two separate transformer units which are bolted together to form a single unit. They are physically some of the largest (and heaviest) items of equipment to be found in a substation. As such, great care must be taken with both the position within the substation, and the ground bearing capacity.

9.14 Manually switched capacitors

9.14.1 Manually switched capacitors – purpose

MSCs are shunt-connected capacitor banks which are switched into service via a CB (i.e. mechanically switched – and not electronically switched). The purpose of MSCs is to provide voltage support in the following circumstances (see also Figure 10.35):

- Voltage support under normal network heavy loading conditions
- Post fault voltage support

Figure 9.18 provides a simple illustration of how an MSC provides voltage support. The impedance Z represents the source impedance of the power system which is essentially inductive.

9.14.2 MSC connection arrangements

Figure 9.19 illustrates how MSCs have traditionally been connected in the United Kingdom. When positioning the MSC on the LV side of the transformer,

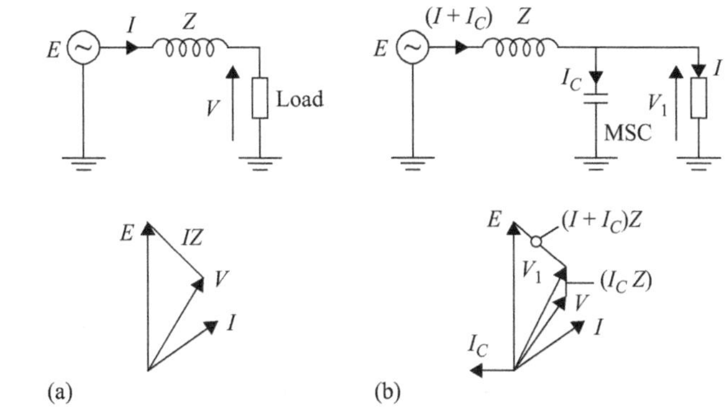

Figure 9.18 MSC voltage support: (a) before MSC and (b) after MSC

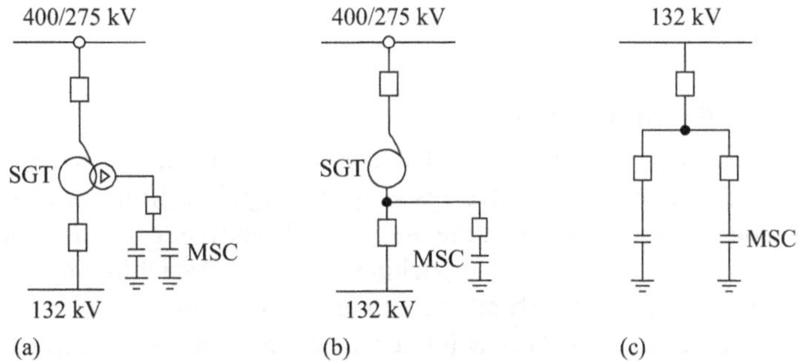

Figure 9.19 Typical MSC connection arrangements: (a) MSC of 2×20 MVAr, (b) MSC of 60 MVAr and (c) MSC of 2×60 MVAr

the voltage is not only raised on the HV network but is also directly raised on the LV busbars.

MSCs are usually manually switched and are installed when manual switching timescales are acceptable. They are also used in conjunction with automatic reactive switching schemes and in association with automatic tap change control schemes, see Section 10.18.1.4. Post-fault switching of MSCs is usually undertaken after a time delay to allow the HV auto-reclose scheme to return any circuits to service (and so avoid any potential over voltage conditions).

It is worthy of note that MSC design usually includes a low inductive reactance in series with the MSC capacitors to limit the energising inrush currents. See also Section 9.6.1.5 on point on wave switching requirements.

A variant of the pure MSC is the mechanically switched capacitor with an associated damping network. As a capacitance source, MSCs have the potential for resonance with the remainder of the power system. To minimise this risk (at salient points in the network as system analysis dictates), the MSC may be fitted with a damping network which includes a damping resistor. This arrangement is usually termed a 'MSCDN'.

9.15 Static VAr compensators

9.15.1 FACTS technology

The abbreviation FACTS stands for 'flexible alternating current transmission system'. It defines a system comprising static equipment used for the AC transmission of electrical energy with the objective of both enhancing the controllability, and increasing the power transfer capability, of the network. FACTS technology is generally power electronics based. The 'SVC', and variants of the SVC, is an important item of FACTS equipment, providing both inductive and capacitive reactive support to the power system in rapid timescales.

SVCs comprise three devices, which are as follows:

- Thyristor-controlled reactor (TCR)
- Filter
- Thyristor-switched capacitor (TSC)

These will be briefly reviewed below.

9.15.1.1 TCR fundamentals

Figure 9.20 illustrates the principle of operation of a TCR. When the gate of a thyristor (usually known as a 'valve') is subject to a pulsed firing voltage, it commences to conduct current through the thyristor. The current continues until it falls to a level close to current zero (the latching current) at which point the thyristor ceases to conduct (i.e. shuts off) and acts as an open circuit.

The performance of the TCR is influenced by the fact that the switching of an inductive circuit results in both a steady-state AC current and decaying DC transient current as outlined in Section 3.8.2. With reference to Figure 9.20(b), if the thyristor is fired at a voltage peak, i.e. an angle of $\alpha = 90°$, this corresponds to a steady-state AC component of current commencing from current zero, with no accompanying DC transient component of current. If however the firing angle, α, is progressively moved beyond 90°, the DC transient component of current becomes greater in magnitude, the impact of which is to produce a total current flow that is of reduced magnitude and duration (i.e. the time between thyristor firing and

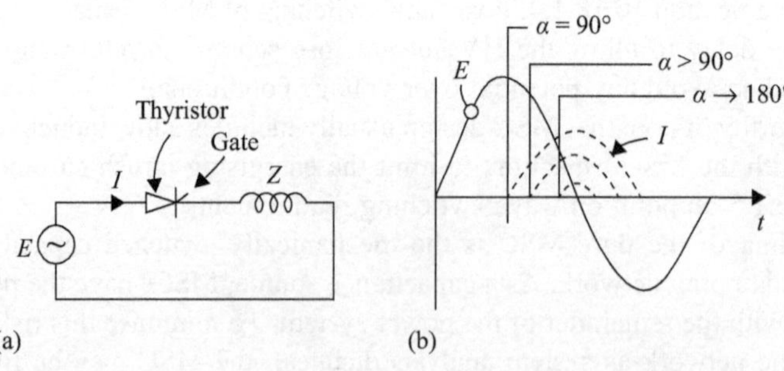

Figure 9.20 TCR fundamentals: (a) thyristor network and (b) voltage and current wave

subsequent current zero is reduced). As the firing angle α approaches 180°, the resultant current magnitude and duration also approaches zero.

TCR in operation

Figure 9.21 illustrates a practical TCR arrangement where the applied voltage is switched by two thyristors, one on the positive half cycle of voltage, and the other on the negative half cycle of voltage. The firing of the thyristors determines the magnitude and duration of the resulting current (as explained in Section 9.15.1.1). This ranges from maximum continuous current at firing angle $\alpha = 90°$, to ever-reducing current magnitude and duration when α is increased beyond 90°, to zero current when $\alpha = 180°$. Figure 9.21(b) illustrates a typical firing current magnitude and duration for a firing angle approximately midway between 90° and 180°. In this instance, the resulting current wave is neither fully sinusoidal in shape nor continuous throughout the half cycle, and as such contains harmonics.

9.15.1.2 TCR harmonic filter

Figure 9.22 illustrates a plot of fundamental frequency as a per cent of the TCR current wave, as the thyristor firing angle progressively moves from 90° to 180°. As the plot shows, as the firing angle moves from 90° to 180°, the per cent of fundamental current decreases, and therefore the per cent of harmonic current increases. Analysis shows that the dominant harmonics are the 5th, 7th and 11th. A filter

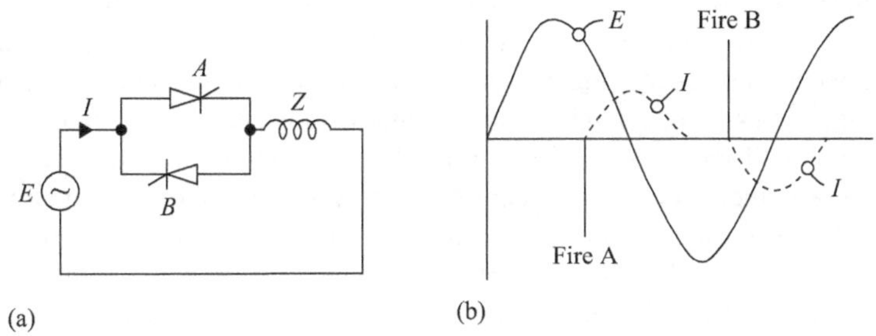

Figure 9.21 TCR in operation: (a) practical TCR and (b) TCR wave forms

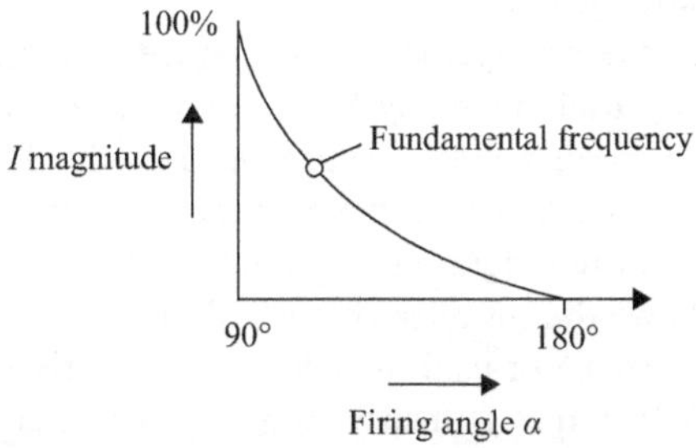

Figure 9.22 TCR fundamental frequency

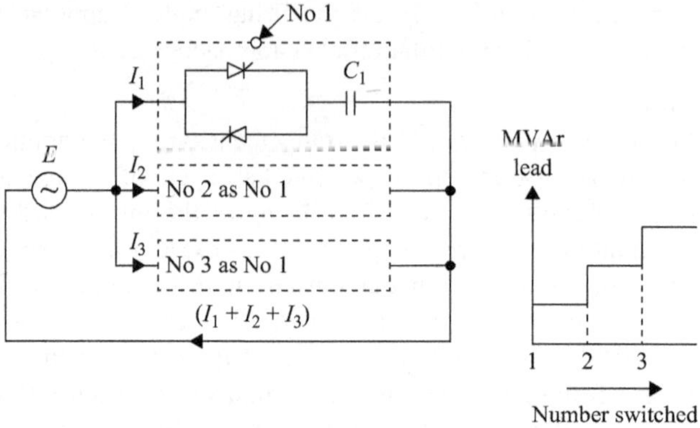

Figure 9.23 TSC arrangement

arrangement located in parallel with the TCR filters out the harmonic currents and therefore prevents them from flowing into the supplying network. Therefore, by varying the thyristor firing angle, varying magnitudes of fundamental frequency inductive reactance current can be arranged to flow from the supplying network into the TCR.

9.15.1.3 Thyristor switched capacitor (TSC) arrangements

Figure 9.23 illustrates the basics of a TSC. The capacitors are switched on and off in blocks. With reference to Section 3.8.3, the very high transient in-rush currents that can occur when energising capacitors make it necessary to choose the point on wave thyristor switching angle. For minimum transient current, point on wave switching should ideally take place at a voltage switching angle of zero degrees. Inductive reactances (not shown) connected in series with the capacitors are often included to further reduce transient inrush currents. Following initial thyristor firing, the thyristors are fired every 180° to produce a continuous steady-state capacitive current, virtually free from transients and therefore harmonics.

9.15.2 SVC alternative arrangements

There are a number of ways in which SVCs may be arranged, and Figure 9.24 shows two common arrangements.

In Figure 9.24(a), the arrangement can produce continually variable lagging and leading VArs. The variable lagging VArs are produced by the TCR only being switched. The variable leading VArs are produced by the TSC being switched (in blocks), and the TCR switched to produce cancelling lagging VArs to achieve resulting leading VArs of the required magnitude.

Figure 9.24(b) shows the instance of an SVC arrangement comprising a variable TCR, and a fixed capacitor bank which is permanently in service – producing leading VArs. The TCR output is then varied to produce lagging VArs, the magnitude of which determines the level leading VArs taken from the network. This arrangement is mostly for the instance of control of leading VArs only. It has the

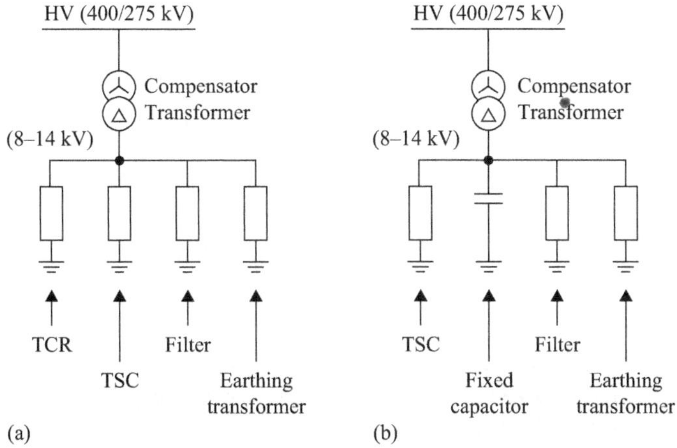

Figure 9.24 SVC alternative arrangements: (a) variable TCR/TSC and (b) variable capacitance

advantage of not requiring thyristors (and the associated cost) for switching the capacitors, and in many instances the filter can be incorporated into the capacitor (with further optimisation of cost). It has the disadvantage of incurring continuous losses since the capacitor bank is always in service (when in use).

SVCs can be directly connected to networks at voltages up to 33 kV, but for most applications, transformers are required to match the optimised thyristor design voltage to the higher voltage of the power network. The transformers are typically termed 'compensation transformers' and are usually a star delta configuration which further reduces the flow of harmonics into the HV network. The SVC voltage is typically in the range 8–14 kV. The TCRs and TSCs are also usually configured as a delta configuration which, again, optimises on harmonic performance.

9.15.3 SVC application

SVCs are termed 'dynamic' devices because of their speed of response (typically tens of milliseconds). Unlike CBs, they can also undertake a very high number of operations without wear or tear. With reference to Figure 9.25, typical SVC applications are as follows:

- Maintain target HV voltage – through the generation or absorption of VArs.
- Power factor correction on the LV side of a transformer to minimise the voltage drop across the transformer by operating in (variable) capacitor mode.
- Increasing the power transfer across an interconnector by maintaining target voltages at defined locations, i.e. maintain the receive end voltage V_R in Figure 3.5(b).
- Maintaining post fault stability through rapid voltage increase at defined points on the network.

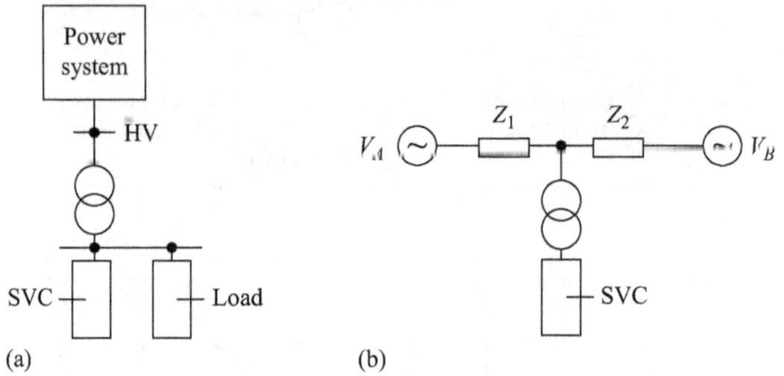

Figure 9.25 SVC applications: (a) (i) maintain target HV voltage and (ii) power factor correction; (b) (i) increase power transfer, (ii) maintain post fault stability and (iii) dampen power oscillations

- Damping power oscillations (between generation groups) by maintaining target voltage at salient points in the network. This application can be likened to that of a generator 'power system stabiliser'.

Generally, SVCs may be set to 'target voltage' mode, the usual application, or 'constant MVAr' mode. In the United Kingdom, the typical SVC ratings may be up to ±75 MVAr. Thyristor valves tend to get hot when operational and frequently require cooling – which may result in continuous and short-duration ratings.

9.16 Voltage transformers

9.16.1 Voltage transformer – overview

The objective of a VT is to obtain a secondary voltage which is an accurate reflection of the HV primary voltage, and of a magnitude suitable for connecting to metering, protection relays, control equipment, etc. VT secondary windings usually have a nominal secondary voltage of 63.5 V between phase and neutral and 110 V phase-to-phase. A VT is a 'parallel connected' transformer, as is a power transformer. However, whereas a power transformer is designed with a specific leakage reactance to limit short-circuit fault levels, VTs are designed with as small a leakage reactance as possible to maximise transformation accuracy.

IEC 60044, instrument transformers, stipulates a range of VT accuracy limits (classes). For example, accuracy class 0.5 covers precision metering with a voltage ratio error of ±0.5%, whereas accuracy class 3P covers protection VTs with a voltage ratio error of ±3%. Accuracy class also specifies a maximum phase displacement between primary and secondary voltages. Metering VTs are generally required to maintain accuracy between 80% and 100% of rated voltage up to stipulated maximum burdens (i.e. CT secondary load) – whereas protection VTs are required to maintain accuracy between 5% and 120% of rated voltage up to stipulated maximum burdens. Some VTs are designed for dual metering and protection purposes.

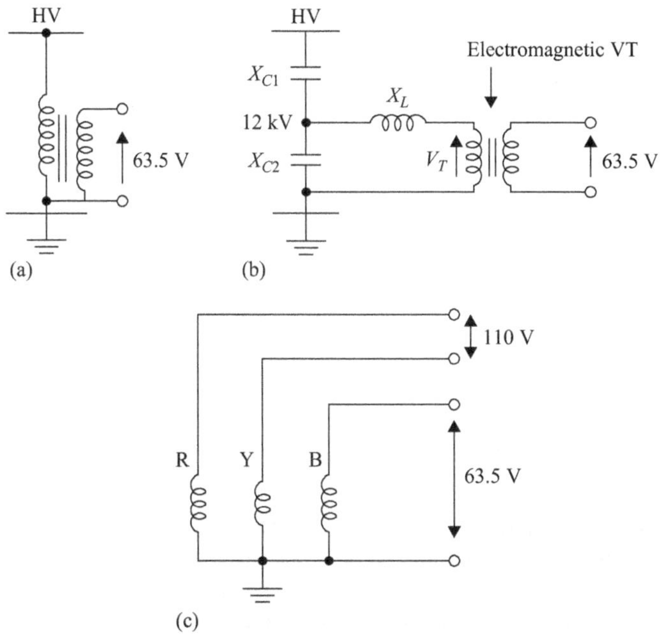

Figure 9.26 Voltage transformers: (a) electromagnetic VT, (b) capacitor VT and (c) usual method of earthing VT secondaries

9.16.2 Voltage transformers – types

With reference to Figure 9.26, VTs generally fall into two types: electromagnetic and the capacitor VT (CVT).

1. **Electromagnetic voltage transformers**
   Figure 9.26(a) illustrates the arrangement for an electromagnetic VT where both the primary and the secondary windings are wound around the same magnetic core. It is usual to apply this type of VT up to nominal voltages of 33 kV, or at higher voltage only when greater accuracy is required. In AIS substations, they are usually provided as three separate stand-alone VTs (one per phase). Similarly, with GIS substations, they are normally provided as an individual unit per phase. With distribution voltage metalclad switchgear, they are usually provided as a single, three-phase unit, capable of being racked in/out of service – these units are usually provide as a five limbed core, which ensures accuracy when transforming zero sequence voltages.
2. **Capacitor voltage transformers**
   CVTs are generally found at network voltages of 132 kV and above where they are more cost effective than electromagnetic VTs. With reference to Figure 9.26(b), the capacitors, in effect, act as a potential divider, typically producing 12 kV at the tapping point to the transformer. The inductance X_L is

arranged such that it interacts with X_{C1} and X_{C2} in a power factor correction (resonant) mode to ensure that V_T remains constant as the load connected to the secondary of the VT varies.

Figure 9.26(c) illustrates the usual method of earthing a three phase VT arrangement (i.e. either three single phase units or one three-phase unit) via the neutral point. However, VTs connected to the 33 kV network, and some VTs in power stations, are often earthed not via the neutral but by directly connecting yellow phase to earth. This must be considered when undertaking phasing out commissioning tests.

9.17 Current transformers

9.17.1 Current transformers – overview

As stated in Section 3.4.7, CTs are used to transform the high levels of HV network current down to lower levels of current suitable for connecting to metering, protection relays and control equipment, etc. CTs differ from both power transformers and VTs in as much as they are 'series connected', the latter two being parallel connected. This has a significant effect on performance characteristics and design considerations.

9.17.2 Current transformer ratings

CTs usually have two ratings, which are as follows:

1. **Rated continuous thermal current**
   This refers to the maximum continuous primary and secondary current that can be tolerated without damage. This current is not always one and the same as the CT ratio, particularly at the higher voltages. For example, a 400-kV network CT may be described as 4,000/2, i.e. a rated continuous primary current of 4,000 A and secondary current of 2 A, but the turns ratio is 2,000/1.
   Most CT secondary current ratings are either 5 or 1 A. The 5 A tends to be used at the lower network voltages essentially for CT cost reasons.
2. **Rated short-time current**
   This is the maximum short-circuit current which the CT can carry for a stipulated duration, without damage. The duration is usually 1 s at 400 and 275 kV network voltages and 3 s for all lower voltages.

9.17.3 Current transformer accuracy

IEC 60044, instrument transformers, defines a range of accuracy classes for CTs, broadly categorised into the following applications:

1. Measurement application
2. General purpose protection application
3. Special purpose application – usually protection

1. **Measurement CTs**
   Measurement (metering) CTs are required to maintain specified accuracy up to 120% of rated continuous current, providing the burden does not exceed the stipulated maximum. Measurement CTs are specified in terms of
   (i) Ratio
   (ii) Rated VA output (which limits the maximum secondary burden)
   (iii) Accuracy class.
   Accuracy classes range from 0.1 to 0.5 where the number defines the plus or minus per cent error, e.g. an 0.5 accuracy class for a CT with a 5-A rated secondary should have a maximum error which does not exceed ±0.5% of 5 A, i.e. ±25 mA. The accuracy class also stipulates a maximum phase angle error.
2. **General purpose protection CTs**
   These CTs are connected to protection relays where greater errors can be tolerated (e.g. IDMTL overcurrent, instantaneous overcurrent, biased differential relays, etc.) in terms of operating quantities, or the ability of a relay to perform when the CT is subject to partial saturation. These CTs are specified in terms of
   (i) Ratio
   (ii) Class (specified as either 5P or 10P)
   (iii) Rated accuracy limit factor
   (iv) Rated VA output
   With reference to the category of class, above, 5P refers to a maximum allowed current error of 5% and similarly 10P to 10%.

   The 'rated accuracy limit factor' is a number which if exceeded would result in the 5P or 10P category errors being exceeded. It is given by:

$$\text{Rated accuracy limit factor} = \frac{\text{Accuracy limit primary current}}{\text{Rated primary current}}$$

   where the 'accuracy limit primary current' is that primary current which if exceeded would result in a loss of the class of accuracy. Within this context, the rated accuracy limit factors have been standardised and include 5, 10, 15, 20 and 30.

   Example:
   Suppose a CT was specified as:
   (i) Ratio = 600/1 (i.e. rated continuous primary current = 600 A, and rated secondary current = 1 A)
   (ii) Class = 5P
   (iii) Rated accuracy limit factor = 10
   (iv) Rated output = 30 VA
   From the above expression, the 'accuracy limit primary current' is equal to the 'rated accuracy limit factor', multiplied by the 'rated primary current' (i.e. 10 × 600 = 6,000 A). Thus, if the CT primary current does not exceed 6,000 A, the maximum allowed error of 5% (of the class 5P CT) would not be exceeded.

Now the reason the CT error commences to increase is that the CT core commences to saturate, which in turn implies that the voltage across both the (parallel connected) magnetising impedance and burden has reached its maximum value consistent with maintaining the accuracy class. If the CT burden is reduced, a higher 'accuracy limit primary current' could be tolerated before the maximum allowed error is exceeded. With this context, the rated burden Z is derived from the relationship.

$$\text{CT rated output (i.e. 30 VA)} = [\text{CT rated secondary current (i.e. 1 A)}]^2 \times Z$$

3. **Special purpose application – protection CTs**
These CTs are termed 'class PX' (formerly just class X). They are required when either CTs need to be precisely matched, or that errors arising from CT saturation need to be avoided.
Section 3.8.2 shows that a network subject to a short circuit results in a fault current that contains a transient DC component. This unidirectional current has the potential to drive the CT magnetic core into saturation, resulting in increased current flowing into the CT magnetising impedance and reduced current flowing out of the CT secondary winding and into the burden (i.e. relay). It can be shown that if the knee point V_K of the CT (see Chapter 15) exceeds the value derived in the relationship below, then CT saturation is avoided:

$$V_K = \frac{I}{N}\left(\frac{X}{R} + 1\right)(R_C + R_B) \tag{9.4}$$

where in the above I is the primary (fault) current, N is the CT turns ratio, X and R are the reactance and resistance of the network between voltage source and fault and R_C and R_B are the CT secondary wiring resistance, and relay burden, respectively.
Class PX CTs are specified in terms of:
(i) Turns ratio
(ii) Rated primary current
(iii) CT secondary rated knee point voltage
(iv) CT magnetising current at rated knee point voltage
(v) CT secondary winding resistance.

Class PX CTs are usually installed at the higher network voltages where the X/R (reactance to resistance) ratio of the network is relatively high, and where some protection applications do require closely matched CTs. They are usually categorised into two types, A and B. Type A is designed to transform accurately without saturation up to maximum network fault current (with due regard to the network X/R ratio), while type B, a less costly design, permits a degree of saturation, and as such has a lower knee point than type A.

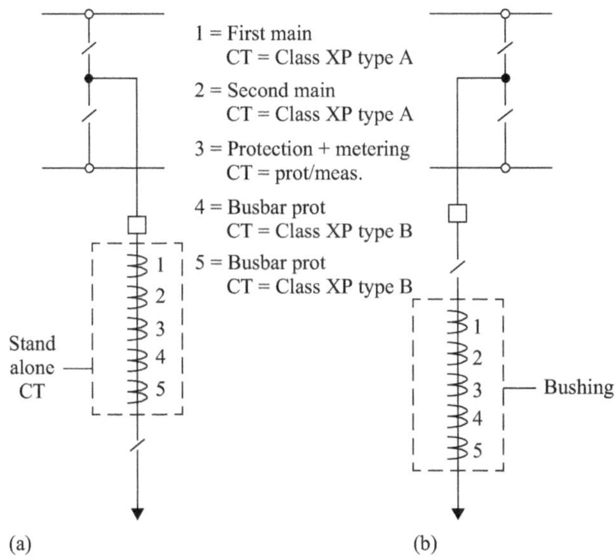

Figure 9.27 Feeder circuit – CT locations – AIS substation: (a) AIS outdoor and (b) AIS indoor

9.17.4 Current transformer – locations

CTs are usually located in close physical proximity to CBs.

With reference to Figure 9.3(a), CTs associated with distribution voltage metalclad switchgear would usually be located in the feeder chamber. As with all CTs, the busbar protection CT and feeder protection CT should be positioned such that their zones of protection overlap.

AIS substations with CBs as illustrated in Figure 9.4(b) and (c) would usually have their CTs positioned in the CB bushings. AIS substations with CBs as illustrated in Figure 9.4(d) would have stand-alone CTs, comprising one CT (stack) per phase. These are usually termed post-type CTs. Some AIS substations are located in-doors and the circuit transition from outdoor to indoor is usually via bushings, in which the CTs are located, see Figure 9.27.

GIS substations with switchgear shown in Figure 9.4(e) would have their CTs located in the GIS housing.

CTs may also be located in transformer bushings or cable slip over CTs located adjacent to cable terminations.

By way of example, Figure 9.27 illustrates CT arrangements for a transmission network feeder circuit, for the instance of an outdoor and indoor AIS substation, respectively.

Chapter 10
Protection and control systems design

10.1 Introduction

Protection systems automatically detect and remove faults from the power system. Equally important are control systems which provide operational information and enable the manual and automatic operation and reconfiguration of the power system. Both are critical to the ongoing functioning and performance of the power system and ensure that it is retained within specified design limits.

With reference to power system construction, it is worthy of note that although the cost of protection and control equipment is much lower than that for the high-voltage (HV) equipment (typically 10% for a new substation), the associated engineering time is much greater (typically 45% for a new substation). This is a reflection both of the volume of equipment associated with protection and control systems and the relative complexity – particularly arising from the high level of equipment interconnectivity.

This chapter will summarise the fundamentals of protection and control systems design and consider the practical application with reference to the requirements of power network construction. The following will be examined:

- Protection systems (Sections 10.2–10.14)
- Protection application (Section 10.15)
- Protection and control settings (Section 10.16)
- Control systems (Sections 10.17–10.22)
- Protection and control accommodation (Section 10.23)
- Protection and control asset replacement (Section 10.24)
- Battery systems and DC supplies (Section 10.25)

10.2 Protection types and operating characteristics

10.2.1 Protection types – introduction

Protection technology is a relatively large subject with a wide range of protection types and operating characteristics. The following sections will limit the discussion to the fundamentals of the most common types of protection, as installed both on the UK power networks and on many networks worldwide. In summary, the following will be examined:

- Protection terminology
- Protection purpose and objectives

- Protection historical development
- Overcurrent (OC) and earth-fault (EF) protection
- Directional overcurrent protection
- High-impedance circulating current protection
- Transformer protection
- Feeder-unit protection
- Power-line carrier protection
- Numeric unit feeder protection
- Distance protection
- Busbar (BB) protection
- Circuit breaker fail protection
- Protection, control and communication channels
- Inter-tripping

10.2.1.1 Protection terminology

The following terms are generally used:

1. **Protection relay**
   A measuring device that detects the existence of a power system fault (e.g. an EF relay)
2. **Protection equipment**
   Protection relays and associated equipment (e.g. fuses, links, test blocks, resistors, trip relays, etc.)
3. **Protection system**
   Protection equipment required to achieve a specific function (e.g. blocked distance protection)
4. **Protection scheme**
   Protection equipment and associated CTs and VTs which enact a specific function [e.g. a first main protection (MP1) scheme]
5. **Protection reliability**
   The ability of a relay to operate when required (often following long periods between faults when the relay is dormant). Reliability is a function of two qualities, which are as follows:
   (i) Protection dependability – the certainty of operation when a fault arises
   (ii) Protection security – the certainty of a relay not operating when subject to currents/voltages for which it is not required to operate (e.g. an out-of-zone fault)
6. **Protection discrimination**
   The characteristic of disconnecting the minimum amount of network to remove the fault from the power system
7. **Protection sensitivity**
   The characteristic whereby a relay operates when the fault condition just exceeds the setting. It additionally defines the minimum setting level to which the relay can be set to detect the fault in question
8. **Protection speed**
   The characteristic of a fast operating time to clear the fault rapidly, both to avoid damage to equipment and minimise power system instability

9. **Protection settability**
   A relay with a range of settings that allows a setting close to the required operating quantity to be chosen.

It is worthy of note that in practice terms 1–4 are often used interchangeably (and therefore imprecisely).

10.2.1.2 Protection purpose and objectives

The purpose of power system protection is to detect a fault on the power system and subsequently instruct a circuit breaker(s) to trip and therefore isolate the faulty item of equipment. This process needs to be accomplished as fast as possible to achieve the following objectives:

- The prevention of power system instability – i.e. the prevention of loss of synchronism of either generators, or parts of the power system, which leads to power system breakup and possible shut down
- Prevention of unnecessary damage to the equipment subject to fault
- Prevention of a risk of damage to healthy equipment– by carrying fault current whose magnitude and duration is in excess of equipment rating
- Prevention of loss of supply to electricity consumers
- Prevention of financial loss arising from both damage to equipment and loss of consumers.

It is worthy of note that protection systems are not legally required to operate at a level of sensitivity, and speed of operation, that would safeguard personnel from the danger of electrical shock. This requirement is impractical from a protection design perspective – although a degree of protection will implicitly be provided.

10.2.1.3 Protection system fundamentals

Figure 10.1 illustrates the fundamentals of a protection system. Salient points are as follows:

- Inputs into the protection relay comprise CTs and/or VTs and in some instances remote-end protection signals.
- There are two fundamental categories of protection – 'main protection' and 'backup protection'.

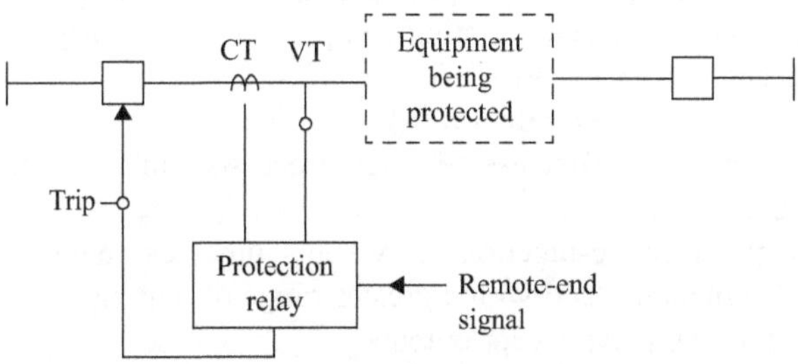

Figure 10.1 Protection system – fundamentals

- Main protection is designed to operate for a fault on the equipment that it is protecting.
- Backup protection is designed to operate if the main protection fails to operate, and after a delay which allows time for the main protection to operate. There are two classes of backup protection: local and remote.
- Local backup protection operates for failure of the associated main protection, i.e. to clear a fault on the local circuit.
- Remote backup protection operates for failure of the main protection on remote circuits. Usually the same relay provides both local and remote backup protection.
- Backup protection is usually provided by inverse definite minimum time lag (IDMTL) OC and EF protection and Zones 2 and 3 of distance protection.

10.2.1.4 Protection relay historical development

Dating from *ca.* the mid-1970s, protection relays have been subject to a number of evolutionary changes, whilst maintaining the core functionality. Since many of the older relays are still in operational service, it is instructive to understand these changes – which in summary are as follows:

1. **Electromechanical relays (up to the 1970s)**
   The earliest protection relays employed 'electromechanical' technology utilising magnetic attracted armatures, induction discs and mixing and summation transformers. The features of this technology were:
   (i) Single-function relays
   (ii) Operation was via moving parts
   (iii) Settings applied manually, on site
   (iv) Relatively large in physical size
   (v) Relatively simple technology to understand – and understood by most engineers
   (vi) Periodic maintenance/testing required
   (vii) Lifespan of at least 30 years.
2. **Static relays (from the early 1970s)**
   These comprised analogue devices utilising discrete electronic components such as transistors. The main features were as follows:
   (i) Single-function relays
   (ii) Virtually no moving parts (apart from outputs)
   (iii) Settings applied manually, on site
   (iv) More physically compact than electromagnetic relays
   (v) More complex to understand than electromagnetic relays
   (vi) Reduced maintenance/testing compared to electromagnetic relays
   (vii) Lifespan up to 25 years.
3. **Digital relays (from the mid-1980s)**
   These comprised the first use of microprocessors in protection relays. The main features were:
   (i) Mostly a single-function relay, but some less complex relays were multifunction – and with a greater range of options
   (ii) No moving parts except outputs
   (iii) Settings undertaken manually, on site, via key pad/LED interface. Number of settings greatly expanded

- (iv) More physically compact than static relays
- (v) Increased complexity and more difficult to understand than static relays
- (vi) Almost maintenance free – self-monitoring
- (vii) Lifespan 15 years.

4. **Programmable numeric multifunctional relay (from *ca.* 2000)**
   These are programmable microprocessor-based relays using algorithms to define functionality. The main features are:
   - (i) Capable of multiple functions e.g. main protection, backup protection and fault recorder in a single relay
   - (ii) All inputs arranged to be digital via A/D convertors
   - (iii) Settings input by file transfer, usually within the factory. Number of settings greatly expanded requiring (ideally) a software-based process for managing settings and settings changes
   - (iv) Physically very compact
   - (v) Further increase in complexity requiring specialist expertise
   - (vi) Lifespan 10–15 years

   NB: Figure 10.2 illustrates the key elements of a numerical multifunctional relay.

Present day numerical (i.e. digitally based) relay technology has become increasingly complex, with the size of a manual/handbook describing relay performance hugely increased from that of earlier technologies. The challenge is therefore how to simply explain the key operating principles. Within this context, the functionality of most numerical protection systems is still based upon that of its electromechanical predecessor, and as such protection performance characteristics are often best and more easily described with reference to this simpler technology.

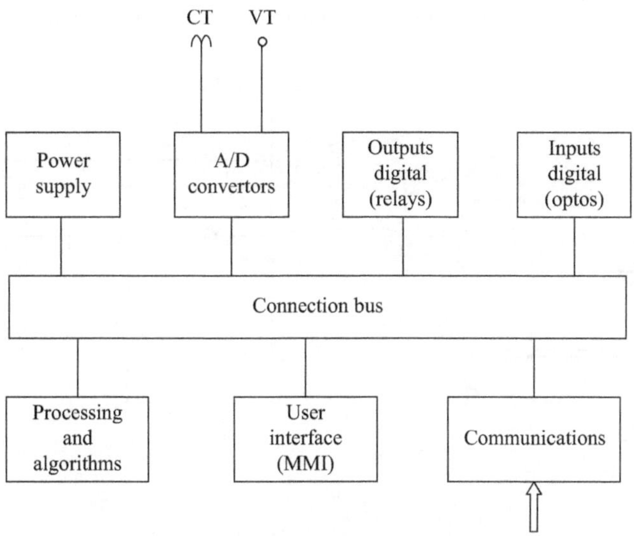

Figure 10.2 Numerical multifunctional relay – key elements

This approach will be adopted in the remainder of this chapter, for example protection performance interpreted in terms of relay contact logic and schematic diagrams, and on the basis of a single functional device.

10.3 Overcurrent and earth-fault protection

10.3.1 Overcurrent and earth-fault protection circuitry

OC and EF protection is arguably the oldest, most basic and most commonly applied form of protection. Figure 10.3 illustrates the required connections with salient observations, which are as follows:

1. **Earth-fault relay**
   Figure 10.3(a) shows the connections for an EF relay and the current flow for a red phase to EF. With a three-phase fault, or a phase–phase fault, the currents circulate around the CTs and do not flow through the relay. Within this context, the PPS and NPS currents summate to zero at the star point, and only the ZPS currents, resulting from an EF, flow through the relay. In Figure 10.3(a), the ZPS current per phase is equal to one-third of the secondary current *i*.
2. **Overcurrent and earth-fault relay**
   Figure 10.3(b) shows the usual connections for a combined OC and EF relay, as frequently used in practice. For the phase–phase fault shown, the currents circulate around the yellow and blue OC relays. However, as only one relay is required to operate to disconnect the fault, one of the relays, usually the yellow, is superfluous and often omitted (and as such shown dotted). Again, PPS and NPS currents (as found in three-phase, phase–phase and phase-earth faults) summate to zero at the star point, and only ZPS currents arising from earth-faults flow through the EF relay.

   OC relays are subject to three-phase load currents and therefore require a setting in excess of maximum load current to avoid unwanted operation. It is

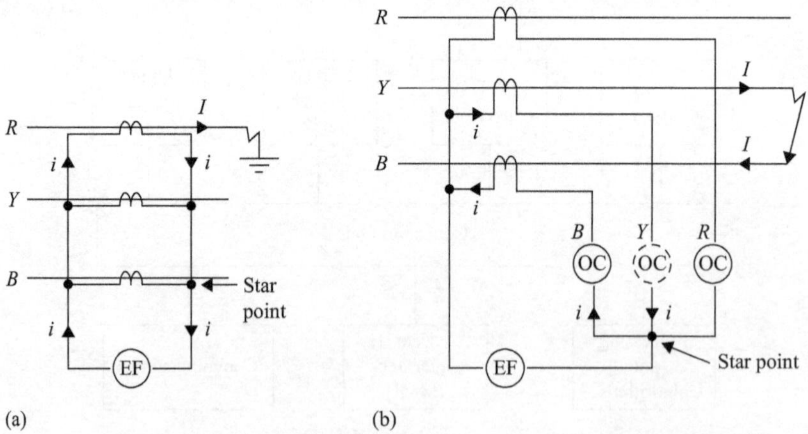

Figure 10.3 Overcurrent and earth-fault relay connections: (a) earth-fault relay and (b) three-phase overcurrent and earth-fault relay

worthy of note that the purpose of an OC relay is to operate and trip a circuit breaker as a result of detection of fault current – and not to prevent marginal circuit overloading.

EF relays are not subject to load current and therefore can be set to a lower (i.e. more sensitive) setting than an OC relay.

Generally, OC and EF relays are used as backup protection – but may be the only form of protection on the lower voltage networks, such as 11 kV.

10.3.2 IDMTL overcurrent and earth-fault relay – operating characteristics

IDMTL (often abbreviated to IDMT) refers to the characteristics of a family of relays whose operating characteristics and origins date back to electromagnetic technology. Figure 10.4 shows the four common operating characteristics (curves) of the IDMT family. It is usual to show the curves on a logarithmic scale. The curves are defined in IEC 60255 and BS 142. Salient observations on Figure 10.4 are as follows:

- The relays have two settings: a current setting (historically termed plug setting or PS), this is the setting at which the relay commences to operate (and below which it will not operate); and a time setting expressed in terms of a 'time multiplier' TM (or sometimes termed 'time multiplier setting' (TMS)).

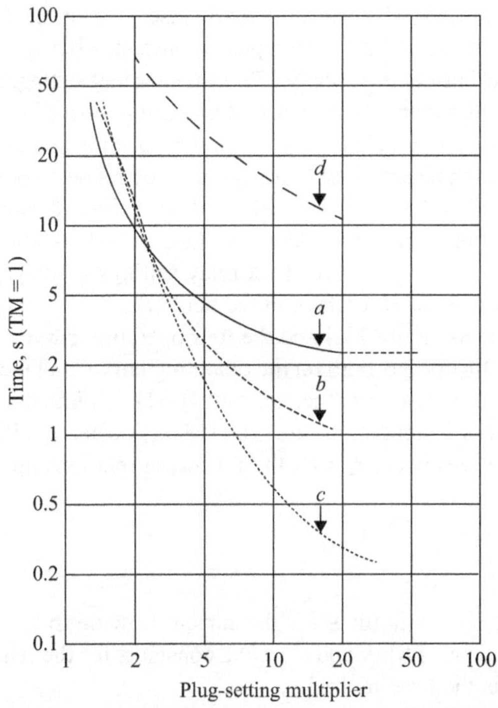

Figure 10.4 IDMTL family of curves: (a) standard IDMT (3/10), (b) very inverse (1.5/10), (c) extremely inverse (0.6/10) and (d) long-time delay (30/5)

- The speed of operation of the relay is dependent upon the multiple by which the current into the relay exceeds the current setting. This is termed the 'PS multiplier' PSM which is given by:

$$\text{PSM} = \frac{\text{Current flowing in relay}}{\text{Current setting of relay}}$$

- Thus, if a relay has a current setting (PS) of 1 A and is subject to a current of 10 A, then the PSM is 10. If the ratio of the CT feeding the relay is 1,000/1, then expressing these values in primary currents, the current setting of the relay is 1 kA, and the current flowing into the relay is equivalent to 10 kA, and again the PSM is equal to 10. With numerical relays, the current setting, in the relay (PS), may be expressed in either primary or secondary values, if the former then the CT ratio is also included as a setting (in the relay).
- The relay operating curves are illustrated in Figure 10.4 and are based upon a TM of 1.0. However, the TM is a variable setting usually between 0.05 and 1.0. Therefore, for example, if a TM of say, 0.4, was selected then the operating times shown in Figure 10.4 would be multiplied by 0.4.
- The relays traditionally (as determined by electromagnetic technology) were specified as having a minimum pickup current of 100% of the current setting (PS) – and a maximum pickup current (i.e. relay creep) of 130% of the current setting; however, with modern numerical relays, the maximum pickup current can be as low as 105%.
- The relay traditionally also had a minimum reset current (i.e. following operation or partial operation) of 70% of the pickup current – but for modern relays, the minimum reset current is typically 95% of the current setting. The reset current is important in determining the chosen current setting (PS), which must ensure that a partly operated relay is able to reset when subsequently subject to maximum load current (following clearance of a through fault by relays nearer the fault).
- Traditionally, IDMT relays also had fixed reset times following full operation and fault current cessation – however, modern relays allow this time to be selectable. It is usually important for relay settings grading for the reset times to be consistent across all relays in the network.
- With reference to Figure 10.4 and the four operating curves, the figures shown in brackets define points both on the operating curve, and the name of the relay characteristic. For example, the standard IDMT is termed a 3/10 relay, which corresponds to a 3-s operating time (at TM = 1) when the PSM is equal to 10.
- Each of the curves is defined by the following relationship (see IEC60255):

$$t = \frac{K}{\left(\frac{I}{I_S}\right)^a - 1} \times \text{TM} \tag{10.1}$$

where t is the operating time, I is the current flowing in relay, I_S is the current setting of the relay (PS), K and a are the constants for the relay characteristic in use and TM is the time multiplier.

The standard IDMT, 3/10, characteristic has values of $K = 0.14$ and $a = 0.02$.

10.3.3 IDMTL relay – settings and grading

Settings applied to IDMT relays in practice are generally tailored to the individual application (e.g. feeder protection, transformer protection, etc. and the network voltage in question), and examples will be covered later. However, virtually all IDMT relays have a settings requirement in common, and that is the requirement to 'grade' with other IDMT relays on the network, i.e. the quality of discrimination.

Figure 10.5 illustrates the principle of IDMT grading, using standard (3/10) relays. Salient points are as follows:

- Figure 10.5(a) illustrates the part of the network to be examined, assume that a three-phase fault takes place at the position shown (close to substation B), with a fault current of 9,000 A.

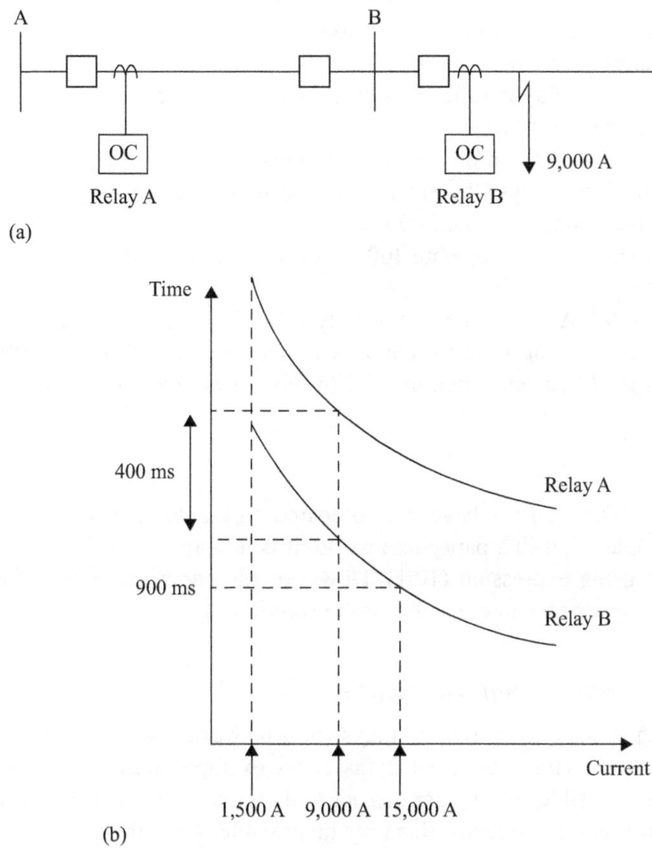

Figure 10.5 Overcurrent grading criterion – example: (a) network to be analysed and (b) grading curves on log paper

- It is assumed that both circuits are subject to a maximum load current of 1,000 A, and the PS of the relay is therefore set to the next highest setting, assume this to be equivalent to 1,500 A. NB: This setting also ensures that the relay will reset following clearance of through fault current – assuming a reset current in this instance of 95% current setting.
- Figure 10.5(b) illustrates the relay curves on a chart (using log–log paper – but not drawn to scale). In the first instance, the curves are drawn such that at ten times the PS (i.e. 15,000 A) the operating time is 3 s (not shown), i.e. the 3/10 relay characteristic.
- Suppose that the TM of relay B is set to grade with IDMT relays at substation C and beyond, and that this requires a TM of 0.3. This would give an operating time of 900 ms (i.e. 0.3 × 3.0 s) at the PS of 15,000 A. The operating curve for relay B is drawn through this point (i.e. PS = 1,500 A, Time = 900 ms).
- The grading criterion between relays A and B to be satisfied is that at the maximum fault current of 9,000 A, the two relays time grade. The traditional time grading is usually 400 ms, comprising the following:
  - Circuit breaker B opening time = 100 ms
  - Relay error = 150 ms
  - Overshoot of relay A (after current cessation) = 50 ms
  - Safety margin = 100 ms
  - Total time discrimination margin = 400 ms

  NB: With modern relays, the errors and overshoot are not as great and the above may be reduced to, say, 300 ms.
- Relay A is therefore set to operate 400 ms slower than relay B at a fault current of 9,000 A.
- The TM of relay A can be determined from the 3/10 operating criterion. By obtaining the operating time for curve A at ten times the PS (i.e. 15,000 A), assume this is '*T*' seconds then the TM of relay A is given by:

  $$\mathrm{TM} = \frac{T}{3}$$
- Historically, IDMT curves have been prepared on log–log paper as illustrated in this example – but in a paper-less world, it is now more usual to determine the settings using expression (10.1). However, the paper solution provides a more easily understandable version of the solution.

10.3.4 IDMT relay – natural grading

It is often advantageous to set IDMT relays (in a network), which need to grade with each other, to the same standard settings. In the example shown in Figure 10.5, this would not be possible (if they are subject to the same fault current). However, in a network with multiple infeeds, this may be possible, since the relays closer to the fault are subject to greater fault current, and grading is facilitated by the 'natural grading' provided by the network. Figure 10.6 illustrates typical, but simplified, network arrangements that facilitate natural grading.

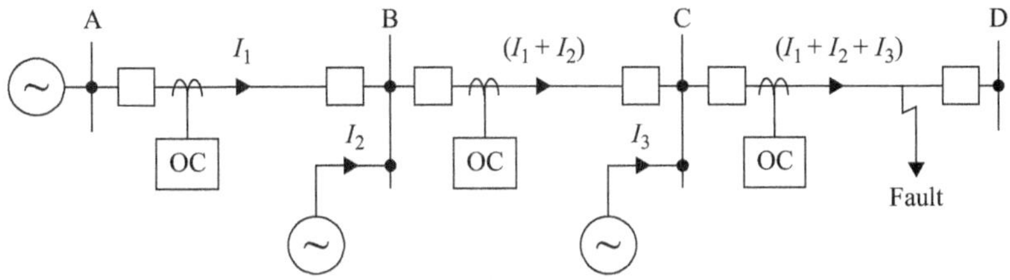

Figure 10.6 Network facilitating natural grading

10.3.5 Instantaneous overcurrent relay

Instantaneous OC relays are operated by current only and have an instantaneous operating time (e.g. 10–20 ms). They are generally utilised as follows:

- In an OC check relay mode, where the contacts are in series with another relay which operates at a lower setting, e.g. pilot-wire protection.
- In conjunction with two-stage OC transformer protection, see Section 10.6.4.
- As a high-set OC (HSOC) relay as part of transformer protection, see Section 10.6.6.

Instantaneous OC relays are usually designed to be immune to the transient component of fault currents.

10.4 Directional overcurrent protection

10.4.1 Directional relays – fundamentals

Directional OC (DOC) relays were first developed when the dominant technology was electromagnetic, and many of the principles developed at that time remain to the present day.

The directional relay which is most in use (as found on the 132 kV and lower voltage networks) is that of the '90° connection – 45° characteristic'. Figure 10.7 illustrates the principle – salient aspects of which are as follows:

- The input to each directional relay comprises the relevant phase current (e.g. *R*) and a voltage which lags by 90° (e.g. the red relay is fed with the *Y–B* voltage).
- The relay characteristic angle – traditionally termed the 'maximum torque angle' (MTA) is arranged to lead the input voltage (i.e. *Y–B*) by 45°.
- The operating zone of the relay is that in which the current is within, plus or minus 90° of the MTA.
- The operating zone ensures positive operation of the relay for any red phase to EF current which is in front of the relay (which usually lags the red phase voltage by at least 45°). The operating time is instantaneous (typically 10–20 ms).
- More detailed analysis would show that the relay also adequately detects phase-to-phase and three-phase faults in front of the relay – and for most

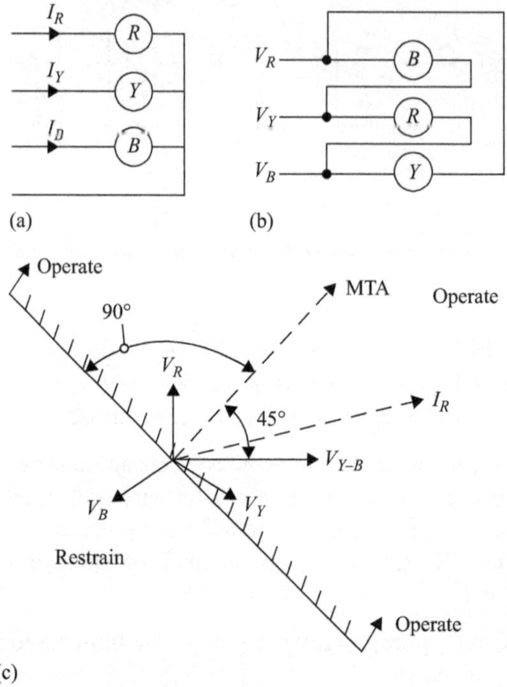

Figure 10.7 Directional relay – 90° connection 45° MTA: (a) current connections, (b) voltage connections and (c) operating characteristic

practical faults, the possibility of mal-operation is virtually non-existent. The relay will of course not operate for a three-phase fault close to the VT terminals, when all three voltages collapse to zero. Fault clearance would then rely on backup protection further back in the network.

10.4.2 Directional overcurrent relay

Figure 10.8 illustrates the essential features of a DOC relay, consisting of the combination of a directional relay and an IDMT relay. Such relays are widely used on the 132 kV and lower voltage networks, with typical applications discussed later in this chapter.

10.4.3 Directional earth-fault relay

Directional EF relays are also available with a residual current input, as shown in Figure 10.3(a), and a residual voltage input derived from a VT secondary whose windings form an open-delta configuration. The VT must be of the five limbed variety (which is expensive) to accurately transform zero sequence voltages. Directional EF relays are used less frequently in practice in the United Kingdom (largely since the DOC relay adequately detects EFs).

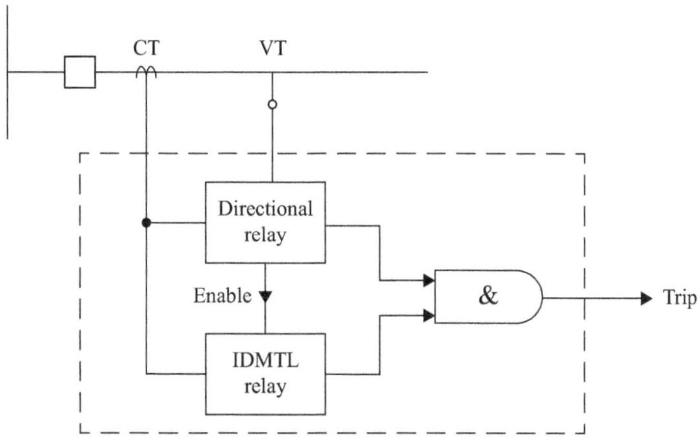

Figure 10.8 Directional overcurrent relay – block diagram

10.5 High-impedance circulating current protection

10.5.1 High-impedance circulating current – background

In the last half of the twentieth century, unit protection of substation equipment was undertaken in many parts of the world, with few exceptions, almost entirely by 'high-impedance circulating current protection'. From the mid-1990s onwards, this type of protection began to be progressively replaced by low-impedance types of circulating current protection, largely for the following reasons:

- The introduction of microprocessor-based relays which not only facilitated the design of low-impedance relays but also provided greater functionality, flexibility and continuous self-monitoring. For some applications, the number of CTs, auxiliary contacts, volume of wiring and relay panel space is also reduced. It is worthy of note that some microprocessor-based relays also operate on a high-impedance scheme principle.
- Virtually maintenance free.
- Cost reduction, although in the early relays this was marginal.

Notwithstanding the above benefits, some power network companies retain high-impedance protection as the protection of choice for some applications, for the reasons given below:

- Simpler technology requiring less expertise
- Good track record of reliability
- Longer service life (typically up to 40 years – with maintenance) than microprocessor-based systems.

The following sections will examine the principles and performance characteristics of high-impedance protection.

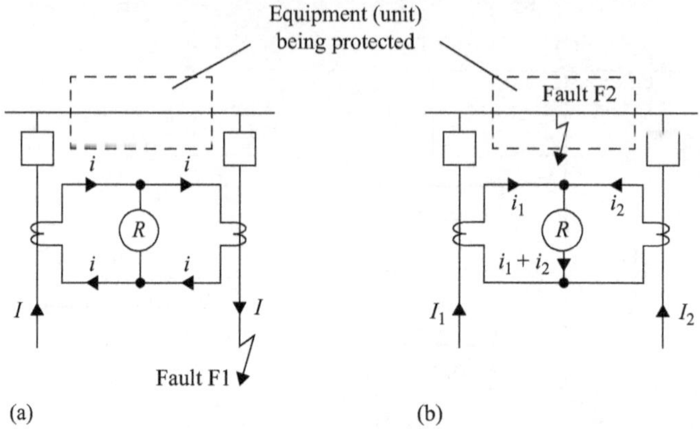

Figure 10.9 Circulating current protection – principles: (a) external fault (stable) and (b) internal fault (operation)

10.5.2 Circulating current protection – principles

Consider the simple arrangement shown in Figure 10.9 where the protected zone of the equipment is as shown and electrically continuous.

Figure 10.9(a) shows the instance of an external, and out of zone, fault 'F1', where the secondary currents are equal in magnitude and phase, and (by Kirchhoff's law) no current flows into the relay 'R'. Figure 10.9(b) shows the instance of an internal (and in zone) fault 'F2', where the secondary currents are not equal in phase (although they may be equal in magnitude), and (by Kirchhoff's law) are additive, and flow into the relay to cause operation.

In practice, primary fault currents contain DC transient components (see Section 3.8.2) which can cause partial CT magnetic core saturation. This results in imperfect CT current transformation, the effect of which may cause mismatch of the secondary currents, which for the external fault shown in Figure 10.9(a) causes spill current through the relay. Measures must therefore be taken to make the relay stable against such spill currents. Two options which exist are as follows:

- The use of a high-impedance protection scheme where the setting of the relay exceeds that of the (effect of) spill current.
- The use of a low-impedance protection scheme where the circulating currents around the CT loop are used to bias the relay towards stability.

10.5.3 High-impedance circulating current – stability voltage

Figure 10.10 illustrates circulating current protection circuitry, which includes the CT equivalent circuit, in which:

Z_A, Z_B = CT magnetisation impedances
R_A, R_B = CT wiring resistance
R_1, R_2 = interconnecting wiring resistance
R_R = relay resistance
R = relay of negligible impedance

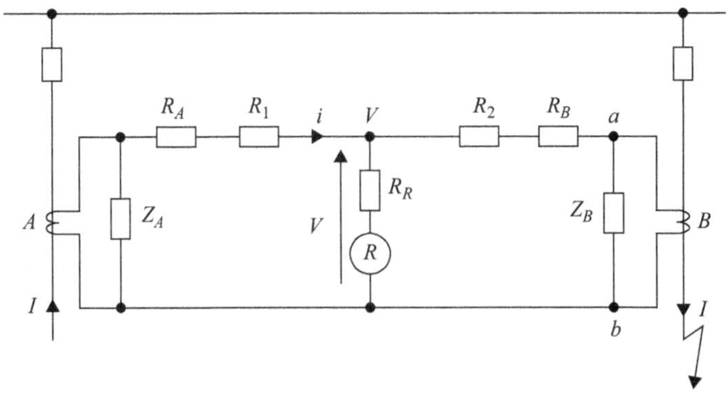

Figure 10.10 Derivation of stability voltage

The most onerous case of CT mismatch arises for the external fault shown in Figure 10.10, when one of the CTs is totally saturated and the other totally unsaturated.

Consider CT B to be totally saturated, the effect of which is to cause the equivalent of a short circuit across points *a*–*b* (i.e. the magnetising impedance appears as a short circuit). Also consider CT A to be totally unsaturated with perfect transformation of primary to secondary current. As a result, the current i arising from CT A circulates around the loop which includes resistors R_2 and R_B and the short-circuit *a*–*b*. In doing so, a voltage V is developed across the resistance R_R (which for the moment is considered to be a high resistance such that the current through it is negligible) and the relay R, such that:

$$V = i(R_2 + R_B)$$

Suppose for example that i was equal to 25 A, and that $(R_2 + R_B)$ was equal to 8 Ω then:

$$V = 25\ \text{A} \times 8\ \Omega = 200\ \text{V}$$

Therefore, if the relay operating current was set to a value such that when 200 V was imposed across the resistor and relay series circuit – it did not operate – relay stability would be achieved for this extreme saturation condition. The voltage V is termed the ‘relay stability voltage’. However, it is usual to set the relay to be immune to a higher voltage than the stability voltage (usually the next highest setting of the relay), say 225 V – this is termed the ‘relay setting voltage’ V_S.

10.5.4 High-impedance circulating current – primary operating current

With reference to Figure 10.11, the total operating current, i_P at the point at which the relay operates (i.e. when the circuitry is subject to the setting voltage, V_S), is that flowing into the parallel combination of the relay and magnetising impedances – i.e.:

$$i_P = i_A + i_B + i_R$$

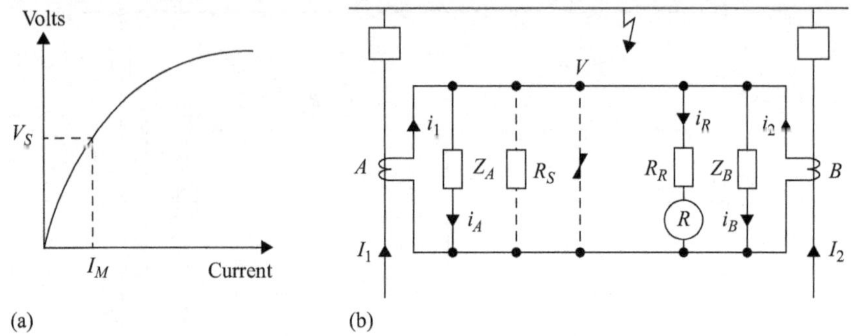

Figure 10.11 Derivation of operating currents: (a) magnetisation curve and (b) relevant circuitry

Figure 10.11(a) shows the magnetisation curve for both of the CTs – and at the setting voltage V_S the corresponding magnetising current is I_M. At V_S equal to 225 V (as determined above), assume I_M has a value of 20 mA (a typical value), then:

$$i_P = (2 \times I_M) + I_R = (2 \times 20\,\text{mA}) + I_R \tag{10.2}$$

At this point, the flexibility of the scheme is demonstrated, since the magnitude of i_R, and hence i_P can be varied by varying the magnitude of R_R. This is perfectly acceptable as long as the relay setting is such that it operates when the setting voltage V_S is applied across the circuitry. Suppose the operating current i_R of relay R is set to 40 mA (typical value), then the value of R_R at the stability voltage is given by:

$$R_R = \frac{225\ \text{V}}{40\ \text{mA}} = 5{,}625\ \Omega$$

And with reference to expression (10.2), the total operating current is given by:

$$i_p = (2 \times 20\,\text{mA}) + 40\,\text{mA} = 80\,\text{mA}$$

If the CT ratio was, say, 1,000/1, then the primary operating current (POC) is given by:

$$\text{POC} = 80\,\text{mA} \times 1{,}000 = 80\,\text{A}$$

With reference to the above, the following is worthy of note:

- The currents through the resistor and magnetising impedances are quadrature to each other and theoretically should be subject to vector addition. However, it is common practice to add them directly. This gives a slightly higher operating current than is the reality, i.e. errs on the side of caution.
- When deriving the stability voltage, the highest through fault current, often termed the 'rated stability limit' current should be considered.
- High-impedance protection at first sight seems counter-intuitive since in the operation mode the CTs are connected in opposition which causes the voltage applied to the whole of the secondary circuit to rise to a high value (as if

the CTs were subject to an open circuit). However, with safeguards, this is a successful type of protection.

- The operating current, as illustrated above, can be quite low, e.g. 80 A. However, the fault current can be thousands of amps. With such high currents, the CT's are driven into saturation for long periods of each current cycle (i.e. 50 Hz cycle). To minimise saturation and to ensure there is sufficient voltage to cause relay operation, it is usual to ensure that the knee point of the CTs is at least twice that of the stability voltage.
- To limit the HVs imposed upon the circuitry, it is usual to fit metrosils, as shown dotted in Figure 10.11(b). The current taken by the metrosil, at the setting voltage, forms part of the total operating current.
- The operating current of the system can be further increased by connecting shunt resistors, termed 'setting resistors', R_S, as shown dotted in Figure 10.11(b).
- In the example shown above, the relay arrangement is that of a current operated relay in series with a relatively high resistance R_R (termed the stability resistor). Relays calibrated directly in voltage may also be used – and in such instances, the operating current at the setting voltage will need to be known to determine the total operating current.

10.5.5 High-impedance circulating current – applications

High-impedance circulating current protection has been very widely used both in the United Kingdom and in many other countries in the following applications:

- BB protection
- Mesh corner protection
- HV connection (HV Conn) and LV connection (LV Conn) protection
- Autotransformer overall protection
- Reactor protection
- Restricted EF (REF) and balanced EF (BEF) protection.

The above will be examined in subsequent sections of this chapter.

10.6 Transformer protection

10.6.1 Two winding transformer – biased differential protection

High-impedance circulating current protection cannot be successfully applied to two winding transformers, and traditionally this has been achieved by low-impedance circulating current protection in the form of 'biased differential protection'. Due to the complexity and cost of this protection, it has been most commonly applied to 132/33 kV two winding transformers (i.e. the higher network voltages) and to generator transformers.

Figure 10.12 illustrates the traditional biased differential scheme circuitry. A modern microprocessor-based relay would comprise a software equivalent of this circuitry. Figure 10.13 illustrates both the essential features of the biased differential relay itself, together with a typical operate/retrain relay characteristic.

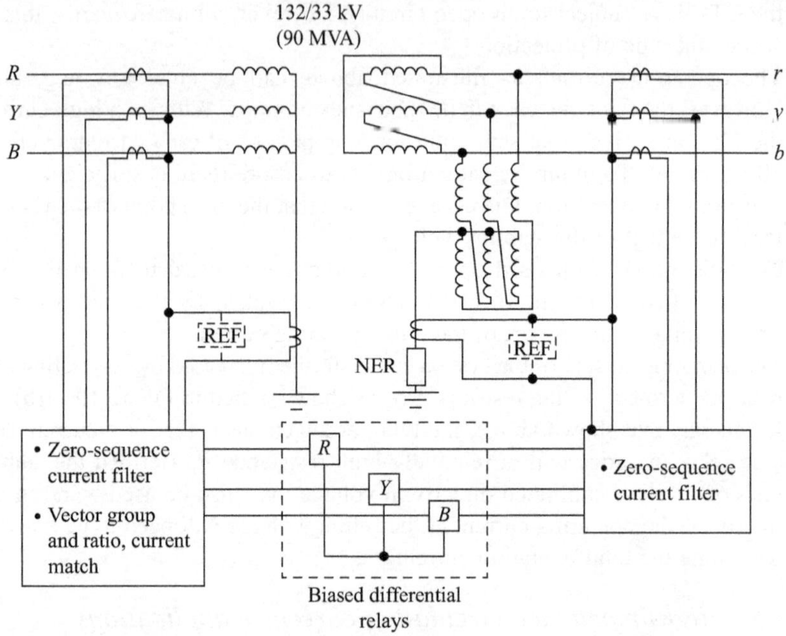

Figure 10.12 Biased differential protection – key features

The setting characteristic, at the boundary of operation, comprises the following relationship:

$$I_D = KI_B + I_S \tag{10.3}$$

where I_D is the 'differential current', I_B is the 'bias current', K is the 'bias slope' and I_S is equal to a minimum operating quantity equivalent to relay friction in older relays. With increasing levels of I_D, the relay is in the operate zone.

The bias current I_B comprises the average of the current in each half of the bias windings 'B', such that:

$$I_B = \frac{I_L + I_D}{2} + \frac{I_L}{2} = I_L + \frac{I_D}{2} \tag{10.4}$$

Inserting expression (10.4) into (10.3):

$$I_D = K\left(I_L + \frac{I_D}{2}\right) + I_S \tag{10.5}$$

Rearranging:

$$I_D = \frac{KI_L + I_S}{1 - 0.5K} \tag{10.6}$$

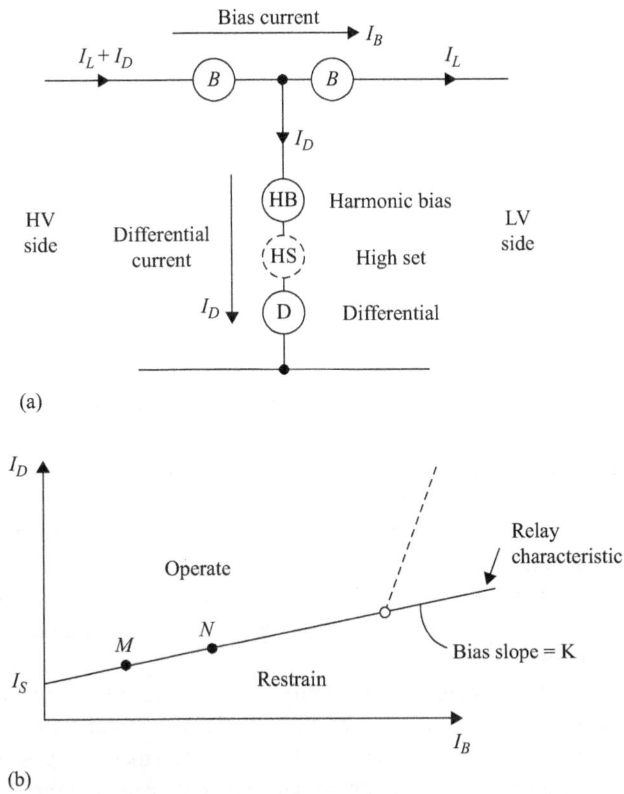

Figure 10.13 Biased differential relay: (a) biased differential relay and (b) operate/restrain characteristic

For the instance of $I_L = 0$ expression (10.6) reduces to:

$$I_D = \frac{I_S}{1 - 0.5K} \tag{10.7}$$

Expression (10.7) represents the minimum operating current of the biased differential relay.

Salient observations relating to Figures 10.12 and 10.13 are as follows:

1. **Biased differential relay**
   With reference to Figure 10.13(a), the biased differential relay comprises a number of elements, which are as follows:
   (i) The differential element '*D*'
   (ii) The bias element '*B*'
   (iii) The harmonic bias element '*HB*'
   (iv) High-set element '*HS*'

(i) **Differential element**
The differential element contains the out-of-balance current between primary and secondary of the transformer.

(ii) **Bias element**
The bias elements are subject to the bias current(s) and contain a setting which determines the bias slope. The bias slope characteristic (depending upon the manufacturer) may be arranged in stages, with a change in the inclination of the slope (e.g. as shown dotted in Figure 10.13) to compensate for magnetising inrush current (so lessening the case for a separate harmonic bias), or to create extra stability for large through fault currents.

(iii) **Harmonic bias element**
Section 3.8.2 explained that when a transformer is energised there is a large, transient flow of current termed the magnetising inrush current. This current only flows in the winding connected to the source of supply and as such it flows into the differential relay as if it was a transformer fault. Magnetising inrush current contains a large second harmonic component of current, and this component is detected by the harmonic bias element and used to bias the relay (i.e. in addition to I_B) such that the relay remains in the restrain zone of the operating characteristic.

(iv) **High-set element**
An HSOC element is frequently provided to ensure independent fast operation for the instance of high levels of internal fault current – particularly a fault coincident with the transformer being switched in to service and magnetising inrush current is present, and the harmonic bias element is operational (thereby causing restraint).

2. **Tap changer**
There is current balance in the biased differential relay (i.e. zero current in the differential element) at one transformer tap position only, usually the nominal tap. At all other tap positions, a spill current flows in the differential element, as if it was an internal fault current. This spill current is greatest at an extreme tap position, coincident with maximum through fault current flowing through the transformer. The bias slope is arranged such that with this extreme condition, the relay remains stable (i.e. in the restrain zone). Satisfactory bias slopes in practice usually range from 20% to 30%.

3. **Zero-sequence filter**
With reference to Figure 10.12, an EF external to the biased differential zone of protection, on either the LV or HV side of the transformer, results in zero-sequence current flowing into the fault – but not flowing across the power transformer star – delta winding (see Chapter 4). As such to ensure there is current balance in the biased differential relay (i.e. no current in the differential element), the zero-sequence current flow in the relay must mirror that through the transformer and must be filtered out – as illustrated in Figure 10.12. The effect of the zero-sequence filter is to desensitise the relay to EFs and make it predominately a phase–phase fault, or three-phase fault, relay.

4. **Vector group/ratio current match**
   To obtain balance for currents that pass right through the power transformer, the biased differential circuitry, self-evidently, must mirror the ratio and vector group of the power transformer, necessitating the arrangements shown in Figure 10.12.
5. **CT considerations**
   The bias elements of the relay provide stability for CT saturation during external fault conditions. As such biased differential relays are capable of satisfactory performance with CTs of a lower knee-point than those required by high-impedance schemes.
6. **Settings**
   With reference to expression (10.6), the protection is typically required to operate at 10%–60% of the transformer HV winding current rating, coincident with a bias current equal to the full load current rating of the transformer.

10.6.2 Restricted earth-fault protection

REF relays are widely used forms of protection on two winding transformers. They are provided to protect the transformer from winding faults to earth. Figure 10.14 shows the most common applications in the United Kingdom, namely 132/33 and 33/11-kV network transformers. This type of protection is also commonly applied to two winding generator transformers. The term 'restricted' largely originates from the fact that the protection is restricted to a zone of the transformer, i.e. the transformer windings. The term 'BEF' is commonly given to the three-CT arrangement as shown in Figure 10.14(b), although this relay serves the same purpose as the REF relay. Salient observations are as follows:

- The relays are only responsive to ZPS currents.
- Figure 10.12 shows the same relays as those in Figure 10.14 incorporated into a biased differential relay scheme. The principle of operation is the same.
- In Figure 10.14(a), the REF on the HV winding is protecting against a winding fault (F1). This fault results in autotransformer action within the faulted winding. As the fault position moves away from the neutral end of the winding to the line terminal, the current flowing in the neutral decreases and the current supplied from the source increases – the current in the relay being the addition of the two (and equal to the current in the fault current).
- With reference to winding fault position (F2) in Figure 10.14(a), minimum fault current arises when the fault position is at the centre of the winding and increases towards the two ends of the winding. The fault current of course only flows though the CT located in the neutral (for a single-end HV-fed fault). Fault current flows into the 132-kV windings but not into the 132-kV REF since zero-sequence currents do not flow through the power transformer.
- REF relays are more responsive to EFs than the biased differential relay shown in Figure 10.12, since the zero-sequence currents are filtered out of the latter.
- The BEF relay shown in Figure 10.14(b) operates for winding fault positions (F3). Again, the fault currents are a minimum for faults at the centre of the

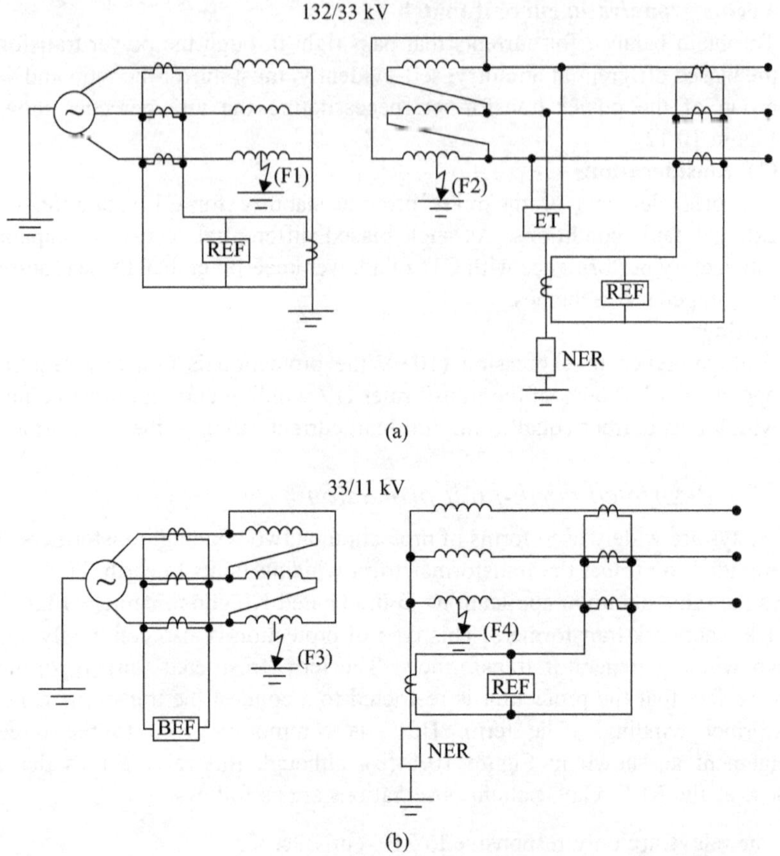

Figure 10.14 Restricted and balanced earth-fault protection – application: (a) REF application to 132/32-kV transformer and (b) BEF and REF application to 33/11-kV transformer

winding. This relay will not detect faults on the LV side of the transformer as the zero-sequence currents circulate around, and are confined to, the delta winding.

- With fault (F4), as the fault moves away from the neutral end of the winding, it increases in magnitude, and only flows through the CT positioned in the neutral (for a single-end HV-fed fault).
- Traditionally, REF (and BEF) schemes have been of the high-impedance type, but more recently biased low impedance relays have been employed.

Settings requirements are typically as follows:

- Relay stability from either a high or low impedance relay perspective is that the relay should remain stable (i.e. un-operative) for an external fault current equivalent to ten times the transformer full load, or the maximum through fault current if this can be determined (taking source impedance into account).

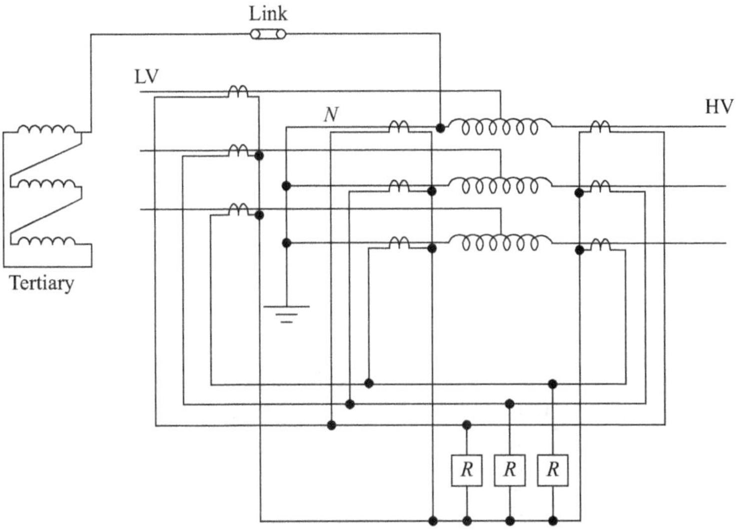

Figure 10.15 Autotransformer – overall protection

- Relay operation is typically required to be between 10% and 60% of the rated current of the protected winding for a solidly earthed neutral, or 10%–25% of the minimum EF current at the transformer terminals for a resistance earthed neutral.

10.6.3 Autotransformer overall differential protection

Figure 10.15 illustrates the protection arrangements for an autotransformer. The relays, *R*, are shown as high-impedance relays but could equally well be a modern low impedance biased differential relay similar to that shown in Figure 10.13 (but without the harmonic bias and HS features). Full protection is afforded against both phase-faults and EFs.

Since autotransformer windings are electrically connected, the currents in the HV, LV and N (neutral) CTs obey Kirchhoff's current law (see Section 3.3.1.1) and therefore the CT secondary windings can be directly connected, and with all three sets of CTs with identical ratios – to achieve current balance. This arrangement is also immune to the effects of tap-changing and magnetising inrush current.

When the transformer is fitted with a delta tertiary winding, but the tertiary is not used, it is usual to connect it into the autotransformer protection scheme as shown. Should the tertiary be used (e.g. to supply a reactor), the link is broken and the tertiary connected to earth via an earthing transformer with the protection for the tertiary and connected equipment being separate to that of the autotransformer.

Relay settings are typically as follows:

- Relay stability is usually required to be 16 times the HV-winding-rated current, or the maximum calculated through fault current, whichever is the greatest.
- Relay operation is usually targeted at between 10% and 60% of the HV-winding-rated current.

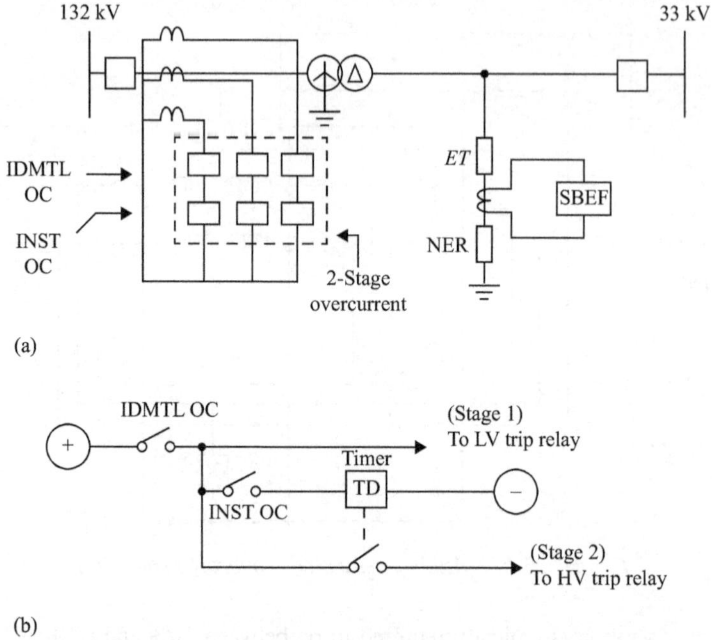

Figure 10.16 Two-stage overcurrent and standby earth fault: (a) two-stage overcurrent and standby earth-fault relay – AC connections and (b) two-stage overcurrent relay – trip circuitry

10.6.4 Two-stage overcurrent protection

The application of IDMTL OC and EF protection varies across the networks. However, the use of a two-stage OC relay arrangement is a very common and a standardised application for the (backup) protection of transformers.

The arrangement is shown in Figure 10.16 for the instance of a 132/33 kV two winding transformer, but it applies equally to autotransformers. The two-stage OC relay is used primarily for technical reasons (i.e. a faster operating time) in preference to separate HV and LV OC relays.

With reference to Figure 10.16, an uncleared fault on the 33-kV side of the transformer will result in operation of the IDMTL OC relay, and tripping of the LV (33 kV) circuit breaker. This is the stage-1 trip.

At the instant of fault inception, the instantaneous OC relay will also operate, which in conjunction with the (subsequent) operation of the IDMT OC relay initiates the timer time delay (TD). Should the operation of stage 1 not clear fault (i.e. the fault is on the transformer side of the circuit breaker), the timer will time out and operate the HV trip relay – which in turn will trip the HV (132 kV) circuit breaker. This is the stage-2 trip. The advantage of this arrangement is should the stage-1 trip clear the fault, the instantaneous OC relay resets the timer immediately (i.e. on current cessation). This is preferable to the longer disengagement time of a separate IDMT OC relay. As such the timer may be set to (typically) between 200 and 300 ms which is a lower and preferable time discrimination than the 400 ms, as given in Section 10.3.3. It is worthy of note that distribution transformers such as 33/11 kV are usually fitted with separate HV and LV OC relays resulting in less complex circuitry.

Two-stage OC settings are typically as follows:

1. **Current setting**
   The current setting (i.e. PS) is typically 150% of the transformer continuous rating which facilitates the following:
   (i) Transformer cyclic rating
   (ii) Matches the tap-changer overload limit – which is often the limiting factor in transformer ratings. NB: Different considerations may apply to 400/275-kV transformers.
2. **Time multiplier**
   The TM setting must satisfy the following;
   (i) Most transformers are specified with a capability of carrying maximum through fault current for a time not exceeding 3 s and are subject to an $I^2t =$ constant criterion. This limit must not be exceeded.
   (ii) Must grade with outgoing feeders on the LV side of the transformer under maximum fault current conditions.

10.6.5 Two-stage standby-earth-fault (SBEF) relay

A standby-EF (SBEF) relay is usually positioned as shown in Figure 10.16(a). This relay is traditionally part of the IDMT family, with a long time delay (TD) characteristic, as shown in curve *d* of Figure 10.4. The characteristic closely matches the thermal rating of, and therefore protects, the neutral earthing resistor (NER), which is commonly of a liquid resistor design. The SBEF relay is also usually arranged as a two-stage device, with circuitry similar to that shown in Figure 10.16(b), but omitting the instantaneous OC relay. Again, with the occurrence of an uncleared fault on the LV (33 kV) network, stage 1 trips the LV circuit breaker – but should the uncleared fault be on the transformer LV connections, stage 2 then trips the HV circuit breaker.

Two-stage SBEF current settings are arranged to match the characteristics of the NER – and typically may be 15% of the maximum EF current of the NER. TMSs are usually set to grade with feeder EF relays.

It is worthy of note that 33/11-kV transformers have traditionally been fitted with a single-stage SBEF.

10.6.6 Transformer HV high-set overcurrent (HSOC)

It is usual with transmission transformers, i.e. network voltages of 400/275, 400/132, 275/132 kV to install an HV HSOC relay, although instances also arise of it being installed on 132/33-kV transformers. The purpose of this relay is to provide high-speed backup protection for faults in the transformer tank in the vicinity of the

HV winding connections, where the fault current is very high. The operating time of the relay is typically 10–20 ms. It is usual to install the relay as a three-pole device, generally fed from the same CTs as the two-stage OC protection. The HSOC setting is typically arranged to be 50% or less than the minimum fault current on the transformer HV terminals, but 150% or more than the maximum through fault current, i.e. must not operate for a fault on the LV BBs.

10.6.7 Buchholz and winding temperature protection

These protective devices are examined in Section 9.11.1.3, as part of the power transformer, and are installed on virtually all transformers (on networks with voltages of 33 kV and above).

10.7 Feeder-unit protection

10.7.1 Feeder-unit protection – principles

Feeder-unit protections usually operate on the current differential unit principle first established by Merz-Price (*ca.* 1910). In essence, this is the circulating current arrangement illustrated in Figure 10.9 excepting that the CTs are now positioned at the opposite ends of a feeder. This presents a number of design difficulties.

- A relay needs to be positioned at either end of the feeder, so the relatively simple arrangement shown in Figure 10.9 is not directly applicable. However, if the current magnitude and phase is transmitted from either end of the feeder to the other via a communication medium, as shown in Figure 10.17, then the principle of operation as shown in Figure 10.9 can again be applied.
- With reference to Figure 10.17, feeders invariably possess shunt capacitance causing a differential current I_C which exists during both load and some fault conditions. This current causes an imbalance which flows in the relay operate (differential) element. The protection characteristic must compensate for this imbalance.
- The communication medium is invariably imperfect with the following design limitations:
  - The electrical characteristics of the communication medium may either limit or corrupt the signal (e.g. modify the magnitude, phase angle or shape).

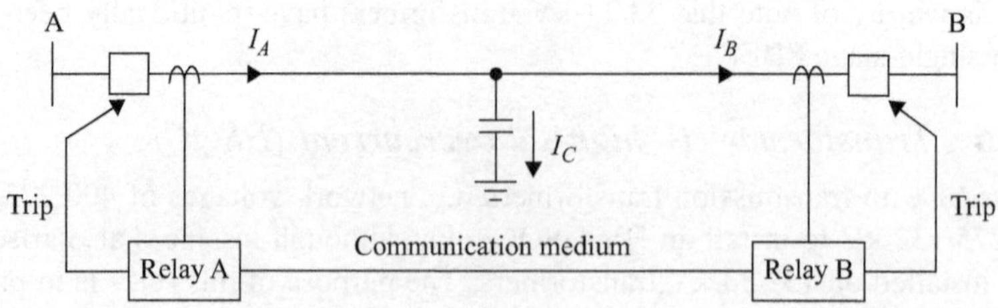

Figure 10.17 Feeder-unit protection – arrangements

- The communication medium may contain a time lag.
- Communication mediums are (more than most equipment) subject to breakdown or in advertent interference, for which conditions, the relay is required to remain stable. As such the communications medium must be subject to monitoring.

- CT saturation effects when subject to fault current DC transients, resulting in spill currents in the differential element, must be eliminated as part of the design.

Taking the above into account, feeder-unit protections usually possess operate/restrain characteristics based upon the following:

- Phase angle comparison of the currents at each end of the feeder
- A combination of phase angle comparison and amplitude comparison of the currents at each end of the feeder (encompassing biased differential).

The following sections will examine three main types of feeder-unit protection, which are as follows:

- Pilot-wire protection
- Power line carrier (PLC) protection
- Numeric protection.

10.7.1.1 Pilot-wire protection

Pilot-wire protection has a long and successful history of being applied to both transmission and distribution networks. As the name suggests, it uses a pilot cable as the communication medium. On the transmission network where the bulk of the network is OHL, the pilot was traditionally rented from a telecommunication company (particularly BT). However, with communication companies increasingly moving to digital communications mediums, pilot-wire protection has almost disappeared (i.e. from the UK transmission networks). The exception is where the pilot cable belongs to the power network company (i.e. private pilot) and installed in the same trench as the HV power cable. This situation is more common on the 33-kV network and below where a greater proportion of the network is HV cable – and therefore pilot-wire protection is still installed.

There are two fundamental categories of pilot-wire protection, which are as follows:

1. **Balanced current**
   With reference to Figure 10.18(a), in this design, the CTs at either end of the feeder are effectively connected in series so that for through fault conditions, the secondary voltages are additive to produce a circulating current around the pilot loop. The relays at both ends of the feeder are connected across the pilot (i.e. parallel connected), and as such are not subject to the circulating current. During a feeder internal fault, the secondary voltages are in opposition causing the secondary currents to flow through the relays – which operate.
2. **Balanced voltage**
   With reference to Figure 10.18(b), this is the opposite arrangement to balanced current, whereby the CTs at either end of the feeder effectively have their

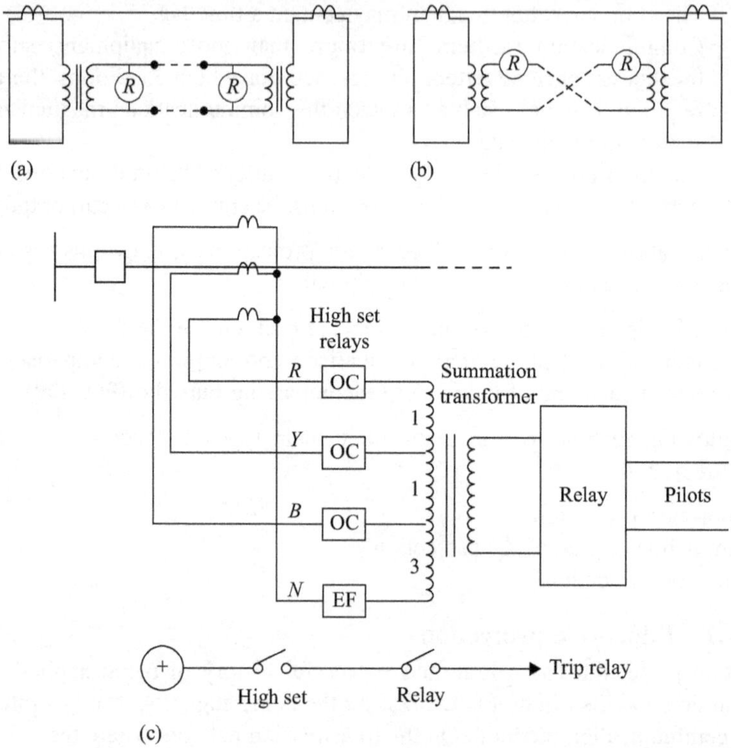

Figure 10.18 Pilot-wire protection – arrangements: (a) balanced current, (b) balanced voltage and (c) summation transformer arrangement

secondary's connected in opposition, so that for through fault conditions, there is no circulating current around the pilot loop. In this instance, the relays at either end are connected in series with the pilot, and as such, for an external fault, there is no current in either relay. For an internal fault, the secondary voltages are no longer equal and opposite resulting in current flow around the pilot loop and through the relays, so causing relay operation.

The connection between the CTs and the pilot is usually via a summation transformer as shown in Figure 10.18(c). This results in a three-phase input being converted into a single-phase output. A consequence of this arrangement is that the relay has different sensitivities for different types of fault. This arises because the inputs to the summation transformer are via a tapped winding, and as shown in Figure 10.18(c), the winding ratio of *R*/*Y*/*B* is 5/4/3, respectively (this ratio being typical). Many relays also include a HSOC and EF feature – the purpose of which is to prevent relay mal-operation in the event of pilot failure (or interference). Settings considerations comprise the following:

- The relay operate setting should exceed (usually by a factor of two) the out of balance current caused by the feeder capacitance. This is assisted by virtue that the relays are usually fitted with a bias element (not shown in Figure 10.18), not dissimilar to that shown in Figure 10.13. It is worthy of note that the relay sensitivity can be very low, typically between 250 and 1,200 A.

- The HSOC is usually set both above maximum load current, and to less than 50% of the minimum phase-to-phase internal fault current. Similarly, the HS EF relay is usually set both higher than ZPS component of maximum load current (arising from natural load unbalance) and lower than 50% of the minimum phase-to-earth internal fault current. The HS relays therefore determine the sensitivity of the protection.

Relay application usually specifies maximum pilot cable electrical characteristics (i.e. resistance and capacitance). The maximum feeder length to which pilot-wire protection can be applied, as result of design limitations, is typically about 35 km.

10.8 Power line carrier protection

10.8.1 Power line carrier protection – background

The most common type of PLC protection is usually that of phase comparison, i.e. a comparison of the phase angles of the currents at each end of the feeder. This type of protection has been successfully applied to OHL transmission feeders of up to, typically, 200 km in length. PLC uses the OHL itself as the communications medium, and as such one of its advantages is that it is under the direct control of the power network company (who own and operate the network) and therefore less susceptible to external interference. PLC was installed extensively and successfully in the United Kingdom until about the early 1990s, and although some installations remain in service, it is rarely installed today for the following reasons:

- Increased availability of digital communication mediums (usually fibre) and protection systems working in conjunction with this medium. It is worthy of note that in those parts of the world where digital communications are limited, that PLC is still installed.
- Comparatively expensive due to the cost of the coupling equipment.
- Diminishing expertise – as the technology is somewhat specialised and no longer a preferred technology (in the United Kingdom and many parts of Europe).

10.8.2 Power line carrier protection – equipment and performance

With reference to Figure 10.19, PLC equipment and performance characteristics comprise the following:

1. **Line trap**
   The line trap is a tuned inductance preventing carrier frequency from entering the substation, and directing it into the OHL feeder. It has negligible impedance to power frequency. It is located at BB level, with a tuning capacitance usually located directly under the inductance. Line traps are usually installed on two phases.

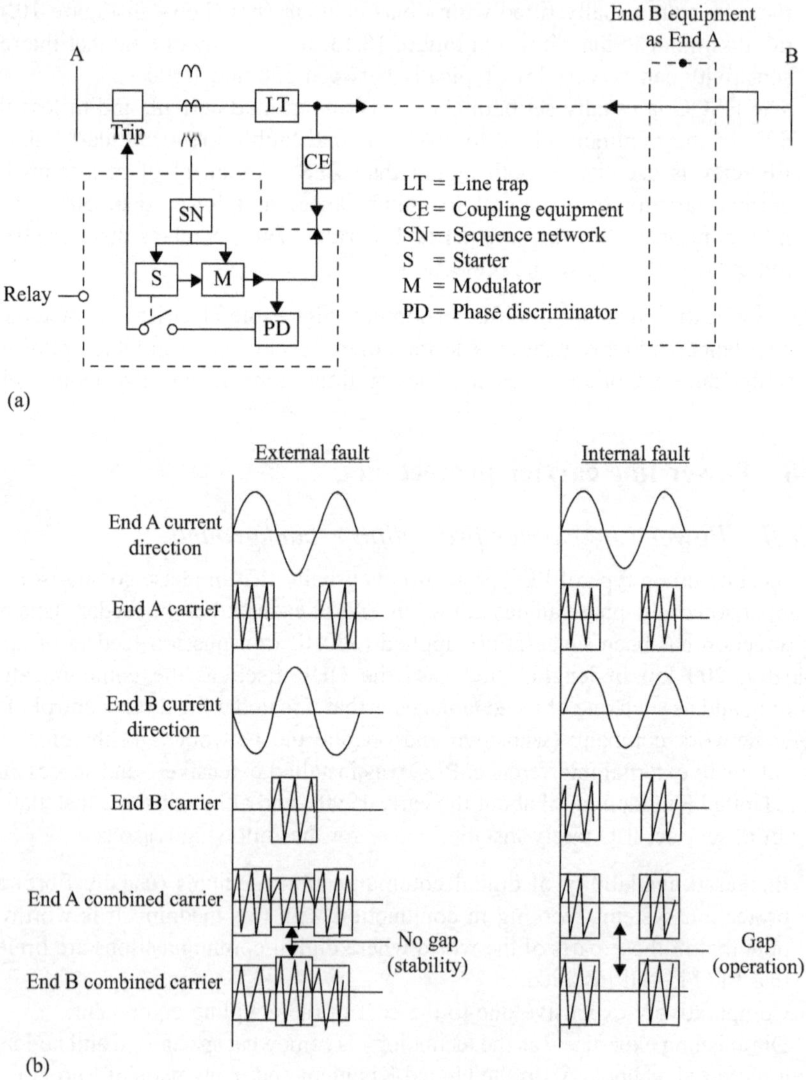

Figure 10.19 Power line carrier protection – operation: (a) carrier protection – equipment and (b) carrier protection – performance (simplified)

2. **Coupling equipment**
   The coupling equipment connects the phase-comparison relay to the OHL. It consists of tuned circuitry, connected to, and working in conjunction with, the capacitors in the capacitor voltage transformers. The coupling equipment prevents power frequency voltages from being impressed on the relay. It is installed on the same two phases as the line traps.

3. **Sequence network**
PLC protection invariably functions by extracting phase-sequence currents. In the first instance, the sequence network derives PPS and NPS currents from the three-phase input current and in doing so filters out ZPS currents.
4. **Starter**
The starters are required to ensure that the carrier signal is only transmitted under fault conditions – and not continuously (a requirement of radio signal transmission regulations). The starters are usually initiated by a step change in the level of PPS or NPS current (i.e. impulse starters) and consist of two levels: 'low-set' (LS) starter and 'high-set' (HS) starter. The LS starter output initiates the modulator for transmission of the carrier signal. The HS starter output is connected in series with the trip signal. Once operated the starters usually remain operated for a duration of 500 ms (typical).
The HS starter has a higher setting than the LS and must not operate until it is assured that the LS at both ends of the feeder have operated (and the carrier signal is transmitted from both ends – and signal evaluation is assured). This usually requires the HS starter setting to be typically 50% higher than the LS. This takes into account, and ensures relay stability for, the unbalance caused by feeder capacitance current.
5. **Modulator**
The modulator generates the carrier signal. It receives both PPS and NPS currents from the sequence network and combines a proportion of each to provide a single-modulating quantity. A frequently used modulating quantity is:
$$5I_2 - I_1$$
where I_2 is the NPS current and I_1 is the PPS current.
Such a modulating quantity ensures a carrier signal output for all fault permutations. The modulator generates blocks of carrier signal as illustrated in Figure 10.19(b) with typical carrier frequencies ranging from 70 to 700 kHz (as determined by regulations). NB: Procedural arrangements are required to determine a suitable carrier frequency – and one which cannot be picked up by and influence an adjacent PLC system.
6. **Phase discriminator**
The phase discriminator receives the local and remotely transmitted blocks of carrier signal and measures the gap in the signal in terms of an angle (where 360° is equal to one cycle at 50 Hz). Should the angle exceed the setting, typically 30°, the phase discriminator operates and outputs a trip signal and, if the HS is also operated, the local end circuit breaker is tripped. It is worthy of note that the carrier signal received from the remote end will suffer attenuation as it propagates along the OHL, as shown in Figure 10.19(b).
Operation of the local phase discriminator usually results in cessation (muting) of the locally transmitted carrier signal – thereby assisting the remote-end phase discriminator to operate.

PLC cannot be applied to cable circuits because the signal attenuation is too great, neither can it be applied to mixed OHL and cable circuits, nor to 'T' circuits, largely due to signal reflection problems with the risk of an unacceptably low level of carrier signal being received at the remote end. PLC has also been used for inter-tripping and protection signalling (e.g. blocking signal) purposes.

10.9 Numeric feeder-unit protection

10.9.1 Numeric feeder-unit protection – considerations

Modern numeric feeder-unit protections invariably communicate via fibre optic or other digital communications medium. Where dedicated fibre is not available, multiplexing techniques are used. An advantage of fibre/digital technology is that comparison of the currents at both ends of the feeder can be carried out on a per-phase basis, whereas with pilot-wire protection or PLC, the three phases are in some way combined. This type of protection may sometimes be classified as ‘teleprotection’.

A difficulty to overcome with fibre/digital communications mediums (which does not exist with older methods such as pilot wire or PLC) is the inherent TD between the sending and receipt of the communications signal, creating a difficulty for the comparison of simultaneous currents at each end. Two widely used methods for managing the TD are as follows:

- Continuous signal measurement
- GPS synchronised time.

These will be briefly evaluated.

10.9.1.1 Continuous measurement

Although the relay at each end of the feeder has an internal clock, these are not accurately synchronised between each end and suffer random drift. One method of compensating for this is that one of the feeder ends, call this end ‘A’, transmits a signal to the remote end, end ‘B’, and this signal contains a time tag of the instant the signal was sent, let this time be *TA1*. This signal is received at end ‘B’ which returns the signal after a time duration, and let this duration be *TBD*, with the returned signal containing the time tags *TA1* and *TBD*. Let the time at which the signal is received back at end ‘A’ be *TA2*, then it follows that end ‘A’ can determine the transmission time (i.e. propagation delay) *TM* between end ‘A’ and end ‘B’ from the following relationship:

$$TA2 - TA1 = TBD + 2TM$$

The end ‘A’ relay can then compare local end current with current data received from the remote end after the delay *TM*. This ensures comparison of the local end current with that current that existed at the remote end at the same instant of time (i.e. real time comparison). NB: The above assumes that the transmission time (*TM*) is the same in both directions.

10.9.1.2 GPS synchronised time

This technique uses a time signal from a GPS satellite to synchronise the clocks within the relays at either end of the feeder. Each end then time tags the per-phase current and transmits to the remote end such that each end compares current with an identical time tag. In this instance, the communication channel propagation delay is not required to be known. This method does however require the additional cost of the communication equipment between substation and satellite – however, the same time signal can be used for other relays and uses in the same substation.

10.9.2 Relay characteristics

Relay characteristics are usually of the biased differential type based upon both the principles shown in Figure 10.13 (but omitting the harmonic bias and HS), and the simplified diagram shown in Figure 15.6. The bias current would usually comprise the feeder load or through fault current. The bias slope can be used to compensate both for CT partial saturation (thereby allowing operation with CTs having lower knee points), and the imbalance arising from feeder capacitance current (resulting in operating current in the differential relay). Phase angle comparison similar to that used by PLC may also be used as a relay characteristic.

10.9.3 Settings considerations

Unlike older types of relays, modern feeder protections are designed to be immune from mal-operation should the communication channel fail. As such OC and EF starters are not required and the operating current setting of the relay can be relatively low. Typical setting criterion include:

- The differential current setting must exceed the feeder capacitance (charging) current.
- The protection must operate for a single end fed fault when the fault is located at the far end of the feeder (i.e. the remote-end circuit breaker is open) – under summer minimum plant conditions.
- The protection must operate for an internal high resistance fault (e.g. 100 Ω to earth) – coincident with a through current whose magnitude is equal to the rated current of the relay. For example, a 1-A relay connected to a CT of ratio 1,200/1 would equate to a through current of 1,200 A.

It is worthy of note that this type of protection is usually designed such that it can be applied to three (or more)-ended feeders.

10.10 Distance protection

10.10.1 Distance protection – introduction

Distance protection has a long and successful history of being applied to all networks at voltages of 33 kV and above. The name 'distance' is derived from the fact that the impedance from the relay location to the point of the fault is directly proportional to distance. A disadvantage of distance protection is that it is relatively expensive, requiring three-phase voltage transformers as well as current transformers. In its basic form, distance protection is not a unit protection, and as such some of its features (i.e. zones 2 and 3) are used as backup protection. However, with the use of an end-to-end communications medium, distance protection can be converted into the equivalent of a unit protection. Distance protection also has the advantage that it can be applied to multi-ended feeders.

The sections that follow will examine the following key features of distance protection design and performance:

- Distance protection – comparators
- Plain distance protection application
- Distance protection impedance measurement
- Distance protection application limits
- Blocked distance protection
- Application to composite feeders

10.10.1.1 Distance protection – comparators

Distance protection operates on the principle of evaluating whether a 'measured impedance' falls within the boundary of operation of a comparator characteristic – where the boundary of operation is determined by the comparator 'setting impedance' (sometimes termed replica impedance). Comparator inputs and outputs are illustrated in Figure 10.20(a). Comparators comprise two classical types: amplitude comparators, and phase angle comparators, either of which may be used to obtain a required relay characteristic. For the purpose of this analysis, phase angle comparators will be assumed.

The phase angle comparator operates on the basis of comparing the phase angle between two input phasors S_1 and S_2, where S_1 and S_2 are derived from the feeder voltages and currents. The comparator operates when the angle θ between S_1 and S_2 (where S_1 leads S_2) satisfies the expression:

$$-90^\circ \leq \theta \leq 90^\circ \tag{10.8}$$

It is usual to show the comparator characteristic on a resistance/reactance diagram. By way of example the following commonly used comparators will be examined.

- Directional comparator
- Impedance angle comparator
- Polarised mho comparator
- Offset mho comparator
- Quadrilateral characteristic

Directional comparator characteristic

With reference to Figure 10.20(b), the directional characteristic comprises:

$$S_1 = V_R \text{ and } S_2 = IZ_R \tag{10.9}$$

where V_R is the relay input (i.e. fault) voltage, I is the relay input (i.e. fault) current, Z_R is the comparator (i.e. relay) setting impedance and Z is the impedance from relay point to fault location.

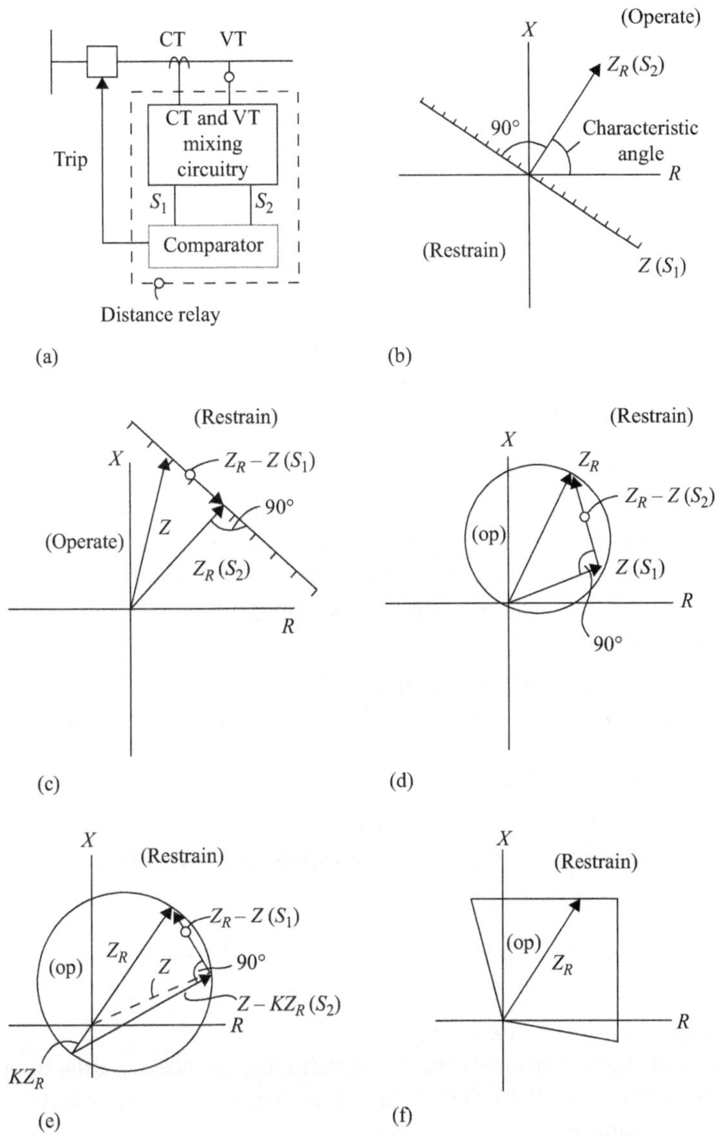

Figure 10.20 Comparator characteristics: (a) distance relay – key features, (b) directional characteristic, (c) impedance angle characteristic, (d) polarised mho characteristic, (e) offset mho characteristic and (f) quadrilateral characteristic

Dividing expression (10.9) by I:

$$S_1 = \frac{V_R}{I} = Z \text{ and } S_2 = \frac{IZ_R}{I} = Z_R$$

For operation, the angle between Z and Z_R must be within ±90°.

This is the same characteristic as that shown in Figure 10.7 – but with a different method of expressing the inputs and presenting the characteristic.

Impedance angle comparator characteristic

This is also known as an 'ohm' characteristic. With reference to Figure 10.20(c), the inputs for this characteristic are:

$$S_1 = IZ_R - V_R \text{ and } S_2 = IZ_R \tag{10.10}$$

Dividing expression (10.10) by I:

$$\text{then, } S_1 = Z_R - Z \text{ and } S_2 = Z_R$$

For operation, the angle between $(Z_R - Z)$ and Z_R be within ±90°.

It is worthy of note that by selecting Z_R as purely inductive then Z_R is positioned on the $+X$ axis, and a reactance characteristic results. Alternatively, by selecting Z_R as purely resistive then Z_R lies along the $+R$ axis and a resistance characteristic results.

Polarised mho comparator characteristic

With reference to Figure 10.20(d), the 'polarised mho' inputs comprise:

$$S_1 = V_P \text{ and } S_2 = IZ_R - V_R \tag{10.11}$$

where V_P is termed the polarising voltage.

Consider initially $V_P = V_R$ and divide expression (10.11) by I,

$$\text{then, } S_1 = Z \text{ and } S_2 = Z_R - Z$$

The resulting characteristic is a circle of diameter Z_R passing through the origin. The operate zone is within the circle.

Offset mho comparator characteristic

The polarised mho characteristic has the disadvantage of not operating for a fault at the relay position, i.e. the origin of the X–R diagram. This is overcome by the 'offset mho' characteristic. Inputs comprise:

$$S_1 = IZ_R - V \text{ and } S_2 = V - KIZ_R \tag{10.12}$$

where K is a number less than one which determines the extent of the offset from the origin.

Dividing expression (10.12) by I,

$$\text{then, } S_1 = Z_R - Z \text{ and } S_2 = Z - KZ_R$$

The resulting characteristic is again a circle which is offset from the origin as illustrated in Figure 10.20(e).

Quadrilateral comparator characteristic

The 'quadrilateral' characteristic (alternatively termed polygonal) comprises an assembly of impedance angle characteristics. Prior to the advent of digital/numeric-based relay, this characteristic was difficult to achieve – but is now the preferred distance protection characteristic for most applications, as will be explained in the following sections.

10.10.1.2 Plain distance protection application

Figure 10.21 illustrates the basic (usually termed as 'plain') application of distance protection. There are usually four zones of protection: Z1 (zone 1), Z2 (zone 2), Z3F (zone 3 forward) and Z3R (zone 3 reverse). Figure 10.21(a) shows the zones associated with substation A. Identical zones exist at substation B but facing towards substation A. Salient points are as follows:

1. **Zone 1**
   Zone 1 comprises an instantaneous comparator (e.g. 15–25-ms operating time) with a setting impedance of typically 80% of the feeder A–B impedance.

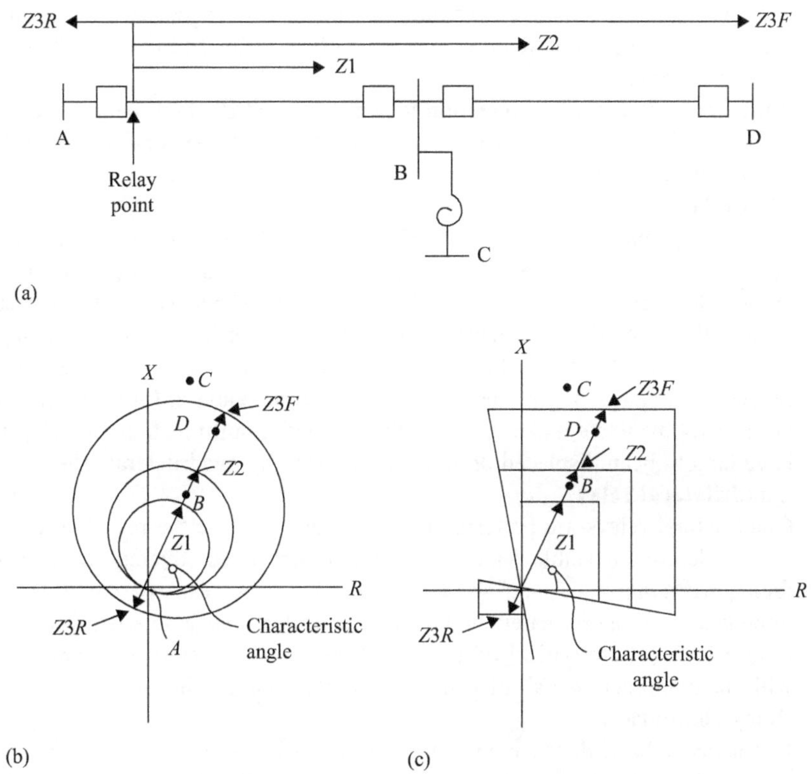

Figure 10.21 Basic distance relay – application: (a) zones of protection, (b) mho relay protection and (c) quadrilateral relay protection

The 80% ensures no overreach to detect a substation B BB fault – which otherwise may arise as result of:

(i) CT errors
(ii) Relay-measuring errors
(iii) Feeder impedance data errors.

2. **Zone 2**
The Zone 2 comparator is typically set to 150% of the feeder A–B impedance, to be assured that it does detect an uncleared BB fault at substation B, thereby acting as backup protection. Zone 2 is time delayed, typically by about 500 ms, to ensure discrimination with main protections. Zone 2 also provides protection for the last 20% of feeder A–B, not covered by zone 1 – this is usually only a requirement for the instance of the feeder being energised from one end only.
3. **Zone 3 forward**
The zone 3 forward comparator is usually employed as remote backup for uncleared faults on the power system in general. Some engineers argue the case for zone 3 existing at all is weak, and care must be taken with the settings to ensure it does not operate unnecessarily. Zone 3 is usually time delayed to discriminate with zone 2 and has a TD setting of typically 1 s. Zone 3 forward should be set to not operate for a fault on the lower voltage network, i.e. on the substation C BBs – this requirement often dictates the setting.
4. **Zone 3 reverse**
Zone 3 reverse provides protection for an uncleared BB fault behind the relay point, i.e. the substation A BBs. The zone 3 forward and zone 3 reverse TDs are usually one and the same.
5. **Mho relay**
Historically, and with reference to Figure 10.21(b), many distance protections consisted of a three-zone relay comprising zones 1 and 2 as polarised mho comparators, and zone 3 an offset mho comparator. The relay settings normally refer to the reach (i.e. boundary) of the comparator at the characteristic angle, which is usually the feeder line angle (as determined by the X-to-R ratio of the feeder), or an angle slightly below. Although these relays are still installed and are usually available as an option within numeric multifunctional relays, they have largely been displaced on new installations by quadrilateral relays.
6. **Quadrilateral relay**
Quadrilateral relays are preferred to mho relays because the resistive reach is independently adjustable – thereby providing improved coverage for resistive faults, particularly EFs on short feeders where the resistive component of impedance is proportionately higher. The quadrilateral characteristics shown in Figure 10.21(c) are typical of a range of possible characteristics (which vary with the manufacturer) all of which are similar, but not identical.
7. **Relay comparators**
In practice, a basic distance protection relay usually comprises 18 comparators, with each of the three zones having comparators for the following six fault conditions: R-E, Y-E, B-E, R-Y, Y-B and B-R (with quadrilateral relays, the reverse zone 3 may be independent comparators).

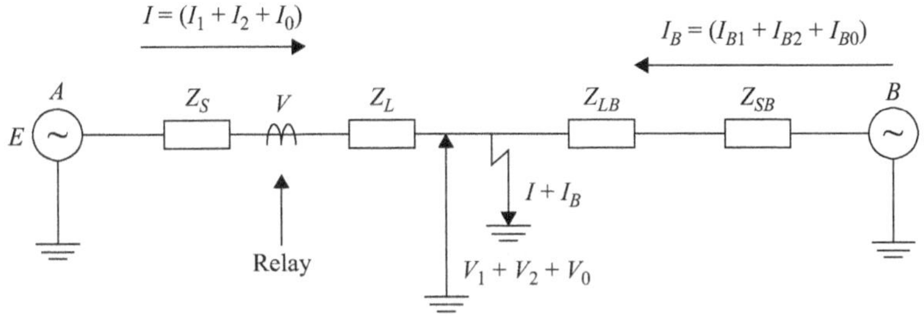

Figure 10.22 Fault current analysis

8. **Switched distance schemes**
   At the lower voltage networks (e.g. 33 kV and some 132 kV), a more cost-effective form of distance protection is a 'switched distance' scheme, in which the voltages and currents are switched, in accordance with which phases are faulted, to a single comparator. A disadvantage of this arrangement is the extra TDs involved with switching relays.

10.10.1.3 Impedance measurement

In the above sections on comparators, the impedance Z from the relay point to fault was taken as being the relay voltage V, divided by the relay current I. However, in practice, this simple analysis is incomplete, and the measured impedance needs to be evaluated in terms of the PPS, NPS and ZPS impedances.

Consider Figure 10.22 where in the first instance the current into a fault on an OHL is from end A only (i.e. end B does not initially exist). Consider the impedance Z measured by the relay in terms of the feeder phase-sequence voltages and currents – based upon the analysis in Figure 4.13, it can be said that:

$$Z = \frac{V}{I} = \frac{(I_1 Z_{L1} + V_1 + I_2 Z_{L2} + V_2 + I_0 Z_{L0} + V_0)}{I_1 + I_2 + I_0} \tag{10.13}$$

Now, at the point of fault, the voltage = 0,
so $V_1 + V_2 + V_0 = 0$ and inserting into expression (10.13),

$$\text{then } Z = \frac{I_1 Z_{L1} + I Z_{L2} + I_0 Z_{L0}}{I_1 + I_2 + I_0}$$

Now, for a single end fed EF, $I_1 = I_2 = I_0$,

$$\text{therefore } Z = \frac{I_1(2Z_{L1} + Z_{L0})}{3I_1} = \frac{2Z_{L1} + Z_{L0}}{3}.$$

However, with reference to Figure 4.6, the sequence impedances of an OHL are not constant but vary with the mutual inductance from an adjacent feeder, with the ZPS impedance being by far the most variable. As a consequence, the impedance measured to the fault position becomes variable.

This situation is made still more variable when the infeed to the fault from end B is considered. In this instance, the value of ZPS current flowing from end A is not the same as that of the PPS and NPS (which are the same) but varies with the values of the respective ZPS impedances of ends A and B.

If, with reference to the current flowing into the fault from end A, the ratio of $\frac{I_1}{I_0} = k$ (where k is variable and a function of the fault infeed from end B – as determined by the values of the sequence impedances).

Then, it can be shown that the impedance Z to fault, measured from end A is given by:

$$Z = \frac{2Z_{L1} + KZ_{L0}}{2 + K}$$

Clearly, for distance protection to function correctly, the impedance measured to a specified fault location must be constant (and proportional to distance), and not vary with the factors cited above. To achieve a constant measure of impedance to the same fault position, distance protection is therefore designed to measure a single impedance, and that is chosen to be the PPS impedance to fault. To achieve this, for EFs, a technique termed 'residual compensation' is used, and for phase–phase faults, a technique termed 'phase-fault compensation'.

1. **Residual compensation**
   It can be shown that the PPS impedance Z_1 to fault can be obtained, if the comparator is fed with a compensating current I_C such that:

   $$Z_1 = \frac{V}{I + I_C} \qquad (10.14)$$

   where V and I are the faulted phase voltage and current, respectively, at the relay point, and where:

   $$I_C = \frac{I_R + I_Y + I_B}{3}\left(\frac{Z_0}{Z_1} - 1\right) \qquad (10.15)$$

   and I_R, I_Y and I_B are the currents in the three phases, and Z_0 and Z_1 are the sequence impedances of the OHL in question from relay point to point of fault.
   From expression (10.15), it is worth recalling that by definition (see expression 4.12):

   $$I_0 = \frac{I_R + I_Y + I_B}{3}$$

   and the quantity $\frac{1}{3}\left(\frac{Z_0}{Z_1} - 1\right)$ is termed the 'residual compensation factor' (RCF) and is a setting on the relay to achieve PPS impedance measurement.
2. **Phase fault compensation**
   Similarly, it can be shown that to measure PPS impedance to fault for a phase–phase fault, the phase–phase voltage (of the faulted phases,) and difference of the faulted phases currents must be measured, e.g. for a Y–B phase–phase fault:

   $$Z_1 = \frac{V_Y - V_B}{I_Y - I_B} \qquad (10.16)$$

10.10.1.4 Relay polarisation

A limitation of any distance protection characteristic that passes through the *X–R* diagram origin is that when the faulted phase voltage drops to zero (i.e. a fault at the feeder terminals, of zero impedance), the comparator cannot function (since there is no voltage to enable comparison to take place). To overcome this, the relay comparator is fed with a voltage that comprises a number of components as follows:

- The faulty phase voltage
- A small per cent of a healthy phase voltage, phase shifted to be in phase with the healthy phase voltage. This is termed the 'cross-polarised' voltage.
- In some instances, a small per cent of the pre-fault voltage. This is termed the 'memory voltage'; this is effective for the instant of a close-up three-phase fault (on the feeder terminals) when both the faulty phase voltage and cross-polarised voltage fall to zero.

Therefore, the voltage fed into the comparator comprises:

$$V_{\text{comparator}} = V_{\text{fault}} + V_{\text{crosspolarised}} + V_{\text{memory}} \tag{10.17}$$

The above comparator voltage comprises V_P as specified in Section 10.10.1.1 with reference to the 'polarised mho comparator characteristic' – and it is also the voltage input into the quadrilateral relay comparators.

An impact of the added voltages is to cause distortion of the relay characteristics for faults which do not coincide with the comparator characteristic angle. The extent of distortion is increased by a high system impedance ratio (SIR) associated with the faulted circuit (see Section 10.10.1.5). The distortion is more pronounced on polarised mho relays and can be helpful in extending the resistive range of overage.

10.10.1.5 Minimum line length

When distance protections are called upon to operate with very small voltages, and currents, relay accuracy decreases. With reference to Figure 10.22, the voltage and current applied to the relay reduces as the source impedance Z_S behind the relay increases. The term SIR is a measure of source impedance compared to relay setting impedance Z_R and defined as follows:

$$\text{SIR} = \frac{Z_S}{Z_R} \tag{10.18}$$

Manufacturers usually produce relay accuracy curves as shown Figure 10.23 which defines the accuracy boundary against increasing SIR. At the boundary of relay operation, the operating time is indeterminate and a more practical boundary is that of the 30-ms operating time. Generally, distance protection is not applied to feeders in excess of an SIR of 30, at which point, with reference to Figure 10.23, relay inaccuracy commences to become unacceptable.

SIR can increase in two ways. The first is that the source impedance behind the relay point is high. The second is that the length of the protected feeder is

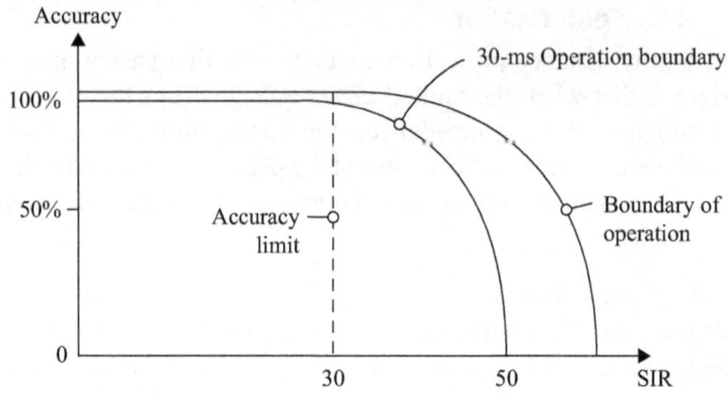

Figure 10.23 Effect of source impedance on operating characteristic

very low. The latter consideration usually results in distance protection not being applied to feeders below a certain length. For the 400 and 275-kV transmission networks, the minimum feeder length to which distance protection is applied is typically 10 km.

10.10.1.6 Zone 3 setting limits

Generally, there are two-setting limits relating to zone 3, which are as follows:

1. **Transformer LV fault**
   With reference to Figure 10.21, zone 3 must not operate for a fault on the LV BBs at substation C, with the maximum number of parallel connected transformers in service (i.e. connecting substations B and C). Any fault uncleared by main protection at substation C should be left to be cleared by backup protections which are electrically nearer to the fault (than zone 3). When calculating how far the zone 3 setting should look into the transformer impedance, it should be noted that the RCF for the substation A to B feeder is unlikely to be suitable for the transformer, as the transformer will have a different ZPS to PPS impedance ratio to that of the feeder. As such a factor of safety should be included in the setting.
2. **Maximum load current**
   Zone 3 must also be set to avoid operation by maximum load current. The latter is usually taken as being at an angle on the *X–R* diagram of between 0° (i.e. on the resistive axis) and 30°. A factor of safety of typically 20% is usually taken between the load impedance and setting impedance. With polarised mho comparators, an option often exists of using a shaped characteristic (often lemon shaped) which allows greater load current. With quadrilateral characteristics, the setting reach into the *R* axis may need to be limited to avoid load encroachment.

A check should be made to ensure that no part of the zone 2 setting characteristic outreaches the zone 3 setting.

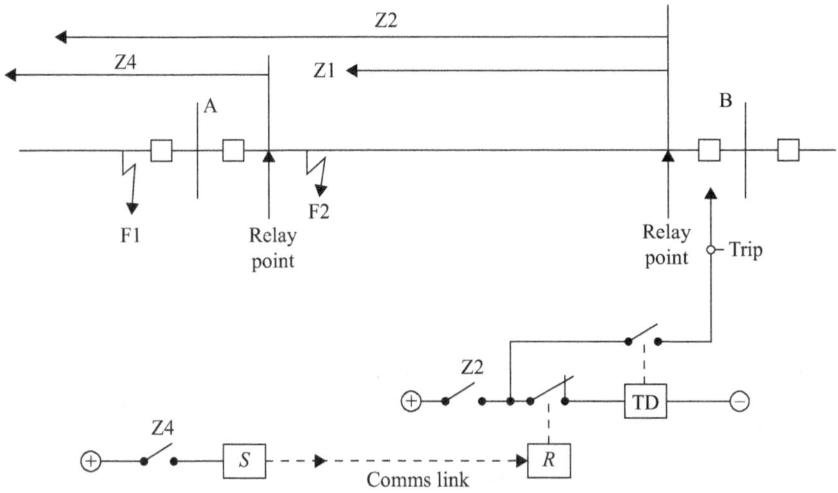

Figure 10.24 Blocked distance protection scheme

10.10.1.7 Blocked distance protection

There are numerous ways in which a plain distance protection can be arranged to approximate to a unit protection – the most common of which in the United Kingdom is a 'blocked' distance protection scheme. With reference to Figure 10.24, both substations A and B are equipped with the basic three-zone distance protection scheme as shown in Figure 10.21, plus an additional zone (i.e. comparator), zone 4 (or other alternative name). For the purpose of explaining the blocking scheme, only the relevant zones are shown.

1. **External fault**
   (i) Consider the fault at position F1, which is external to feeder A–B.
   (ii) The fault is detected by both Z2 at substation B, and Z4 at substation A.
   (iii) At substation B, Z2 starts the TD.
   (iv) At substation A, Z4 sends a block signal via the communications link.
   (v) At substation B, the block receive signal opens the contact to de-energise the timer – and no trip occurs at end B.
2. **Internal fault**
   (i) Consider the fault at position F2, which is internal to feeder A–B, but beyond the reach of Z1 at substation B.
   (ii) The fault is detected by Z2 at substation B, which starts the timer TD.
   (iii) At substation A, Z4 does not detect the fault, and there is no block signal send.
   (iv) At substation B, the timer TD times out and trips the circuit breaker at substation B.

The circuitry shown also exists at the other end of the feeder to discriminate for faults beyond substation B. The timer TD must be set in excess of the longest possible time for receiving the block signal, typically between 30 and 60 ms. If the Z2 comparator operating time is, say, 15 m, the overall time to output the trip signal would be between 45 and 75 ms. If the blocking channel becomes faulty, the blocking circuitry is usually disabled and the protection then reverts to the plain three-zone distance protection. Since blocking schemes use zone 2 comparators, and not zone 1, a lower SIR is achieved, and therefore (with reference to Figure 10.23 and Section 10.10.1.5), blocking is applicable to shorter feeder lengths, than the plain distance scheme, possibly as low as 4 km.

10.10.1.8 Three-ended circuits

Blocked distance protections are frequently applied to protect three-ended circuits, although on the 132-kV network, this may extend to four-ended circuits. With reference to blocked distance protection being applied to the network shown in Figure 10.25, the main considerations are as follows:

1. **Zone 1 setting – 'T' point coverage**
   Where the 'T' point cannot be covered by zone 1 from all three ends then blocked distance protection should be applied.
2. **Zone 1 setting – electrically closest substation**
   The zone 1 setting reach at each substation is set to 80% (typically) of the electrically shortest feeder length, i.e. zone 1 at substation A is set to 80% of feeder A–B, or feeder A–C, whichever is the lesser impedance (in this instance, it is feeder A–C).

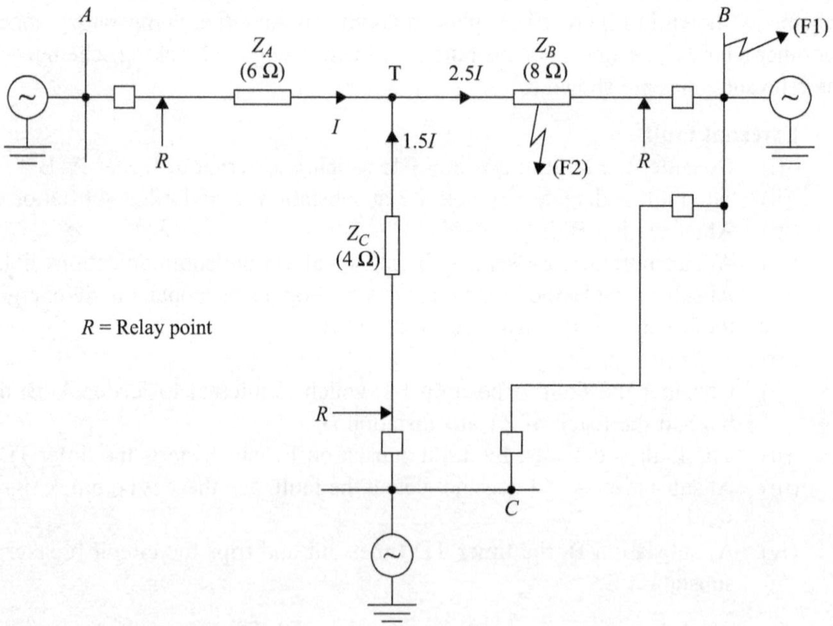

Figure 10.25 Blocked distance protection – three-ended circuit

3. **Zone 1 setting – parallel path consideration**
   Consider the zone 1 at substation A and fault position (F1). The zone 1 must not detect this fault when considering the reduction in impedance to fault, arising from the parallel path – which comprises the circuit from the 'T' point to substation C in series with the feeder from substation C to B (assuming no generation at substation C). If this fault is detected, the zone 1 setting at substation A must be reduced to 80% of the parallel path impedance.
4. **Zone 1 setting – fault current direction**
   Consider a fault at position (F2) between substation B and the 'T' point, for the instance of no generation at substation C. In such an instance, the current from substation A to the fault may divide with some flowing via substation C. This would result in a block signal being sent from substation C to substations A and B so preventing the fault being cleared by the blocking circuitry. In such an instance, the fault must be detected by zone 1 from at least one substation end which would then result in sequential tripping.
5. **Zone 2 setting – throttling**
   Consider a fault at position (F1). The impedance measured by the relay at substation A is the ratio of the measured voltage and current at the relay point. For the currents shown in Figure 10.25, then:

   Voltage, V, at substation A $= (6 \times I) + (8 \times 2.5I) = 26I$

   And impedance, Z, seen by relay at substation A $= \dfrac{V}{I} = \dfrac{26I}{I} = 26\ \Omega$

   The effect of the in-feeding current from substation C is to cause the apparent impedance (i.e. 26 Ω) seen at substation A (on the A–B feeder) to be higher than the actual impedance (i.e. 14 Ω). This effect is termed 'throttling'.

   The zone 2 setting at each substation must therefore be able to detect a fault on the remote-end BBs under worst throttling conditions. NB: This is required both for the blocking scheme to function correctly and for zone 2 to clear an otherwise uncleared BB fault.

In most applications, the above prove not to be problem, but they must be considered – and in some instances, it may be concluded that it is impractical to apply distance protection.

10.10.1.9 Transformer feeders

Figure 10.26 shows the instance of distance protection being applied to a transformer feeder. The distance relay at substation A may be set with a zone 1

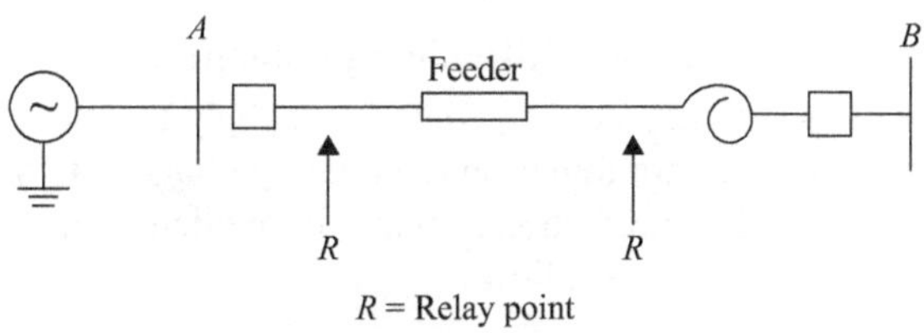

Figure 10.26 Application of distance protection to a transformer feeder

reach which looks part way into the transformer. This serves the following purposes:

- Provides 100% zone 1 protection of the feeder. Within this context, the distance relay at substation B may not operate for a feeder fault due to:
  - High SIR arising from the transformer high impedance
  - Little or no system infeed at substation B.
- Acts as fast acting back up to transformer protection.

Care must be taken in choosing the setting, since as explained in Section 10.10.1.6, the ZPS to PPS impedance of the transformer differs to that of the feeder, so impacting on the value of the residual compensation, and therefore to ensure that the distance protection does not operate for a fault on the LV BBs, a factor of safety must be included in the setting impedance. The factor of safety would apply to all three zones.

10.10.1.10 Composite feeders

Distance protection is applied widely to OHL feeders, and to some cable feeders. Prior to the advent of numeric/digital relays, application to cables was difficult due to the limited range of the residual compensation setting (the required setting for a cable being lower than that for an OHL). Distance protection has also been successfully applied to mixed OHL and cable circuits with care being required to select an appropriate value of residual compensation – to avoid overreach or excessive under-reach.

10.11 Busbar protection

10.11.1 Busbar protection – overview

The design and satisfactory performance of BB protection is of critical importance since it concerns the simultaneous tripping of large numbers of circuit breakers in a substation. Within this context, a particular concern with BB protection is that it should be immune from mal-operation, arising from either inadvertent interference or design deficiencies. The following sections will examine common designs of BB protection, namely:

- Distribution network(s) BB protection
- High-impedance circulating current
- Numeric

10.11.1.1 Distribution network busbar protection

At distribution voltages, it may not be economic to apply the more comprehensive forms of BB protection that are found on the higher voltage networks. Within this context, two designs of BB protection commonly applied to the 33-kV network (and some lower voltages) are as follows:

- Frame earth
- Rough balance

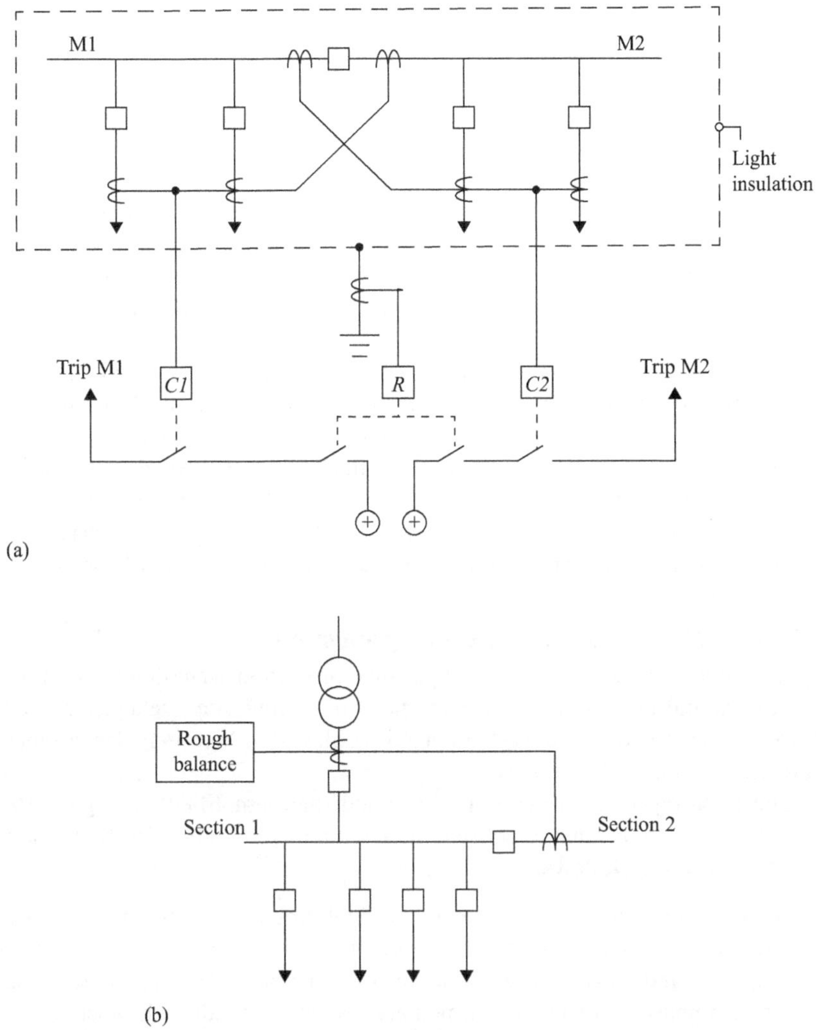

Figure 10.27 Busbar protections – distribution networks: (a) frame earth busbar protection and (b) rough balance busbar protection

- **Frame earth BB protection**
  'Frame earth' is a simple form of EF BB protection, for metalclad switchgear. It is achieved by lightly insulating the metalclad framework from earth. If the framework is then connected to earth at only one point, as shown in Figure 10.27(a), and a current transformer fitted over this connection, to which a relay *R* is connected, any fault to the framework would result in operation of the relay. This is the principle of frame earth protection.

  However, in practice, it is usual to add security to the system by the addition of an EF check zone for each section of BBs, such that it requires the operation of both the frame earth relay and the check zone to trip all circuit breakers connected to the faulted zone. It is assumed of course that the design of the switchgear is such that a phase-to-phase fault is unlikely. Variations to Figure 10.27(a) exist with the check zones available as either a high-impedance or biased differential scheme.

- **Rough balance busbar protection**
  With reference to Figure 10.27(b), the rough balance BB protection measures only the infeeds into a faulty zone, i.e. in this instance the incoming transformer and the bus-section. For faults on section 2 (i.e. an out of zone fault), the transformer current would be matched by the current flowing through the bus-section circuit breaker, and no current, would appear in the rough balance relay, i.e. relay stability.

  The rough balance relay usually comprises a HS element and an IDMT element. For faults on the feeders, the HS element is either time delayed to discriminate with the feeder protection operation, or blocked by a feeder protection starter relay. For faults involving section 1 BBs, the HS element would operate and trip all circuit breakers connected to the BBs. The IDMT relay caters for a stuck feeder circuit breaker condition. The rough balance relay is usually a three-phase device, i.e. individual CTs per phase (although the IDMT may have an EF element).

10.11.1.2 High-impedance busbar protection

BB protection schemes using high-impedance protection were dominant until the advent of digital protections. However, many high-impedance schemes remain in service, and some power network companies still install high-impedance schemes in preference to digital alternatives.

The high-impedance scheme is illustrated (and simplified) in Figure 10.28, with reference to a double BB substation (and circuits initially selected as shown). Salient points are as follows:

- With reference to Figure 10.28(a), the high-impedance scheme comprises a check zone which is common to all circuits, apart from the bus-sections and couplers – and separate discrimination zones for each BB. Separate relays exist for each phase (not shown) – which are responsive to all fault types.
- The CTs are directed to the appropriate discrimination zone by disconnector auxiliary contacts which reflect the position of the BB selector disconnectors, as illustrated in Figure 10.28(b). Care must be taken with the timing of the auxiliary contacts to ensure they make/break at the appropriate point in the disconnector movement. When the disconnectors are fully open, the CTs are usually short-circuited by the auxiliary contacts (not shown).
- Tripping of each respective BB zone i.e. M1, M2, R1 and R2 requires operation of both the zone discrimination relay, e.g. M1D, and the check relay. This two out of two design provides a level of security against mal-operation.
- Figure 10.28 shows one discrimination relay per BB zone, however, with distributed relay rooms (blockhouses), it is usual to provide one discrimination and check relay per circuit (to minimise long CT wiring and tripping circuits), which trip their own circuit breaker.

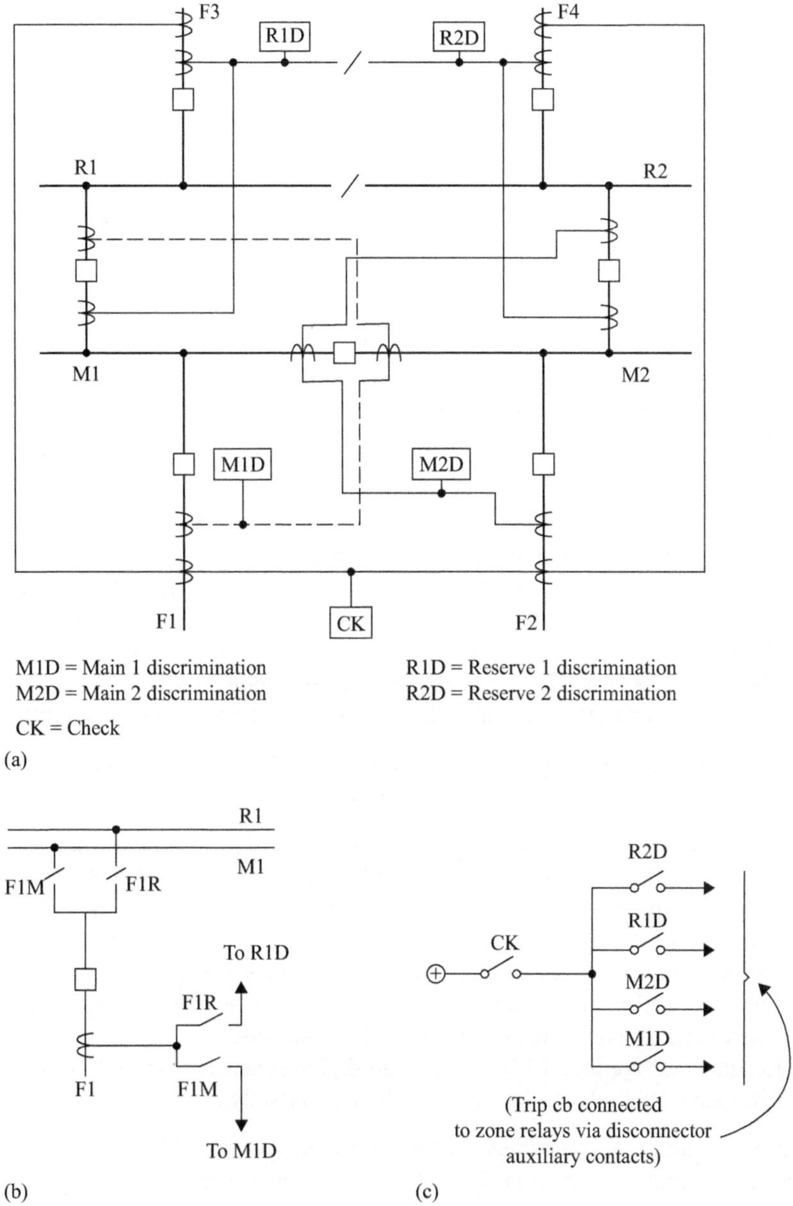

Figure 10.28 Busbar protection scheme – high impedance: (a) busbar protection scheme – outline, (b) busbar selector disconnector circuitry and (c) tripping arrangements

- High-impedance BB protection operation settings vary with the number of connected circuits. Typical setting requirements are as follows:
  - Operating currents for both discrimination and check zones to be between 50% and 100% of BB-rated current with confirmation that such a setting would detect a fault on the BBs under minimum system in-feed conditions (with a factor of safety of at least two).
  - Stability voltage to be based upon BB maximum short-circuit current.

  Instances have arisen of the settings being difficult to achieve for all operating conditions.

10.11.1.3 Numeric busbar protection

Dating from the mid-1990s, a rigorous debate took place in the much of the United Kingdom, and elsewhere, about the merits of moving away from the high-impedance BB protection to that of low-impedance BB protection. The case for the latter was facilitated by the advent of digital protections. The case against the high-impedance scheme was (and is):

- Requires extensive wiring, auxiliary contacts, relays, resistors and metrosils.
- Extensions to the BB protection (i.e. additional circuits) was not an easy task, involving additional wiring, significant commissioning, settings calculations and possible changes to settings.
- The scheme was not fully self-monitoring, and some components particularly settings resistors may become faulty without this being monitored and alarmed.

A digitally based BB protection resolved most of the above difficulties but also carried some disadvantages, which are as follows:

- The digital scheme is invariably a one-out-of-one system (i.e. no check zone), such that if the BB protection fails, the BB protection for the whole substation is lost – which was not the case with high-impedance schemes. This was a major concern for system operators. Later digital designs were to have two independent operating characteristics – but in the same relay.
- The high-impedance scheme was readily understandable to most engineers, whereas the digital scheme is more highly specialised.
- Repair times are invariably longer with digital schemes than high impedance.
- Lifespan of the relay is likely to be shorter, possibly 50% shorter.

Most power network companies (in the United Kingdom), although not all, now install digital/numeric low-impedance BB protection. Salient aspects of the design are as follows:

- Input is usually from a single CT per circuit.
- Protection tripping/stability logic is provided by two different algorithms within the one relay as stated above, requiring the operation of both to cause a protection trip output signal. One of the algorithms may operate on a check zone principle as shown in Figure 10.28 where the infeeds of all the feeders are integrated.

- CT and switchgear position inputs are usually fed into a per circuit 'bay unit' which retransmits the data to a BB protection central unit which contains the scheme logic algorithms, and the appropriate trip output signals, see Figure 15.9. The trip outputs signals are transmitted back to the bay units – which in turn trip the associated circuit breakers.
- Critical 400 kV sites may be designed with a backup central unit in a separate cubicle – with arrangements in place for rapid changeover from the main to the backup unit, should the main unit fail.
- Relay functionality in both algorithms is usually that of a biased differential relay, similar to that illustrated in Figure 10.13. Settings are heavily influenced by manufacturers criteria but must be capable of operating for a BB fault under minimum in-feed conditions usually coincident with a specified through load bias current – which influences the selection of the bias slope setting.

10.12 Circuit breaker fail protection

10.12.1 Circuit breaker fail – requirements

Circuit breaker fail protection, in the United Kingdom, has been installed since the early 1980s at network voltages above 132 kV and at some critical 132 kV sites. The reason for its introduction was to ensure that the risk of an uncleared fault, arising from the failure of a circuit breaker to trip (i.e. open) was discriminatively cleared. This followed actual experience of a number of 'stuck' circuit breakers resulting in indiscriminate clearance by backup protection. The reasons for the circuit breakers failing to trip were either a failing of the trip coil circuitry or circuit breaker mechanism failure.

Prior to 1980, circuit breaker fail protection was dependent upon backup protection detecting the fault. However, with the tendency towards ever-increasing maximum load currents, without a corresponding increase in minimum fault currents, backup protection settings could not be depended upon to detect all (stuck breaker) faults. In addition, backup protection suffers from the disadvantage that it is relatively slow in operating – so risking system stability.

Figure 10.29 illustrates the principle of application of circuit breaker fail protection to a double BB substation – only two circuits are shown to simplify the explanation. There is more than one form of circuit breaker fail scheme but the one shown was by far the most common application – prior to the advent of numeric relay schemes (which largely operate on the same principle). For ease of explanation, operation of the protection will be via contact logic. Salient points are as follows:

- Circuit breaker fail protection operates for the instance of a circuit breaker failing to trip following a trip command from an associated trip relay.
- Figure 10.29(a) depicts the circuitry that exists for each circuit in the substation, a current-operated relay, CCK (current check), obtains a DC input via the operation of a trip relay, and providing the current input also exceeds the setting (typically 200 mA), so indicating continuing current flow through the circuit breaker, the relay operates.

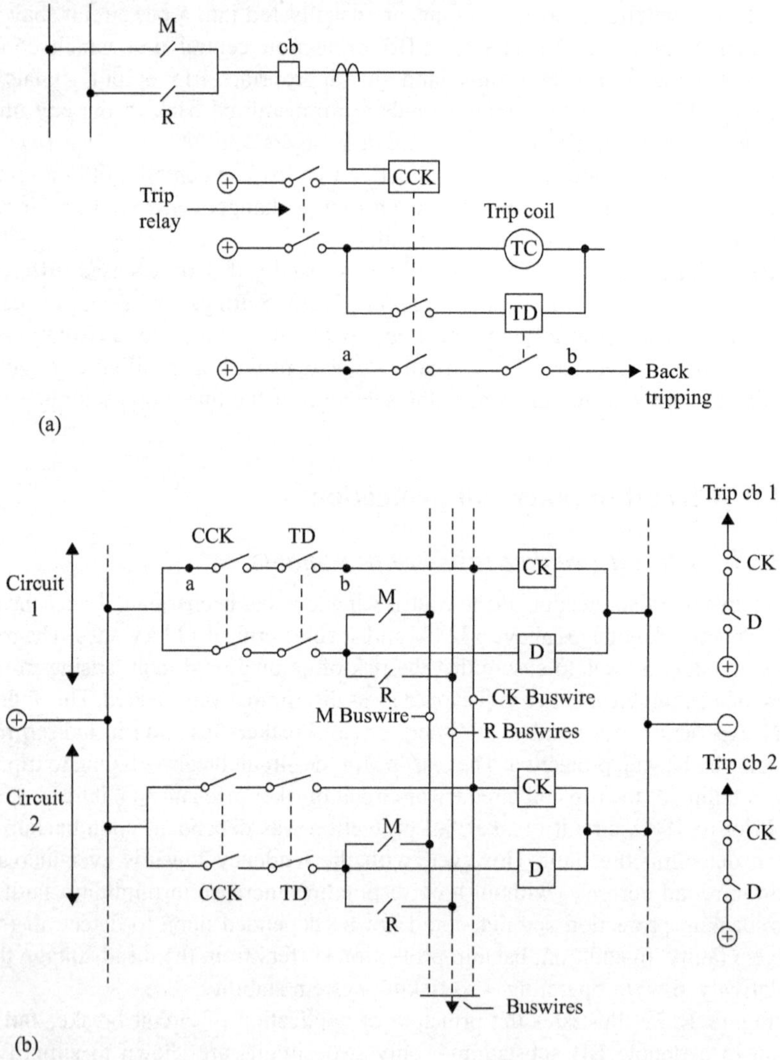

Figure 10.29 Circuit breaker fail protection – circuitry: (a) back tripping – initiation circuitry and (b) back tripping – trip circuitry

- The CCK relay energises a timer, TD, which is positioned across the circuit breaker trip coil.
- If at the expiration of the TD (typically 150 ms) the current is still flowing in the CCK relay (implying that the circuit breaker is stuck), the 'back tripping' of all other circuit breakers connected to that BB will commence.
- In Figure 10.29(b) (which shows only two circuits – to illustrate the principle of operation), the circuitry between points 'a' and 'b' is one and the same as

that in Figure 10.29(a). As a result, with the closing of the CCK and TD contacts, current flows into the D (discrimination) and CK (check) relays associated with circuit 1, which in turn close contacts to again request the circuit breaker associated with circuit 1 to trip.

- If circuits 1 and 2 are selected to the same BBs, say the main BB, M, such that the, M, auxiliary contacts in circuits 1 and 2 are closed, then current flows through the, M, auxiliary contacts associated with circuits 1 and 2 to operate the D (discrimination) relay associated with circuit 2.
- Similarly, current also flows from circuit 1 onto the check buswires to operate the, CK, relay associated with circuit 2.
- Closure of the D and CK contacts on circuit 2 results in a 'back trip' of the circuit breaker associated with circuit 2.
- In addition (and not shown), it is frequently arranged for the busbar protection discrimination and check relays to trip into the circuit breaker fail protection discrimination and check buswires, as an additional means of tripping a busbar.
- The setting of the CCK relay is required to monitor circuit breaker resistor interrupter currents, which dictates a setting as low as 200 mA (i.e. 400 A at 400 kV on a 2,000/1 CT ratio).
- The setting of the timer TD should exceed the longest circuit breaker arc interruption time and should take account of the CCK and TD relay reset times.
- The time from fault inception to the circuit breaker fail protection clearing the fault (i.e. arc interruption of all in-feeding circuit breakers connected to the BB in question) is typically in the order of 200–250 ms.
- Similar circuitry to that examined above applies to mesh and single-switch substations.

10.12.2 Circuit breaker fail protection – numeric relays

It is common practice with numeric multifunction relays to incorporate the BB protection and circuit breaker fail protection into the one relay. This is because the two protection schemes share many of the same inputs and outputs, i.e. CTs, disconnector auxiliary contacts, trip outputs, etc., and for both types of protection, tripping occurs on a per BB basis. This is cost effective compared to hard wired older technology schemes (of which many still remain in service) and (in principle) results in simpler commissioning and circuit extensions.

10.13 Protection, control and telecomms – communication channels

10.13.1 Communication channel requirements

The communication channel requirements are generally as follows:

1. **Protection**
   (i) Unit protection communication channels – to enable comparison of electrical quantities at both ends of a feeder

(ii) Protection signalling channels – e.g. distance protection blocking communications channel
(iii) Inter-tripping communications channel

Protection signalling equipment operating times are typically 15 ms, and inter-trip equipment operating times typically 40 ms. This does not take into account the communications medium propagation time. The longer inter-tripping equipment operating time is usually associated with higher levels of equipment security.

2. **Control**
   (i) SCS (substation control system) (SCADA) communication channel between substation and electricity control centre
   (ii) Commercial metering communication link between the substation and metering settlement point.
3. **Telecommunication**
   (i) Operational telephony communication channel
   (ii) Public telephony communications channel
   (iii) Internet and email access communications channel

   Although an increasing number of substations have all of the above facilities, some may not have internet and email as a result of being remote from a wideband network – and the cost of installation is not justified.

10.13.2 Communication channel mediums

Communication mediums are in a state of change as digital communications become more available. They generally comprise the following:

1. **Private pilots**
   Private pilot cables are usually laid in the same trenches as the HV power cable. These can be used for both protection communications (for pilot-wire protection) and inter-trip signalling.
2. **Rented pilot wires**
   Rented pilot wires can be rented from telecommunications companies and are generally utilised as follows:
   (i) Pilot-wire protection
   (ii) Voice frequency protection signalling and inter-tripping
   (iii) Operational and public telephony

   Rented pilot wires are generally being phased out and replaced by rented digital communication systems.
3. **Power line carrier**
   PLC uses the power line itself as the communications medium. It is generally used for:
   (i) Protection communications
   (ii) Inter-tripping
   (iii) Protection signalling

   Again, this medium is rapidly being displaced (at least in the United Kingdom) by digital communication mediums.

4. **Radio**
   A limited amount of radio has been used in the United Kingdom using aerials and satellite dishes, e.g. for GPS time synchronisation.
5. **Optical fibres**
   Optical fibre technology is increasingly becoming the dominant form of communication medium for all applications. The most common form of transmission uses synchronous digital hierarchy which is an international standard protocol that transfers multiple digital bit streams, synchronised over optical fibre. This technology requires the installation of multiplexor equipment in the substation. The equipment, together with the fibre optic connecting cable, is usually provided by, and rented from, a telecommunication company.

10.13.3 Communications equipment – requirements

At an early stage, in a scheme (usually at time of development), it is essential that the communication requirements relating to the construction work are evaluated for purposes of design and costing. Depending upon the scale of the construction work and the communications infrastructure that is either in existence or needs to be installed, the design work and related costs can vary significantly. The range of work includes:

- Addition of extra hardware into the substation-based multiplexor.
- Additional multiplexor either because the existing multiplexor is full to capacity, or it is a new substation.
- New incoming fibre optic cables, again because the current cable is full to capacity or that it is a new installation. NB: At transmission substations, it is usual to duplicate all protection facilities with each facility carried in a separate fibre cable – taking different physical routes.

10.14 Inter-tripping

10.14.1 Inter-tripping – application

Inter-tripping is the operation of protection at one of a circuit causing a signal to be transmitted to trip a circuit breaker at the other end. It is termed a 'direct transfer trip'. It differs from protection signalling, such as blocking, in as much as it is an unconditional tripping signal. Inter-tripping is commonly applied as follows:

1. **Transformer feeders**
   With reference to Figure 10.30(a), operation of the transformer protection results in an inter-trip send to trip the remote-end circuit breaker. In addition, operation of the feeder protection would result in an inter-trip send to trip the transformer LV circuit breaker. At transmission voltages the inter-tripping and protection are duplicated. Where private pilots laid in the same trench as the HV cable are used for inter-tripping, a design of inter-tripping termed

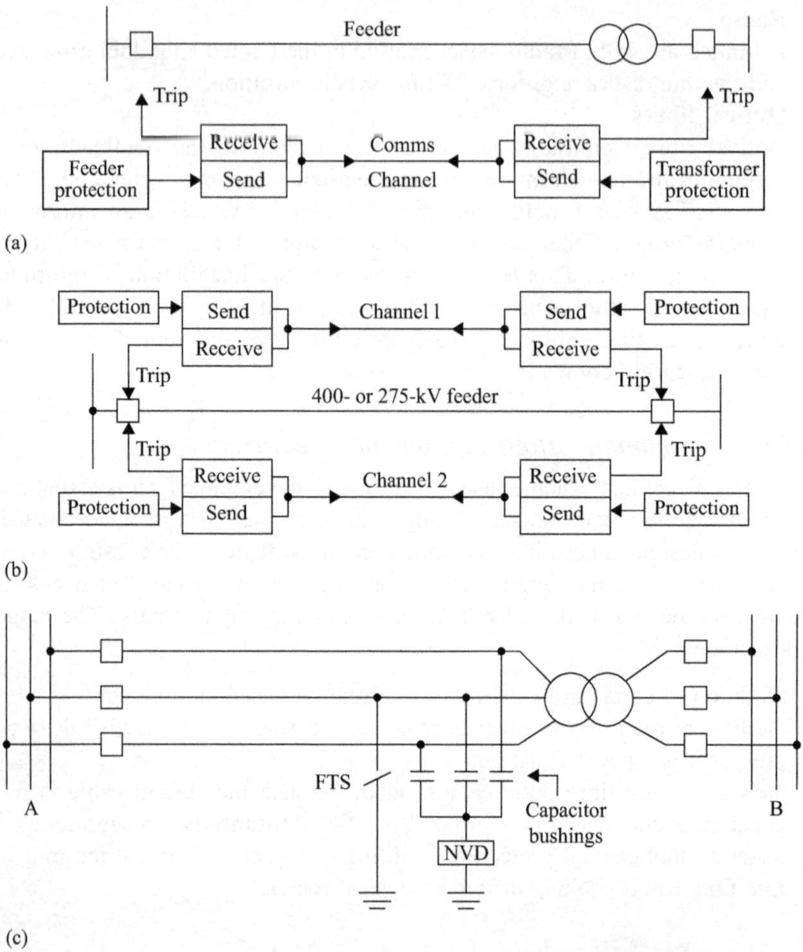

Figure 10.30 Inter-tripping application and methods: (a) inter-tripping application to a transformer feeder, (b) inter-tripping application to a transmission feeder and (c) NVD protection – and fault thrower switch

'surge proof' is generally used – this is usually a DC system, insulated and arranged to be immune from impressed voltages.

2. **Transmission network feeder**
   Figure 10.30(b) illustrates the inter-tripping arrangements associated with a transmission network feeder. In this instance, the inter-tripping is duplicated. Prior to the advent of numerical multifunction relays (containing an integral inter-tripping facility), there would have been only one inter-trip, associated with the non-unit main protection. The inter-trip send may also be initiated from BB protection and circuit breaker fail protection.

10.14.2 Unstabilisation, fault throwers and neutral voltage displacement protection

Two forms of inter-tripping that do not employ inter-trip send and receive equipment are unstabilisation and fault throwers, which are as follows:

1. **Unstabilisation**
   Unstabilisation consists of a feeder-unit protection at one end of a circuit being caused to operate for an out-of-zone fault at the other end of the circuit.
   For example, the operation of a BB protection may result in the tripping of a circuit breaker of an in-feeding feeder circuit, but because the fault is between the CTs and the circuit breaker, the fault continues to be fed from the remote end of the feeder. In this instance, a contact from the BB protection trip relay is applied to the local feeder protection, which modifies the data transmitted to the remote end of the feeder such that it concludes that an in-zone fault has occurred and therefore trips the associated (remote end) circuit breaker.
   Similarly, with a transformer feeder, operation of the transformer protection results in unstabilisation of the feeder protection to cause tripping of the remote-end circuit breaker.
   This method of inter-tripping has the advantage that an additional inter-tripping equipment and communications channel are not required.
2. **Fault thrower**
   This method of inter-tripping consists of applying a single phase to earth fault to the power system via a 'fault thrower switch'. This switch is designed for this purpose. It is initiated by protection. With reference to the network in Figure 10.30(c), a typical application is when a transformer fault occurs and the transformer protection initiates a fault-throwing switch, which shorts one of the phases to earth. The fault then causes the feeder protection at the remote end to operate and trip the associated in-feeding circuit breaker. This method of inter-tripping is used on the 132- and 33-kV systems, especially when it is difficult to obtain a conventional inter-trip channel (i.e. absence of telecommunications infrastructure). It additionally saves on the cost of inter-tripping equipment.
3. **Neutral voltage displacement protection**
   Neutral voltage displacement (NVD) protection is frequently employed on the 33-kV network when an inter-tripping communication medium between source substation and a transformer feeder is not available. Figure 10.30(c) shows the principle of operation. It is usually applied as follows:
   (i) Should a feeder fault arise the protection at substation A operates to trip the in-feeding circuit breaker. However, without conventional inter-tripping, the circuit breaker at substation B remains closed and, in the instance of a parallel feed, back-feeds the fault (but with a fault current so low that no current-operated protection would operate). The NVD detects an imbalance in the voltage to earth caused by the fault and trips the LV circuit breaker. NB: In some applications, a two-stage NVD may be required.

(ii) With a broken OHL conductor falling to the ground, on the transformer side of the break, the NVD protection operates both to trip the LV circuit breaker and closes the fault thrower to cause the protection at substation A to operate.

10.15 Protection application

10.15.1 Protection application considerations

Each power network company usually has a protection application policy document – detailing the protection to be applied to the most common types of circuits (e.g. feeder, transformer, BBs, etc.). This usually leads to the formation of standard protection application diagrams – which greatly simplify the design requirements during construction. In addition, standardisation provides confidence that the design will function properly.

The extent to which a circuit is protected is dependent upon the network voltage. The higher the voltage, the more critical the equipment is to the scale of security of supply, and the greater the economic justification for increasing the level and standard of protection. Within this context, the 400- and 275-kV transmission networks are designed in accordance with a protection policy that stipulates

> No credible failure of protection and associated systems shall prevent a fault from being removed from the transmission system. The term 'failure' means fails to operate within a maximum specified operating time. This is specified as 40 ms for main protection operating times, to achieve the target fault clearance times specified in Figure 5.10.

To implement this policy, on the 400- and 275-kV networks, all faults are required to be detected by at least two independent and discriminative high-speed protection systems. Each protection is allocated to an independent tripping system, fed from separate DC supplies. Circuit breakers are supplied with two trip coils, each of which connects to one of the two tripping systems. This is in effect a one-out-of-two-trip system arrangement.

The 132- and 33-kV networks are normally designed with a protection policy that stipulates

> Sufficient redundancy shall be in-built such that failure of a single item of protection will not lead to restrictions in availability of equipment.

Both the 132- and 33-kV networks are therefore generally provided with a single high-speed main protection, and a backup protection. Tripping is via a single battery into a single trip coil, i.e. essentially a one-out-of-one system, supported by busbar protection. Main protection operating times are generally required to be 50 ms or less, such that when the circuit breaker opening time is added, the target fault clearance times specified in Figure 5.10 are not exceeded.

It is worthy of note that in practice, modern main protection operating times for protections installed on all networks have operating times under 40 ms with some as low as 15 ms.

The following sections will summarise typical protection application arrangements for the commonly installed equipment of feeders and transformers.

10.15.2 Protection application – feeders

The following feeder applications will be summarised:

- 33 kV
- 132 kV
- 400 kV (275 kV is similar)

- **33-kV feeders**
  Figure 10.31(a) illustrates typical protection application to a 33-kV feeder. This arrangement shows distribution voltage metalclad switchgear with CTs positioned as shown. The main protection is mainly unit protection, communicating via pilots or fibre. In rural areas where communication connections are not readily available, distance protection may be used. Backup protection is usually IDMTL two-pole OC and EF protection. Some rural areas with a closed 33-kV ring circuit are protected by directional relays with time-graded settings to achieve discrimination around the ring.
- **132-kV feeders**
  Typical 132-kV feeder protection is shown in Figure 10.31(b). This shows a double BB AIS substation with high-impedance BB protection using discrimination and check zones. The main protection will again be either unit protection via pilot or fibre cables, or distance protection, with the backup protection being that of IDMTL two-pole OC and EF relays. In some instances, directional relays may also be used as mix of second main protection (MP2)/backup protection for ring feeders (not shown).
- **400-kV feeder**
  Figure 10.31(c) shows the protection application for a 400-kV feeder (and equally a 275-kV feeder). As stated in Section 10.15.1, the main protection is duplicated and termed MP1 and MP2, respectively. MP1 is usually required to be a unit protection (mostly communicating via fibre), and the MP2 a non-unit protection, i.e. usually distance protection. Very short-length feeders may dictate the use of two-unit protection, and if this is the case, the hardware platforms and algorithms of each are usually required to be different, to de-risk common mode failures – but may be supplied from the same manufacturer. Duplicate inter-tripping would usually be specified. The backup protection would usually comprise an IDMTL EF relay to protect against high-resistance EFs. The setting of this relay would need to be higher than the standing ZPS component of the load current of the feeder (arising from natural unbalance of the network). The BB protection assumes a numeric relay with a single CT input. Circuit breaker fail protection is provided.

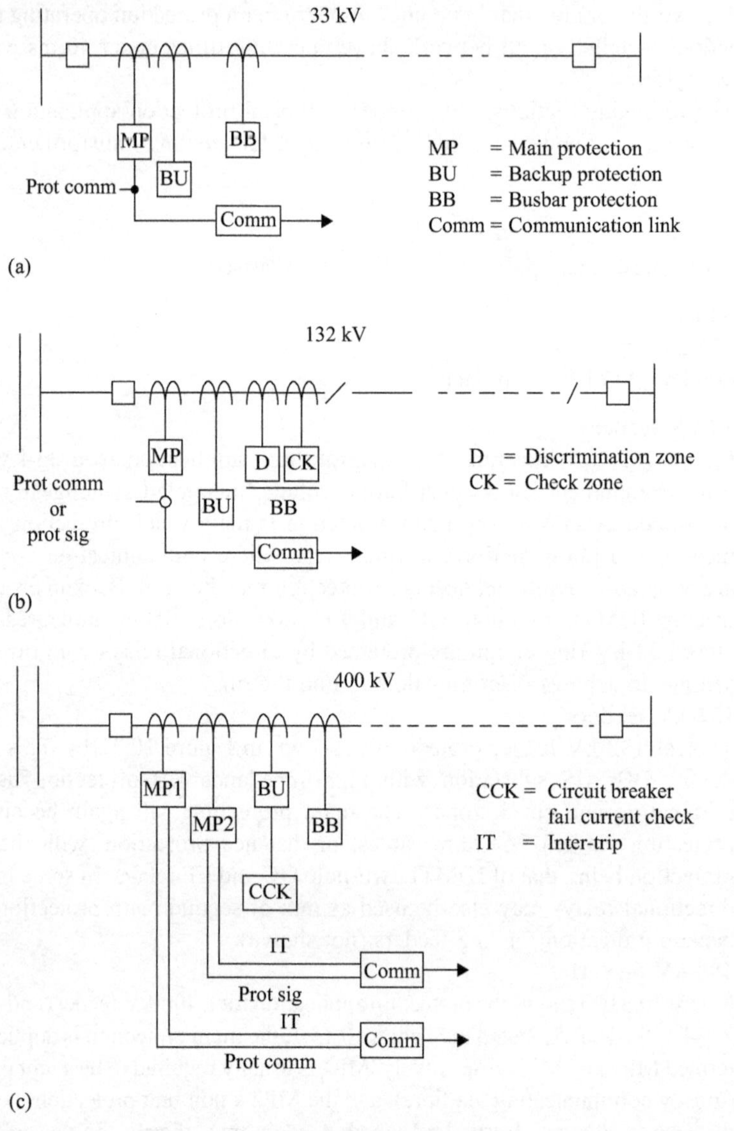

Figure 10.31 Feeder protection – application: (a) 33-kV feeder protection, (b) 132-kV feeder protection and (c) 400-kV feeder protection

10.15.3 Protection application – transformers

The following transformer protection will be briefly reviewed:

1. 33/11-kV transformer
2. 132/33-kV transformer
3. 400/132-kV transformer

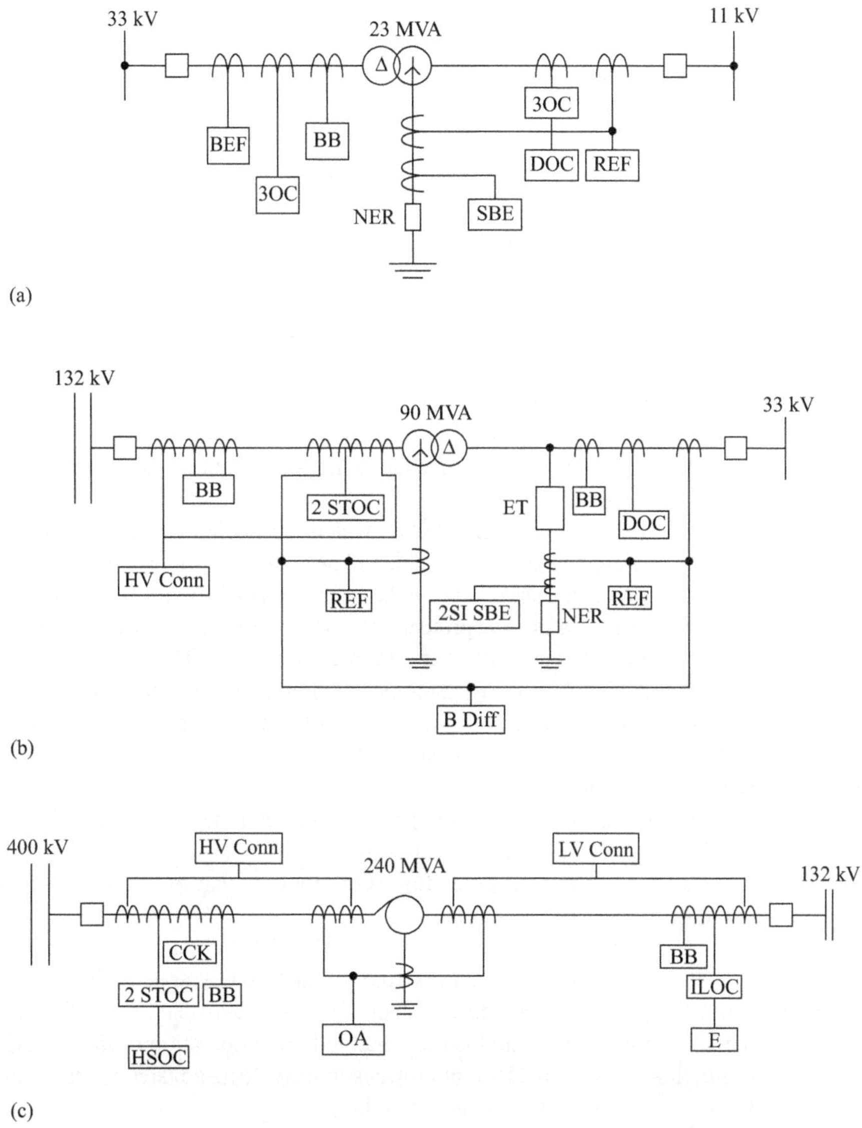

Figure 10.32 Transformer protection – application: (a) 33/11-kV transformer protection, (b) 132/33-kV transformer protection and (c) 400/132-kV transformer protection

1. **33/11-kV transformer**

   With reference to Figure 10.32(a), the main features are as follows:

   (i) Main protection of the transformer is provided by both the HV BEF and LV REF protection together with the HV IDMTL three-pole OC relay (3OC).

(ii) The LV IDMTL OC relay (3OC) together with the SBEF relay provide backup protection to the 11-kV network and trip the LV circuit breaker. The HV OC relay will time grade with the LV OC relay.
(iii) The DOC relay is applied as in (2), below.

2. **132/33-kV transformer**
With reference to Figure 10.32(b), the key protection features of a 132/33-kV transformer are as follows:
(i) Transformer main protection is provided by a biased differential relay (B Diff), and HV and LV REF protection.
(ii) The two-stage OC relay provides backup protection as described in Section 10.6.4, and the two-stage SBEF relay provides backup protection for EFs as described in Section 10.6.5.
(iii) Where the distance between the HV BBs and the transformer is relatively long, it is usual practice to provide HV connections (HV Conn) protection. This may operate on either a high-impedance or low-impedance protection principle.
(iv) The DOC relay is a three-pole relay, which is installed when two or more transformers are operated in parallel, and provides discriminative protection clearance for an uncleared fault on the transformer LV terminals. This relay also provides discriminative protection for faults on the transformer HV side, if the transformer is connected as a transformer feeder. The relay both trips the LV circuit breaker, and initiates the LV circuit breaker auto-reclose relay.
(v) BB protection is illustrated as the high-impedance type with discrimination and check zones (at 132 kV).

3. **400/132-kV transformer**
With reference to Figure 10.32(c), the protection arrangements associated with a 400/132-kV transformer circuit are as follows:
(i) The transformer overall protection (OA) may be that of a biased differential or high-impedance relay.
(ii) The operation of the two-stage OC (2 STOC) protection and the HV HSOC are described in Sections 10.6.4 and 10.6.6, respectively.
(iii) The LV-interlocked OC relay protects for a fault between the LV CT and the LV circuit breaker and is only enabled upon operation of the LV BB protection. It trips the HV circuit breaker and affects a faster trip than the second stage of the two-stage OC relay.
   - Both HV Conn and LV Conn protection are usually applied. The LV Conns protection provides discriminative clearance for the instance of an LV fault when two or more transformers are connected in parallel (an alternative would be to extend the transformer protection to include the LV Conns – which is usually not preferred).
(iv) The HV and LV BB protection are assumed to be numeric protections with single CT input.
(v) The LV IDMT EF relay (E) provides backup protection for faults on the 132-kV network. It is located on the LV side of the transformer to enable a more sensitive current setting to be obtained.

All transformers are fitted with Buchholz and winding temperature protection as described in Chapter 9.

10.15.4 Protection application – complex circuits

Complex circuits may be defined as:

- Three or more circuit ends
- Composite circuits (e.g. mix of OHL, cables and transformers)
- Weak infeeds

- **Circuits with three or more circuit ends**
  At transmission network voltages of 400 and 275 kV, it is unusual to have a circuit with more than three ends – but this may be the case at the lower voltages. Considerations are as follows:
  - Feeder-unit protections are usually not designed to be applicable to more than three circuit ends.
  - IDMTL protection is usually difficult to apply to three or more circuit ends as sensible discrimination under all running conditions is problematic to achieved, as such compromise settings may be accepted.
  - The limits of application of distance protection have been discussed earlier in this chapter – but instances have arisen of distance being applied to as many as five ended circuits at 132 kV (usually transformer feeders). When application proves difficult, the circuit under consideration may need to be sectionalised through the introduction of additional circuit breakers, and in some instances, a new substation.
  - Protection application must consider maximum and minimum fault levels and all possible operational running arrangements.
- **Composite circuits**
  Composite circuits may involve combinations of OHL, cables, transformers, quadrature boosters, series compensators, etc. Such combinations may prove problematic for the application of distance protection and IDMTL protection, particularly when considering maximum and minimum fault-level conditions. Complex circuits may also be subject to ferroresonance. In some instances, mid-circuit protection may be required, e.g. covering a mid-circuit cable, for the purpose of signalling to the remote ends to lock out auto-reclose.
- **Circuits with weak infeeds**
  Weak fault-level infeeds are often associated with intermittent generation, or wind generation where the contribution to fault current is low. Situations may arise where the maximum load current exceeds the minimum fault current. Consideration may have to be given to the impact of planned outages on fault levels, or to modifying protection settings in accordance with operational circumstances, or to the application of operational tripping schemes – see Section 10.21.

10.16 Protection and control settings

10.16.1 Protection and control settings policy

It is critically important that each power network company (or organisation acting on their behalf) develops and maintains a protection and control settings policy document, which details the settings requirements for every relay on the network. The policy document should also be supported by the technical criterion which underpins the choice of setting. It is advantageous both to power network construction and the security of the power system in general, if the relay settings across a particular network (or networks) are standardised. In many instances, this objective is achievable – but usually with exceptions. Settings policy generally exists at two levels, which are as follows:

- Power system level
- Relay-specific level

With reference to the protection and control procedural requirements covered in Chapter 20, the above will be briefly reviewed.

10.16.1.1 Settings policy – power system level

The power system level of settings policy considers the required protection operating and stability levels with reference to power system currents, voltages, impedances, etc. Operating times are also considered as integral to this level since time impacts directly on the performance of relays and the status of the power system. Typical considerations which fall into the power system level of policy comprise:

- IDMTL OC and EF relay operating currents (usually set around maximum load currents), specified in terms of primary (i.e. power system) currents – and target relay operating times for defined fault current conditions.
- Biased differential protection minimum POC, together with the through fault stability current for determining the bias slope setting.
- High-impedance protection minimum POC and through fault stability current.
- Distance protection settings in terms of the impedance to be measured (e.g. PPS), the feeder length to be covered by each zone, the feeder line angle and the zone operating times.
- System maximum load currents (usually short-term overloads) to which the relays may be subject.
- System maximum and minimum fault currents to which the relays may be subject.
- Protection target operating times to achieve the target fault clearance times specified in Section 5.4.1.5.

The technical data required to calculate or determine the levels of current, voltage, impedance, etc. should be obtained from:

- Maximum and minimum fault level studies – see Section 4.5.1
- Impedances database, i.e. OHL, cables, transformers and wound items of equipment, generators, etc. – see Section 4.2.8
- System studies – usually undertaken for non-standard applications or unusual network-operating conditions.

10.16.1.2 Settings policy – relay-specific level

This comprises the requirements for setting the individual relay(s) such that they accord with the power system level settings. For example, if an OC relay has to be set to a primary operating current of 1,000 A – the relay-specific level defines how the relay is to be configured to operate at a secondary current equivalent to 1,000 A. This may involve the positioning of switches or multipliers, or the software equivalent. With reference to the latter, some relays allow settings to be entered directly in primary quantities. In all instances, the CT and VT ratios must be taken into account. Most relays now possess more functionality than that required for a particular application, and as a result, some settings need to be either disabled or set to level at which they will not operate.

Manufacturers usually provide a settings guidance document for each relay. However, it is incumbent upon the power network company to agree the settings format with the manufacturer for their specific application, and this in effect becomes the relay-specific level policy. Agreement on the settings format may be achieved at the time of the type tests, as described in Chapter 21. However, it may be achieved/agreed it is essential that the settings policy is held in a documented policy suite and subject to change control. With programmable numeric relays, the settings policy will usually be held as a software file reference. Section 20.3.3 discusses the procedural requirements for relay settings.

Computer-generated relay settings

Most relay settings are capable of determination by computer-based systems, in such instances, only the technical parameters of the circuit in question need to be input. Increasingly, this will be the means by which relay settings are determined in future. Within this context, it is critical that any computer-based solution is trialled and confirmed to be error free. In addition, it is essential that some engineering expertise is retained in undertaking and understanding relay settings – since reliance on computer-based solutions usually result in a steep decline in expertise!

Multifunctional relays

As previously stated, the modern tendency is to incorporate individual relay functionality into a single relay which carries multiple functions. A typical example of integration for a 400-kV feeder as illustrated in Figure 10.31(c) is as follows:

1. **First main protection unit relay**
   Comprising the following functions:
   (i) MP1 (unit protection)
   (ii) First inter-trip
   (iii) Backup protection (IDMTL)
   (iv) Fault recorder
   (v) Communications medium interface
2. **Second main protection unit relay**
   Comprising the following functions:
   (i) MP2 (distance protection)
   (ii) Second inter-trip

   (iii) Backup protection (IDMTL)
   (iv) Communications medium interface
3. **Circuit breaker unit relay**
   Comprising the following functions:
   (i) Circuit breaker fail CCK
   (ii) Circuit breaker fail timer
   (iii) Synchronising
   (iv) Auto-reclose
   (v) Circuit breaker control (trip/close)
   (vi) Communications interface
4. **Substation common unit relay**
   Comprising the following functions:
   (i) BB protection
   (ii) Circuit breaker fail protection

Some multifunction relays are virtually universal in their functionality in as much as they carry most, if not all, circuit functionality and only that which is required is selected – with an acceptance of significant redundancy of functions. The selection of which functionality is selected for use in effect becomes a relay setting itself. A feature of multifunctional relays that has to be diligently managed is the very high number of settings that have to be set (or disabled).

10.17 Control systems

10.17.1 Control systems – introduction

As stated at the outset of this chapter, control systems are largely concerned with the manual and automatic operation of equipment usually under normal loading conditions. This differs from protection systems which are concerned with power system fault conditions – although the two systems often work in close collaboration. The following key control systems will be briefly reviewed:

- Voltage control
- Synchronising
- Auto-switching
- Operational tripping
- SCADA systems

10.18 Voltage control

10.18.1 Voltage control – introduction

The purpose of the power system is to deliver power from sources of generation to consumers, and in doing so, deliver it within a defined voltage range. As the power flows from generator to consumers, it does so via a range of equipment containing reactive (inductive and capacitive) and resistive impedances which modify the voltage profile. A required characteristic of the power system is therefore the automatic control of voltage between generation and the consumer terminals to ensure it is retained within the defined limits. This is accomplished by a combination of reactive compensation (shunt reactors and shunt capacitors) and transformer automatic tap change control (ATCC) schemes. Within this context, an overview of voltage control will be undertaken with reference to the following:

- ATCC performance characteristics
- ATCC types
- Voltage control strategy
- Automatic reactive switching (ARS) schemes

10.18.1.1 ATCC performance characteristics

When a transformer carries load current, a voltage is dropped between HV and LV terminals (essentially across the transformer leakage reactances as typically shown in Figure 3.14), resulting in a variation in transformer LV voltage. Automatic adjustment of the transformer tap position enables a constant secondary voltage to be maintained as the load through the transformer varies. This is achieved by an ATCC scheme, also known as an AVC (automatic voltage control) scheme. The heart of the scheme is an AVC relay, which some literature refers to as an AVR (automatic voltage relay) – which must not be confused with the generator AVR as described in Chapter 3, which serves a similar, but not fully identical, purpose.

In simple terms, and with reference to Figure 10.33, the AVC relay is supplied with the transformer LV voltage via a VT. The relay has a nominal voltage setting with a dead-band voltage range either side of nominal. Any voltage excursion outside the dead band for a defined period of time results in a signal to the tap changer to adjust the transformer tap position (either up or down as appropriate) to bring the LV voltage back within the dead band.

10.18.1.2 ATCC types

ATCC operation must satisfy two main requirements, which are as follows:

- Maintain the transformer LV voltage within specified limits
- Ensure that when transformers are operated in parallel, the taps are aligned to minimise transformer circulating current, as described in Chapter 9 and Figure 9.15.

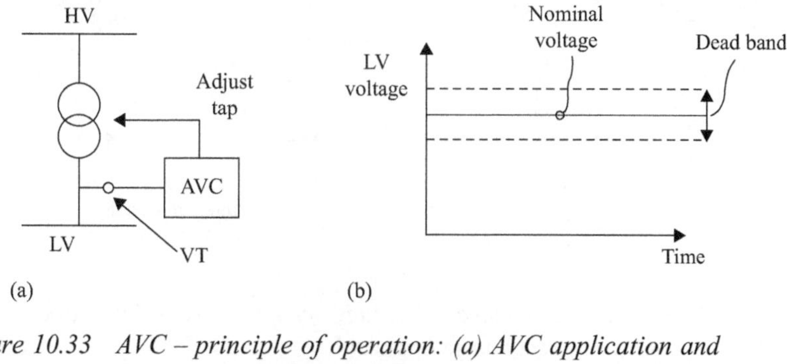

Figure 10.33 AVC – principle of operation: (a) AVC application and (b) AVC relay characteristic

There are three traditional designs for achieving the above, which are as follows:

1. **Master – follower system**
   This system comprises selecting one of the transformers in the substation to 'master' and the other transformer(s) to 'follower'. The AVC relay associated with the master takes control and makes the decision to tap. Once the master transformer has tapped, the follower is arranged to tap in the same direction, maintaining all transformers on the same tap. Prior to the advent of computer-based relays, this system involved very significant inter-transformer wiring, particularly if three transformers were involved.
2. **Circulating current system**
   This system is suited to a radial network as illustrated in Figure 10.34(a). If the transformers are initially on the same tap, there is no circulating current between the transformers. Should one of the transformers be instructed by its AVC relay to tap, so the transformers are on different taps, a circulating current arises around the transformers. If the transformer CT secondary currents are compared the circulating current appears as an imbalance, and the imbalance current is fed into the AVC circuitry in such a way that the two transformer AVC relays are presented with voltages that require them to tap towards each other. This system is arranged so that the transformers are capable of running one tap apart. Generally, the circuitry involved with the circulating current scheme is simpler than that for the master follower, but the circulating current scheme is only suitable for sites with two parallel connected transformers. Both the circulating current scheme and the master-follower scheme can be arranged to incorporate a 'line drop compensation' feature to enable voltage at a point remote from the transformer to be maintained at a constant level.

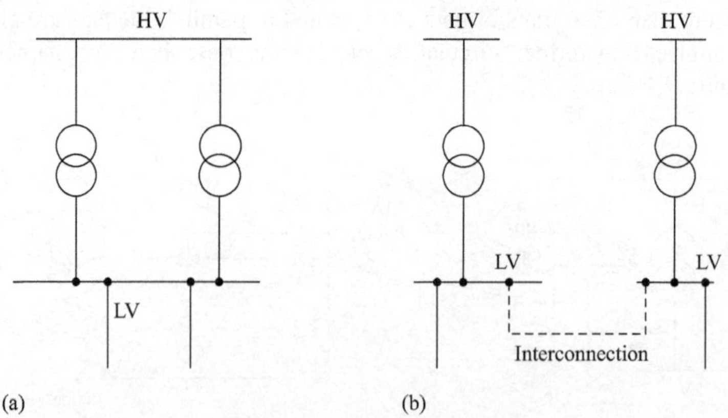

Figure 10.34 Transformer connection arrangements: (a) radial network and (b) interconnected network

3. **Reversed reactance system**
   The reversed reactance system is applicable to an interconnected network as illustrated in Figure 10.34(b), although it can equally be applied to a radial system. The design dates back to the days of electromagnetic technology. Each transformer LV current is fed into its AVC relay via a CT. By passing the CT current through an inductance, in a reverse direction, the voltage seen by the AVC relay, as a result of transformer circulating current, is such as to make the transformers tap together. The advantage of this system is that it requires no connecting circuitry between each transformer AVC. A slight inaccuracy in voltage measurement arises from the fact that the design assumes the load current to be at a defined power factor.

Modern computer-based relays use algorithms to mimic all of the above methods – and with greater options and functionality.

10.18.1.3 Voltage control strategy

Clearly, a coordinated strategy is required in the control of voltage across the whole power system, both to avoid hunting between the various AVC relays, and to maximise the contribution that each device may make. Figure 10.35 provides a

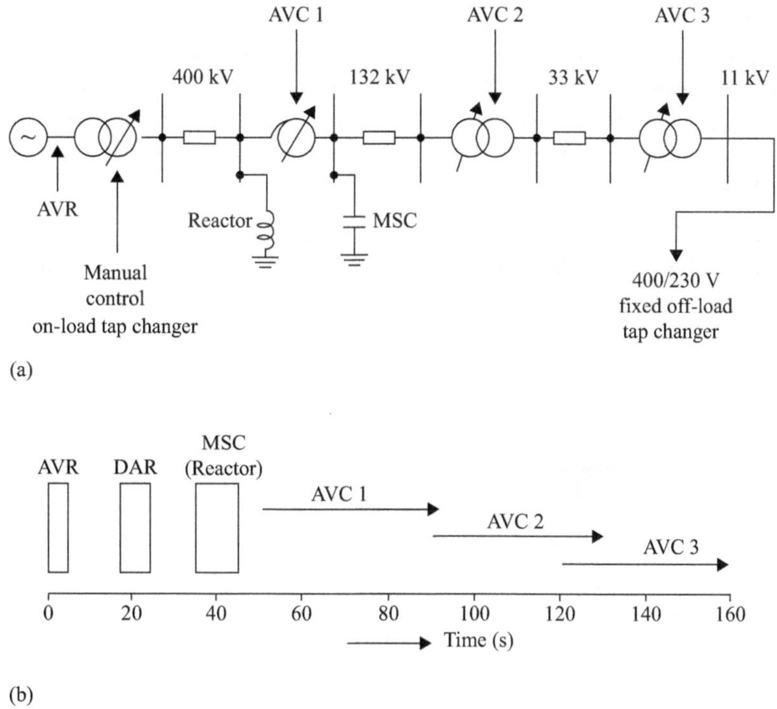

Figure 10.35 Voltage control – equipment and coordination: (a) power system voltage control equipment and (b) voltage control – time coordination

simplified overview of a power system wide voltage control strategy. Salient observations are as follows:

- The positioning of the reactors and MSCs is an example only. In practice, both could be either on the HV or LV network.
- The generator AVR is fast acting to hold the generator terminal voltage constant.
- Voltage control of the power station 400 kV BBs is via manual control of the generator tap changer.
- For the instance of network voltage being reduced as a result of a circuit fault and subsequent trip, time is allowed for delayed auto-reclose to operate and restore the circuit, and with it restoration of voltage levels.
- Time is allowed for reactive compensation e.g. reactors and MSCs to operate to maintain the transmission network voltage within specified limits.
- AVCs are used to regulate voltage for load flow through the transformers.
- Evaluation has determined that the order of priority of AVC tap changing should be at the higher voltage first. This minimises the risk of AVC relay hunting and repeated tap changing to either over, or under, correct.
- A strategy is also required for the AVC dead bands. Analysis shows that generally a wider dead band at the higher voltages provides optimised performance.

10.18.1.4 Automatic reactive switching (ARS) scheme

With reference to Figure 10.36, ARS systems are generally associated with the interface between the 400 (or 275) and 132-kV networks. They provide a coordinated voltage control solution to the maintenance of both the transmission HV voltage and the transformer LV BB voltage. The ARS comprises a single unified control device with the following inputs and outputs:

- VT and CT inputs from the LV of each transformer
- Circuit breaker and disconnector auxiliary contacts inputs from each substation
- Outputs to control each MSC, each reactor and each transformer tap changer.

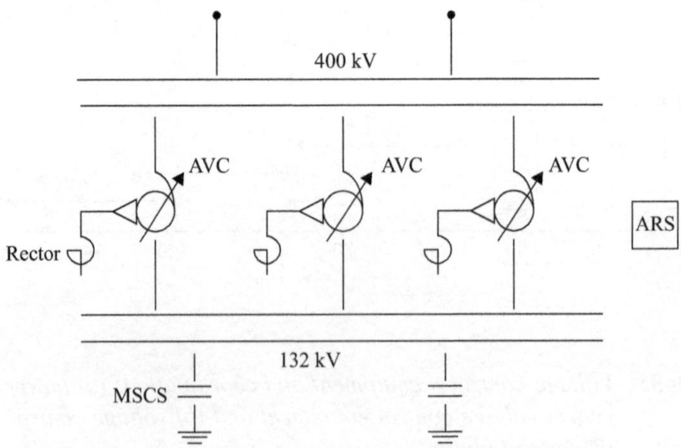

Figure 10.36 Main equipment – ARS system

The ARS relay replicates the running arrangements of each substation to determine which transformers, reactors and MSCs are connected to which BBs. It usually contains an AVC relay which is associated with each LV BB. When switched, the MSCs and reactors influence the voltage on the HV BBs in addition to the LV BB, whereas the AVC relays only influence the voltage on the LV BBs. Instances arise of reactors being connected to the HV BBs. The voltage control regime would follow the strategy outlined above. The ARS system, although relatively complex provides a high standard of voltage control. It is time consuming to commission.

10.19 Synchronising

10.19.1 Synchronising relay – purpose

With reference to Figure 10.37, a synchronising relay is used to ensure that two AC supplies which are required to be connected in parallel are aligned within a specified range of settings, and therefore classified as being in 'synchronism', thereby allowing a circuit breaker to be closed to parallel the supplies. The settings limits associated with synchronism comprise:

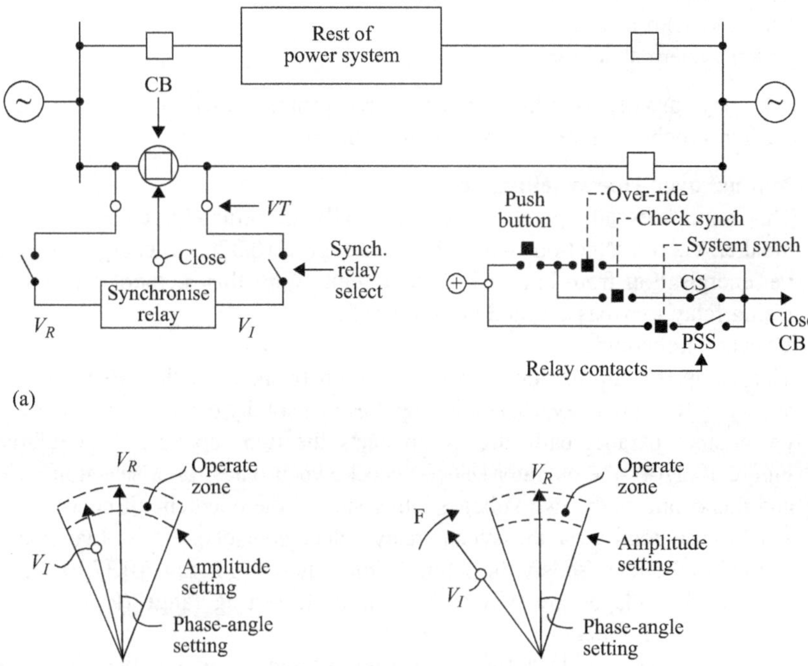

Figure 10.37 Synchronise relay inputs and operating settings: (a) synchronise relay and selection switch, (b) check synchronise and (c) power system synchronise

1. **Voltage amplitude**
   Both voltages must be within a defined voltage amplitude range, close to nominal voltage
2. **Voltage phase angle**
   Both voltages must be within a defined phase angle relative to each other
3. **Supply slip frequency**
   Both supplies must be within a defined 'slip' frequency range relative to each other.

Circuit breaker closure without satisfying the synchronise relay settings criterion is termed an 'out-of-synchronism' closure and risks the following:

- Power system instability and possible loss of electricity supplies
- Damage to generators, and other equipment, particularly as a result of pole slipping
- Operation of (non-unit) protection relays arising from large out-of-synchronism current flows and voltage depressions.

Generally, there are three different ways in which a circuit breaker can be closed, which are as follows:

1. Manual over-ride synchronism (deadline)
2. Check synchronise
3. Power system synchronise (PSS)

Modern relays usually have both the check synchronise function and PSS function in the same synchronise relay. These will be briefly reviewed:

1. **Manual over-ride synchronise**
   This comprises an operator, located locally or remotely, closing a circuit breaker, via the selection switch shown in Figure 10.37(a) to energise a circuit, i.e. energisation from one end of the circuit. With this method, the synchronising relay is bypassed (and not required).
2. **Check synchronise**
   This comprises an operator, closing a circuit breaker via the selection switch and synchronising relay, to parallel supplies in a solidly connected power system i.e. another parallel path already connects the two supplies. With reference Figure 10.37(a), the operator selects 'check synchronise' on a selection switch, and this connects the two voltages either side of the open circuit breaker to the synchronise relay (via the synch. relay select contacts). If the two voltages applied to the relay satisfy the settings limits shown in Figure 10.37(b), i.e.:
   (i) Both voltages are within the amplitude setting range (typically 85% nominal voltage)
   (ii) The phase angle between the two voltages is within the phase-angle setting (typically ±25°)
   (iii) The above two criteria remain in place for a defined TD (typically 2 s) to ensure that the power system is solid (and not split).

Then if all of the above settings limits are met, the check synchronise function contact 'CS' closes. The operator may then press the close push button and a signal is sent to the circuit breaker close coil to close the circuit breaker and parallel the supplies.

3. **Power system synchronise**
   PSS comprises the paralleling of two independent (i.e. split) power systems with slightly different frequencies. With reference to Figure 10.37(a), an operator selects PSS on the selector switch, which, via the relay select contacts, connects the voltages either side of the open circuit breaker to the synchronise relay. The settings criterion, as shown in Figure 10.37(c), for the PSS function to operate comprises:
   (i) Both voltages must be within the amplitude setting range (typically 85% nominal voltage)
   (ii) The slip frequency of the two voltages must be within the slip frequency setting range (typically within 0.125 Hz)
   (iii) The voltages angle must be within the phase-angle setting, which is determined by the circuit breaker closure time and the slip frequency setting – with the objective of circuit breaker interrupter closure when the two voltages are in phase (angle setting is typically 3°).

   Closure of the PSS contacts results in a signal being sent to the circuit breaker close coil to close the circuit breaker and parallel the two supplies – independently of operator involvement.

10.19.2 Voltage selection scheme

Figure 10.37(a) shows two voltages connected, via the synch. relay select contacts, to the synchronise relay: V_R (the BB running voltage) and V_I (the incoming voltage). The arrangement shown is for explanatory purposes only. In practice, a 'voltage selection' scheme exists for directing the appropriate voltages to the synchronise relay. The voltage selection scheme is fed from all VTs connected to the same BB (involving only one phase – usually yellow phase) and replicates the BB running arrangements and BB voltage. When selecting a synchronising function, the voltage V_R is applied to the synchronise relay (through the synch. relay select contacts) from the voltage selection scheme, and the incoming voltage V_I from the incoming VT.

Voltage selection schemes also exist for synchronising circuit breakers at mesh substations. However, the circuitry is more complex by virtue that two voltage selection circuits must exist, representing the voltage both in the clockwise and anti-clockwise directions respectively around the mesh. This is required to enable manual over-ride synchronise (deadline) to be performed – which could require a circuit breaker to energise equipment in either direction around the mesh, but with a voltage (i.e. running voltage) behind the open circuit breaker, no matter the direction of energisation.

10.20 Auto-switching

10.20.1 Auto-switching – introduction

The term auto-switching is an umbrella term describing four categories of activity, which are as follows:

1. **Auto-reclose**
   This is the automatic reclosing of a circuit breaker following a trip caused by a protection operation.
2. **Auto-opening**
   This comprises the automatic opening of a circuit breaker for operational reasons (e.g. part of an operational tripping scheme).
3. **Auto-close**
   This is the switching into service, via a circuit breaker, equipment that has been on standby – for operational reasons.
4. **Auto-isolation**
   This comprises the automatic opening of disconnectors for the purpose of isolating faulty equipment, following tripping of all in-feeding circuit breakers.

The following aspects of auto-switching will be examined:

- Auto-reclose application
- Single-shot auto-reclose
- High-speed auto-reclose (HSAR)
- Delayed auto-reclose (DAR)
- Mesh substation auto-switching
- Trip relay reset and persistent inter-trip

10.20.1.1 Auto-reclose application

Auto-reclose comprises the automatic reclosure, and return to service, of a circuit breaker following tripping of that circuit breaker by protection. Auto-reclose is commonly applied to OHL feeder circuits on the basis that the fault is likely to be transient (due to lightning, encroachment of foliage in severe weather, conductor clashing in high winds, ice dropping away, etc.), and therefore, the circuit can be successfully returned to service with the fault having disappeared, i.e. not permanent. Auto-reclose is usually not applied to cable circuits since cable faults are likely to be permanent. It is often applied to mixed OHL and cable circuits if the majority of the feeder is OHL. Auto-reclose has proved essential to maintaining continuity of supply to consumers, avoidance of fragmentation of the network and retention of system stability. Auto-reclose application in the United Kingdom is generally as follows:

- Single-shot auto-reclose mostly found on the 33-kV network and some parts of the 11-kV network
- HSAR – limited application in the United Kingdom, but extensive application worldwide
- DAR – applied extensively in the United Kingdom at network voltages of 132 kV and above.

10.20.1.2 Single-shot auto-reclose

With reference to Figure 10.38, single-shot auto-reclose consists of the following sequence:

1. Protection operation, auto-reclose initiated and circuit breaker trip
2. Short TD until it is assured the circuit breaker has fully cleared the fault
3. Short TD to be assured that the trip relays have reset
4. Close pulse sent to reclose the circuit breaker
5. If on circuit breaker closure the fault reoccurs within a specific time of the close pulse output, then the fault is taken as being permanent and the auto-reclose relay locks out – otherwise the auto-reclose relay is available to reclose for further faults (i.e. after the reclaim time).

The time from the auto-reclose being initiated to the output of the close pulse is usually termed the 'dead time'. The time from the close pulse being output to the time when the relay commences to be locked out (i.e. for the occurrence of a permanent fault) is termed the 'reclaim time'.

The total reclose time i.e. dead time plus reclaim time for a fast operating and resetting protection relay may be in the order of a few seconds (shorter times are theoretically possible but may be more onerous for the circuit breaker) – but with IDMTL type of relays where both the relay operation and reset is much slower the total auto-reclose time may exceed 15 s.

10.20.1.3 High-speed auto-reclose

HSAR systems are applied to 11-kV rural distribution networks – the application is well described elsewhere and out of the scope of this book. In the United Kingdom, HSAR was widely used on the 275-kV network but has largely been removed and replaced by DAR. The existing installations on the 275- and 400-kV networks are limited and usually remote from sources of generation. HSAR is widely used on transmission networks in many parts of the world. The basis of application is that the circuit is reclosed at high speed before synchronism is lost. Application usually exists in following two forms:

1. **Three-pole trip and HSAR**
   This comprises protection operation followed by tripping of all three poles (phases) of the circuit breaker with rapid three-phase auto-reclose (without a check for synchronism), typically in about 400 ms. This ensures that the system is reconnected at high speed to aid system stability and may reconnect a split system before synchronisation (and stability) is lost.
2. **Single-pole trip and auto-reclose**
   This system requires the detection of single-phase-to-earth faults by the feeder protection with subsequent tripping of the corresponding phase of the circuit breaker (i.e. single-pole trip) – followed by high-speed reclose of the single phase. This arrangement is superior in ensuring retention of power system stability – but adds significant complexity to the protection arrangements and circuit breaker trip/close circuitry and mechanism.

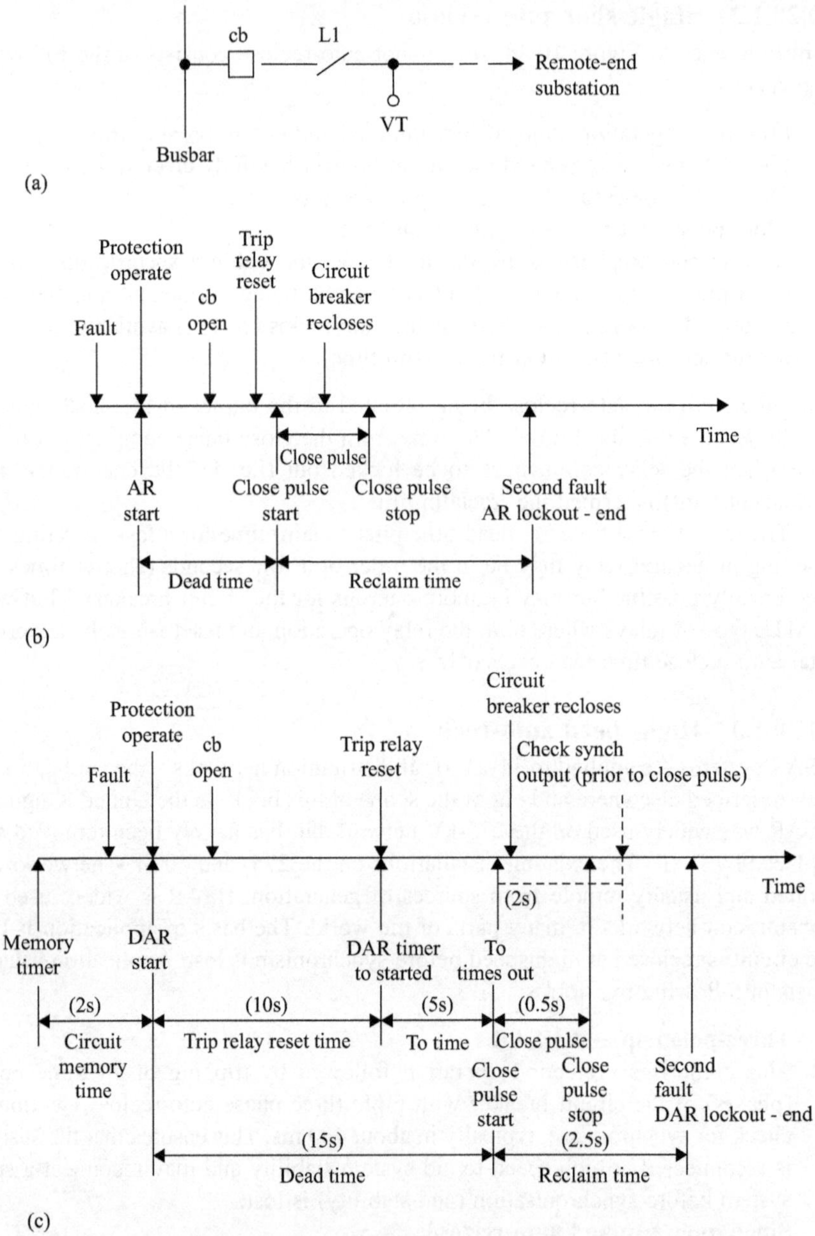

Figure 10.38 Auto-reclose arrangements: (a) feeder arrangement, (b) single-shot auto-reclose and (c) DAR

10.20.1.4 Delayed auto-reclose

The highly interconnected nature of the transmission networks in the United Kingdom is such that if HSAR was applied, the occurrence of a second fault on reclosure (i.e. a persistent fault) could result in power system instability. This risk is overcome by the application of DAR, which delays the auto-reclose until steady state system conditions are likely to have been re-established. With reference to Figure 10.38(c), salient points relating to DAR performance are typically as follows:

1. The circuit must be in service (typically for a minimum of 2 s) prior to the fault arising. This is confirmed by monitoring that the circuit breaker and disconnector auxiliary contacts were in the closed position prior to the protection operation, and that the VT was energised.
2. A feeder fault causes protection operation, which initiates the DAR, and tripping of the feeder circuit breaker. It is assumed that the remote-end circuit breaker is also tripped.
3. The trip relay reset scheme resets the trip relays, in typically 10 s. This causes the DAR internal timer TD to commence.
4. Assuming that dead-line auto-reclose is to take place (with check synchronise closure at the remote end), on completion of the TD time (typically 5 s), the auto-reclose relay initiates a close pulse to reclose the circuit breaker. NB: Prior to reclose, the relay checks that the BB voltage behind the open circuit breaker is ‘live’, and the feeder voltage in front of the open circuit breaker indicates ‘dead’.
5. At the instant, the close pulse is output, the reclaim timer is commenced (typically with a 2-s drop off time after close pulse cessation) during which time the relay is locked out and will not allow further reclosures. NB: Should a second fault occur in the reclaim time, it is considered that the feeder is subject to a permanent fault – for which reclosure is not permitted (i.e. lockout).
6. Typical overall time for a dead-line reclosure is 15 s.
7. If alternatively the feeder was dead-line reclosed from the remote end, the local end VT becomes live, and after completion of the time-delay TD, the synchronise relay is selected by the DAR relay – and if the BB and incoming voltages (see Figure 10.37) are within the setting limits of the synchronise relay, the check synchronise function sends a signal to the DAR relay, which outputs the close pulse to reclose the circuit breaker.

NB: In practice, there may be slight variants to the above explanation – albeit achieving the same end result.

10.20.1.5 Mesh substation auto-switching

The most complex application of auto-switching to be found on the UK networks is probably associated with mesh substations. This requires interconnected circuitry for the auto-reclose of all circuit breakers in the mesh and all transformer LV circuit breakers, together with circuitry for the auto-isolation of the feeder and transformer disconnectors associated with each mesh corner.

An example of the performance characteristics, and significant added value, of mesh substation auto-switching equipment is as follows. With reference to the mesh arrangement illustrated in Figure 20.15, should a trip on auto-reclose onto a permanent feeder fault on Line 1 occur, this would result in the automatic opening of disconnector 103, to isolate the faulty feeder, and automatic reclosure of circuit breakers 420, 120, and the transformer T1 LV circuit breaker, to restore the mesh and return healthy items of equipment back to operational service.

10.20.1.6 Trip relay reset and persistent inter-tripping signal

Trip relay reset and inter-tripping circuitry is integral to the performance of auto-switching schemes and DAR in particular. For example, with reference to the feeder shown in Figure 20.11, should feeder circuit breaker X105 DAR close onto a persistent feeder fault (causing it to lock-out), an inter-trip signal will be initiated from the operated (feeder protection) trip relay at substation A and sent to substation B. This will hold off the DAR at substation B from reclosing circuit breaker X405. Detection, by the DAR, of the persistent fault at substation A will result in the auto-isolation (i.e. opening) of disconnector X103, which in turn will delay the resetting of the trip relays at substation A (by typically 75 s). The operated trip relays at substation A will maintain the inter-trip signal to substation B, and if the duration of the inter-trip signal exceeds an inter-trip timer setting at substation B (typically set at 60 s), the DAR at substation B is permanently locked out, and disconnector X403 will auto-isolate (i.e. open). Thus, the feeder is fully isolated and the DAR relays at both ends of the feeder reset.

10.21 Operational tripping

10.21.1 Operational tripping – requirements

The term 'operational tripping' (alternatively termed 'operational inter-tripping' or in some instances 'system to generator tripping') describes special and bespoke auto-switching arrangements which are required to prevent either of the following:

- Power system or generator instability – requiring a typical maximum operating time of 200 ms
- Network overloading – requiring a typical maximum operating time of 5 s.

Examples of operational tripping schemes are given with reference to Figure 10.39, which are as follows:

1. **Example 1**
   - (i) Consider the situation where the no. 1 circuit between substations C and D is out for maintenance, and the nos. 1 and 2 circuits between substations C and E are removed from service via a double-circuit fault. This leaves only circuit no. 2 between substations C and D as the remaining connection to the rest of the power system.
   - (ii) Consider as a result of (i) the transfer impedance (see Section 3.2.5) between the sources of generation and the rest of the power system, when all generation is in service, to be so large, that system instability arises.

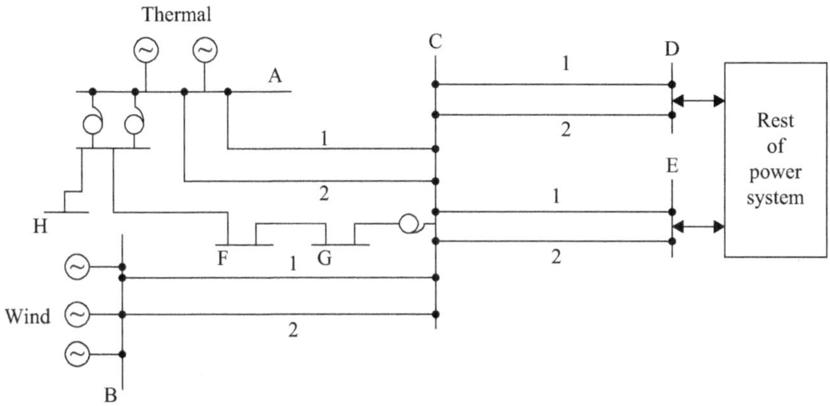

Figure 10.39 Example system for operational tripping

(iii) As a result, a number of generators need to be tripped and removed from service until the power transfer levels are such that power system stability is maintained. This requires an operational tripping scheme.

(iv) An operational tripping scheme is therefore designed to be installed at substation C. This is usually a programmable logic device.

(v) The operational tripping inputs to the logic device are as follows:

(a) At substation C, circuit breaker and disconnector auxiliary contacts from the two circuits to substation D, and the two circuits to substation E, to reflect whether these circuit ends are in or out of service – and feeder protection trip relay contacts on the same four circuits.

(b) At substation D and E, on the circuits to substation C, circuit breaker and disconnector auxiliary contacts, and associated feeder protection trip relays, to feed into inter-trip send circuits for each feeder at each respective substation. The inter-tripping equipment sends a signal to substation C, as an input to the operational tripling scheme, to indicate whether the circuit end in question is in or out of service, or in the process of being tripped.

(vi) The operational tripping scheme outputs comprise inter-tripping signals to trip the generators at power stations A and B, as specified by the system designers. NB: generator status inputs may also be required.

(vii) The operating logic within the operational tripping logic device at substation C may comprise a very wide range of permutations of generators to trip depending upon the operational arrangements pertaining at the moment of loss of the last relevant circuit. A typical time from loss of the last relevant circuit (i.e. operation of trip relay) to tripping of generation may be 180 ms.

(viii) The operational tripping logic device will be subject to comprehensive factory acceptance tests before being dispatched to site. It is usual to equip the logic device with test inputs both to facilitate site testing and to provide inputs to the device when equipment is out for maintenance.

2. **Example 2**

   Consider the situation whereby if both of the higher voltage circuits between substations A and C are out of service, there is overloading of the LV network from substations A–F–G–C. This requires the following:

   (i) An operational tripping scheme located at either substation A or C (as appropriate).

   (ii) Inputs to the operational tripping logic device are required from both ends of the circuits between substations A and C to indicate the circuit ends are either in or out of service. Inputs are required from auxiliary contacts but not protection trip relays (speed of operation is not critical).

   (iii) Logic device output comprises an inter-trip signal to trip a circuit breaker to severe the substations A–F–G–C circuit at a point(s) defined by the system designer. Circuit breaker tripping time is required to be within 5 s of loss of both of the substations A and C circuits.

With increasing power system complexity, the number of operational tripping schemes has steadily increased. A requirement of the scheme designer is to ensure that operation of one operational tripping scheme does not (unintentionally) cascade into other scheme, with severe and unintended implications for the power system.

10.22 SCADA system

10.22.1 SCADA system – overview

The term 'SCADA' is a generic term which stands for 'supervisory, control and data acquisition'. SCADA systems invariably comprise programmable computer-based devices for monitoring and controlling equipment used by many industries. With reference to electrical power networks, SCADA systems are concerned with four key requirements, which are as follows:

1. **Control**

   Control is the opening and closing of circuit breakers, disconnectors, earth switches, etc. It also includes in/out switching functions such as protection and inter-tripping, etc., and transformer tap changing.

2. **Metering**

   Metering refers to operational metering (i.e. not commercial metering) including circuit voltage, current, MW and MVArs – to assist operational control of the power network(s).

3. **Indications**

   Indications refers to information on the status of HV power equipment e.g. open/closed position of circuit breaker, disconnector, etc., and the position of protection and control in/out switches. It additionally also usually includes transformer winding temperature relay temperatures and tap position.

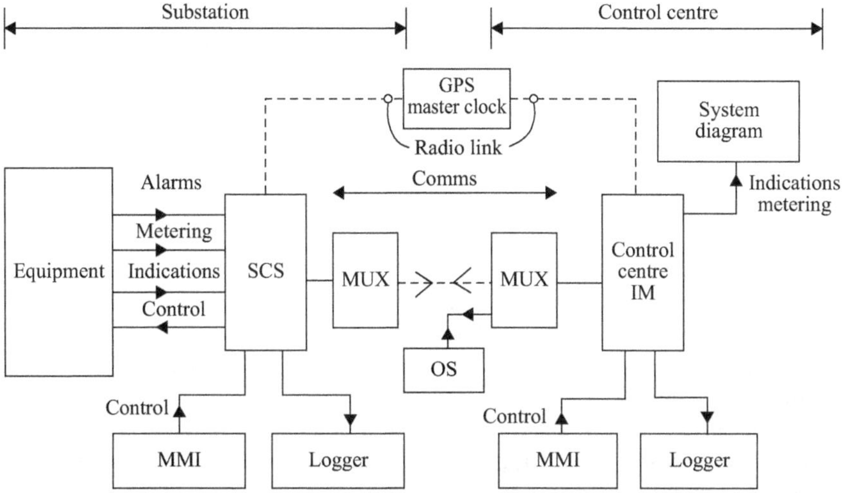

- MMI = Man–machine interface
- IM = Intelligent machine
- OS = Other substations
- SCS = Substation control system
- MUX = Multiplexer

Figure 10.40 SCADA system– typical

4. **Alarms**
   This comprises the status of all equipment-related alarms in a substation – of which there are many, and usually ever increasing as equipment becomes more sophisticated or complex.

10.22.2 SCADA system – design

With reference to Figure 10.40, the typical design of a SCADA system comprises the following:

- The equipment comprises all power system equipment within the substation. It is usually grouped on a per-bay basis. It may also include other equipment such as that associated with security systems. The data routes may be directly into the SCS via copper connections, but increasingly this is being undertaken via fibre, sometimes comprising a local area network which may be connected either radially or as a ring.
- The SCS is the central processing unit within a substation to which data flows to/from the equipment. Alternative terms may be used to SCS. The SCS also communicates with the control centre.
- The control centre interfaces with all other substations. Some power network companies may divide the control across a number of control centres.
- The means of communication between SCS and control centre is via a communications protocol. Perhaps one of the best known in the United Kingdom is GI74, employed by National Grid, and although an old design, and in the

process of being replaced, it has proved to be highly reliable and effective over many years. IEC 870 is often now the preferred protocol for new installations – possessing superior time tagging and the ability to handle increased volumes of data – particularly the ability to handle discrete alarms as opposed to common alarms required by older systems.

- The 'MUX' comprises a multiplexor unit for substation to control centre communications. Usually, two separate communications routes are required to aid reliability with separate MUXs per route.
- The man–machine interface (also termed HMI, human–machine interface) allows both local (SCS) and remote (control centre) control of the equipment – usually presented as a substation single-line diagram graphical layout on a HMI screen.
- The 'logger' logs the changing status of all alarms and indications and records all control actions.
- It is usual for the control centre to display a 'system diagram' of the whole network indicating the status of the equipment – so providing the control engineer with an overview to aid operational decisions.
- The 'GPS master clock' is required for the purpose of synchronisation of (precise) time.
- The 'control centre intelligent machine' provides the interface with all substations – it is self-evidently a critical item in the control infrastructure – and is usually supported by a backup machine.

10.22.3 SCS design requirements

A power network company will usually provide a generic technical specification for the SCS and peripheral equipment – which will be validated at time of type tests – see Section 21.3.1.1. The essential design task for a specific substation under construction is the collation of all requirements, i.e. the substation and equipment-specific control, metering, indications and alarm requirements. Careful attention must be paid to names (e.g. circuit names) and nomenclature. Many power network companies develop generic equipment models to ensure consistency and standardisation. Once collated and agreed as correct, the data must be programmed into the SCS prior to factory tests. It is essential that the data in the SCS and control centre machine is identical.

10.23 Protection and control accommodation

10.23.1 Relay panel requirements

Traditionally, protection and control equipment including fuse/links and test blocks and associated interconnecting wiring and cable terminal blocks have been housed in panels (indoors) and weather-proof kiosks (outdoors). The panels historically comprised a sheet steel front which had to be accurately cut to accommodate the dimensions of the relay. Within the panel, terminal blocks were mounted on either side to accommodate cable terminations and routing of wiring. In some instances, the cables were terminated in a wall box behind the panel, with wiring between wall box and panel. This type of accommodation has largely been replaced by protection and control cubicles, usually facilitating detachable modules (within a standardised 19-in. wide frame), mounted in vertical rows (i.e. one above the other) – with each module serving a different function and mounted on a front-opening swing door to provide good access to the rear of the modules. Relay panels traditionally comprised almost one panel per function, but with modern, physically small, multifunction relays, the number of cubicles required is far fewer. However, at transmission voltages, it is still common practice to segregate MP1 and MP2 into adjacent cubicles.

Usually substations contained one suite of panels for the control of circuit breakers and disconnectors etc., (i.e. open/close) termed the control panel suite – and another (usually) separately located suite of panels for protection and control functions, termed the relay panel (or cubicle) suite. The control panel suite has now largely been replaced by the SCS as shown in Figure 10.40. However, a circuit breaker cubicle (containing open/close interpose relays, synchronising relay, trip circuit supervision, etc.) is often included in the relay cubicle suite.

Relay cubicles/panels are usually located in a central relay room housing all cubicles and physically arranged to reflect the layout of the substation (as far as practical). In some instances, they may be split over two relay rooms. However, some substations because of their physical size may have small individual relay rooms per circuit located adjacent to the HV equipment for that circuit. These are termed ‘distributed relay rooms’ or ‘blockhouses’, and each usually contains their own individual battery and DC distribution board.

10.23.2 Cubicle design and practical considerations

With reference to the installation of relay cubicles, it is preferable to carry out as much as possible in the factory under factory-quality-controlled installation and test conditions. In some instances, whole blockhouses worth of equipment (including all panels but usually not the battery) have been assembled and tested in the factory and transported to site as complete, only requiring the termination of the cables/fibre on site. The following should be considered with reference to protection and control cubicle (and kiosk) design:

- The designation/nomenclature/labelling of each circuit, and associated item of protection and control.
- Cubicle access arrangements including accessibility to all relays, etc.
- Location of test blocks and test arrangements.
- Cable termination locations. Some installations have a central marshalling area where all external cables are terminated and inter-cubicle or inter-circuit wiring connections made – and internal cabling to each cubicle terminated.
- Termination types and security.
- Location of fuses/links.
- Cubicle earthing arrangements.
- Cubicle lighting and illumination.
- Logical layout of cubicles and relays.

10.24 Protection and control asset replacement

10.24.1 Asset replacement considerations

The lifespan of the HV equipment is nominally 40 years. That of older protection and control equipment is typically 25 years, but modern computer-based protection and control equipment may only be 10 years or less. Therefore, protection and control assets will require replacing numerous times in the lifespan of the HV equipment.

An ongoing objective of protection and control asset replacement policy is to engineer arrangements that minimises the work content and time to undertake the work. Installations possessing significant standardisation would appear to be the obvious solution. However, despite numerous endeavours to achieve this objective, changes in technology, equipment design and relay interconnectivity make this objective elusive. Within this context, a strategy looking both at the short term, where the technology may be reasonably frozen, and the longer term where technology change may be anticipated, may be an optimum way forward.

General considerations relating to asset replacement are as follows:

- All asset replacement needs to balance cost of replacement with the benefits of introducing new types of equipment and new technologies.
- Consideration needs to be given to replacing either all protection and control equipment on a circuit bay – or just equipment that is currently a source of risk.
- With reference to feeder asset replacement, both ends of the feeder will usually require replacing at the same time – particularly when unit protection is involved.
- Protection and control asset replacement is usually ideally undertaken coincident with that of the HV equipment.
- Standardisation of equipment design is important in terms of familiarisation of operational personnel, training and spares holding.
- Careful consideration must be given to the replacement of common equipment such as BB protection and SCSs, and how the substation continues to operate to the required design and operational standards during the change-over period.
- Circuit outages need to be considered planned and agreed – and must be part of a long-term replacement strategy. In the ultimate, outages can be the major constraint on replacement.
- Consider the potential impact of the proposed asset replacement on the next asset replacement (perhaps only 10 years away) and where economies could be delivered.
- Consider the lifespan of the substation or any connecting circuits. It may be advantageous to delay work or even not undertake it at all.
- Protection and control asset replacement strategy needs to optimise and takes account of:
  - Technical performance benefits
  - Cost
  - Specialist resource availability
  - Outage availability
  - Timing

– Volume of new construction work coincident with potential asset replacement
– Drawings availability and management
– Economies of large-scale asset replacement.

10.25 Batteries and DC supplies

10.25.1 Battery systems

The performance of protection and control equipment is self-evidently highly dependent upon a source of power. This is invariably provided by a stand-alone battery system, independent of the power system itself, and therefore immune from loss of AC supply. A fundamental design decision surrounding the choice of battery system is the operating voltage to which the operating range of relays, circuit breaker trip coils, etc. must be aligned – and which determines the size of inter-connecting cables and associated voltage drops. The following sections will examine the following key requirements of batteries and DC supplies:

- The lead–acid Plante battery
- Battery types and performance characteristics
- DC supply arrangements
- Relay and trip coil voltage limits
- Battery sizing

10.25.1.1 The lead–acid Plante battery

The lead–acid Plante type of cell is by far the most common form of cell found in substation batteries. The cell comprises a positive terminal (connected to lead peroxide plate), a negative terminal (connected to a lead plate) and an electrolyte of dilute sulphuric acid contained in a glass/plastic container, into which the plates are immersed. The nominal voltage of the cell is 2 V. In the United Kingdom, sub-station batteries are usually 110-V nominal (comprising 55 cells) or 48-V nominal (comprising 24 cells).

The cell is described as 2-V nominal since if disconnected from a battery charger, and with no load connected, the cell terminal voltage will settle at 2 V. The ideal voltage to which a cell should be charged is 2.25 V per cell i.e. 123.75/54 V (for 110/48-V nominal batteries). The maximum voltage to which a battery should be boost charged is 137.5/60 V (for 110/48-V nominal batteries).

The capacity of a battery is usually expressed in ampere–hours (A h) for a 10-h discharge rate. For example, a 300-A-h battery is capable of providing 30 A for 10 h without suffering any deterioration. After the 10 h, the battery terminal voltage will have dropped to approximately 104.5/45.6 V (for 110/48-V nominal batteries). The higher the discharge rate (e.g. a 6-h rate, 1-h rate, 30-min rate, etc.), the lower the terminal voltage at the end of the discharge. Manufactures produce a suite of discharge curves for information purposes. Typical substation battery sizes range between 100 and 400 A h.

10.25.1.2 The recombination cell

In the mid-1980s, an alternative to the Plante cell, the 'recombination cell' appeared on the market. This is more accurately termed a 'lead–acid valve-regulated cell' or more commonly a 'sealed cell'. Although the Plante cell suffers loss of electrolyte during the charging process and requires the electrolyte to be periodically topped up, and the electrolyte-specific gravity maintained within limits, the sealed cell has none of these disadvantages. Instead the charge and discharge process is retained entirely within the enclosed cell.

Sealed cells are usually provided in a 'monobloc' of six cells, resulting in a nominal monobloc voltage of 12 V (i.e. 6 × 2 V/cell). The equivalent of the 110 V nominal Plante battery is therefore a 108-V nominal battery (i.e. nine monoblocs), but the 48-V nominal battery is retained via four monoblocs.

Salient characteristics of the sealed cell are as follows:

- Sealed cells have a superior (higher) voltage discharge profile for the same discharge current, than a Plante battery.
- Sealed cells are relatively maintenance free compared to Plante.
- Estimated lifespan of a sealed cell is 10–15 years compared to 20–30 years for Plante.
- Sealed cells are cheaper than Plante.
- Experience has generally shown that sealed cells are not as reliable as Plante.

Based on the above, the common practice in the United Kingdom has been to use sealed cells for smaller battery installations, such as blockhouses, e.g. less than 100 A h, and Plante batteries for the larger centralised relay room arrangements.

Traditionally, 48-V batteries were used for all telecommunications, alarms, indications and protection relay power supplies – with 110-V batteries being used for trip circuits. However, the modern practice is to standardise more on 110 V, as it is less susceptible to interference voltages. In some instances, 110/48-V DC to DC convertors have been used.

A more recent development in DC supplies is an uninterruptable supply comprising an AC/DC convertor (e.g. 230-V AC to 110-V DC) as the main source of DC supply – but on failure of the AC input, there is an instantaneous changeover to a DC generator – which is powered by bottles of compressed air. Capacitor banks ensure no loss of output during the changeover. This system removes the requirement for batteries. The compressed air bottles would typically supply the DC system for 6 h, by which time the AC supply should be restored. This is an attractive solution for transmission substations in particular where the AC supplies are backed up by diesel generators.

10.25.1.3 DC supply arrangements

Figure 10.41(a) shows typical arrangements for a substation DC supply, employing batteries. A battery charger maintains the cell voltage at an optimum level of charge at 2.25 V/cell i.e. 123.75 V for 110-V nominal Plante battery. The DC supply arrangement is usually equipped with high and low-voltage alarms and a battery EF detection system. The battery EF system is usually centre point resistor tapped, providing equal sensitivity to positive and negative pole EFs, with an ability to detect EFs down to typically 50 kΩ. Batteries of 48 V are usually

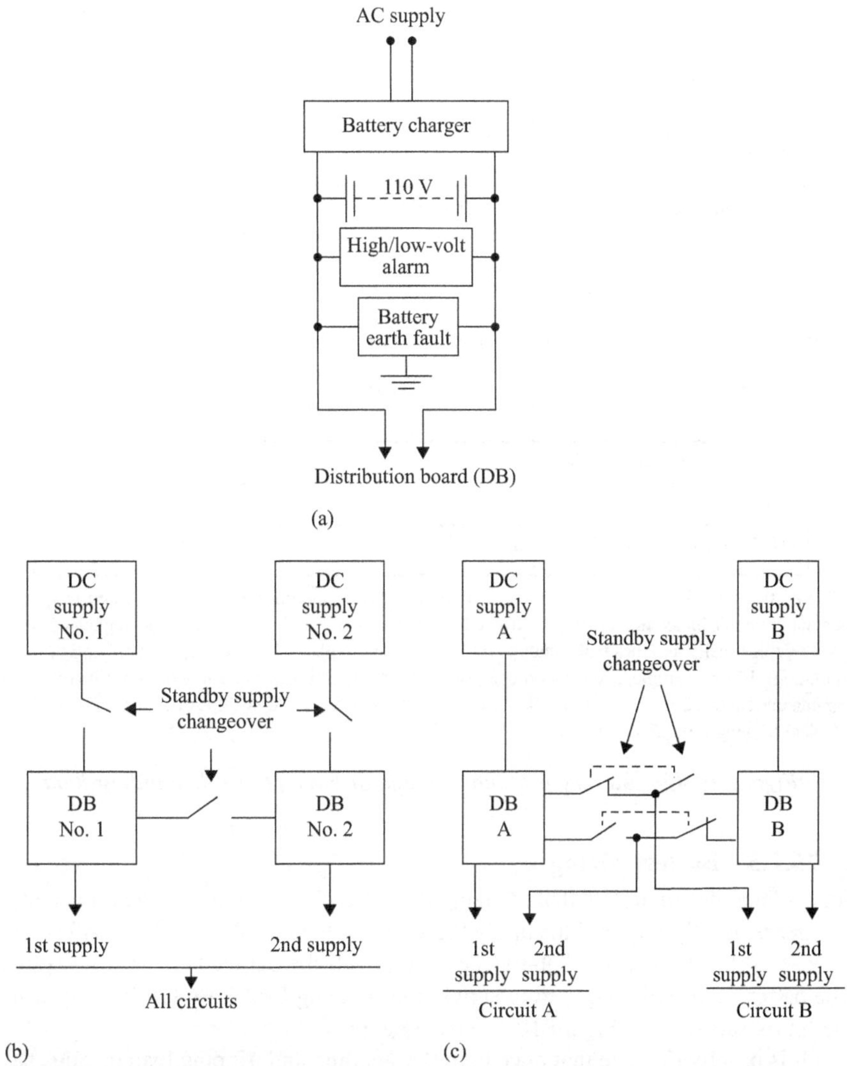

Figure 10.41 Transmission substation – 110-V nominal supply arrangement: (a) DC supply arrangement, (b) centralised DC supplies and (c) blockhouse DC supplies

operated with the positive pole directly earthed with fuses in the negative pole, which is telecommunications technology standard practice.

Figure 10.41(b) and (c) shows typical DC supply arrangements for the transmission networks which require an independent MP1 and MP2.

10.25.1.4 Relay and trip coil voltage limits

The rated voltage and working range of relays and circuit breaker trip coils are usually as given in Figure 10.42. See BS 142 and ENA standards.

DC system nominal voltage (V)	110*	48
Relay-rated voltage (V)	125	54
Maximum operating voltage (V)	137.5	60
Minimum operating voltage (V)	87.5	37.5

NB: *Relays which are either not double-pole switched or whose wiring leaves the relay room, or is extensive, are susceptible to mal-operation from cabling capacitance discharge should an earth-fault arise on certain parts of the wiring. These relays must be designed to withstand a discharge from a 110-μF capacitor charged to 150 V.

Figure 10.42 Relay, switchgear controls and trip coil operating voltage limits

48-V nominal battery (V)	46
110-V nominal central battery (V)*	102
110-V nominal blockhouse battery (V)*	93

NB: *At the end of the 6-h period, the above voltage limits should not be reduced when tripping all associated circuit breakers that may be tripped simultaneously (i.e. busbar fault). For duplicated battery systems, this usually assumes that standby arrangements are in place (i.e. one battery supplying the whole substation). If, for example, a substation comprised of ten circuit breakers each with two trip coils with a tripping current of 12 A per circuit breaker, then the tripping load would be 240 A (i.e. 10 × 2 × 12 A), for a typical tripping time of 40 ms.

Figure 10.43 Battery terminal voltage at the end of 6-h standing load

10.25.1.5 Battery sizing

An essential design task is that of sizing the battery. Within this context, the battery charger is usually required to supply the standing load (with a factor of safety), but on loss of the charger, the battery must supply all the demands of the DC system. The battery is usually required to supply the standing load for 6 h following which the limits specified in Figure 10.43 usually apply.

It is usually the case that once both the standing and tripping load profiles have been determined, the battery manufacturer with more specialist battery data would be requested to determine the battery size. IEEE document 485 provides guidance on this matter.

Given that, Figure 10.43 specifies the minimum battery terminal voltage, and Figure 10.42 the circuit breaker trip coil minimum operating voltage, a calculation needs to be undertaken to calculate the voltage drop across the wiring interconnecting the battery to the circuit breaker at time of the circuit breaker tripping (at the end of the 6-h period) to ensure that the voltage appearing at the trip coil is within limits.

Chapter 11
Impressed voltage

11.1 Impressed voltage – composition

Impressed voltage (IV) is an umbrella term used by many in the electrical power industry to encompass the following:

- Capacitive coupling (arising from the electric field and voltage source)
- Inductive coupling (arising from the magnetic field and current source)
- Conductive coupling (arising from current flow through a connection with earth)
- Trapped charge (arising from residual charge left on the capacitance of an item of equipment).

The term IV relates to the existence of a voltage on an item of equipment, or metallic object, which is not directly connected to the energised power system, but which has arisen as a result of either an electrical coupling mechanism, or a residual charge at the time of circuit de-energisation. Such voltages can be dangerously high and a source of danger and therefore must be eliminated before contact is made with the equipment/metallic object. IV is a significant safety hazard to be managed and controlled on any electrical construction site.

11.2 Permanent and temporary works

IVs may arise on either permanent works or temporary works. Examples of IV on permanent works include inductive coupling from a cable phase conductor to the cable metallic sheath, and conductive coupling giving rise to both touch and step potentials associated with earthing systems. In all instances of permanent works, the designer must specify a design that is both not a safety hazard and optimises the performance of the equipment. Temporary works that may be subject to IV include the following:

- Site construction, or maintenance, of equipment
- Cranes, MEWPS, winching or lifting equipment
- Scaffolding and metallic fences
- Long metallic objects in general.

The severity of IV arising on temporary works is usually more difficult to determine than that of permanent works. That is because the magnitude of IV on temporary works varies greatly with site specific arrangements, i.e. the proximity and orientation of the HV system and the dimensions of the equipment or object subject

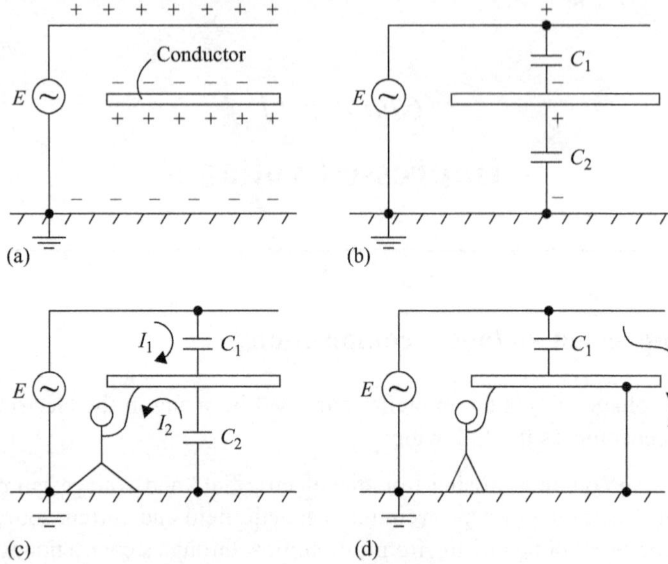

Figure 11.1 Capacitive coupling – principles: (a) Conductor subject to electric field, (b) equivalent circuit capacitance, (c) person subject to current flows and (d) person protected by earthing

to the IV. In many instances, empirical methods are used for protecting people from IV, on temporary works, such as standard earthing methods defined in safety rules. In other instances, calculations need to be undertaken to determine the severity of the IV and the optimum methods of protection.

11.3 Impressed voltage – principles

11.3.1 Capacitive coupling – principles

In Figure 11.1(a), an energised power system produces an electric field (from an energised OHL or busbar, etc.) that induces charges on any conductor that exists within the field. The charges dispose themselves as shown, since the nature of a conductor is that no field can exist within it. Figure 11.1(b) shows the electric field interpreted in terms of the equivalent capacitance. As illustrated in Figure 11.1(c), should a person touch the conductor then the person is subject to current flows I_1 and I_2. I_1 is a continuous current, but I_2 is a transient current until capacitance C_2 discharges. Figure 11.1(d) shows that the application of an earth to the conductor short-circuits and protects the person.

Figure 11.2(a) shows the capacitances described in terms of their capacitive reactances X_{C1} and X_{C2}, respectively. As the length of the conductor increases, the magnitudes of both X_{C1} and X_{C2} decrease. However, the ratio of X_{C1} to X_{C2} will remain constant irrespective of the length of the conductor. The magnitude of

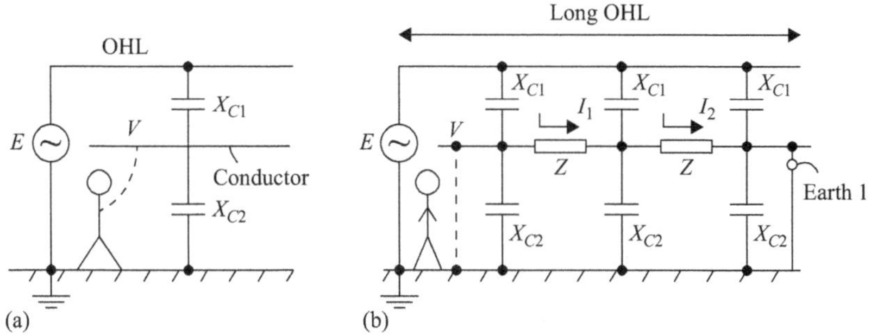

Figure 11.2 Capacitive coupling – practicalities: (a) Equivalent circuit – short conductor and (b) equivalent circuit – long conductor

V will depend upon the ratio of X_{C2} to $(X_{C1} + X_{C2})$, and therefore, the voltage, V, of the conductor will remain constant irrespective of the physical length of the conductor. It is to be noted that as X_{C1} and X_{C2} decrease with increasing conductor length, the magnitude of current through a person touching the conductor will increase.

Very long conductors such as an OHL or cables have an equivalent circuit based on distributed shunt capacitive reactances X_{C1} and X_{C2} and distributed circuit impedance Z (mostly inductive). This is shown in Figure 11.2(b). The connection of an earth, i.e. earth 1, at one end of the circuit will result in a voltage V at the other end of the circuit, given by:

$$V = (2I_1 + I_2)Z$$

If a person then touches the conductor, the magnitude of V could be dangerously high – therefore requiring the closing of earth 2 (shown dotted) to protect the person. This requirement would be unusual for the relatively short lengths of conductors found in a substation – but would be required for long OHL or cables. If work is to be undertaken part way along the conductor – other earths may be required at the point of work to afford satisfactory protection of the person. It is worthy of note that when opening and closing earth switches at either end of a circuit,

- The first earth switch to close is subject to maximum capacitive closing current.
- The last earth switch to open is subject to maximum capacitive breaking current.

11.3.2 Inductive coupling – principles

Figure 11.3(a) shows a current of magnitude I flowing through the power system (e.g. along an OHL or busbar) giving rise to a magnetic flux around the power conductor. Figure 11.3(b) shows the instance of a conductor which is shorted (earthed) at both ends to form a continuous loop, encompassing a proportion of the magnetic flux of magnitude Φ. Faraday's law states that a changing flux (e.g. from

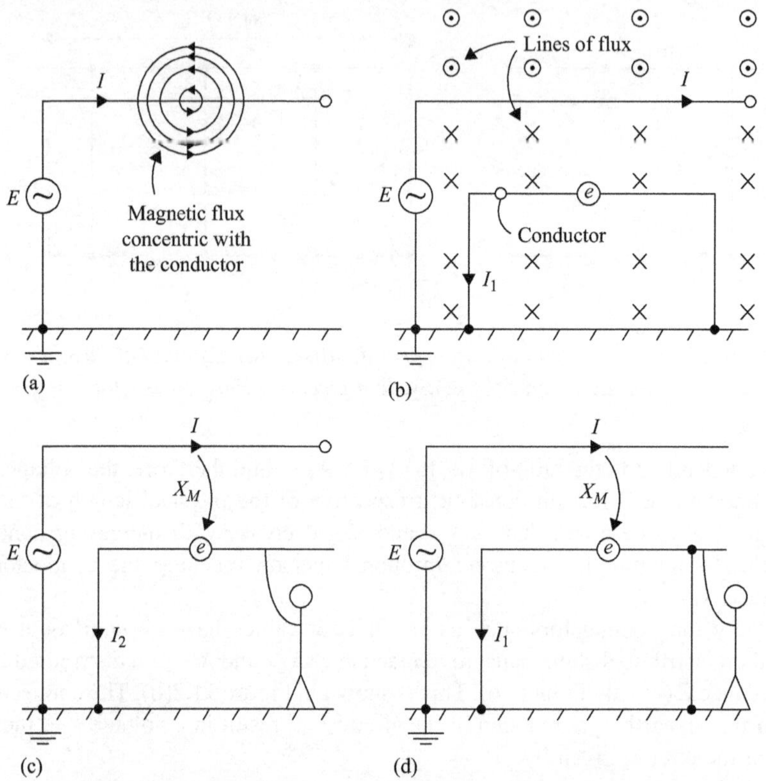

Figure 11.3 Inductive coupling – principles: (a) Magnetic flux around a conductor, (b) flux encompassed by a conductor loop, (c) person subject to induced emf and (d) Person protected by earthing

an AC source) within a closed loop induces an instantaneous emf into that loop given by the expression:

$$e = N\frac{d\Phi}{dt}$$

where Φ is the flux enclosed, and N is the number of conductors (i.e. one in this instance).

With reference to Figure 11.3(c), the magnitude of e can also be given in RMS terms as:

$$e = I\,X_M$$

where I is the power system RMS current, and X_M is the mutual inductive reactance between power system and conductor.

Figure 11.3(c) also shows that where a person forms part of the closed loop, the induced emf e gives rise to a current I_2 through the person, which could be dangerously high. Applying a short-circuit (earth) as in Figure 11.3(d) protects the person.

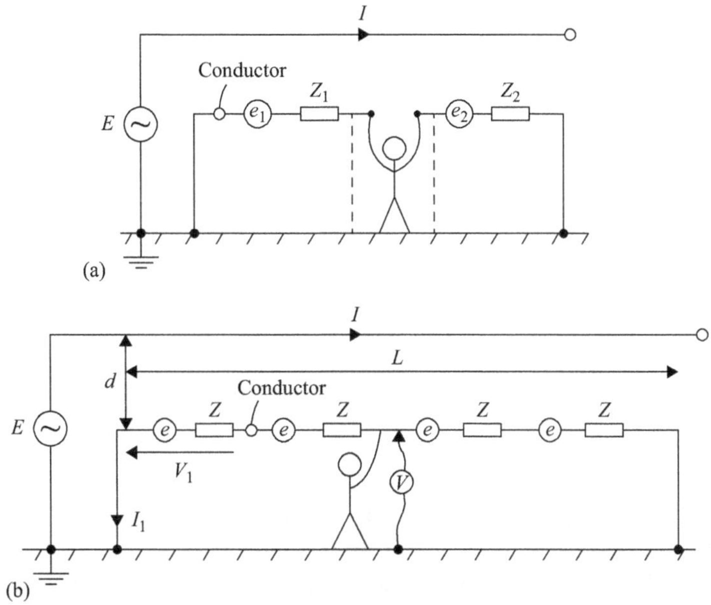

Figure 11.4 Inductive coupling: practicalities: (a) Induced emf either side of a break in a conductor and (b) magnitude of emf along a conductor

Figure 11.4(a) shows the instance of a conductor (e.g. a parallel OHL) subject to an induced voltage arising from power system current I. Should a break occur in the inductively coupled circuit the person becomes subject to voltages e_1 and e_2. To protect the person, earths (shown dotted) must be applied. With reference to Figure 11.4(b), it can be shown that the magnitude of an induced voltage into a conductor varies as follows:

- Increases with length of parallelism L
- Decreases with increasing separation d
- Increases with increasing current I.

Figure 11.4(b) shows the instance of a conductor earthed at two substations. The conductor equivalent circuit comprises a distributed induced voltage per unit length of e, and a distributed impedance per unit length of Z. In the figure shown, the circulating current I_1 is:

$$I_1 = \frac{4e}{4Z} = \frac{e}{Z} \tag{11.1}$$

Thus, I_1 remains constant no matter the length of parallelism (i.e. for a specific combination of OHL dimensions). Now, the voltage V_1 along any unit length of the conductor shown in Figure 11.4(b) is given by:

$$V_1 = e - I_1 Z$$

But from (11.1) $I_1 = \frac{e}{Z}$.

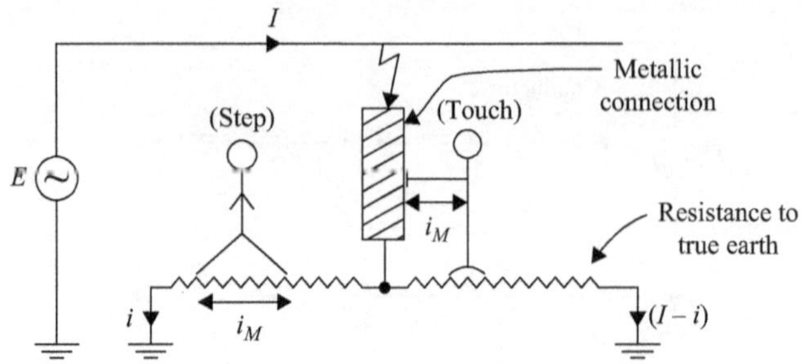

Figure 11.5 Step and touch potential

So $V_1 = e - \left(\frac{e}{Z}\right)Z = 0.$

So the voltage, V, measured at any point along the conductor, for the simple case illustrated, is zero. In practice due to the resistance of the earth, voltage V_1 will have a finite value.

Again, it is worthy of note that when opening and closing earth switches at either end of a circuit, the opposite of capacitive coupling occurs, i.e. inductive coupling results in

- The last earth switch to close is subject to maximum inductive closing current
- The first earth switch to open is subject to maximum inductive breaking current.

11.3.3 Conductive coupling – principles

Conductive coupling arises as a result of current flow through the earth (usually fault current) creating a voltage difference between two points, which may be remote from the path of current flow. Should a person come into contact with these points danger from electric shock may arise. There are two main forms of conductive coupling:

1. Step and touch potential
2. Transferred potential

These will be considered below:

- **Step and touch potential**
  With reference to Figure 11.5, step and touch potential are usually associated with fault current flowing into a substation earth (although the concept applies equally to an OHL tower subject to fault current flowing through it to earth), creating a voltage in the earth which can be imposed upon a person, i.e. conductively coupled. Touch potential is the voltage difference between two points bridgeable between a person's hand and feet. The horizontal distance is standardised at 1 m (arm–body–leg). Alternatively, step potential is the voltage difference between two points on the surface of the ground bridged by a

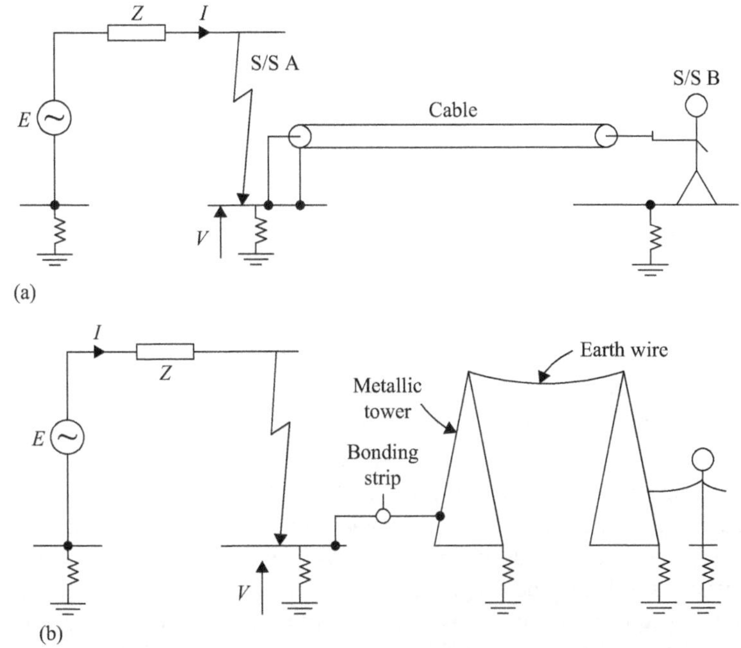

Figure 11.6 Transferred potential: (a) Transferred potential along a cable and (b) transferred potential along an OHL

person's feet. The step is again standardised as 1 m (leg–leg). Usually, touch potential is higher than step potential and more dangerous from an electric shock perspective. Both touch and step potential should be designed to be within safe limits as part of the permanent works design.

- **Transferred potential**
  Figure 11.6 gives two examples of transferred potential. In both instances, a rise of potential, V, is transferred to the point of work some distance from the substation. In the case of Figure 11.6(a), the person would be protected if either the cable was earthed at the point of work, or the cable (and its metallic sheath) was not connected to earth. In the case of Figure 11.6(b), the rise of potential, V, at the substation is transferred along the tower aerial earth wire. Safety from the transferred voltage, if touching the tower, is assisted by virtue that each tower will be earthed, so lowering the voltage at the tower that is being touched. Safety when working on the OHL phase conductors would be achieved by earthing the phase conductors to the tower. It is worthy of note that if the OHL phase conductors had also been earthed at the substation, they would also be subject to voltage V, which would be transferred along the OHL.

11.3.4 Trapped charge – principles

With reference to Figure 11.7, when circuit breaker CB is opened, the current I is interrupted at current zero. This coincides with E and V being at peak voltage – and

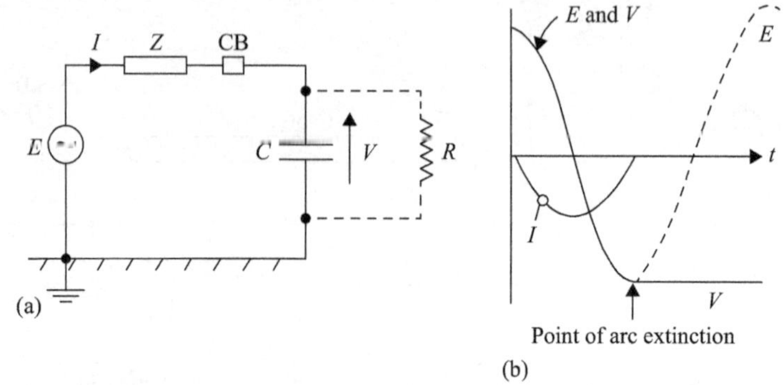

Figure 11.7 Trapped charge: (a) Circuit subject to trapped charge and (b) voltage waveforms arising from trapped charge

thus, the capacitance remains charged at this voltage. This condition is termed 'trapped charge'. All capacitances have a leakage resistance R in parallel through which the trapped charge discharges and thus V eventually reduces with time. In some instances, this may be a significant time. Typical examples of trapped charge remaining on equipment for extensive lengths of time (i.e. hours to days) following circuit breaker opening include MSCs, GIS busbars and cables. Overhead lines can exhibit trapped charge following circuit de-energisation – but the trapped charge usually decays in 20–60 s through insulator leakage resistance. Dissipation of trapped charge can be through a wound VT if connected to all three phases of the circuit, otherwise the equipment will have to be earthed.

Cable HV pressure tests can also result in trapped charge on completion of the test. It is therefore usual to discharge the trapped charge, on test completion, in a controlled manner, usually using a discharge resistor. However, the dielectric performance of XLPE cables in particular is such that charge can re-appear on the cable following discharge through the resistor. The default position following discharge is therefore to earth the cable at both ends by closing of earth switches thereby ensuring the complete discharge over time. Similar discharge procedures may be required after pressure testing GIS equipment.

11.4 Inductive coupling – analysis

11.4.1 Current transformer analogy

Inductive coupling between a power system conductor and another isolated conductor (not part of the power system, e.g. an isolated busbar) may be considered to be the equivalent to a current transformer (CT) with a one-to-one turns ratio, but without a magnetic core (and therefore a very low value of magnetising impedance). With reference to Figure 11.8, the CT analogy is underpinned by the fact that the current in the power system conductor (the primary) is determined by the power system load, and not the isolated coupled conductor (the secondary), i.e. the equivalent of a series connected transformer, see Section 3.4.7.

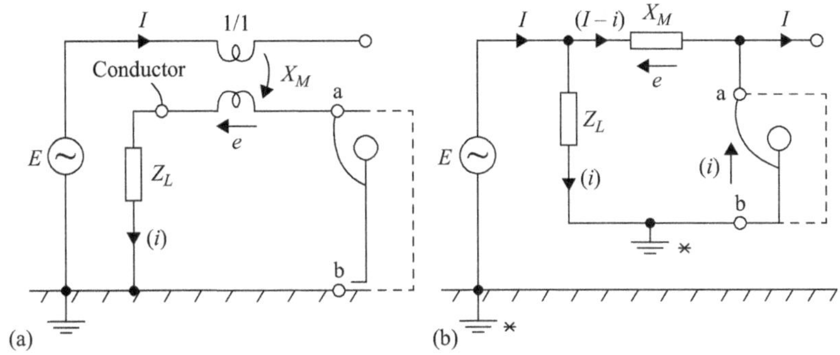

*Figure 11.8 Inductive coupling – CT analogy: (a) circuitry, (b) equivalent circuit NB: *earth to be considered as not electrically connected*

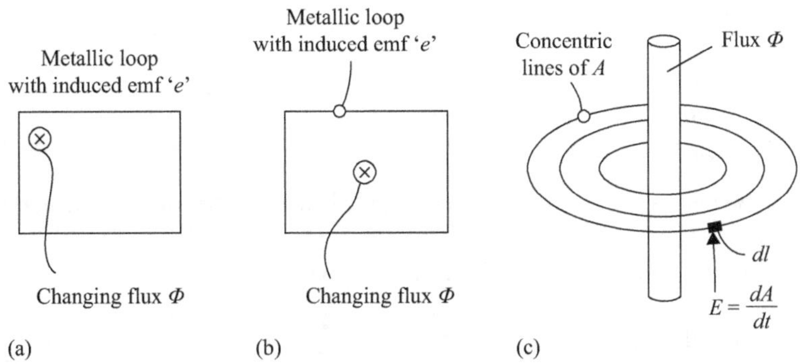

Figure 11.9 Magnetic vector potential 'A': (a) Flux located close to loop, (b) flux remote from loop and (c) disposition of field 'A'

Figure 11.8(b) shows the equivalent circuit (drawn as % impedances) for the inductively coupled conductors. As stated above, the magnetising impedance X_M (which is one and the same as the mutual impedance between primary and secondary) is relatively small in magnitude and therefore has little influence on the magnitude of current I in the power system conductor. However, the body impedance of the person shown is relatively large compared with both X_M and Z_L (which comprises mostly of the leakage reactance of the isolated conductor). This results in the voltage imposed across the person approximating to that of the open circuit voltage across a–b (i.e. the effect of the current flow, i, has negligible effect on the voltage to which the person is subject).

11.4.2 Inductive coupling – position of induced emf

Faraday's law states that a changing magnetic flux Φ within a closed metallic loop causes an induced emf, e, around the loop. However, Faraday's law says nothing about where exactly the emf is located around the loop, and with reference to Figure 11.9(a) and (b), it seems to make no difference to the magnitude of e (around the closed loop) whether the flux is concentrated close to a point on the loop perimeter as opposed to being in the centre of the loop. It also begs the question that if a portion of the loop was removed, whether an induced emf still exists in the remaining portion of the loop – and if so, what determines the magnitude of that emf, and where is it located?

To answer this question, there is a requirement to briefly revert to the physics of power engineering. It is reasoned that a more fundamental field than the magnetic flux Φ exists, termed the 'magnetic vector potential', 'A'. In the same way that the flux Φ forms concentric circles around a conductor arising from current I, the 'A' field forms concentric circles around the flux Φ, see Figure 11.9(c). As such it can be shown that:

$$\int A \cdot dl = \Phi \tag{11.2}$$

i.e. integrating $A{\cdot}dl$ around the closed loop.

And, with reference to Figure 11.10, at any point dl, $A \propto (I/r)$

Now from Faradays Law:

$$e = N\frac{d\Phi}{dt} \tag{11.3}$$

where in the above e is the induced emf, and N is the number of turns, which in this instance = 1.

Placing Φ from expression (11.2) into (11.3):

$$e = \frac{d(\int A \cdot dl)}{dt} \tag{11.4}$$

Now from basic theory, it is well documented that:

$$e = \int E \cdot dl \tag{11.5}$$

where E is the electric field at point dl in V/m.

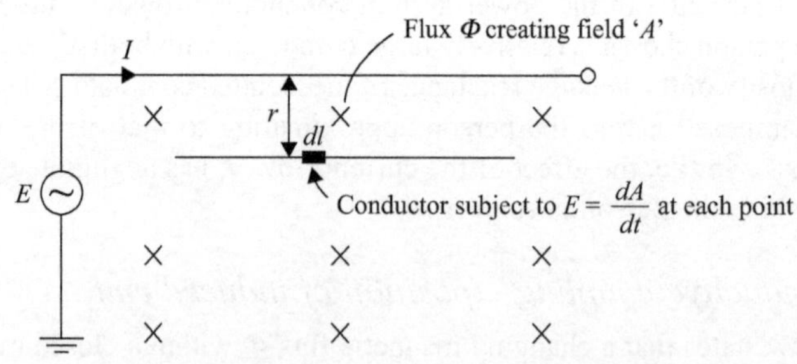

Figure 11.10 Electric field and emf induced on a conductor

Equating expressions (11.4) and (11.5), it can be shown that:

$$E = \frac{dA}{dt} \tag{11.6}$$

With reference to Figure 11.10 and expression (11.6) each point dl on the conductor is subject to an electrical field E, depending upon the changing magnitude of A at that point. And the total induced emf into the conductor is given by expression (11.5). This somewhat simplified explanation shows that an electric field and induced emf does exist in a conductor even if the conductor does not form a closed loop. Furthermore, that the magnitude of E at a particular point (and hence e along the conductor) is a function of A which in turn is a function of Φ. Magnetic vector magnetic potential is largely a mathematical concept which gives a greater understanding of the existence and position of an induced emf than does Faraday's law alone. As such it provides a better explanation for the origin of induced emf's experienced on site.

11.5 Physiological effects of electricity

11.5.1 Electric shock

An electric shock occurs when a person's body becomes part of an electrical circuit causing electric current to pass through the body. The current can cause a wide range of harmful effects from minor skin sensation to death. A number of factors influence electric shock, comprising the following:

- The magnitude of current through the person's body.
- The pathway of current inside the body (e.g. hand–hand or hand–foot).
- The duration of the current.
- The surface area of contact. The greater the surface area, the less the required touch voltage for the same impact.
- Touch voltage magnitude. A greater touch voltage causes a reduced body impedance.
- The physical characteristics of the person. People have different body impedances.

11.5.2 Impedance model of the human body

A single impedance model of the human body is difficult to define since the impedance varies with a number of factors as stated above. However, as a starting point, Figure 11.11 provides a generalised model. This model is taken from IEC 60479 (2005). It is based upon total skin contact area of 100 square cm and a touch voltage of 25 V and applies to 95% of the population. At a touch voltage of 1,000 V, the impedances reduce to approximately 33% of those at 25 V. The impedances are mostly resistive but with some shunt capacitance.

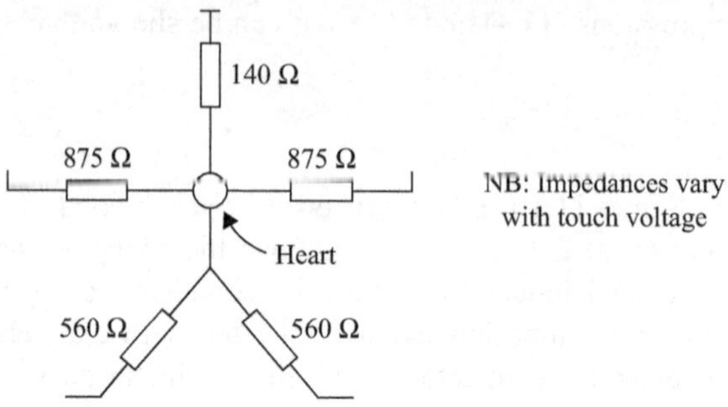

Figure 11.11 Human body – generalised impedance model

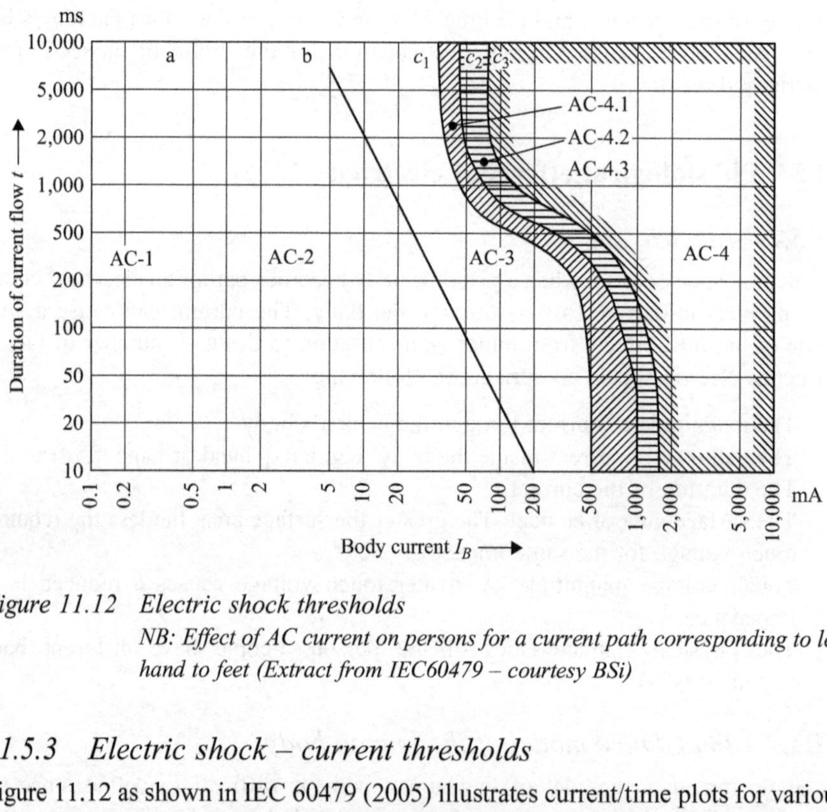

Figure 11.12 Electric shock thresholds
NB: Effect of AC current on persons for a current path corresponding to left hand to feet (Extract from IEC60479 – courtesy BSi)

11.5.3 Electric shock – current thresholds

Figure 11.12 as shown in IEC 60479 (2005) illustrates current/time plots for various levels of electric shock. This authoritative document is used by many in the electrical industry. It is worthy of note that authoritative American documentation although similar does not have fully identical electric shock thresholds. Figure 11.12 is applicable to virtually the entire population. The various thresholds are as follows:

1. **Startle threshold**
   This is given as 0.5 mA continuous, as defined by line 'a'. This is also known as the threshold of reaction. As the threshold is exceeded, there is increasingly

significant pain as the level of current and the duration increase, but usually no lasting harmful effects arise in the area designated AC-2.

2. **Immobilisation (cannot let go) threshold**
The immobilisation threshold is given as 5 mA continuous with a current/time relationship threshold for lower time durations, as defined by line 'b'. This threshold covers virtually the whole population, although an adult male may usually withstand 10 mA continuous. For the area designated AC-3, increasing pain, immobility and breathing difficulties may occur, but usually no organic damage.
3. **Ventricular fibrillation threshold (fatal)**
Ventricular fibrillation (VF) of the heart is a fatal condition since the heart no longer pumps blood and oxygen. VF may continue after the electrical current has ceased. Areas AC-4.1, AC-4.2 and AC-4.3 show probabilities of VF occurring as being up to 5%, up to 50% and exceeding 50% of the population, respectively.

11.5.4 Electric shock – voltage thresholds

It is frequently a requirement to translate the current thresholds in Figure 11.12 into a voltage threshold since IV in the first instance results in the creation of a voltage. This is particularly the case for both inductive coupling and conductive coupling who's coupling voltage emanate from a current source. The current sources and their impact are categorised as follows:

Current source	Impact
• Three-phase balanced load current	Inductive coupling
• Single phase to earth fault current cleared by protection systems in a maximum time of 200 ms	Inductive and conductive coupling

The following sections will derive voltage thresholds for electric shock relating to the two current sources.

11.5.5 Touch potential – safe voltage threshold

Historically, and based upon the requirements of international telecommunications directives (CCITT directives), safe limits of substation rise of earth potential were specified as either 430 V, or if the protection cleared the fault in 200 ms, then 650 V. These voltages are applicable to transferred potentials where it is assumed that the person at the remote end of the transferred voltage (and subject to the electric shock) is in direct contact with the earth (in practice, this is not always the case). In light of an improved understanding of electric shock criteria, the 430/650 V limits are no longer the threshold voltages. Nonetheless they remain the design threshold to which UK substation earth systems should be designed (see ENA Technical Specification 41-24).

It is currently recognised that situations may arise when substation rise of earth potential may significantly exceed the 430/650 V threshold and where these impact on third parties mitigation measures would need to be undertaken.

ENATS 41-24 recognises that higher voltage thresholds are acceptable from an electric shock perspective for both touch potentials and step potentials. In general, touch potentials are a higher magnitude than step potentials, therefore only touch potentials are considered. On the assumption that the substation earth comprises a metallic mesh, the touch potential is usually the greatest between any equipment connected to the peripheral edge of the metallic mesh and a point on the ground 1 m away from it. The following will determine the maximum voltage for this touch potential criterion.

BSEN 5022 (Earthing of power installations exceeding 1 kV AC) specifies in Annex NA the criterion to be used in the United Kingdom for determining maximum touch potential voltage. This criteria is based upon the body current limits and body impedances specified in IEC 60479. The model used utilises the following data as provided in Figure 11.12.

- A body current I_B of 600 mA corresponding to a duration of 200 ms on VF curve C2 in Figure 11.12. Curve C2 depicts up to a 5% probability of VF occurring. Therefore, this criterion accepts a low risk of VF on the basis that the coupling voltage only lasts for 200 ms. In practice, this is a statistically low eventuality.
- A body impedance Z_T not exceeding 5% of the population, which is based upon an expected (i.e. anticipated) touch voltage exceeding 1 kV. Based upon this criterion, IEC 60479 Table 1 gives Z_T as 575 Ω hand to hand.
- For a required electrical contact of left hand to both feet. IEC 60479 specifies that the corresponding Body Factor (BF) for converting the ohmic impedance from hand to hand, to hand to both feet as 0.75.
- Persons boot resistance of 4 kΩ per foot, and substation chippings (150 mm) resistance of 2 kΩ per foot.

With reference to Figure 11.13, the touch voltage expression (from BSEN 5022) for V_T is:

$$V_T = I_B[(Z_T \times BF) + (R_F + R_C)] \tag{11.7}$$

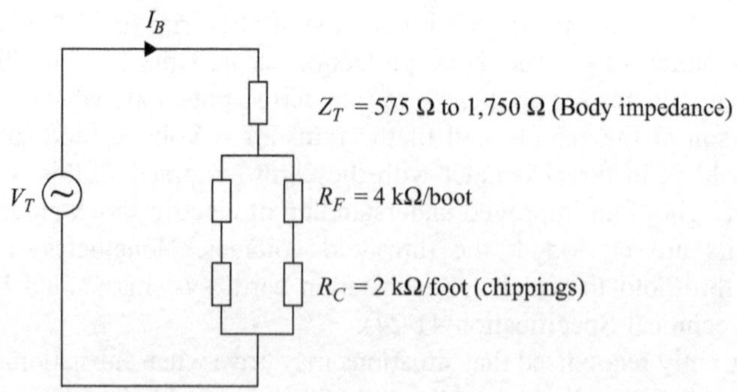

Figure 11.13 Touch voltage threshold model

Inserting the values from above.

$$V_T = 0.6[(575 \times 0.75) + (2{,}000 + 1{,}000)]$$
$$= \quad 260 \quad + \quad 1{,}800 \quad = 2{,}060\ \text{V}$$
$$\text{(Person)} \quad + \text{(Boots + Chippings)}$$

Therefore, the design threshold touch voltage in a substation which is not to be exceeded is:

$$V_{T(\text{Substation})} = 2{,}060\ \text{V} \tag{11.8}$$

This is shown graphically in BSEN 5022 Figure NA.2.

If the person is not in a substation but, say in a field, standing on the soil surface and the chippings are replaced by foot resistance to earth (typically 330 Ω), then inserting the values into expression (11.7):

$$V_{T(\text{Soil surface})} = 1{,}559\ \text{V} \tag{11.9}$$

NB: In practice, this is usually assumed to be 1,700 V.

11.5.6 Single phase to earth fault – inductive coupling safe voltage threshold

A single phase to earth fault would cause the highest level of inductive coupling between the power system conductor subject to the fault and other de-energised objects such as busbars, OHL, cables, etc. However, unlike the subject of touch potentials little exists in the way of standards documents for determining the safe voltage threshold. However, the criterion that applies to touch potentials logically applies equally to voltages arising from inductive coupling. Therefore, the voltage limits given in expressions (11.8) and (11.9) apply equally to inductive coupling and temporary works designers should ensure that either these limits are not exceeded or that appropriate safety control measures are applied.

It is to be noted that the above thresholds are applicable to a construction workforce working in either a substation, or a field, or a cable trench, etc. However, third parties such as the general public may have shoes with a lower resistance than the boots defined above, or may be standing on soil of a different resistivity to that assumed above. In such instances and depending on the circumstances, lower values than the above thresholds may be required to ensure personal safety.

11.5.7 Balanced load current – inductive coupling safe voltage threshold

Unlike fault currents cleared in 200 ms or less, balanced load currents are continuous. The question therefore arises of the safe voltage threshold to be adopted when a person is subject to a continuous inductively coupled voltage. With reference to a construction workforce undertaking work, a cautious approach would be to adopt the startle threshold, since the work would be unpleasant and potentially unsafe to undertake if the workperson was constantly being startled. With reference to Figure 11.12, the continuous startle threshold current I_B is 0.5 mA which will be the assumed threshold for this text. It is acknowledged that an adult male will tolerate a higher threshold than 0.5 mA and therefore that some organisations might choose different criterion and a higher threshold level. Furthermore, if an equipotential zone is established between the equipment and the person when work is being undertaken, a higher equipment voltage threshold may be acceptable, see Figure 11.24 for an example of an equipotential zone.

To determine the touch voltage, expression (11.7) must be used with the value of $I_B = 0.5$ mA. However, at lower touch voltages, the body impedance increases, and with reference to IEC60479, at the voltages likely to be experienced then $Z_T = 1{,}750\ \Omega$. Then for the instance of work in a substation involving a person standing on chippings:

Recalling (11.7) then, $V_T = I_B\,[(Z_T \times BF) + (R_F + R_C)]$
On inserting values $= 0.0005\,[(1{,}750 \times 0.75) + (3{,}000)] = 2.16$ V
so $V_{T(\text{Substation})} = 2.16$ V
and it can be similarly calculated that:

$$V_{T(\text{Soil surface})} = 1.7\ \text{V}$$

Clearly, with such a low voltage, any light insulation would render a workperson safe. Again, it needs to be noted that the above applies to a construction workforce and may not be applicable to third party situations (e.g. general public) requiring threshold calculations to be undertaken relevant to the situation.

NB: See comparative figures in IEC 61201.

11.6 Capacitive coupling – IV magnitude calculations

11.6.1 Conducting plane images

With reference to Figure 11.14, if a charged conductor such as an OHL or busbar is positioned at height h above a conducting plane such as the earth, the conducting plane has the same effect on the electric field, emanating from the conductor, as an equal and opposite charge (i.e. an image conductor) placed at distance h below the conducting plane surface.

Using the method of images, it can be shown that the voltage induced, V_C, on a de-energised conductor, C, as a result of a single charged conductor and its image, as shown in Figure 11.14 is given by:

$$V_C = \frac{V_L}{\sqrt{3}} \frac{\ln\left(\frac{d_2}{d_1}\right)}{\ln\left(\frac{2h}{r_1}\right)}\ \text{V} \tag{11.10}$$

where V_L is the line voltage.

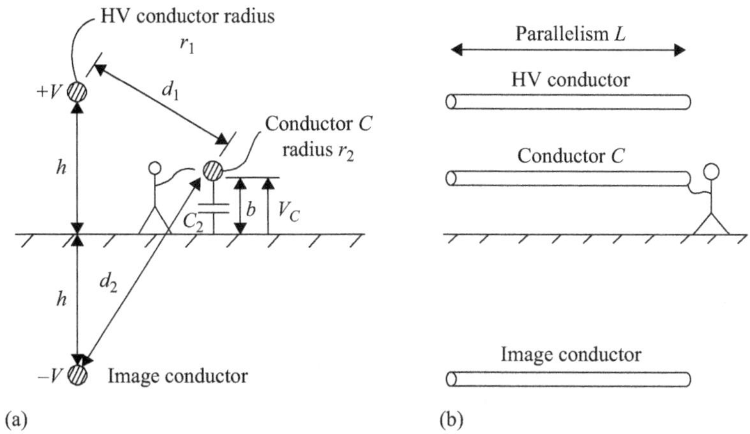

Figure 11.14 Concept of an image conductor arising from capacitive coupling: (a) cross-section view and (b) parallel view

It can also be shown, again using the method of images, that the value of capacitance C_2 is given by:

$$C_2 = \frac{2\Pi\varepsilon}{\ln\left(\frac{2b}{r_2}\right)} \text{ Farads/m}$$

$$\text{or } C_2 = \frac{2\Pi\varepsilon}{\ln\left(\frac{2b}{r_2}\right)} L \text{ Farads}$$

where ε is the permittivity $= 8.85 \times 10^{-12}$ Farads/m and L is the length of the parallel conductor in metres.

Now capacitive reactance X_{C2} is given by:

$$X_{C2} = \frac{1}{2\Pi f C_2} = \frac{\ln\left(\frac{2b}{r_2}\right)}{(2\Pi f)(2\Pi\varepsilon L)} = \frac{\ln\left(\frac{2b}{r_2}\right)}{4\Pi^2 f\varepsilon L}\ \Omega$$

where f is the frequency $= 50$ Hz.

$$\text{Inserting values, then } X_{C2} = \frac{57\ln\left(\frac{2b}{r_2}\right)}{L}\ \text{M}\Omega \tag{11.11}$$

Example: With reference to Figure 11.14, calculate X_2 (assume $b = 2$ m; $r_2 =$ 0.1 m; $L = 20$ m).

Inserting the values into expression (11.11):

$$\text{Then } X_{C2} = \frac{57\ln\left(\frac{4}{0.1}\right)}{20} = \underline{10\ \text{M}\Omega}$$

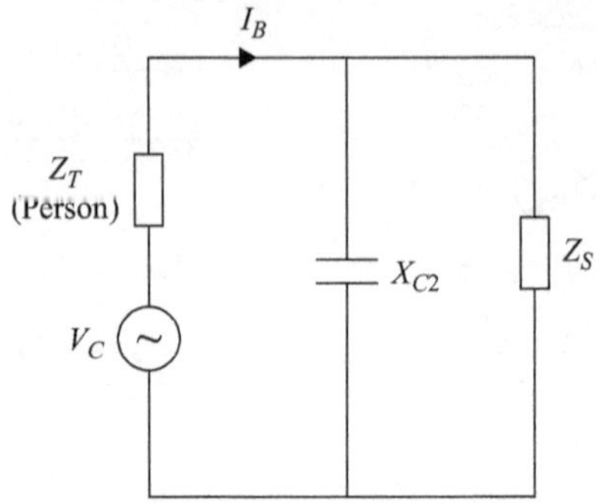

Figure 11.15 Determination of current flow through a person due to capacitive coupling

If with reference to Figure 11.14 a person then touches the conductor, the method for determining the resulting current flow through the person is as follows (see Figure 11.15):

- Using expression (11.10) determine the voltage on the conductor prior to the person touching it i.e. the open circuit Thevenin voltage V_C
- Apply the open circuit voltage V_C to the circuit shown in Figure 11.15, i.e. the Thevenin equivalent circuit, to determine the current through the person. In Figure 11.15, Z_S is the Thevenin equivalent source impedance between the source of generation and conductor *C*.

As determined from Section 11.5.5, the impedance of the person to ground will be approximately 3.5 kΩ (or less). In addition, if the location of conductor *C* is outside of 'safety distance' (see Chapter 18) which will generally be the case in practice, the calculated voltage induced into conductor *C*, from expression (11.10), will be significantly less than the HV network voltage, thereby implying a high source impedance Z_S which, because it is significantly higher than X_{C2}, can be ignored in Figure 11.15. Based upon these assumptions and with reference to Figure 11.15, the current through the person is to good approximation:

$$I_B = \frac{V_C}{X_{C2}} \tag{11.12}$$

As explained in Section 11.3.1, on touching the conductor there will also be transient current flow from X_{C2} until X_{C2} is fully discharged. The transient current usually decays in micro-seconds, having negligible electric shock impact, unless significant capacitance such as that in a long OHL is concerned.

11.6.2 Three-phase coupling

Figure 11.16 shows the instance of a conductor *C* capacitively coupled to live three-phase busbars. Again, using the theory of images, the voltage V_C induced into

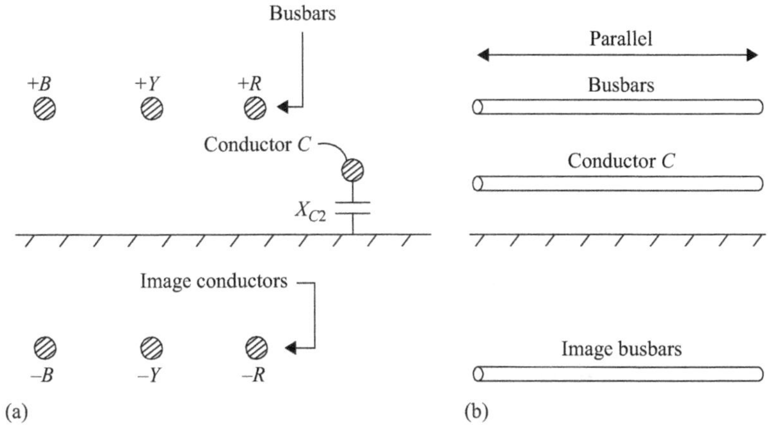

Figure 11.16 Three-phase capacitive coupling: (a) cross-section view and (b) parallel view

conductor C from each phase is given by expression (11.10) resulting in voltages V_{CR}, V_{CY} and V_{CB} arising from red, yellow and blue phases, respectively. Using the superposition theorem, the total voltage on conductor C is the vector addition of the three voltages:

$$\text{i.e. } V_C = V_{CR} + a^2 V_{CY} + a V_{CB} \tag{11.13}$$

And the current I_B flowing through any person touching the conductor is again given by expression (11.12). Three-phase coupling will result in a lower voltage on conductor C than single phase coupling due to the partial cancelling effect of the three balanced voltages.

11.6.3 Example of capacitive coupling

With reference to Figure 11.17, calculate the current through the person on touching the conductor. Details are as follows:

Voltage = 400 kV
Height above ground of R phase = 13 m
of Y phase = 22 m
of B phase = 32 m
of E = 43 m
of conductor C = 3 m
separation D = 10 m
Impedance of person to ground = 3,300 Ω (including boots)
Radius of phase conductor and conductor C = 0.1 m
Parallelism (L) of conductor C with OHL = 20 m

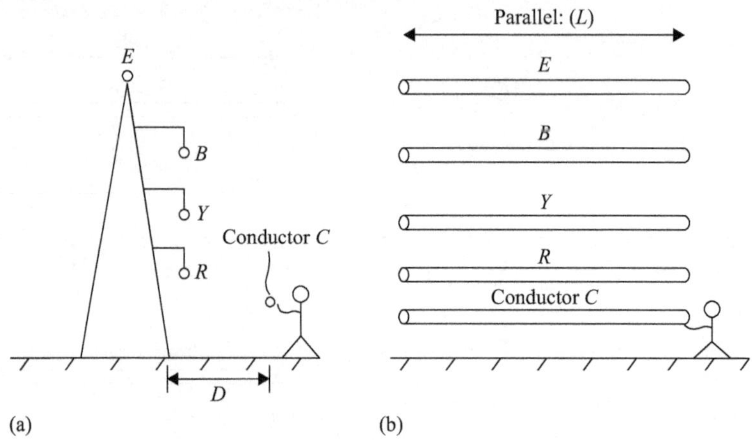

Figure 11.17 Capacitive coupling – worked example: (a) cross-section view and (b) parallel view

Solution

Using expressions (11.10) and (11.13) and inserting the values above, the conductor voltage is calculated to be:

$$V_C = 5.8\,\text{kV}$$

Using expression (11.11) and inserting the values above, the capacitive reactance is calculated to be:

$$X_{C2} = 11.7\,\text{M}\Omega$$

Therefore, from expression (11.12):

$$I_B = \frac{V_C}{X_{C2}} = \frac{5.8\,\text{kV}}{11.7\,\text{M}\Omega} = \underline{0.49\,\text{mA}}$$

i.e. just below the startled reaction current of 0.5 mA.

It is worthy of note that should the other circuit of the twin OHL be energised that capacitive coupling from that circuit should also be added to that of the first circuit. Depending on the phasing of the two circuits, the voltage V_C may approach double that of a single circuit. To avoid complex calculations, a simple approach to estimating V_C may be to consider red phase alone (i.e. the nearest phase) from the single circuit (resulting in an overestimate of voltage and current), i.e. $V_C =$ 11.9 kV.

$$\text{Therefore, } I_B = \frac{V_C}{X_{C2}} = \frac{11.9\,\text{kV}}{11.7\,\text{M}\Omega} = \underline{1\,\text{mA approx}}$$

It needs to be noted that both the conductor *C* and the phase conductors also exhibit capacitance to the aerial earth conductor *E*. A fully rigorous analysis must take these capacitances into account – however, the impact is small and within the limits of acceptable accuracy (and acceptable factor of safety) this consideration can usually be ignored.

NB: Computer-based calculations, if available, would provide greater accuracy.

11.7 Inductive coupling – IV magnitude calculation

11.7.1 Conducting plane images

If a current carrying conductor such as a phase conductor on an OHL or busbar is positioned at a height h above a conducting plane such as the earth (assumed to be zero resistivity and perfectly conducting), the conducting plane has the same effect on the magnetic field, surrounding the conductor, as that of an equal and opposite current carrying conductor positioned at distance h below the conducting plane's surface. This is shown in Figure 11.18.

Using the method of images, it can be shown that the mutual impedance Z_{PC} between current carrying phase conductor and conductor C as shown in Figure 11.18 is as follows (where in expression (11.14) it is assumed that the radius of conductor C is much less than dimension d_1 in Figure 11.18):

$$Z_{PC} = \mu f \ln\left(\frac{d_2}{d_1}\right)\ \Omega/\text{m} \tag{11.14}$$

where f is the frequency = 50 Hz and μ is the permeability = $4\Pi \times 10^{-7}$ Henries/m.

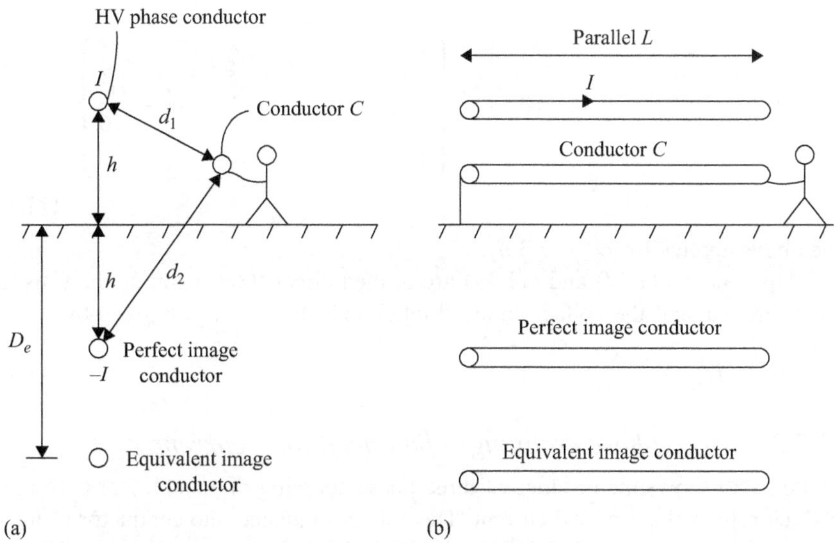

Figure 11.18 Concept of image conductor arising from inductive coupling: (a) cross-section view and (b) parallel view

And the voltage V_C induced into the conductor is:

$$V_C = IZ_{PC} = I\mu f \ln\left(\frac{d_2}{d_1}\right) \text{ V/m} \tag{11.15}$$

In practice, the finite resistivity of the earth causes the image conductor current to penetrate well below the surface to an equivalent depth D_e. The concept of, and formula for, D_e was first determined by Carson in 1926. The concept applies equally to current flowing in the earth, for phase to earth fault return current. The magnitude of D_e is given by:

$$D_e = 659\sqrt{\frac{\rho}{f}} \text{ m} \tag{11.16}$$

where ρ is the earth resistivity in ohm metres and f is the frequency = 50 Hz.

So, if for example $\rho = 100\ \Omega$ m (typical for farmland), then:

$$D_e = 659\sqrt{\frac{100}{50}} = 932 \text{ m}$$

In light of the above, more specialist texts give the mutual impedance as being:

$$Z_{PC} = \Pi^2 10^{-4} f + j4\Pi 10^{-4} f \ln\left(\frac{D_e}{d_1}\right) \ \Omega/\text{km} \tag{11.17}$$

where D_e and d_1 are given in Figure 11.18 – and the above applies for $D_e > 7.3\ d_1$, or alternatively:

$$Z_{PC} = \Pi^2 10^{-4} f + j4\Pi 10^{-4} f \sqrt{\left\{\left[\frac{\ln\left(1 + 1.382\left(\frac{D_e^2}{d_1^2}\right)\right)}{4}\right]^2 - \frac{\Pi^2}{16}\right\}} \ \Omega/\text{km} \tag{11.18}$$

The above applies for $D_e < 7.3\ d_1$.

Expressions (11.17) and (11.18) are termed either the Carson or the Carson–Clem formula, and the voltage induced into conductor C is again given by:

$$V_C = IZ_{PC} \text{ V} \tag{11.19}$$

11.7.2 Three-phase coupling – balanced load current

Figure 11.19 shows the instance of three-phase coupling of a conductor C from an OHL carrying balanced load current. The voltages induced into conductor C due to currents in each phase of the OHL is determined from expression (11.19), i.e. voltages V_{RC}, V_{YC} and V_{BC} arising from the currents in red, yellow and blue phases, respectively. To this must be added voltage V_{EC} arising from current I_E flowing in

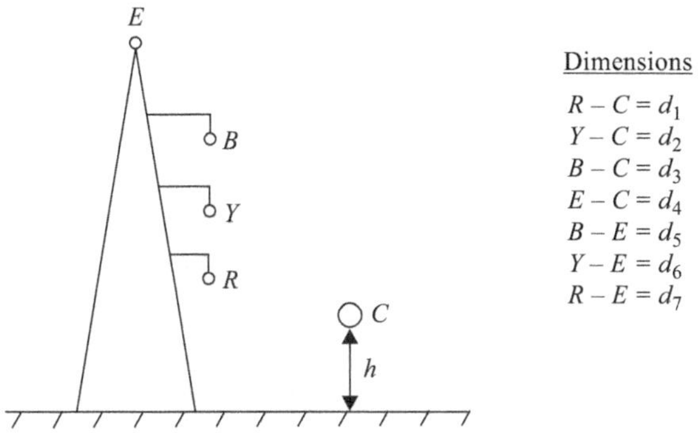

Figure 11.19 Example of three-phase inductive coupling

the tower earth wire. The tower earth wire current is given by the following expression:

$$I_E = \frac{I_R Z_{RE} + a^2 I_Y Z_{YE} + a I_B Z_{BE}}{Z_{EE}} \text{ A} \tag{11.20}$$

where Z_{RE}, Z_{YE} and Z_{BE} are the mutual impedances per km between the phase conductors and the earth conductor, and Z_{EE} is the self-impedance of the tower earth wire per km. Again, using the superposition theorem, the induced voltage V_C into conductor C is given by:

$$V_C = I_R Z_{\mathrm{RC}} + a^2 I_Y Z_{\mathrm{YC}} + a I_B Z_{\mathrm{BC}} + I_E Z_{\mathrm{EC}} \text{ V/km} \tag{11.21}$$

where in the above Z_{RC}, Z_{YC}, Z_{BC} and Z_{EC} are the mutual impedances from the phase and earth conductors, respectively, to conductor C, and I_E is from expression (11.20).

Clearly, for the instance of inductive coupling arising from a busbar where there is no equivalent to the tower earth wire, the term $I_E Z_{\mathrm{EC}}$ disappears from expression (11.21).

Specialist texts also define the three-phase inductive coupling in terms of distances between conductors and to earth. This is given by the expression below and with reference to the distances given in Figure 11.19 and is derived from expressions (11.17) and (11.19):

$$V_C = 4\Pi 10^{-4} fI \left\{ \frac{\sqrt{3}}{2} \ln\left(\frac{d_3}{d_2}\right) + j\ln\left(\frac{\sqrt{d_3 d_2}}{d_1}\right) + Z_e \right\} \text{ V/km} \tag{11.22}$$

where I is the magnitude of the balanced load current, and f = frequency.

NB: Expression (11.22) only applies for the instance of $D_e > 7.3\ d_1$:

$$\text{and where } Z_e = -\frac{Z_{\mathrm{EC}}}{Z_{\mathrm{EE}}} \left\{ \frac{\sqrt{3}}{2} \ln\left(\frac{d_7}{d_6}\right) + j\ln\left(\frac{\sqrt{d_7 d_6}}{d_5}\right) \right\} \Omega/\text{km} \tag{11.23}$$

For the instance of the aerial earth conductor not being present, as with substation busbars, then expression (11.23) disappears. Although expression (11.22) looks cumbersome it is less arduous than expression (11.21) as there is less vector manipulation involved.

11.7.3 Single phase to earth fault inductive coupling

The inductive coupling from a phase conductor, subject to a single phase to earth fault current, to a de-energised conductor, similar to the arrangement shown in Figure 11.19 (when, say, red phase is the faulted phase) can be determined from expression (11.21). In these instances, two of the phases (say, yellow and blue) carry negligible current. The magnitude of the induced voltage into the conductor is much greater than for load current. This is because both fault currents are usually much greater in magnitude than load current, and the cancelling effect of the other two phases (which are present for balanced load current) are absent.

11.7.4 Example of inductive coupling

With reference to the 400-kV busbar arrangement given in Figure 11.20, determine the current through the person's body on touching the conductor, for the instance of:

1. A three-phase balanced load current of 4 kA
2. An external red phase to earth fault current of 20 kA.

Details are as follows:

Parallelism between busbar and conductor $C = 20$ m
Earth resistivity $\rho = 100\ \Omega$ m
Impedance Z_T of person's body to ground (including boots and chippings) is:

- 3,750 Ω for touch voltage on conductor C approximately equal to 100 V
- 4,300 Ω for touch voltage on conductor C approximately equal to 25 V or less.

Frequency = 50 Hz

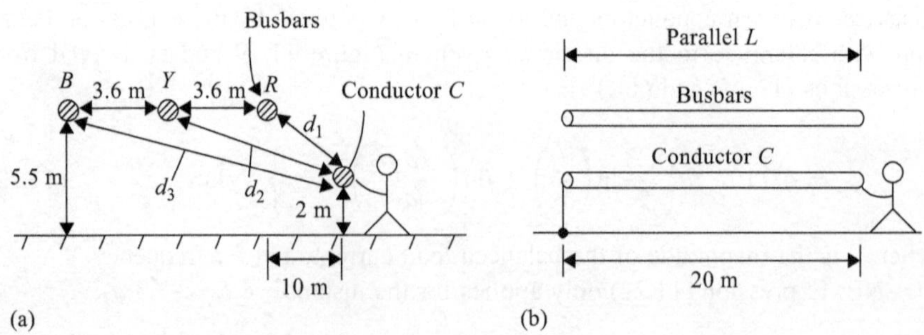

Figure 11.20 Inductive coupling – worked example: (a) cross-section view and (b) parallel view

Solution to (1)

From expression (11.16) and inserting relevant values $D_e = 930$ m.

From Figure 11.20, the distances d_1, d_2 and d_3 can be readily calculated. These distances can be inserted into expression (11.22) to calculate conductor voltage V_C noting that there is no earth wire and therefore no Z_e term.

On inserting the values into (11.22), $V_C = 2$ V.

And the current through the person's body is therefore given as:

$$I_B = \frac{V_C}{Z_T} = \frac{2}{4{,}300} = \underline{0.46\ \text{mA}}$$

i.e. just under the startle reaction current of 0.5 mA.

Solution to (2)

Insert the values into expression (11.17) to determine the mutual impedance Z_{RC} between the red phase conductor and conductor C:

then $Z_{RC} = 5.4$ mΩ

Using expression (11.19) to determine conductor voltage V_C:

i.e. $V_C = IZ_{RC} = 20{,}000 \times 0.0054 = 108$ V

$$\text{Therefore, } I_B = \frac{V_C}{Z_T} = \frac{108}{3{,}750} = \underline{29\ \text{mA}}$$

With reference to Section 11.5.5, the fault will be cleared in under 200 ms, as such the current through the person's body of 29 mA is well below the VF level of 600 mA, in the same way that the conductor voltage of 108 V is well below the VF threshold voltage of 2,060 V. For the VF values to be exceeded, the parallelism between busbar and conductor would need to be approximately 380 m. Again, computer-based systems, if available, would provide a more accurate result.

11.8 Practical considerations

11.8.1 Combined effect of capacitive and inductive coupling

So far capacitive and inductive coupling have been considered separately – but in practice, they both exist at the same time. This begs the question of the magnitude of their combined effect. Figure 11.21 illustrates a de-energised conductor subject to both capacitive and inductive coupling. Consider earth switches *ES1* and *ES2* are initially open where the resistance to true earth from *ES1* and *ES2* is negligible. If a person touches the conductor that person is subject to flow of current through his body arising from capacitive coupling only. On closing *ES1*, the capacitive coupling is short circuited and the person rendered safe from capacitive coupling; however, the person is now subject to inductive coupling. Closing *ES2* now renders the person safe from inductive coupling. Therefore, the effects of capacitive and inductive coupling can usually be considered separately. With reference to Section 11.3.1,

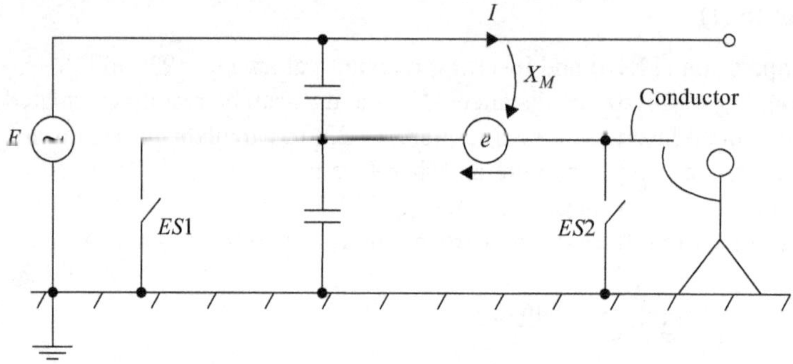

Figure 11.21 Combined effect of capacitive coupling and inductive coupling

where very long lengths of parallelism exist, both capacitive and inductive coupling effects do exist simultaneously, and the coupling summated – but for many practical circumstances to good approximation – they can be taken to exist separately.

11.8.2 Capacitive vs inductive coupling – severity

A relevant question from an electric shock perspective is which in practice tends to be the most severe and dangerous: capacitive coupling or inductive coupling? The answer is very dependent upon a range of factors including physical position, distances and current magnitudes. Generally, when the calculations for three-phase balanced voltage and current (up to maximum load current), for the same physical arrangements, are carried out, capacitive coupling tends to be slightly more severe. However, when single phase to earth fault currents are considered, inductive coupling results in very high levels of induced voltage and the most onerous and severe condition.

11.8.3 Working in an air-insulated substation

When work is undertaken under power network company safety rules (i.e. on the existing power network) prescribed and well-proven methods of working and earthing are utilised and under such circumstances IV situations which may give rise to electric shock are controlled and negated.

However, when work is being undertaken under contractors' safety rules, the potential for electric shock is often greater. This is because the equipment is being assembled from scratch, and under such circumstances, the rules governing when to earth (or to take other precautions) to combat IV can be less easy to define, with concomitant greater risk. Moreover, earthing connections on the equipment may not be readily available in the early stages of construction. A fundamental question that frequently arises is how far (i.e. distance) does a person, and the work, need to be from sources of IV before no safety precautions from IV need to be undertaken. This could occur for example in a contractor's CDM zone within an operational substation, which may be, say, between 40 and 80 m from the nearest source of IV (i.e. a reasonably significant separation). The answer of course is that it depends upon a mix of factors: the separation distance between power system and the equipment, the length of parallelism, the height above (or under) the ground, the voltage of the power system, the currents flowing in the power system and the resistivity of the earth. The response to this question is that unless a contractor is in possession of empirical rules governing such situations, then calculations using the formulae described in the preceding sections must be undertaken, or alternatively a computer-based solution.

11.8.4 Rules and guidance for controlling IV

Rules and guidance worthy of consideration for controlling IV when undertaking construction work on an operational construction site comprise the following:

- Wherever possible, always earth equipment under construction to remove the effects of capacitive coupling. It is preferable to earth midway along the equipment. In substations, this should be considered as being a default position.
- Equipment containing lumped capacitance such as cables and capacitor banks, and windings with inter-turn capacitance should always be earthed to remove the effects of capacitive coupling and possible stored charge.
- When equipment is either under, or on the surface of, the ground, earthing to mitigate capacitive coupling is usually not necessary.
- When lifting equipment, under construction, attach a 'trailing earth' to the equipment – which should be long enough and flexible enough to stretch from an earth bar to the point of work – to mitigate capacitive coupling.
- When a break is inserted into a piece of equipment or long objects, earth should be placed either side of the break.
- Cranes, MEWPS, etc., should be earthed upon reaching the work area and prior to being used. Scaffolding should always be earthed.
- Consider the requirement for equipotential zones.
- When considering the protection afforded by earthing, bear in mind that an earth rod sunk in the ground has a finite resistance and does not constitute a solid connection to true earth.
- Temporary metallic fences should always be earthed.
- Always use equipment that has been 'approved' for use in controlling IV.
- For temporary works, consideration should be given to the preparation of a set of empirical rules identifying situations where IV will not be a danger (e.g. distance from source and length of parallelism is too great) and control measures to be taken when IV will be a problem.
- If in doubt, employ a reputable computer-based solution and consult a professional with appropriate competence.

11.8.5 Microshocks

With reference to Figure 11.22, when a person approaches a charged conductor, the gap becomes progressively smaller to the point where a spark jumps across the gap. This is concentrated on a small area of skin – and the sensation can range from tingle to pain. This is termed a microshock. Once the person touches the object, then the contact area increases and the sensation disappears.

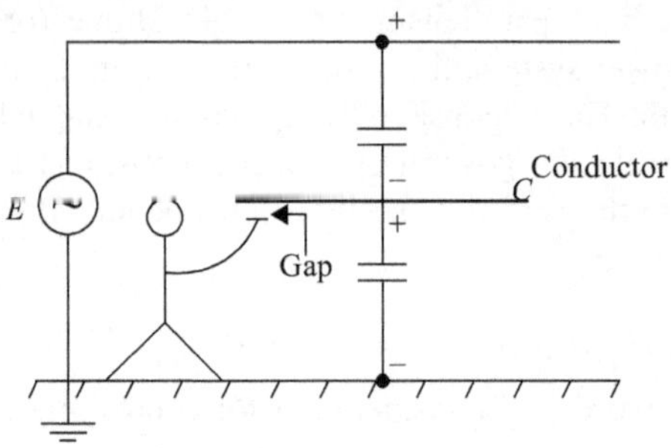

Figure 11.22 Microshocks

Microshocks result in an impulse current of short duration which (transiently) lowers the capacitive reactance of the body, which in turn increases the current flow. If the gap is closed very slowly, several microshocks can arise since the applied voltage is alternating i.e. reoccurring peaks.

Microshocks have no long-term health effects but do cause startled (and slightly painful) reactions. They may occur when touching motor vehicles, which are partially insulated from ground, in either a substation or above an OHL, or when touching structures, or even people shaking hands. They can be eliminated by earthing the conductor, or minimised by screening, or wearing well insulated footwear or gloves.

11.8.6 Overhead lines

Work on OHLs that are part of the power system would be undertaken under power network company safety rules and IV is usually well managed and controlled through stringent application of the rules. New OHL constructed under contractor safety rules would need to take account of both capacitive coupling, and inductive coupling that might arise from other OHLs that may run parallel to the OHL on which the work is undertaken.

When restringing one side of a tower-based OHL, the levels of both capacitive and inductive coupling at the higher voltages and currents can be significant. For this reason, the winching equipment is usually located within an 'equipotential zone' which bonds all the equipment together to a common earthing bar which in turn is connected to an earth spike. In some instances, bonded conductor plates are positioned under the equipment to ensure no difference in potential. The equipotential zone also ensures that should a fault occur between the in service circuit and the tower, all within the equipotential zone are subject to the same rise in potential and therefore safeguarded.

11.8.7 Long metallic objects

Long objects such as pipes, fences, railway lines, scaffolding and de-energised OHL and cables may be subject to significant IV from either energised OHL or busbars, or single core cables. If a long object is buried either in, or in close

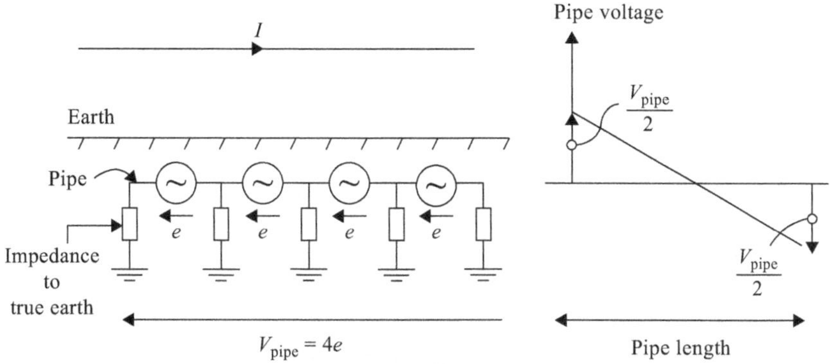

Figure 11.23 Inductive coupling into a pipe

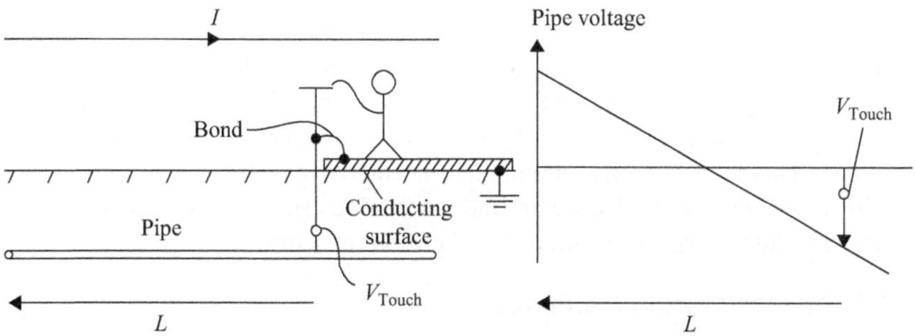

Figure 11.24 Protection of a person via an equipotential zone

proximity to the ground, the capacitive coupling effects are negated. Inductive coupling can however penetrate the ground, the depth of coupling being dependent upon the resistivity of the ground, and in turn the (Carson) equivalent earth depth. Capacitive coupling effects are removed by earthing the object – but as shown in Figure 11.2, if the object is very long, such as a parallel, energised OHL, then several earths dispersed along the object may be required to mitigate the potentially dangerous effects of IV.

When a long object is buried in the earth, i.e. a pipe, and its surface is in continuous contact with the earth, any induced voltage into the object from inductive coupling (e.g. from an OHL) results in circulating currents in the earth so maintaining the pipe at earth potential throughout its length. However, such circulating currents may lead to pipe corrosion and it is therefore usual to coat the pipe with an insulating material. Figure 11.23 shows the instance of an insulated pipe being subject to inductive coupling, and the resulting voltage profile which arises (due to the distributed capacitive impedance between the pipe and earth). The midpoint of the pipe is at earth potential. A person touching the pipe, e.g. via a valve as illustrated in Figure 11.24, will be subject to the touch potential shown. This situation can be protected against by the creation of an equipotential zone as

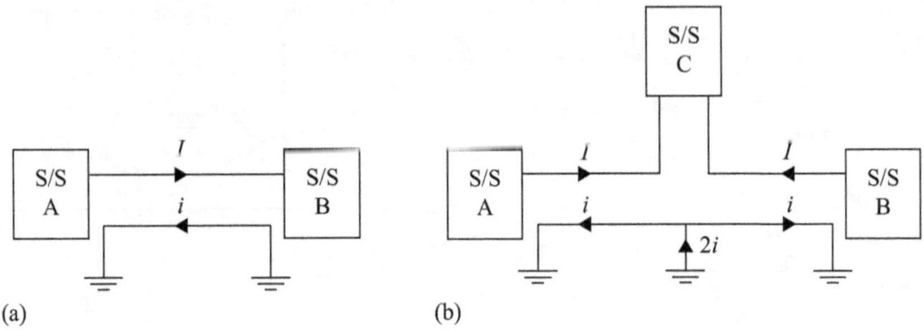

Figure 11.25 OHL circulating currents: (a) circuit earthed at both ends and (b) circuit earthed at both ends and at mid-point

shown in Figure 11.24. Generic methods of protecting individuals from inductive coupling voltages on long objects comprise the following:

- Inserting insulating sections in the object to reduce the voltage in each section.
- Ensure the pipe and all connections to it are fully insulated at all times.
- Run a 'mitigation wire' in parallel with, and periodically connected to, the object. This partly short-circuits the object therefore reducing the voltage along the object. This solution causes circulating currents and potentially unwanted heating.
- Create an equipotential zone around the work as shown in Figure 11.24.
- Use insulted working techniques at the point of work.

11.8.8 Operational considerations

Figure 11.25(a) shows the instance of a double circuit lattice steel tower OHL with one circuit in service and the other removed from service and earthed at both ends of the circuit. The in service circuit subjects the out of service circuit to inductive coupling which in turn results in a circulating current i, the magnitude of which may be between 8% and 12% I, and which varies slightly from phase to phase. It is worthy of note that the magnitude of I is not dependent upon the length of parallelism but is constant for a particular OHL configuration – see Section 11.3.2. A typical 400 kV OHL with a balanced load current of 4 kA in the in service circuit (e.g. current I in Figure 11.25(a)) causes, by inductive coupling, a circulating current i in the earthed circuit of the order of 450 A. Figure 11.25(b) shows the same OHL configuration but with the in-service circuit turned in to substation C (mid circuit). Should the earthed circuit be now subject to a midpoint earth, that earth is subject to a worst case scenario of $2i$, i.e. 900 A. If the midpoint earth is not rated for 900 A make/break (and some may not be), then the earthing switching sequence must ensure that the midpoint earth is the first to close and the last to open.

11.8.9 HV cables

The following summarises some of the main IV considerations with reference to HV cables:

1. **Three core cables**
   Figure 11.26 illustrates the key features of a typical three core cable in which each core has an insulation layer but no metallic screen around the core

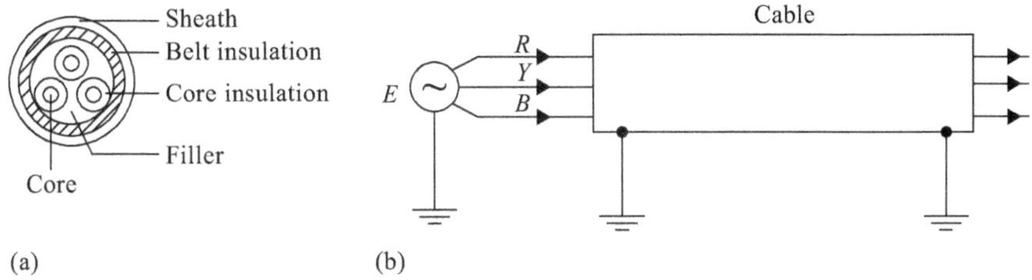

Figure 11.26 Three core cable: (a) Cable cross-section and (b) cable earthing arrangements

(sometimes termed a belted cable). Each core has a capacitance to each other core and to the earth sheath. For balanced load (voltage) conditions, the capacitive currents to earth cancel and there is no flow of current from the sheath into the earth. Furthermore, since the sheath is at earth potential there is no electric field external to the sheath.

With reference to inductive coupling, the three balanced load currents shown in Figure 11.26(b) result in the induced voltages into the cable sheath summating to zero, so there is no circulating current in the sheath to earth loop. Furthermore, the three fluxes arising from the balanced load current summate to zero, so there is no (or little) magnetic flux external to the cable sheath. Should a single phase to earth fault from a core conductor to the cable sheath arise, the currents in core and sheath will be equal and opposite and there will be little magnetic flux external to the sheath and virtually no circulating current in the sheath. A similar argument applies to a phase to phase fault. Thus, under all operational conditions, the circulating current in the sheath and magnetic flux external to the sheath is virtually zero.

2. **Single core cables**

Single core cables are mostly used at the higher voltage of 400, 275 and 132 kV. This is because the greater insulation thickness favours single core cables in preference to three core – both to facilitate manual handling of the cable, and to minimise cable overheating which otherwise would lead to cable de-rating (i.e. limiting the current rating).

Figure 11.27 shows a single core cable (i.e. one cable out of three single core cables, which comprise a three-phase circuit) solidly earthed at both ends. A leakage current flows to earth through the cable to sheath capacitance, and since the sheath is maintained at earth potential no electric field exists outside the sheath. The current flowing into the cable's inner conductor causes inductive coupling into the sheath to earth loop, resulting in a circulating current around the sheath and earth. This causes heating loss in the sheath, which in turn limits the rating of the cable. It is unusual for a solidly earthed cable, as shown, to be used above 66 kV.

Unlike three core cables, single core cables result in a magnetic flux which is external to the sheath, the magnitude of which is proportional to the currents $(I - i)$ as shown in Figure 11.27. There is a partial cancelling of the flux emanating from the three cables, i.e. one cable per phase, when balanced

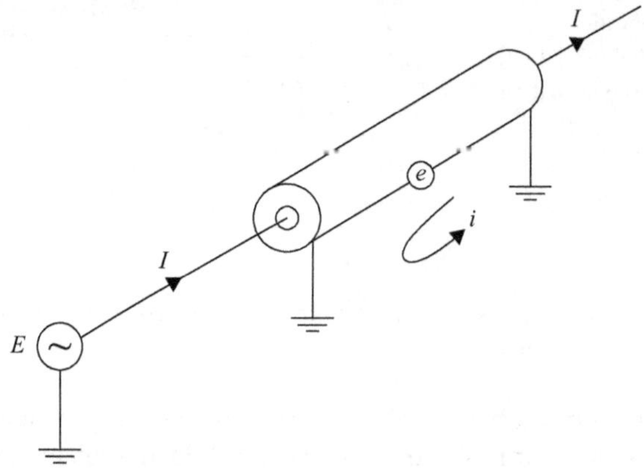

Figure 11.27 Single core cable

three-phase load current flows (more so when the cable is laid in trefoil). The greatest magnitude of flux external to the cable arises for the instance of an external earth fault where the fault current does not return via the cable sheath.

Specialist texts are available (see bibliography) providing calculations to determine the voltage induced into the sheath, and in turn to determine the sheath circulating currents. The sheath to earth loop physical dimensions would again be determined by the Carson equivalent earth as described in Section 11.7.1. It is again worthy of note that the induced current into a cable sheath is a constant magnitude for that cable arrangement, no matter the length of the cable, for the reasons explained in Section 11.3.2.

3. **Single point bonding cable systems**

The current rating of a single core cable can be improved by eradicating the sheath circulating current. This may be achieved by single point bonding systems. The simplest example is to solidly earth the cable at one end, and to earth the cable through sheath voltage limiters (SVL) at the other end. The SVLs are required to limited transient over voltages usually arising from either circuit switch in, or lightning. This arrangement effectively results in an open circuit for inductively coupled voltages into the cable sheath. Historically, it has been the practice in the United Kingdom to limit the sheath standing open circuit voltage to 65 V at 132 kV, or 150 V at 275 and 400 kV. Although the linkage with electric shock currents is not well articulated, statistically, over the years these voltages have not proved problematic. The voltage limits usually result in a typical cable length of about 300 m. This arrangement is sometimes termed end point bonding.

In most practical situations, a return path for earth faults beyond the end of the cable is provided (to prevent interference with communication circuits) by provision of an earth continuity conductor (ECC) see Figure 11.28. The ECC is usually transposed to minimise inductively coupled circulating currents. In some instances, the cable sheath is interrupted mid circuit with an insulated barrier joint (known as split sheath single point bonding) – which can increase the length of single point bonding solutions up to a maximum of about 1,000 m.

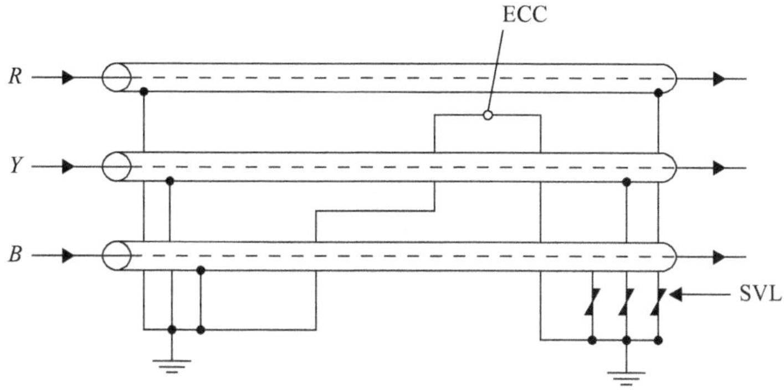

Figure 11.28 Practical single point bonding system

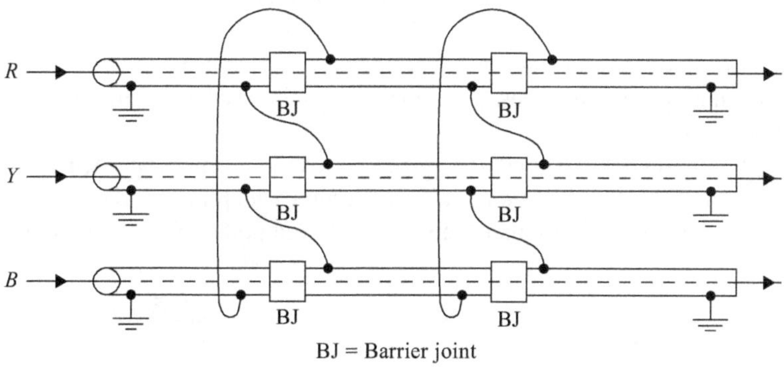

Figure 11.29 Cross-bonded cable system

4. **Cross-bonded cable system**
The cross-bonded cable system overcomes the limitations of cable length suffered by the single point bonding system. With reference to Figure 11.29, the cable sheath is sectioned by the insertion of non-conducting barrier joints. Over the three sections shown, any one sheath is connected in series to the sheaths of the two other phases, thus the inductively coupled voltage across the three sections comprises $V_R + V_Y + V_B$ which summate to zero, or virtually zero, therefore resulting in little to no circulating current to de-rate the cable. With reference to the flat cable formation shown in Figure 11.29, the cable sheaths are not subject to the same inductive coupling when considering the mutual coupling from the two other phases, and therefore to compensate for this, the cable sheaths are physically repositioned (transposed) so that across the three sections they all occupy the top middle and bottom positions (this is not illustrated in Figure 11.29). This further reduces the magnitude of inductively coupled voltages into the sheath.

5. **Working on cables**
When laying a new cable, the safety hazard is primarily that of inductive coupling from parallel OHL or cables. However, as long as the ends of the cable are insulated then the IV danger is removed. Once a new cable is installed in the ground, it can still be subject to inductive coupling from parallel OHL, cables or busbars and therefore jointing of the cable may require insulated working techniques. This involves the jointer working from an insulated platform (usually 10 kV minimum insulation) within a joint bay with insulated sides – thereby ensuring the jointer is insulated and protected from any IV effects on either the cable sheath or cable phase conductor. Insulated working may also be required when constructing cable sealing ends. Once a cable sealing end is constructed, capacitive coupling can take place from any adjacent energised busbar or OHL which in turn can charge the capacitance of the cable. Thus to render safe, the cable sealing end should be earthed on completion.

All work on previously commissioned cables under the jurisdiction of a power network company safety rules should only be undertaken following an assessment of IV. The chosen safety precautions will usually involve one of the following earthing regimes, a number of which may be required in the course of work on a cable:

- Figure 11.30(a) shows the instance of circuit earths at each end of the cable. This results in a circulating current which lowers the voltage at the point of work – but also enables transferred potential to the point of work.

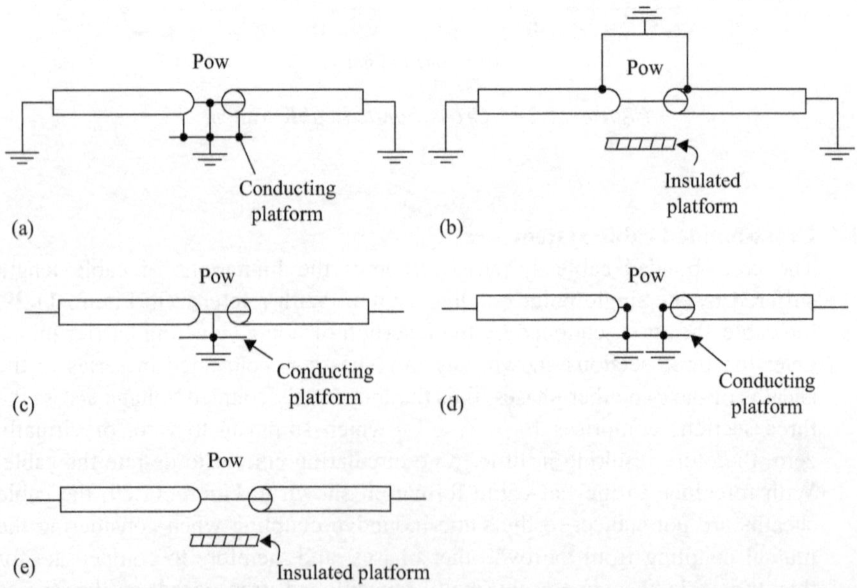

Figure 11.30 Example methods of working on cables: (a) cable earthed – earthed working, (b) cable earthed – insulated working, (c) cable unearthed – earthed working, (d) cable unearthed and split – earthed working and (d) cable unearthed – insulated working

Earths at either end also protect against inadvertent circuit energisation. Working from the earthed conducting platform (i.e. earthed working) provides safe working via an equipotential zone. Earthing at the point of work provides the necessary protection. This would usually be the default working arrangement.

- Figure 11.30(b) shows the instance of where earthing at the point of work is not easy to achieve so requiring insulated working from an insulated platform.
- Figure 11.30(c) would be advantageous for the instance of a cable located mid circuit and connected at either end to an OHL, where there are no circuit earths located at the cable ends. In such an instance, it may be considered advantageous to disconnect the OHL.
- Figure 11.30(d) builds on Figure 11.30(c) where there is break in the cable requiring earths either side of the break. Figure 11.30(e) is an option to Figure 11.30(c). In Figure 11.30(c)–(e), it is assumed that the IV in the cable metallic sheath is either negligible, or that the sheath is fully insulated at the point of work, otherwise sheath earthing as in Figure 11.30(b) would apply.

The above work methods falls into two categories: either earthed working or insulated working. Typical cable circulating currents arising from IV may be as high as 200 A when induced from balance load currents – or 2 kA when induced from single phase fault currents. The above methods may also be considered for circuits under contractor's safety rules at various stages of construction.

11.8.10 Electric field shielding

Both equipment and people can be screened from electric fields, i.e. from capacitive coupling. The classical method is by use of a 'Faraday cage' as illustrated in Figure 11.31. The electric field in the cage is zero (may be thought of as

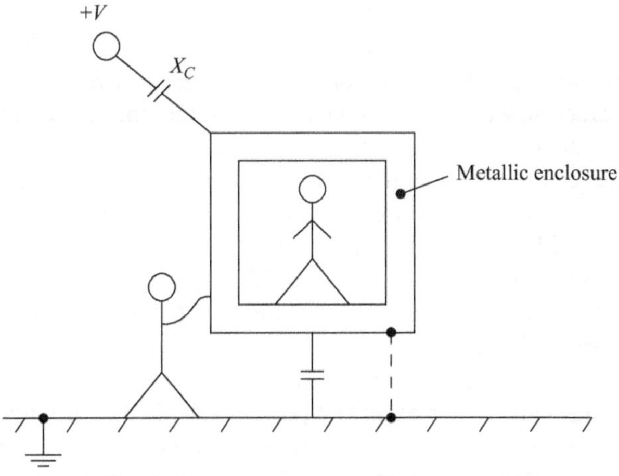

Figure 11.31 Electric field screening

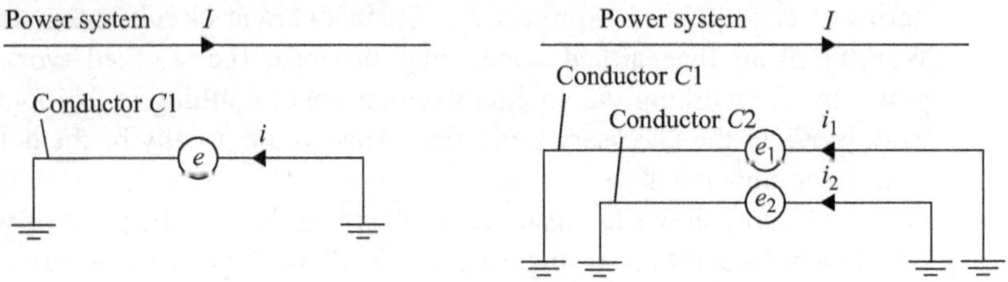

Figure 11.32 Magnetic field screening

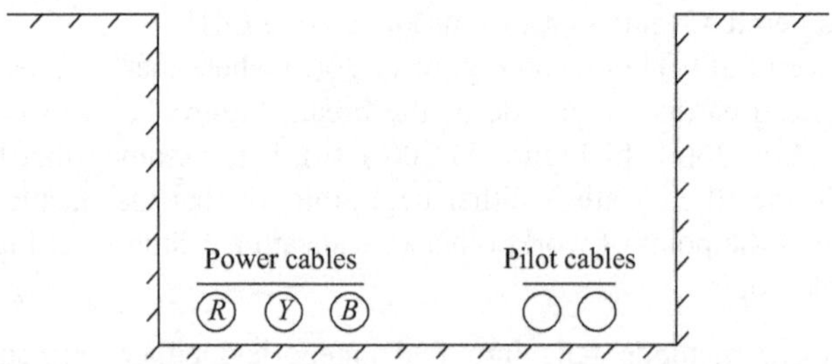

Figure 11.33 Pilot cable screening

short-circuited capacitance within the cage). A person touching the outside of the cage would be subject to electric current flow through their body, so to prevent this, in practice, the cage must be earthed. Earthed scaffolding around equipment forms a reasonably effective Faraday cage. Partial screening is also afforded by an earthed conductor in close proximity to equipment/conductor to be worked upon.

11.8.11 Magnetic field screening

Consider Figure 11.32, where a power system conductor carries a current *I*, which via inductive coupling induces an emf into conductor *C1*, resulting in circulating current *i*. Consider now a second conductor *C2* brought into close proximity to *C1*. Conductors *C1* and *C2* are each subject to an induced emf and resulting circulating currents. However, the available magnetic flux from *I* to cause an induced emf into conductor *C1* is reduced by the existence of i_2. This may be likened to a CT (see Section 11.4.1) with the power system conductor as the primary, with two secondaries represented by conductors *C1* and *C2*. As such, current flow in conductor *C2* reduces current flow in conductor *C1*. Conductor *C2* is therefore said to 'screen' conductor *C1* and vice versa.

A practical example of screening is illustrated in Figure 11.33 which shows pilot cables in the same trench as an HV power cable. If the earthing arrangements are such that there is no circulating current in the power cable sheath, then the sheath provides no screening from inductive coupling from the power cable core

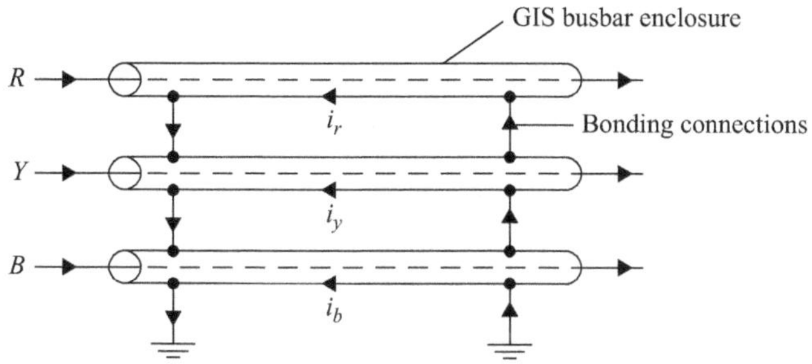

Figure 11.34 Circulating currents in GIS enclosures

conductors. However, the pilot cable sheaths are usually solidly earthed at both ends so inductive coupling from the power cable to the pilot cable sheath takes place causing circulating currents in the pilot sheath, which advantageously provides screening from induced voltages into the pilot cable cores. Typical magnitudes of induced voltages into the pilot sheath are:

- 25 V/km – arising from three-phase balanced power cable load of 2 kA
- 1,500 V/km – arising from power cable fault current of 63 kA.

11.8.12 GIS enclosures

With reference to Figure 11.34, GIS busbar enclosures are subject to inductive coupling from phase conductors. To avoid rise of potential, the enclosures are usually bonded together resulting in circulating currents. The magnitude of the circulating current may be as high as 90% of the busbar current, e.g. for a busbar continuous load current of 4 kA, the enclosure circulating current could be as high as 3,600 A. The three enclosures are not physically symmetrical, so causing spill currents in the earth connection. GIS enclosures when subject to internal and external fault currents are in turn subject to high touch potentials which must be designed to be within touch potential voltage limits.

11.8.13 GIS switchgear – trapped charge

During normal switching of GIS substations, a trapped charge will remain on de-energised busbars. This may remain for many hours due to the high busbar support insulations' levels (low conductivity). In addition, the power system voltage can couple through the grading capacitors of an open circuit breaker to superimpose an AC voltage on the trapped charge. On the 400 kV system, this may result in peak voltages as high as 900 kV. Under these conditions, a GIS busbar support insulator can be subject to levels of electric field stress which lead over time to a low energy flashover across the insulator, or alternatively a flashover on circuit re-energisation. The presence of any particulate contamination contributes to the breakdown of GIS insulation under trapped charge conditions. The risk of trapped charge causing flashover is low, but even so it is essential that GIS exposure to trapped charge is minimised, which can be accomplished through the following measures:

- Connection of power transformer/reactors with earthed star windings to the busbar
- Electromagnetic voltage transformers connected to all three phases of the busbar
- Connection of air insulated equipment such as bushings to the busbar.

When none of the above are available and the period of exposure to the trapped charge is likely to exceed 60 min before the busbar is re-energised, then an earth switch should be closed on to the busbar.

11.8.14 Air cored reactors

Air cored reactors generate powerful magnetic fields which permeate significant distances from the reactor. As such, the field results in inductive coupling with any closed metallic loop. Of specific concern is the reinforced concrete base on which the reactor stands, in which there are steel reinforcing bars (rebar) which form closed interconnecting loops, and therefore subject to induced voltages and circulating currents. These currents can cause the rebar to heat which in turn heats the concrete to unacceptable temperatures. Such heating can be prevented by the following special measures:

- Insulation at all bar crossings
- Use of insulating coating on the bars.

11.8.15 Electric and magnetic field design limits

There is ongoing public concern over the physiological effects on the general public of electric and magnetic fields arising from power systems. Permitted levels in the United Kingdom are as follows:

- Magnetic fields = 360 μT (i.e. flux/m^2)
- Electric fields = 9,000 V/m

These levels apply when the time of exposure is extensive.

Overhead lines in the United Kingdom are designed and constructed to meet the above limits. The levels are influenced by the height and spacing of the OHL, and the phasing of the conductors on a double circuit OHL. Cables do not cause the same high levels of fields. The electric field is contained entirely within the cable, and the magnetic field is greatly cancelled due to the close proximity of the three-phase conductors, see Section 11.8.9. Electricity substations produce fields (affecting the general public), but well below the limits. Currently, there are no limits on emf grounds on how close a house may be built to a substation.

11.9 The management of impressed voltages

The management of IVs is a very significant factor in power network construction both in terms of the design of the permanent works and the on-site safety management of the temporary works. IV relating to permanent works is mostly underpinned by standard and well-proven calculations and methods (e.g. voltages induced into cable sheaths). IV relating to temporary works is more frequently site specific, but which, none the less, is best managed through empirical methods of control, even if significant factors of safety are in built. Within this context, it is essential to ensure that robust safety rules/quality management procedures are prepared to define the required IV control measures. It is equally important to ensure that engineers who are competent in understanding IV and its dangers, are available to apply those safety rules/procedures – or otherwise refer the work situation to an engineer who is competent in undertaking the IV severity level calculations. The understanding and management of IV is one of the fundamental competencies for all who undertake power network construction.

Chapter 12
Substation earthing design

12.1 Introduction

Substation earthing is critical both to the operational performance of the power system, and the provision of a safe working environment. It is a salient area of design when either constructing a new substation or extending or modifying an existing substation.

Substation earthing is a highly specialist subject in its own right, and therefore this chapter will provide an overview of the basic theory and key requirements, relevant to power network construction, as summarised below:

- Objectives and regulations
- Resistivity
- Earth electrode systems
- Earthing conductors and earth mats
- Earthing system and earth fault current
- Voltage limits
- Practical considerations
- Earthing design considerations

12.2 Objectives and regulations

12.2.1 Historical

A fundamental question is why power systems are earthed at all? Common practice with early power systems (up to *ca.* 1915) was to insulate equipment and operate with the power system unearthed. However, as power networks expanded, the phase-to-earth capacitance of the networks increased, such that when a phase-to-earth fault occurred, arcing and over-voltages on the healthy phases was experienced, resulting in equipment damage. Furthermore, rapid detection and isolation of the fault via protection systems was difficult to achieve.

To overcome these problems, power systems, worldwide (with some, but few, exceptions), were subsequently designed as earthed systems, with the neutrals of generators and transformers connected to substation earth mats, which were purposely designed to be as close to earth potential as practical. An earth mat also has the added advantage of providing a safe working environment – from the danger of electrical shocks.

12.2.2 Regulatory requirements

Regulatory requirements relating to power system earthing include the following:

1. **ESQCC regulations**
   The electricity safety, quality and continuity regulations (see Chapter 2) stipulate the following with reference to earthing:
   'A generator or distributor shall ensure that
   (i) As far as reasonably practical, his network does not become disconnected from earth in the event of any foreseeable current due to a fault.
   (ii) In respect of any HV network which he owns or operates – that the network is connected with earth at, or as near as reasonably practical, to the source of voltage, but where there is more than one source of voltage, the connection with earth need only be made at one such point'.
   With reference to the above 'connected with earth' means:
   (iii) 'Connected to earth in such a manner as will at all times provide a rapid and safe discharge of energy'.
   Other sections of the regulations deal with the earthing of LV networks.
2. **Electricity at work regulations**
   Regulation 8 of the electricity at work regulations stipulates that:
   (i) 'Precautions shall be taken, either by earthing or other suitable means, to prevent danger arising when any conductor (other than a circuit conductor) which may reasonably foreseeably become charged as a result of either the use of a system, or a fault on a system, becomes so charged'.
   Regulation 8 goes on to say:
   (ii) 'Conductors which, although not part of a system, are within electrostatic or electromagnetic fields created by a system may be subject to this regulation. Appropriate precautions are necessary if the induced voltages or currents are large enough to give rise to danger'. See Chapter 11 on impressed voltages.
   (iii) 'A conductor shall be regarded as earthed when it is connected to the general mass of earth by conductors of sufficient strength and current carrying capability to discharge electrical energy to earth'.

In summary, the above regulations stipulate two requirements, which are as follows:

- HV networks must be connected to earth (close to the voltage source)
- Earthing arrangements must be put in place to prevent danger (i.e. electrical shock effects), which may arise directly, or indirectly (i.e. via induction), from the electrical network.

12.2.3 Substation earthing objectives

In general, substation earthing should satisfy the following objectives:

- The earth system must be capable of carrying earth fault currents without overheating, mechanical damage or unduly drying out the surrounding soil.

- The resistance of the earth system must be sufficiently low to ensure operation of protection relays for earth faults both at the substation, and on connected OHL and cable feeders, etc.
- The resistance of the earth system must be sufficiently low to ensure that a person in the vicinity of the earthed facilities is not subject to excessive voltage and therefore not exposed to the hazards of electrical shocks. Specifically, with reference to health and safety, the earthing design must achieve the following:
  - Limit the substation 'rise of earth potential' (ROEP) to within defined voltage levels to limit 'transferred potential'. NB: ROEP is alternatively termed 'earth potential rise'.
  - Limit the 'touch potential' and 'step potential' to within defined voltage limits.
  - Limit 'third-party infrastructure threshold' voltage to within defined voltage limits.
- The earth system must be isolated from services entering the substation, so that any ROEP is not transferred to telephones, water mains or any other metallic connections entering site.

The means by which the above is achieved will be examined in the following sections.

12.3 Earth resistivity

12.3.1 Resistivity

All materials, including the earth, present a resistance to the flow of electricity. The magnitude of the resistance is dependent upon an intrinsic characteristic of the material termed its 'resistivity'. A material with a low resistivity readily allows the passage of electrical current. The definition of resistivity is:

> The longitudinal electrical resistance, 'R', of a uniform rod of material, 'l', of length 1 m, with a cross sectional area (CSA), 'A', of 1 m^2. This is expressed mathematically as follows:

$$\rho = R\frac{A}{l}\ \Omega\,\text{m} \tag{12.1}$$

The resistance, R, of a sample of any material can then be obtained by rearranging expression (12.1) as follows:

$$R = \rho\frac{l}{A}\ \Omega \tag{12.2}$$

where ρ is the resistivity of the material in ohm-metres, l is the length of the material in metres and A is the CSA of the material in square metres.

Typical values of earth resistivity are provided in Figure 12.1. Suffice it to say that the earth is essentially resistive (i.e. not inductive or capacitive). It is also

Earth type	Resistivity (Ω m)
Sea water	0.1–10.0
Concrete	1.5
Marshy soil	10.0 30.0
Clay	10.0–100.0
Farm land	60–200
Sand	250–500
Slate	300–3,000
Rock	1,000–10,000
Sandstone	100,000

Figure 12.1 Material resistivity

worthy of note that that there are slight variations in the figures given, depending upon the literature consulted – those given are therefore an average. Actual values can be obtained from field tests.

Invariably, some sites will have earth comprising layers of different resistivity and in those instances, resistivity correction factors would have to be included in any calculation of resistance.

12.4 Earth electrode systems

12.4.1 Types of electrodes

The term 'earth electrode' refers to a conductor or group of conductors that provide an electrical connection to the mass of earth. Within this context, although the huge mass of earth is in itself an excellent conductor of electricity of virtually zero resistance, small volumes of earth tend to be relatively poor conductors (depending upon the resistivity of that part of the earth in question). Therefore, a high magnitude of current, such as fault current, flowing into the earth will cause a high potential gradient in the area immediately surrounding the point of entry into the earth.

The purpose of an earth electrode system is therefore to lower the potential gradient by lowering the resistance between the point of entry of the fault current and the point of 'remote earth' (alternatively termed the 'reference earth' or 'true earth') – the point of remote earth being the point of zero potential (i.e. zero voltage).

NB: Virtually, all of the huge mass of earth will stand at zero potential, and the point of remote earth that is relevant is that point nearest to the earth electrode system. Commonly used earth electrode systems are as follows:

- Driven vertical rods
- Buried horizontal and vertical plates
- Buried horizontal strip
- Buried grid or mesh
- Multiple-driven vertical rods in a hollow square.

The above will be briefly reviewed.

12.4.2 Driven vertical rods

With reference to Figure 12.2(a), the resistance R_E of a vertical rod to remote earth is given by:

$$R_E = \frac{\rho}{2\pi L}\left[\text{Ln}\left(\frac{4L}{r}\right) - 1\right]\ \Omega \qquad (12.3)$$

where ρ is the resistivity of the earth and expression (12.3) is valid when the length L is much greater in magnitude than that of radius r.

Figure 12.2(b) illustrates the value of resistance R_E as a function of the depth of the rod (i.e. L) for an earth resistivity of 100 Ω m (i.e. typical farm land) and a commonly used rod radius of 8 mm. As can be seen, the resistance begins to level out at a rod depth of about 3 m with a resistance of approximately 20 Ω. A rod depth of 15 m would result in resistance of approximately 7 Ω.

Applications of driven vertical rods apply to earthing of LV systems, fence earthing, general field earths, etc.

12.4.3 Buried horizontal or vertical plate

With reference to Figure 12.2(c), the resistance R_E of a buried vertical or horizontal plate to remote earth is given by:

$$R_E = \frac{\rho}{8r}\left(1 + \frac{r}{2.5h + r}\right)\ \Omega \qquad (12.4)$$

where ρ is the resistivity of the ground, r is the radius of an equivalent circular plate, and h is the plate centre depth.

For example, a plate with a centre depth of 1.0 m, and a radius equivalent to 1.5 m, in a ground of resistivity 100 Ω m (i.e. farm land), would have a calculated resistance of 11.4 Ω.

Applications using plates are usually to be found on the 11 kV and LV systems. They are advantageous where the soil contains rock and it is difficult to drive vertical rods.

12.4.4 Buried horizontal wire (or strip)

With reference to Figure 12.2(d), the resistance R_E to remote earth is given by:

$$R_E = \frac{\rho}{2\pi L}\left[\text{Ln}\left(\frac{2L^2}{dh}\right) - K\right]\ \Omega \qquad (12.5)$$

where ρ is the resistivity of the ground, d is the diameter of the wire or width of strip and K is the obtained from tables (and is 1.3 for round wire, and 1.0 for a strip of conductor.)

For example, a round wire conductor in ground of resistivity 100 Ω m (farm land) of length 20 m, diameter 0.02 m, at a depth of 0.5 m, with $K = 1.3$ would have a resistance of approximately 8 Ω.

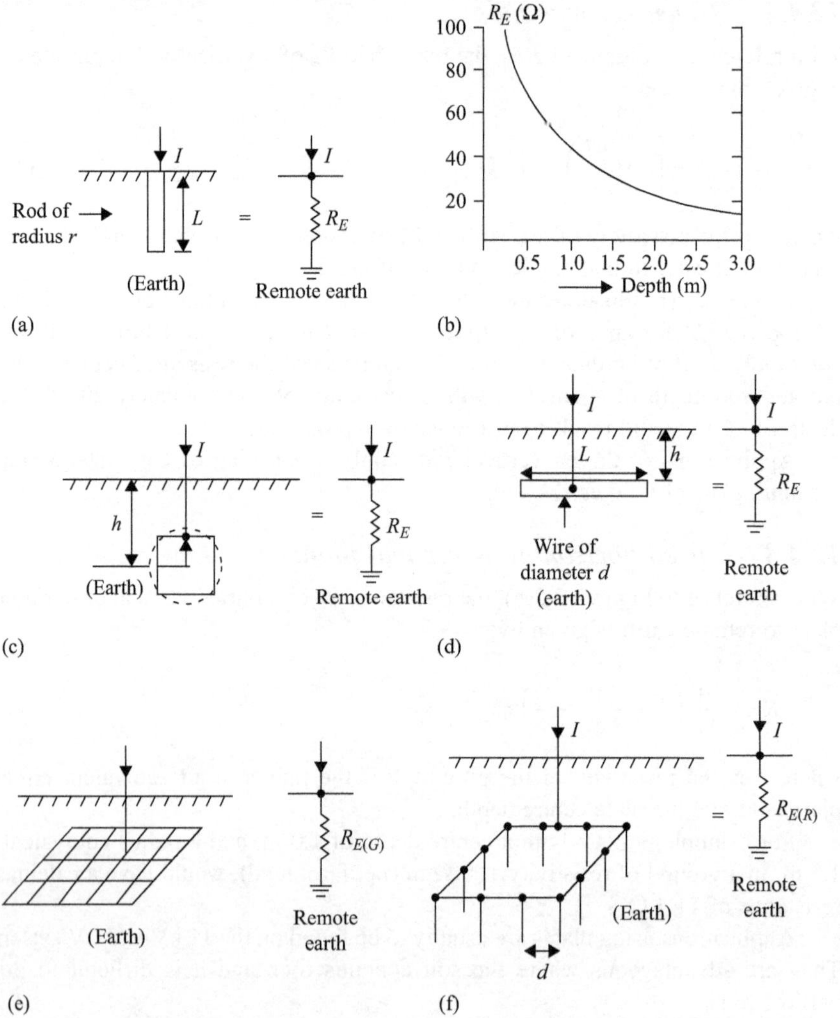

Figure 12.2 Earth electrode systems: (a) driven vertical rod, (b) R_E as a function of depth with driven vertical rod of 8 mm, (c) buried vertical/horizontal plate, (d) buried horizontal wire, (e) buried horizontal grid and (f) multiple-driven vertical rods

12.4.5 Buried horizontal grid

With reference to Figure 12.2(e), the resistance, $R_{E(G)}$ of the grid, to remote earth, based upon a grid buried at a depth of between 0.3 and 1.0 m is given by:

$$R_{E(G)} = \rho\left(\frac{1}{4r} + \frac{1}{L}\right)\ \Omega \tag{12.6}$$

where $r = \sqrt{\frac{A}{\pi}}$, A is the area of the grid, and L is the sum of the lengths of all buried conductors.

For example, a 171 × 171 m^2 grid (alternatively termed a mesh), that comprises 20 × 20 conductors spaced 9 m apart, buried in earth of 100 Ω m (farm land) would result in a calculated resistance to remote earth of approximately 0.27 Ω.

This type of earth electrode system is generally installed at network voltages of 132 kV and above (and some 33 kV substations), which are subject to high fault currents, consequently, requiring a relatively low earth resistance to minimise ROEP. This type of earth electrode system is frequently termed an 'earth mat'.

12.4.6 Multiple-driven vertical rods

With reference to Figure 12.2(f), the resistance, $R_{E(R)}$, of a multiple-driven rod arrangement, (comprising a hollow square) to remote earth, is given by:

$$R_{E(R)} = \frac{1}{N}\left[R_E + \left(\frac{\rho}{2\pi d}\alpha\right)\right]\ \Omega \tag{12.7}$$

where R_E is the resistance of a single isolated rod as given in expression (12.3), N is the number of rods, ρ is the ground resistivity, α is a factor which depends on the number of rods along each side of the square (as specified in BSEN 7430), and d is the distance between adjacent rods.

For example, for the instance of a hollow square of driven vertical rods, in ground of resistivity 100 Ω m (i.e. farm land), where the distance, d, between adjacent rods is 10 m, $\alpha = 9.4$ and the number of rods, N, is 20, then the calculated value of $R_{E(R)}$ is approximately 2.36 Ω.

Multiple-driven rods are mostly used in conjunction with another electrode system (e.g. horizontal grid) to lower the overall resistance to remote earth. It is worthy of note that when piles have been installed, they may also be employed as part of the electrode system, acting as if they were rods.

12.4.7 Combined horizontal grid with vertical rods

The total equivalent resistance, R_T, of an earth electrode system comprising a combination of a horizontal grid and vertical rods is given by:

$$R_T = \frac{\left[R_{E(G)}R_{E(R)} - (R_M)^2\right]}{R_{E(G)} + R_{E(R)} - 2R_M}\ \Omega \tag{12.8}$$

where $R_{E(G)}$ and $R_{E(R)}$ are given by expressions (12.6) and (12.7), respectively, and where:

$$R_M = R_{E(G)} - \frac{\rho}{\pi L}\left[\ln\left(\frac{\pi L}{W}\right) - 1\right]\ \Omega \tag{12.9}$$

where L is the length of vertical rods in metres and W is the width of strap in metres.

Generally, with transmission substations where the resistance of the horizontal grid is relatively small, the addition of the vertical rods often makes very little reduction in overall resistance – but can assist reducing earth resistance in distribution substations where horizontal rod earth resistance is higher.

12.5 Earth conductors and earth mat

12.5.1 Earthing conductor – considerations

Earthing conductors are usually made from either copper or aluminium. Bare copper is generally used for the below ground earth electrode system (e.g. plates, wire strip or horizontal grid) and buried at a typical depth of 0.6–1.0 m. This provides mechanical protection and is below the frost line. The conductors should ideally be surrounded by non-corrosive, good conducting, fine textured soil, firmly rammed (typically 150 mm coverage). Driven rods are also usually made from copper and driven to a minimum depth of 3.0 m and typical maximum depth of 15 m. Aluminium is often used for above ground connections – and could be used below ground if the soil type will not lead to corrosion, although this is not a common practice.

Earthing conductors must of course be sized in accordance with the fault current they will have to carry, the duration of the fault current and the maximum allowable temperature rise of the conductor. The latter requirement is usually stipulated as 405°C for copper conductors and 325°C for aluminium-based conductors – assuming an initial temperature of 30°C. Tables are published (e.g. ENATS 41-24) specifying the conductor size against fault current and duration.

12.5.2 Earth mat connections

With reference to Figure 12.3, the following is usually connected to the earth mat (i.e. earth electrode system).

- All HV equipment metallic tanks, connections and supports that are required to be at earth potential (e.g. circuit breaker tanks, transformer tanks, earth end of surge diverter, earth switches, insulator metallic supports, etc.).
- OHL terminal towers (usually via two connections from the earth mat to two separate tower legs).
- Transformer neutral connections, required for earthing the HV system, which may be via a neutral earthing resistance.
- HV cable sheaths.
- Earthing mats which are positioned under manually operated equipment (e.g. earth switch, disconnector, etc.) to provide an equipotential zone.
- CT and VT neutral connections.
- Multicore and multi-pair cable armouring.
- All normally de-energised metalwork in the substation (e.g. panels, cubicles, kiosks, lighting columns, compressed air equipment, oil tanks, water pipes, etc.).
- Adjacent substations (depending upon the technical merits of joining the earthing systems of two substations). If connected, this is usually via two separate connections.

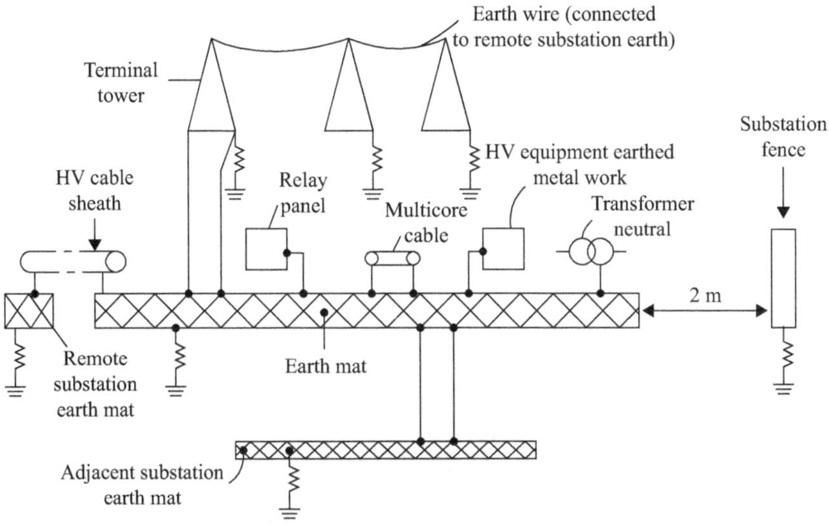

Figure 12.3 Earth mat – connections

12.6 Earthing system and earth fault current

12.6.1 Earthing system

Figure 12.3 illustrates that if an HV network fault take place within the substation, the flow of fault current to remote earth is not only via the substation earth mat (i.e. earth electrodes) but also via OHL (i.e. towers) (that are both individually earthed, and carry an earth wire that connects to a remote substation) – and via HV cable sheaths which are usually connected to earth at both the local and remote substation (for both solid and bonded type cables). The combination of the local substation electrodes and the connecting OHL and HV cable earth return paths is usually termed 'the earthing system' of the substation. It is worthy of note that transformer neutrals connected to the earth mat are also part of the earthing system which may also include the resistance of a neutral earthing resistor. To determine the flow of earth return current via the transformer (and via the earth mat), it is essential to analyse the network with reference to a sequence network diagram (i.e. symmetrical components) as detailed in Chapter 4. It is also worthy of note that most 33 kV OHL have wood pole supports and are of an unearthed construction, and therefore unlike towers provide no earth return current path.

The characteristics of OHL and cable earth return paths will be briefly reviewed below:

1. **OHL earth return path**
   Figure 12.4 shows the instance of an OHL comprising an aerial earth conductor connected to each tower and each tower individually earthed. The OHL can be represented by an equivalent ladder network, in which the earth wire impedance, Z_S, has both an inductive and resistive component, and the tower footing

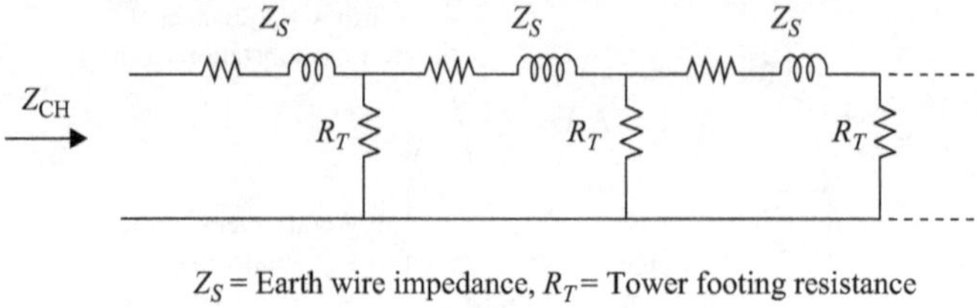

Figure 12.4 OHL ladder network

only a resistive component, R_T. It can be shown that the OHL input impedance, Z_{CH} (known as the 'chain impedance'), looking into the ladder network (of indefinite length) is given by:

$$Z_{CH} = \frac{Z_S}{2}\left(1 + \sqrt{1 + \frac{4R_T}{Z_S}}\right) \tag{12.10}$$

For example, consider a 132-kV single-circuit OHL with three spans per kilometre of OHL, with an earth wire self-impedance of (0.18 + j0.72) Ω/km, and a tower footing resistance of 10 Ω. Applying these parameters to expression (12.10) results in a chain impedance of (1.7 ∠41°) Ω approximately. It is of interest to note that expression (12.10) is not a function of OHL length (and is valid for an OHL of infinite length); however, its validity is dependent upon a certain minimum length of OHL (typically in excess of 20 spans), as more specialists texts will confirm.

2. **HV cable return path**
   A model similar to Figure 12.4 can be derived for HV cable sheaths, with the tower footing resistance replaced by the cable capacitance to the earthed sheath. However, for short cable lengths, the capacitive reactance will be relatively high and usually ignored when determining the chain impedance. A cable system comprising three single-core cables will of course result in each sheath being subject to mutual inductive coupling from the adjacent cable, and this must be included in the derivation of chain impedance.

12.6.2 Earth system impedance

With reference to Figure 12.5, the total impedance, Z_E, of the earth system is given by the parallel combination of earth mat resistance, R_E and the chain impedances of the OHLs, Z_{CH}, and HV cables, Z_{CS}. Mathematically, this is expressed by:

$$Z_E = \left[\frac{1}{R_E} + \frac{1}{\Sigma\left(\frac{1}{Z_{CH}}\right)} + \frac{1}{\Sigma\left(\frac{1}{Z_{CS}}\right)}\right]^{-1} \tag{12.11}$$

where $\sum()$ is the summation of all impedances comprising the parallel connected OHL and HV cables, respectively. Z_E, of course, contains both resistive and reactive components, and thus the constituent parts of expression (12.11) need to be summated vectorially.

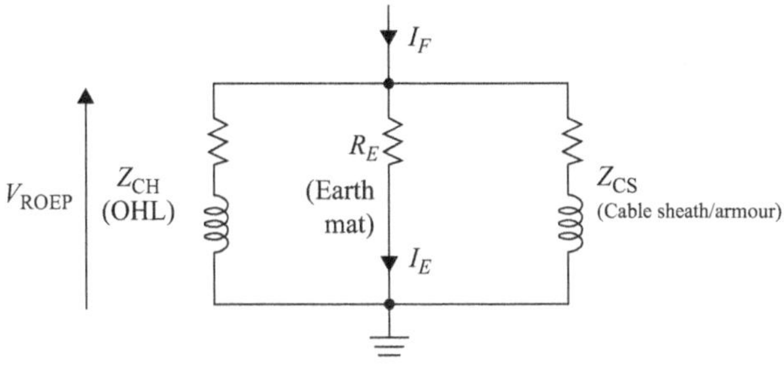

Figure 12.5 Earth system impedances

In the above calculations, no account was taken of the remote substation earth resistance. This is because it is usually relatively small compared to the chain impedances. However, if the remote substation is very close, such that the chain impedance is relatively small, then the earth resistance of the remote substation may need to be included in the determination of Z_E.

Some texts produce tables of chain impedances for individual network OHL and cables.

It is worthy of note that with reference to Figure 12.5, some texts define the impact of OHL and cable chain impedances on the flow of earth fault current through the earth electrode system in terms of a screening factor, K, such that:

$$K = \frac{\text{Fault current through the earth electrode system}}{\text{Total fault current}} = \frac{I_E}{I_F}$$

12.6.3 Earth fault current

To determine the maximum value of current passing through the substation earth electrode system, the more onerous of two situations needs to be considered as follows:

- An earth fault in the substation itself – where return fault current back to the source of supply flows via the substation earth electrode system to the earthed neutrals of all the connected transformers in the network. This is the more usual situation. NB: Account needs to be taken of any OHL and HV cable chain impedances in determining the magnitude of earth electrode current.
- An earth fault external to the substation under consideration, whereby (most of) the fault current returns to the source of supply by flowing via the substation earth electrode system to the earthed neutrals of the transformers in the substation. This condition is much less likely and requires large concentrations of transformers with earthed neutrals.

It is always the case that the magnitude of current flowing into an earth fault is determined by the impedances of the power system equipment itself, as opposed to the earthing system, since the impedance of the latter (by design) is relatively small, and can be ignored. The earth fault current to be considered, when calculating

ROEP, is therefore the maximum earth fault current that can arise, taking all practical operating conditions into account, plus a factor of safety (say 20%), to allow for future increases in network fault levels. For networks that are relatively close to sources of generation, the fault current considered is usually the winter maximum sub-transient current (see Chapter 5) for an earth fault on the substation busbars. Where a network is resistance earthed, the value of the neutral earthing resistance will of course limit and determine the magnitude of earth fault current.

12.7 Voltage limits

12.7.1 Substation ROEP – voltage design limits

With reference to Figure 12.5, the rise of substation earth potential, $V_{(\mathrm{ROEP})}$, is given by:

$$V_{(\mathrm{ROEP})} = I_F Z_E = I_E R_E$$

where Z_E is the total earthing impedance, as defined in expression (12.11).

As pointed out in Section 12.2.3, earth mat design must evaluate ROEP with reference to three voltage limits as follows:

- Transferred potential
- Touch potential
- Third-party infrastructure potential.

These will be briefly reviewed in the following sections, which should be read in conjunction Sections 11.3.3 and 11.5.5 of the chapter on impressed voltages.

12.7.2 Transferred potential

'Transferred potential', see Figure 11.6, defines the instance of a substation ROEP being transferred to a remote location via a metallic connection (e.g. cable) connecting the local site to a remote substation. In the United Kingdom, and based upon international telecommunications regulations, the transferred potential from a substation earth mat is limited to 650 V for a maximum duration of 200 ms (i.e. cleared by main protection), or 430 V for a duration in excess of 200 ms (i.e. cleared by backup protection). These voltage levels are not based upon present electrical shock criterion – but are long standing and take into account economic considerations.

All substations at a network voltage of 33 kV and above are likely to be equipped with fast acting busbar protection (i.e. operating time under 200 ms), and therefore the design ROEP limit for these substations is 650 V. If earthing calculations reveal a ROEP in excess of 650 V then in the first instance, consideration should be given to reducing the earth electrode resistance by increasing the density of the earth mat. If this proves impractical (taking into account the cost) then the higher ROEP is accepted and the site designated as a 'hot site', requiring all incoming metallic connections (other than those comprising the power network itself) to be isolated or insulated from the site – and an evaluation made of whether the level of touch potential is within acceptable limits. An evaluation should also be made on the impact on surrounding infrastructure, see Section 12.7.4.

NB: If the ROEP is less than 650 V then the design is satisfactory requiring no further action.

It is instructional to note that to maintain the ROEP design limit to 650 V, that:

- An earth electrode current of 10,000 A (say) would necessitate an earth electrode resistance of 0.065 Ω.
- An earth electrode current of 50,000 A (say), would necessitate an earth electrode resistance of 0.013 Ω.

12.7.3 Touch potential

With reference to Figure 11.5, 'touch potential' defines the instance of a person touching the tank/body of an earthed item equipment, positioned 1 m away, and being subject to the voltage difference between the position of the hand and the position of the feet. It is worthy of note that all points on, and metallic connections bonded to, the earth mat remain at the same potential (i.e. voltage), and therefore the touch potential relates to a person standing upon the ground (i.e. close to but not directly on the earth mat). Calculations show that touch potential is more severe than step potential (see Section 11.3.3), and therefore only the level of touch potential needs to be assessed. Calculations also show that the greatest level of touch potential usually takes place at the peripheral edge of the earth mat (e.g. metallic grid). Although touch potential also arises within a metallic grid, these levels are of a lesser magnitude.

Formula exists for determining the magnitude of touch potential at the external edges of an earth electrode (e.g. see ENA TS 41-24), however, the expression is complex and readers are directed to more specialists texts should a greater understanding be required.

It is worthy of note that the maximum touch potential will be lower than the ROEP (usually significantly lower) since ROEP is a measure of voltage gradient between the earth electrode system and the remote earth, whereas touch potential is only a measure of voltage across a distance of 1 m from the earth electrodes.

If, unusually, the calculated touch potential exceeds the maximum allowed level of touch potential of 2,060 V (in a substation with chippings), see Section 11.5.5, then measures must be taken to reduce the value to below the maximum. Options to achieve this comprise:

- Increase the density of electrodes in the earth mat
- Drive rods into the ground at the periphery of, and connect to, the earth mat
- Expand the area of the earth mat. This may require the substation boundary fence to be bonded to the earth mat, see Section 12.8.1.

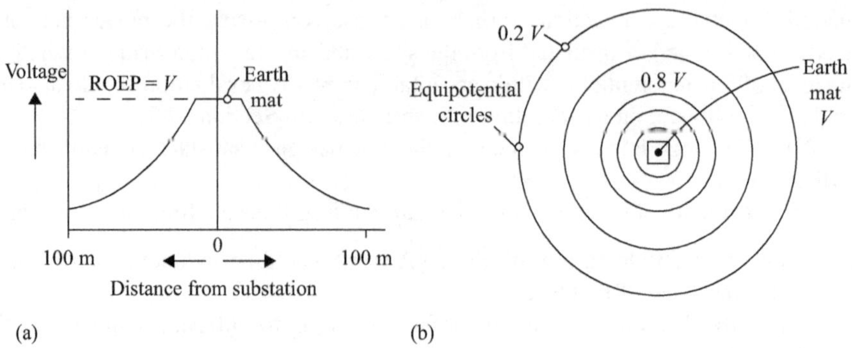

Figure 12.6 Voltage profile around a substation: (a) side explanation view and (b) plan view

12.7.4 Third-party infrastructure

As stated earlier, the flow of earth fault current through the substation earth mat results in the earth mat being subject to a ROEP with reference to remote earth. As a consequence, a voltage gradient exists between the earth mat and point of remote earth (some distance away). Figure 12.6 illustrates the voltage profile from both a side elevation and plan view perspective. The location of remote earth is of course when the voltage is zero (approximately). In most instances (depending upon earth resistivity), the voltage drops to 20% of ROEP at a distance of about 100 m from the substation.

Mathematical relationships exist for a plot of voltage profile, for the various types of earth electrode systems. For the instance of an earth grid, the voltage, V_P, at a distance, x, from the centre of an earth grid is given by:

$$V_P = \frac{\rho I_E}{2\pi r} \sin^{-1}\left(\frac{r}{x}\right) \tag{12.12}$$

where ρ is the earth resistivity, I_E is the earth electrode current, see Figure 12.5, r is the radius of a circular grid of an equivalent cross-sectional area as the actual grid and $\sin^{-1}\left(\frac{r}{x}\right)$ is expressed in radians.

The voltage profile invariably impinges upon third-party infrastructure, e.g. domestic metallic water pipes, metalwork in hazardous industrial plants, commercial premises, etc. As such, the relevant power network company (or contractor) has a duty to assess who will be affected by the voltage profile, and as such subject to unacceptable levels of voltage.

NB: For hazardous industrial plants, the maximum allowable voltage level is generally taken to be 650 V and for domestic/commercial premises 1,700 V. Where necessary, mitigation measures will need to be undertaken to minimise the risk of danger from electrical shock. Within this context, a risk assessment will need to prepared and may involve the statistical likelihood of electrical shock arising, and 'as low as reasonably practical' criterion utilised, as described in HSE and other documentation – as the mitigation measures need to be balanced against 'reasonable' cost.

12.8 Earthing design – considerations and requirements

12.8.1 Substation fence earthing

It is common practice to position substation fences at least 2 m beyond the substation earth mat and to earth the fence separately. By this means, the touch potential (of the fence) is less than if it was bonded to the substation earth mat. The 2 m is a distance in excess of the reach of a person with outstretched arms (and thus the distance cannot be bridged). Common practice (certainly in the United Kingdom) is to separately earth the fence with earth rods of 3 to 5 m depth and at intervals of 50 m and at all corners of the fence and directly under OHL crossings of the fence. As with the substation earth mat, mathematical expressions are also available (see EATS 41-24) to determine the touch potential of the fence.

Fences internal to the substation, either permanent or temporary, and contained within the substation earth mat, are usually bonded to the substation earth mat (typically at 50-m intervals).

12.8.2 High-frequency earths

Where a part of an item of HV equipment is earthed, and is likely to be subject to high frequencies, it is usual to earth the equipment via a 'high frequency earth', to ensure that the high frequency voltage is conducted directly to earth and not imposed upon the equipment. High-frequency earths are usually applied to the earth end of surge diverters and capacitor voltage transformers, particularly where power line carrier is utilised in conjunction with a CVT – see Chapter 9. The high-frequency earth must have no bends as a bend presents a high-impedance to high frequencies, and surge voltages in particular. It usually comprises a rod driven vertically into the ground to a depth of typically 5 m.

12.8.3 Indoor substations

Indoor substations are usually built on a concrete raft. This necessitates a surface laid earthing conductor loop, with spur connections to earth the various items of equipment. It also requires an external (to the building) earthing conductor at a distance of approximately 1 m from the building, at a typical depth of 0.6 m and connected to the internal earthing conductor every 20 m. To further lower the earthing resistance, driven rods connected to the external conductor may be required, or even a second external buried conductor, typically 10 m beyond the first.

12.8.4 Reactive compensation compounds

Special care is required with reference to reactive compensation compounds containing air cored reactors, since they can cause high-induced circulating currents in close metallic earth mat loops – leading to overheating of the earth mat. Therefore, the earthing system in the vicinity of the compound should either not comprise closed loops, or comprise plastic coated insulation at crossing points to prevent circulating currents.

12.8.5 Test facilities

Earthing systems should be provided with test facilities, including access points, to enable periodic testing of the earthing system.

12.8.6 Earthing design method

When undertaking site installation, the earthing system is one of the first tasks to be undertaken – and therefore early consideration should be given to earthing design. Considerations should include the following:

- Whether the substation will be outdoor or indoor
- Evaluation of the earth through analysis of samples, i.e.
  - earth resistivity
  - chemical/physical nature of the earth
- Land space required and availability
- Whether any reinforcing steelwork within the substation, or any earth stability piles, can be used as part of the earth electrode system
- The value of maximum earth fault current to be used in determining ROEP (including factors of safety)
- Determination of any chain impedances to be used in the ROEP calculation – see Section 12.6.1
- Whether the earth mat is to be directly connected to any adjacent substation earth mat (or power station earth mat)
- Design of the earth mat (see Sections 12.4.1–12.4.7), including specification of conductor material and sizing
- Determination of the earth system impedance
- Determination of voltages with required limits
  - ROEP, and transferred potential
  - Touch potential
  - Voltage contour plots and impact on third-party infrastructure. NB: Voltage contours should be plotted on an ordnance survey map, or similar, to assess the infrastructure affected.
- Mitigation measures where required, to contain voltages within limits
- Preparation of drawings for installation and commissioning

Earthing calculations must be retained, since they form the basis of future earth mat development.

12.8.7 Computer-based solutions

Modern earthing design is invariably assisted by the utilisation of computer-based solutions. One solution that is widely used (in the United Kingdom) is Current Distribution, Electromagnetic Field, Grounding and Soil Structure Analysis.

12.8.8 Standard documentation

The following specialist documentation is relevant to substation earthing:

- ENA TS 41-24: Guidelines for the Design, Installation, Testing and Maintenance of Main Earthing Systems in Substations. 1992
- ENA Engineering Recommendation S34: A Guide for Assessing the Rise of Earth Potential at Substation Sites. 1986
- ENA Engineering Recommendation S5/1: Earthing Installations in Substations. 1966
- BS EN 50522 Earthing of Power Installations Exceeding 1 kV AC 2012.

Chapter 13

Civil, structural and building engineering design

13.1 Introduction

At first sight, power network construction may be thought of as being predominately concerned with the design, installation and commissioning of power network assets (e.g. circuit breakers, protection systems, OHL, HV cables, etc.). However, a fully operational power network can only be achieved with the supporting infrastructure of what may be termed 'civil, structural and building engineering' (CSBE) – that is the civil engineering oriented assets concerned with groundworks, roadways, support structures, buildings, etc. Such assets may typically comprise of the order of 45% of a major power network construction project and are therefore a significant consideration – and some projects are of course almost entirely civil engineering orientated.

It has frequently been suggested that CSBE is the 'Cinderella' activity in power network construction. It is usually the last design task to be considered (following determination of the power network requirements), but the first task to be undertaken on-site – so providing the least available time to resolve and prepare the design. As one eminent civil engineer once (metaphorically) reminded the author, 'power engineers must always remember that CSBE is the foundation upon which the power network stands'.

An aspect of CSBE worthy of note is that it usually involves significantly more temporary works than those concerned with power engineering assets. Within this context, the following salient CSBE design tasks will be reviewed:

- Earthworks, drainage, trenches and landscaping
- Structural engineering
- Substation infrastructure
- Environmental works
- OHL and cable civil design
- Civil engineering work standards
- Substation LV AC supplies.

Clearly, CSBE is a huge subject in its own right – and therefore of necessity, this text is limited to an overview and indication of salient design requirements.

13.2 Earthworks, drainage, trenches and landscaping

13.2.1 Earthworks

The following will consider the requirements for the construction of a new substation. The suitability of the ground upon which to build a substation is arguably the first consideration in civil engineering design. This invariably commences with a number of surveys, typically comprising the following:

1. **Geotechnical**
   This comprises intrusive ground investigations to take samples for the evaluation of soil types, and ground bearing capacity for foundation design. The investigations will also establish the position of the water table and any earlier ground activities such as mine shafts, tunnels, contaminated land, etc. This is a highly technical and specialist task.
2. **Topographical**
   This consists of an on-site survey, often employing GPS, to survey topographical details, e.g. ground elevations, buildings, roads, railways, trees, rivers, streams, etc., with a view to assessing the impact on design options.
3. **Underground services**
   This survey usually employs detection equipment such as ground penetrating radar systems, or similar, to detect any existing underground services passing through the proposed site, e.g. gas pipes, water pipes, sewers, HV cables, communications cables, etc.
4. **Access routes**
   This survey consists of an evaluation of the access routes from factories to site – especially the access route(s) and suitability for entering site – particularly when considering physically large and heavy vehicles. The crossing over and under of bridges should also be evaluated.

Following completion of the surveys, the design can be prepared which should broadly address the following:

- Methods of dealing with soil contamination, waterways, old workings and other obstacles, etc.
- Ground levelling – although some sites of necessity will contain slopes.
- Ground improvement and stabilisation measures.
- Piling works may be required when the existing ground, either in its own right, or following stabilisation works, still cannot satisfactorily support the loads that will be imposed. Piles are usually metallic columns driven into the ground (although concrete-based materials may alternatively may be used), and held in position by the friction of the compacted soil around the pile (termed a friction pile). Alternatively, they can be driven into the ground until they reach a predetermined depth of bedrock (termed a column pile).
- Building or strengthening of access roads to the site, or the laying of temporary roads – and other civil engineering work arising, e.g. strengthening bridges, diverting streams, thinning out woodland, etc.

- It is usual to specify that the substation ground is covered with a layer of chippings to minimise the impact of 'touch potential', see Section 11.5.5.

13.2.2 Drainage

Drainage essentially falls into two categories consisting of uncontaminated and contaminated water, respectively. Key aspects of each category are as follows:

1. **Uncontaminated water**
   Drainage of uncontaminated water comprises the removal of surface and sub-surface water (usually arising from rain water) from an area, to prevent waterlogged ground – by, in effect, lowering the water table. This usually involves the design of a trench system in which materials conductive to water flow away are inserted in the ground. This frequently involves porous tiles or pipes. Drained uncontaminated water is usually arranged to flow into a river stream or soak away – or into a water utility clear water drainage system. The drainage systems must be designed to accommodate worst storm conditions (e.g. one in a hundred years).
2. **Contaminated water**
   This drainage system usually relates to the transfer of sewage away from the substation for discharge into a waste water system, and eventual transfer to a sewage treatment works.

Integral to the design of any drainage system are measures that allow discharges from the site to be controlled and, in the case of major sites, stopped, to prevent any possible environmental incident. This is usually achieved by use of a 'penstock valve' arrangement, which is in effect a sluice for controlling the flow of water. At some major substations, it may also be necessary to install pumps to facilitate effective drainage.

NB: All discharges in the United Kingdom are subject to environmental regulations.

When designing the drainage system, the location of the drains must take account of the positioning of the power equipment and associated foundations, and other infrastructure such as buildings and roads. The drainage system is one of the earliest requirements during site installation.

13.2.3 Trenches

Trenches may be required in a substation for a number of reasons, the most common of which are as follows:

1. **Multicore cable trenches**
   Careful consideration needs to be given to the design of multicore cable trenches, i.e. both the size of the trench and the location. Once excavated, the trench is usually constructed from concrete – either cast in situ or precast. Consideration also needs to be given to the traffic to which the trench may be subject – which in turn determines the strength of the trench and trench cover.

In those instances where either shorter runs of trench are required or low capacity trenches with few multicores, ducting may alternatively be specified. This is usually cheaper than trenching but less accessible.

2. **Substation earthing**
   As described in Chapter 12, the substation earth mat is usually buried at a depth of between 0.6 and 1.0 m. This involves the excavation of an earthing trench and subsequent backfill.
3. **HV cables**
   HV cables crossing a substation are usually installed in either a cable trench or cable ducts.

13.2.4 Landscaping

Public acceptance of a power network installation is influenced by visual impact. In some instances, this may necessitate landscaping. That is, the improving of the aesthetic appearance of the installation by changing surrounding contours and adding ornamental features such as plants and trees, the latter being used to screen the installation. Landscaping may also include locating a substation in an ornamental building which is in keeping with the surrounding area. In some instances, it may result in the height of the ground being lowered to lower the profile of the substation, assuming flooding is not a risk.

13.3 Structural engineering

13.3.1 Structural engineering – considerations

Structural engineering is essentially concerned with the design of support systems. In the context of power engineering, it is mostly concerned with supporting the HV equipment, but in a wider sense would include supports for buildings, bridges, etc. The following will be reviewed:

- Equipment foundations
- Equipment support structures
- Gantries and anchor blocks.

13.3.1.1 Equipment foundations

The design of equipment foundations is a significant task, and ranges from foundations for busbar supports to foundations for transformers. It also includes foundations for OHL towers. The location of foundations in a substation together with their physical dimensions is critically important – since they must fully align with both the HV equipment, and busbar, physical dimensions and spacing. Foundation position design must also ensure that access is possible both for construction and to enable maintenance. Equipment foundations must take account of the compressive forces arising from the equipment, shear forces arising from the weight of the equipment (at the equipment boundary), turning moments and horizontal shear forces arising from both short circuit and wind forces, and torsional forces arising from the rotational movement of equipment. Although not generally a consideration in the United Kingdom, some countries must also take account of seismic forces – although UK nuclear power stations and adjacent substations do consider the impact of seismic forces. Foundations are usually of either reinforced concrete (usually the larger and heavier foundations) or unreinforced concrete.

The largest foundations, also termed plinths, are usually associated with transformers (although the term plinth may be used for any discrete foundation immaterial of size). Before finalising the position of this foundation, consideration needs to be given to accessibility – both when bringing the transformer on to site, or removing a faulty transformer. Consideration also needs to be given to the method of moving the transformer on to the foundation without damaging the plinth. Typical methods involve the use of winches and steel bar rollers, or for lighter transformers, the use of power lifting equipment. In some instances, steel runner plates may form part of the foundation design to enable skid-mounted transformers to be slid into position.

13.3.1.2 Equipment support structures

Equipment support structures range from supports for insulators to supports for circuit breakers, disconnectors, etc. In height, such structures may range from 2 m up to possibly 25 m – although the modern tendency is to lower the height. Equipment support structures have historically been constructed from reinforced concrete – or galvanised steel or, in some instances, aluminium. Metallic structures are usually of lattice construction or solid. In recent years, steel has been used in preference to concrete – but with recent improvements in durability, concrete is again being considered.

Of particular importance are the fixing arrangements between the support structure and the foundation. This is usually accomplished by fixing bolts, either fixed into the concrete foundation, or inserted into pockets left in the foundation for grouting in on-site. In some instances, the design requires a protective cement-based grout to be fixed to the structure base plate. The location of the fixings together with the type and level of tightness are critical to the overall design.

Structures are subject to all the forces to which the foundation is subject, and transfer those forces to the foundation – as such structures and foundations comprise an integrated design.

13.3.1.3 Substation line entry gantries and anchor blocks

Substation line entry gantries provide the structural support interfaces between an OHL terminal tower and the substation equipment. A simplified illustration is shown in Figure 13.1(a). The gantry is constructed from lattice steel, although it could equally be formed from other steel configurations, e.g. tubular. Gantries have also been constructed from reinforced concrete, although modern practice is mostly steel. The design of the gantry and its foundations must take account of the down lead short circuit, wind and ice forces, together with similar forces on the down dropper. Successful gantry design is highly dependent upon information flow and coordination between the OHL designer responsible for the down lead (and usually

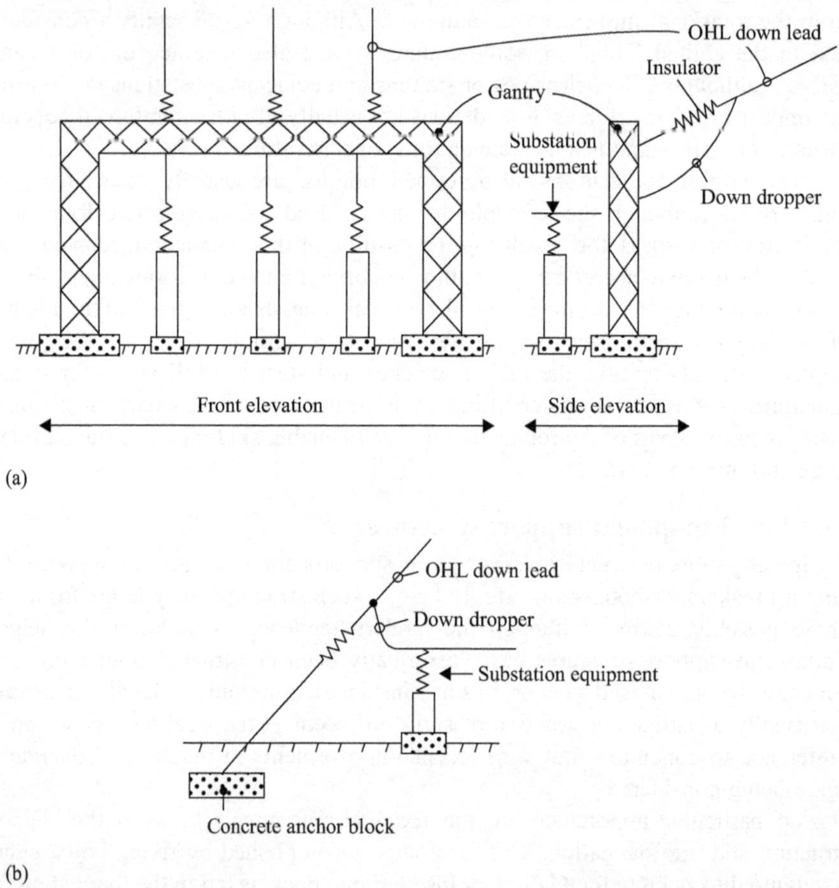

Figure 13.1 Substation gantry and anchor block arrangement: (a) OHL gantry arrangement and (b) 'anchor block' arrangement

the down dropper) forces, and the substation designer with responsibility for the gantry and foundation design. Many power network companies have a range of standard gantry designs to match defined OHL and terminal tower designs and positions.

An alternative to the gantry is the lighter weight construction of the 'anchor block' arrangement as illustrated in Figure 13.1(b).

13.4 Substation infrastructure

13.4.1 Substation infrastructure – general

A substation of necessity must contain infrastructure and facilities both to enable construction and maintenance work to be undertaken, and to secure the substation from external interference. Within this context, the following will be briefly reviewed:

- Buildings
- Building services
- Roads and footpaths
- Vehicle parking
- Substation lighting
- Substation fencing
- Substation security

13.4.1.1 Buildings

Although substation buildings come in a wide variety of shapes and sizes, they contain certain minimum common requirements, with a certain level of standardisation within each power network company. In general, substation buildings contain the following (for 33- to 400-kV network substations):

- Switch gear hall – where the substation is either GIS or distribution voltage metalclad. NB: Some GIS are installed outside
- Standby control room
- Relay room – housing relay cubicles/panels
- Permit and drawings record office. This may be integral to the standby control room or relay room
- Transformer bay – typically for 33 kV substations and below
- 0.4-kV AC supplies room. For transmission substations, this would include a diesel generator room
- Battery room
- DC distribution board room
- Telecommunications equipment room (with smaller substations this may be integral to the relay room)
- General equipment store room
- Portable earthing equipment room – AIS substations only
- Toilet and washing facilities room
- Changing and shower facilities room – where equipment is filled with SF6 on-site
- General mess room with cooking facilities, table chairs, etc. – often incorporating first aid facilities.

In some instances, there may be more than one building complex, e.g. on large transmission substations, the control room, relay room and battery room may be in one building complex, the 0.4-kV facilities in another and the general facilities in a third.

13.4.1.2 Building services

Integral to the design of a substation building is the services that make the building functional. Typical requirements are as follows:

- Lighting – in those instances where the AC supply source is not duplicated, the lighting may include emergency lighting which is supplied from a battery

- Heating – relay and telecommunications rooms in particular must be retained within defined temperature limits
- Ventilation, including air conditioning or air flow design. Battery rooms in particular must be ventilated
- Electrical power, i.e. socket outlets, cooker supplies, etc.
- Floor arrangements for incoming protection and control cables. In some instances, computer floors are installed
- Dehumidification, to remove water vapour from switch rooms and relay rooms. These may enable rooms to operate with lower temperatures
- Water, for drinking, washing, toilets, etc. – and emergency washing facilities in battery rooms
- Waste pipes and sewerage
- Equipotential earth bonding to satisfy BS7671 IET Wiring Regulations
- Fire detection and suppression facilities
- Intruder alarms
- Sound control facilities, e.g. where extraction fans or motors may be installed in a building

13.4.1.3 Substation roads and footpaths

The number and location of access roads and footpaths in a substation is an important design consideration. Generally, the following should be considered:

- Road and footpath positioning with reference to construction, operation, inspection, maintenance, replacement and demolition, especially when considering the positioning of MEWPS and cranes
- Road positioning and strength with reference to the movement of large items of equipment such as transformers, reactors and quadrature boosters, which are both physically large and especially heavy
- The ability to accommodate all types of vehicles to which the road is likely to be subject
- Road widths with reference to vehicle wheel loads and widths. Typical design widths are as follows:
  - Wide heavy load roadway = 5 m
  - General vehicle access road = 3 m
  - Footpath = 1 m
- The arrangements for a vehicle turning
- The proximity of passing vehicles and wide loads to live HV equipment (see Figure 8.8), i.e. adequate clearances to be maintained
- The proximity of passing vehicles and loads to none live equipment and general infrastructure i.e. risk of impact.

13.4.1.4 Vehicle parking

Consideration must be given to the design of the vehicle parking arrangements within a substation – both whilst undertaking work on the equipment and general parking, including that for visitors. With reference to the latter, a car parking area is usually arranged close to the main building(s), and away from the power network.

13.4.1.5 Substation lighting

Lighting columns have traditionally been installed in substations to provide illumination of the following:

- Roadways and footpaths
- Access gates
- Around buildings, particularly control buildings
- Around AIS sites, in general, to facilitate operations during hours of darkness.

Certainly, those who have been called to a substation, on standby, during the night have had their task eased by virtue of substation lighting. In recent years, substation lighting has also been installed for security reasons. Considerations relating to substation lighting design are as follows:

- Colour, glare factor, lifetime and cost
- Level of illumination at defined locations
- On/off switching arrangements
- Proximity to live equipment
- Accessibility and maintenance.

Suffice it to say that substation lighting is a specialist subject in its own right.

13.4.1.6 Substation fencing

The purpose of a substation fence is to provide a physical barrier and deterrent both to prevent danger to third parties and to stop intruders. The ESQC Regulations, see Section 2.4.2, stipulate that for AIS substations, a fence or wall is required with a minimum height of 2.4 m. As a result, most substations are surrounded by a 'palisade' type of fence in accordance with BS EN 1722, although some are of an unclimbable wall-type construction with intruder deterrent features on the top. Fence design considerations are as follows:

- The fence must be designed in conjunction with the concrete foundations, and must take account of wind loading, temperature range, etc.
- Fences must be earthed in accordance with the requirements specified in Chapter 10.
- Some fences are equipped with security and deterrent features, e.g. an electrical fence positioned behind and extending above the main fence, often working via electrical pulses. When armed, these give an uncomfortable electrical shock if touched and generate an alarm if any wire is cut.
- The design should consider the possibility of intruders burying under the fence.
- Vehicle access gates in the fence should be typically 1 m wider than the largest vehicle required to enter site. In many larger substations, the gates are electrically powered and maybe either double leaf or a sliding type.

13.4.1.7 Substation security

With increasing risk of terrorism and organised theft, many substations, particularly those critical to the security of the network, are equipped with substation security systems. These usually consist of cameras (CCTV) and detection devices mounted on structures within the boundary of the substation fence. Output from the cameras is usually routed to recording devices in the substation – and linked to a central control room. Triggering of the detection system usually also switches on targeted lighting, as a further deterrent to the intruder. Such systems can be expensive to install although costs are reducing. In addition, building 'hardening' designs are being implemented to make buildings more secure.

13.5 Environmental works

13.5.1 Environmental works – overview

A number of civil engineering-based activities fall within the category termed 'environmental works' – because of the potential interaction with, and impact on, the wider environment. In summary, this includes:

- Oil containment
- Fire protection
- Flood defences
- Equipment noise attenuation

These will be reviewed in the following sections.

13.5.1.1 Oil containment

Substation equipment containing oil has great potential to cause water pollution arising from oil incidents – so requiring the design and installation of oil containment measures. Furthermore, it is an offence under UK law to pollute 'controlled waters' (i.e. water courses, lakes, coastal waters, etc.) as required by the Environmental Act 1990. Within this context, a key document for shaping the design of oil containment works is The Control of Pollution (Oil storage) England Regulations 2001.

The main sources of potential oil spillage within a substation are transformers, reactors, quadrature boosters and oil storage tanks. Oil circuit breakers, many of which are still in service, also contain significant volumes of oil. In addition, substation access roads associated with oil handling may also need to be included in the oil containment zones.

Figure 13.2 provides an overview of the key requirements associated with the design of an oil containment system and typical of the arrangements to be found at network substations ranging from 33 to 400 kV. Salient design features are as follows:

1. **Oil containment tank**
   Figure 13.2(a) illustrates the instance of a transformer and associated coolers, known as the 'primary containment system', standing on a concrete plinth. This is surrounded by a secondary oil containment system comprising:
   (i) A bund wall arrangement with a typical oil and water tight holding capacity of 115% (110% minimum) of the total oil content of the transformer and cooler arrangement, with a typical wall height of about 0.4 m.

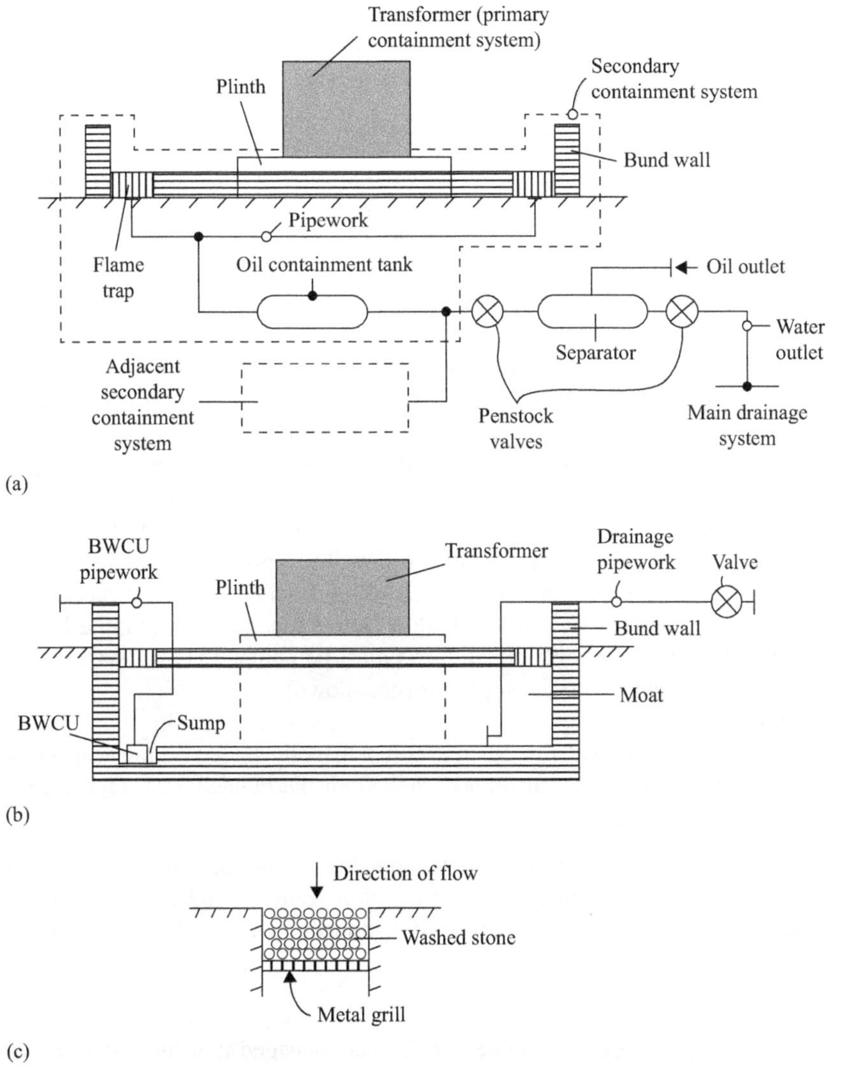

Figure 13.2 Oily water drainage system: (a) oil containment tank, (b) moat type containment system and (c) flame trap example

(ii) An oil containment tank (dump tank) connected by a free draining closed gravity system for rapid removal of the oil, i.e. away from the spillage area surrounded by the bund wall.

(iii) A flame trap to extinguish any fire that may exist in the oil that is within the bunded area, to ensure it does not enter the oil containment tank.

The oil is pumped from the oil containment tank into a closed gravity system which flows to an oil separator. This separates any oil and water mix and operates on the principle that oil being less dense than water, positions itself at the top of the separator. Most separators contain a coalescent unit which facilitates oil and water separation by causing oily water particles to separate out into droplets of oil and water, respectively. Penstock valves are usually positioned on either side of the separators. The separators are usually also fitted with automatic closure devices to prevent further flow into the unit if excessive oil levels arise. An alarm usually provides a warning before maximum capacity of oil or water is reached to enable time for the oil or water to be emptied from the separator. The water outlet from the separator is clean enough to enter the main drainage system.

2. **Moat type containment system**
   An alternative to the oil containment tank is the moat type containment system as shown in Figure 13.2(b). In this instance, the underground containment is directly under the bunded area (whereas the oil containment tank is usually some distance away). Pipework as shown, which usually extends beyond the boundary of the potential fire damage zone, facilitates removal of the oil via a pump.
   An ongoing problem for the moat is the removal of rainwater. This is usually accomplished through the installation of a 'bund water control unit' (BWCU) located in the sump of the moat. The BWCU comprises an automatic sensor and pump to remove water, but is deactivated should the sensor detect oil in the sump. The output from the BWCU flows into a separator. NB: Similarly, water accumulation in oil containment tanks must be pumped via a BWCU into the pipework connected to the separator (not shown).
3. **Flame trap**
   A flame trap(s) is usually installed to ensure that oil which is burning cannot enter either the moat or oil containment tank from the bunded area. Figure 13.2(c) illustrates a typical arrangement.

On some sites, the volume of oil to be contained does not warrant either of the solutions illustrated in Figure 13.2, and therefore, only a bunded area is provided. Clearly, the design of oil containment works must take account of the volumes of oil that exist at the substation in question.

13.5.1.2 Fire protection

Fire is a significant, albeit infrequent, risk to be managed in a substation, necessitating careful design consideration from a minimisation and control perspective. Fire risks may be categorised as:

- Risk to life – of both personnel undertaking work in the substation and to the general public
- Risk to the environment – generally arising from fire effluent discharge, and fire-fighting foam and water run off
- Risk to security of supply – in the event of the fire damaging critically located equipment – particularly if the fire also affects adjacent equipment
- Financial risk – in replacing fire damaged equipment
- Third-party damage – including damage to heritage sites
- Reputation risk – i.e. confidence in the technical and operational ability of a power network company, and possibly contractors, which may affect share price, etc.

Power network equipment which presents the greatest fire hazard risk are those items which contain oil, namely: transformers, reactors, oil-circuit breakers, oil-storage facilities, diesel generators, oil-filled cables, cable-sealing ends, etc. Within this context, the equipment, etc., most at risk from fire damage and of most concern comprises:

- Equipment containing oil in which the fire commences, as listed above
- Adjacent equipment which if damaged would cause consequential circuit outages
- Protection, control and communications equipment, including interconnecting cables – which are essential to the operation of the substation
- Substation buildings – particularly if occupied
- Third party property

A preferred hierarchy for fire protection and containment within a substation is as follows:

- Physical separation of equipment to remove equipment at risk from the fire hazard.
- Installation of fire barriers between the equipment creating the fire hazard and equipment at risk.
- Installation of automatic fire-protection systems.

The above-mentioned points will be briefly reviewed.

1. **Physical separation**
   Significant physical separation of items of equipment ensures that they are removed from the 'fire damage zone' of other equipment which is on fire. The fire damage zone is taken to be the area in which items of equipment are liable to be subject to damage due to the effect of radiated heat and flame impingement from the fire. However, the separation distance must be balanced against the increased substation space required, and as a compromise fire barriers may be required.
2. **Fire barrier**
   A fire barrier wall may be erected to shield and safeguard an item of equipment which is in the fire damage zone of another item of equipment. The design considerations are illustrated in Figure 13.3. The material used for the fire barrier typically requires a minimum fire resistance of 4 h. Fire barrier walls are used extensively in the United Kingdom as a shield against transformer fires. Reinforced concrete is usually the easiest and cheapest type of barrier wall, but other materials such as 'durasteel' may be more appropriate if time/space is an issue.
3. **Automatic fire-protection systems**
   Automatic fire-protection systems are usually installed in indoor substations and basements, and generally operate by removing oxygen from, and cooling, the fire. The most common systems comprise:
   (i) Water spray types – this requires heat detectors, deluge valves, pumps and water storage tanks.

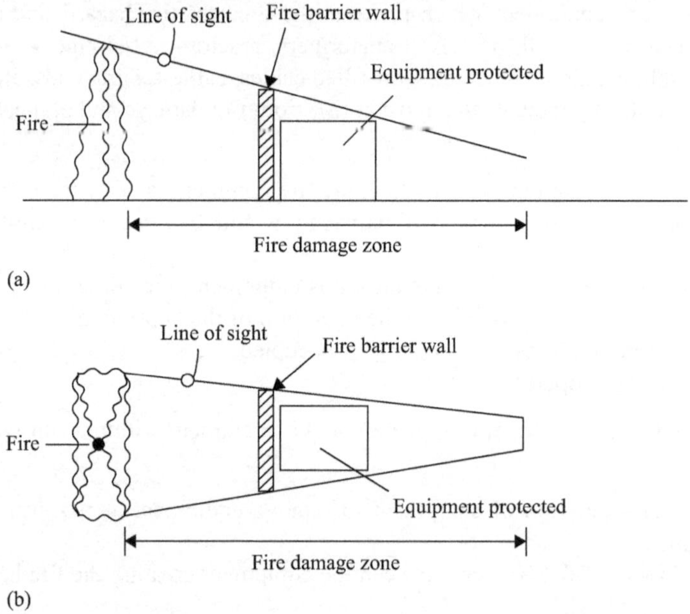

Figure 13.3 Fire barrier – principle of protection: (a) side elevation and (b) plan elevation

(ii) Gas type – this uses a non-toxic gas system supplied from gas cylinders. Older systems used carbon dioxide as the gas, but this has largely been replaced by gases such as 'argonite'. Entry points into the protected area are usually interlocked, such that entry can only be gained when the extinguishing medium has been isolated.

Other design precautions relating to fire comprise the following:

- Protection and control cables should be run in separate routes to power cables.
- Protection and control cables should be segregated wherever possible on a per circuit basis, and routed away from oil filled equipment.
- Adequate means of escape should be provided for all persons on the site, which where appropriate must be signed, maintained and kept free of obstacles.
- Fire resistant construction materials should be used to contain and restrict fire.

Numerous health and safety documents stipulate requirements relating to fire, the most significant of which are:

- ESQC Regulations, see Chapter 2, which state '... for every substation, generators and distributors shall take all reasonable precautions to minimise the risk of fire associated with equipment'.
- The Regulatory Reform (Fire Safety) Act 205 – a salient requirement of which is to undertake a site assessment to minimise the consequences of fire by prevention or by protection systems and procedures.

All substations should contain a fire plan, as integral part of an emergency plan, which should be agreed with the fire brigade.

13.5.1.3 Noise enclosures

Section 9.11.1.6 pointed out that transformers emit a noise (hum) due to the extension/contraction of the transformer core laminations arising from the alternating magnetic flux. The hum takes place twice per cycle and at a frequency of 100 Hz. The hum can be an environmental nuisance, particularly in urban areas, and on certain sites this may require the installation of a 'noise enclosure' to reduce the level of hum external to the closure.

Noise enclosures were traditionally brick based but are now more likely to be a steel frame with the walls and roof fabricated from panels comprising acoustic material sandwiched between steel plates, and acoustic panels for noise absorption within the enclosure. Design considerations relating to noise enclosures are as follows:

- The enclosures usually only enclose the transformer and not the coolers, except in instances where the coolers are integral to the transformer.
- The enclosure is usually required to be weatherproof, so requiring a roof.
- Two doors are usually required for purposes of access and egress.
- The enclosure must maintain the required clearances to any transformer bushings.
- Enclosures must be designed to reduce the noise to defined levels, typically 20 dB(A) insertion loss at 100 Hz.
- Enclosures must enable easy and safe access for maintenance and inspection.
- Enclosures reduce the heat loss from a transformer and must be considered when evaluating thermal ratings.

Noise enclosures may also be applied to reactors and quadrature boosters.

13.5.1.4 Flood defences

From the beginning of the twenty-first century, there has been an increase in flooding, both in the United Kingdom and many other parts of the world. This has arisen both as a result of over-flowing rivers due to excessive rainfall, and coastal flooding caused by storms, often coincident with high tides. Such flooding is largely attributable to global warming and is likely to either continue or increase in the short to medium term. Some substations in the United Kingdom have been subject to flooding (or have been at risk of being flooded), resulting in equipment damage, particularly protection and control equipment located nearer the ground, so making the substation non-operational. All new substations must therefore be designed from a flood defence perspective, with remedial measures applied at existing substations, at risk.

The UK Environment Agency (EA) defines a 'flood plain' as 'the area that would naturally be affected by flooding if a river rises above its banks, or high tides and stormy seas cause flooding in coastal areas'. Substations within flood plains are at greatest risk of flooding. The EA produces 'Flood Maps for Planning', and define two probability scenarios as follows:

- Areas that could be affected by flooding, with no defences, assuming:
  - With reference to the sea, a one in two hundred or greater probability of flooding, occurring each year
  - With reference to a river, a one in a hundred probability of flooding, occurring each year
- The additional extent of an extreme flood affecting further outlying areas with a one in a thousand chance of flooding, occurring each year (river or sea)

The above provides data for identifying the extent of flooding, and the probability of flooding arising – as a basis for considering flood defence measures. As such, early in the construction process and using the above data a flood risk assessment should be undertaken. Typically, a one in a hundred (for a river) and one in two hundred (for the sea) risk would be used for substations connected at lower network voltages, and a one in a thousand risk for higher voltage networks (where the impact on the power system of losing the substation is greater). In addition, the potential height of the flood water needs to be ascertained, and again, the EA or other specialist service providers can provide such figures. As part of the risk assessment, a factor of safety would usually be added, typically 20%.

A cost benefit analysis should then be undertaken to determine optimum flood defence measures. Options include the following:

- Reposition the substation out of the flood plain.
- Raise the height of the equipment that may be subject to flooding – this will apply particularly to equipment which would normally be close to ground level – particularly cubicles, panels, kiosks, etc.
- Construct a water-tight wall around the substation – using either flood gates or ramps to gain access. Consideration must be given to making drains, etc. water tight, and may require the installation of water pumps to control the rise of ground water.
- Construct water tight walls around critical assets only. This will require the installation of a water tight base, or the use of pumps to control the rise in ground water.
- Use of temporary flood barriers around the substation – where it considered that the response time is adequate.
- Where the height of the flood waters is not excessive, equipment may be protected by:
  - Installation of flood proof doors, etc. – with the possibility of raised thresholds
  - Raising or sealing all access holes
  - Sealing of cable troughs
  - Raising the height of walls to prevent water ingress

ENA Engineering Report 138 provides authoritative guidance on flooding with reference to substations, including flood zone information and a method of risk assessment.

13.6 OHL and HV cable civil design requirements

13.6.1 OHL civil design

Civil engineering design requirements relevant to OHL generally include the following, and contain significant temporary works:

- Tower (and pole) foundation design as described in Chapter 7
- Erection of scaffolding over roads, railways, etc.
- Laying of temporary roads or trackways to erection sites
- Creation of a site office, with car parking, etc.
- Establishment of lay-down areas
- Diversion of streams, etc.

13.6.2 HV cables civil design

Civil engineering design requirements for cables generally encompass the following, which again involves significant temporary works:

- Removal and replacement of cable trench soil
- Trench de-watering (if required)
- Cable trench shuttering and barriers
- Cable sealing end compounds and support structures – including protective scaffolds
- Where required, the establishment of a site office including parking
- Laying of temporary roads
- Establishment of lay-down areas
- Diversion of streams, waterways, etc.

In addition, an increasing number of cables are being installed in tunnels, particularly in metropolitan areas such as London – and in such instances, the civil engineering challenges are of a different order of magnitude, and include:

- The tunnel excavation arrangements
- Permanent removal of the tunnel waste
- Sealing and strengthening the tunnel to make it both water tight and mechanically sound
- Establishing ventilation and access shafts
- Provision of tunnel services:
  - Electricity
  - Ventilation
  - Communications
  - Transport arrangements within the tunnel

13.7 Civil engineering work standards

13.7.1 Civil engineering standards

Civil engineering design differs from power engineering design in two distinct ways:

- Power engineering equipment is underpinned by national/international standards such as IEC to a far greater extent than civil engineering requirements for power network construction. Within this context, civil engineering is more bespoke.
- Whereas much of power engineering equipment is factory assembled, civil engineering equipment/works is mostly assembled on-site (e.g. buildings, drainage, plinths, etc.), as such, exposed to the vagaries of the weather.

As a result, it is usual for a power network company (or alternatively a contractor or consultant) to also prepare a generic technical specification for CSBE with typical content as follows:

- Standards and codes of practice for governing the level (quality) of workmanship (e.g. standard/quality of concrete for certain applications)
- Hold and notification point arrangements
- Level and reference points requirements (usually ground related)
- Site sampling, evaluation and testing arrangements
- Site quality management arrangements.

13.8 Substation LV AC supplies

13.8.1 LV AC supply – requirements

Although substation 400/230 V supplies lie essentially in the domain of power engineering, rather than civil engineering, they are included in this chapter since they mostly form part of the substation support infrastructure, rather than the power network itself. Substation LV AC supplies are critical to substation operation and are required for the following purposes:

- Substation battery chargers
- Protection and control equipment AC/DC convertors
- Transformer pumps and fans
- Motorised disconnectors and earth switches
- Outdoor kiosk heating
- Substation buildings heating, lighting and power supplies
- Air conditioning
- Substation lighting
- General motor loads, e.g. pumps, compressor motors, etc.

13.8.2 LV AC supply arrangements

Figure 13.4(a)–(c) illustrates typical means for obtaining 400 V AC supplies from the HV networks. Figure 13.4(a) and (b) is now only used in the UK subject to

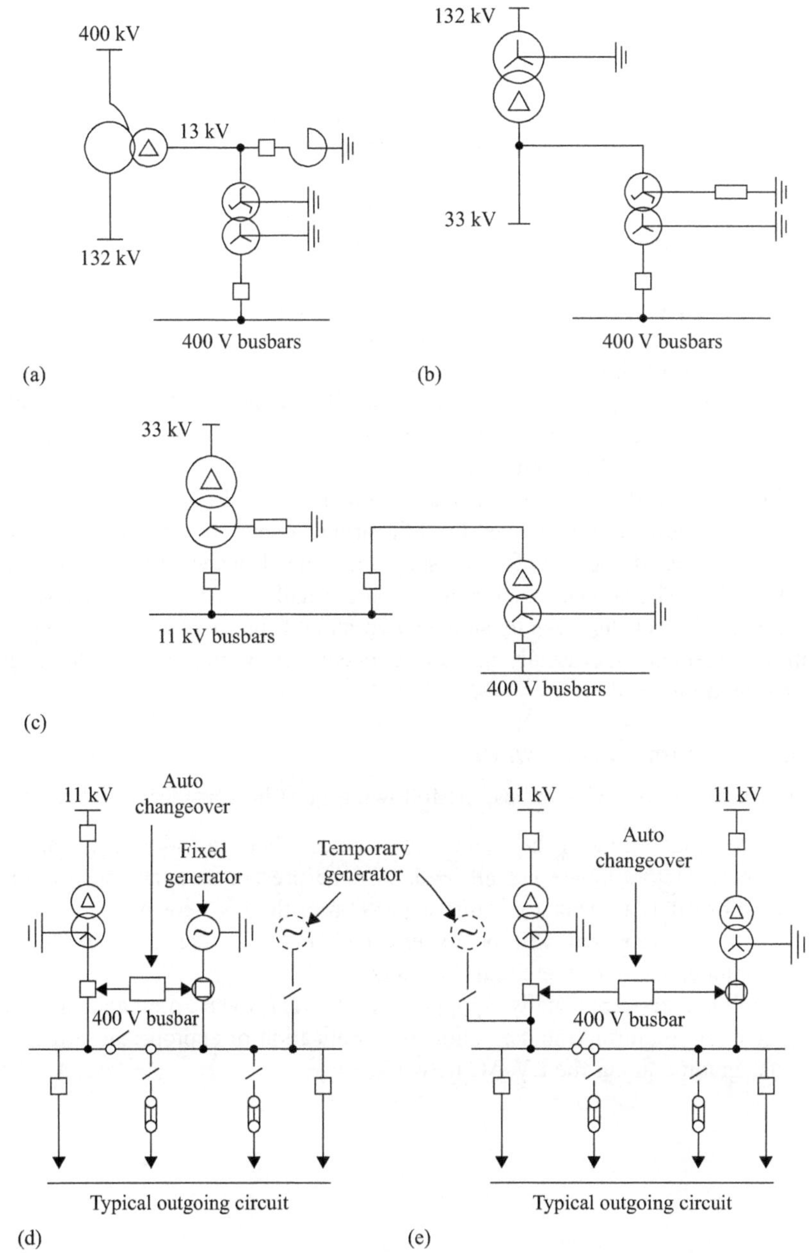

Figure 13.4 LV AC supply arrangements: (a) SGT tertiary connection, (b) earthing transformer connection, (c) distribution transformer connection, (d) transmission substation and (e) distribution substation

satisfactory LV AC power metering arrangements (i.e. who is the energy supplier and at what tariff).

Figure 13.4(d) shows typical arrangements for the provision of AC supplies in transmission substations. It comprises an incoming 11-kV supply from a distribution company, whereby the 11/0.4-kV transformer may be located either within, or eternal to, the substation boundary. Arrangements are usually made, via insulating glands, to ensure the transformer substation rise in earth potential is not transferred on to the 11-kV network. The 11-kV supply is the preferred source of supply with automatic changeover to the permanently installed generator should the 11 kV supply fail. The generator is usually diesel powered requiring a fuel tank of sufficient size to power the diesel and enable it to supply full load for, typically, up to 1 week. Once started, the diesel is usually arranged to run for a minimum time of 30 min. Facilities are also provided to readily connect a fully rated mobile generator – for the instance of both loss of incoming 11 kV supply and the permanently connected generator.

Figure 13.4(e) illustrates a typical arrangement for a 33 kV network substation. Incoming supplies could be derived from the arrangements shown in Figure 13.4(c). In some instances, the source of 11 kV supply may be via a switch fuse rather than a circuit breaker. The number one incomer would usually be from the main source of supply with auto changeover to number two should the number one supply fail. Again arrangements are usually made to connect a fully rated mobile diesel generator should the need arise.

13.8.3 Design considerations

When designing LV AC supplies, the following must be considered and resolved:

- The full load rating of the equipment to be installed, including application of diversity factors (since not all loads will be present simultaneously), and a factor of safety to cater for future expansion of the LV network.
- The fault levels that will be experienced and the associated short-circuit breaking capacity of the circuit breakers.
- Circuit breakers are usually equipped with IDMTL overcurrent and earth fault protection, requiring the derivation and application of appropriate settings.
- Fuse grading along the LV AC network.

Chapter 14

Detail design, manufacture and site installation

14.1 Introduction

The three project stages of detail design, manufacture (including procurement) and site installation are primarily in the domain of the contractor. The earlier stages of detail design and manufacture are concerned with the detailed engineering requirements of the project in which every nut, bolt and wire, etc., is identified. In contrast, the later stage of site installation is concerned with the erection of the manufactured equipment (and construction of the associated site infrastructure) – in accordance with the requirements of the detail design (i.e. drawings and schedules). This chapter will examine the technical considerations and requirements relating to these three key stages.

14.2 Detail design

14.2.1 Detail design – overview

Detail design commences after release of the project contract, integral to which are relevant sections of the construction design specification (the SDS and technical specifications, see Figure 1.3), that provide the starting information for commencing the detail design. The technical stages comprising detail design generally consists of the following:

1. **Detail design specification (DDS)**
   With reference to Section 20.2.5, the DDS provides the final level of design specification for obtaining both a technically compliant design and one that satisfies health, safety and environmental requirements. In summary, it comprises the following:
   (i) Expansion of the SDS to a final level of detail which enables the preparation of project drawings and schedules to commence. This will include the identification of all relevant technical specifications.
   (ii) Specification of the method and means of technical implementation of the project, i.e. site installation and commissioning. Consideration should also be given to identifying operation and maintenance requirements, etc., for later inclusion in related documentation.
   (iii) Specification of temporary work requirements.

(iv) Specification of the calculations that need to be undertaken for reasons of equipment design and settings, and subsequent production of the calculations.

(v) Identification of health, safety and environmental hazards and method of resolution/control.

It is usual to provide one SDS per substation bay or OHL or HV cable feeder.

2. **Drawings and schedules production**
This comprises the production of all drawings and schedules against which the equipment and site infrastructure will be installed and commissioned. Drawings may be based upon existing standard drawings or unique to the project in question. It is essential that all required drawings are identified, with particular attention focused on interface boundary drawings both between technologies (e.g. civils to substation HV equipment) and at the boundaries of the project. Drawings must be produced correctly and first time, to avoid unwanted re-work. This is dependent not only upon the content of the DDS but the competencies of the drawing practitioner. Drawings are usually subject to engineering assurance/design acceptance (as outlined in Section 20.2.4) before being forwarded to site for site installation to commence.
3. **Manufacture/procurement interfaces**
During detail design, both the DDS and the production of drawings must be aligned with the equipment interfaces of the manufactured/procured equipment – particularly in terms of physical dimensions and input/output interfaces.

The following sections will examine detail design considerations (much of which should be included in the DDS) with reference to the main technologies, namely:

- Substation HV equipment
- Protection and control equipment
- Civil, structural and building engineering
- OHL
- HV cables.

14.2.2 Substation HV equipment

With reference to Chapters 8 and 9, the following summarises the salient technical considerations at the time of detail design:

1. **Single-line diagram**
A single-line diagram (SLD) is the reference document from which virtually all other design flows. Although initially prepared prior to contract release, it must be finalised and frozen at the outset of detail design. A SLD should comprise the following:
   (i) Single-line representation of the power network including all electrical equipment relevant to the project
   (ii) The nomenclature of all equipment

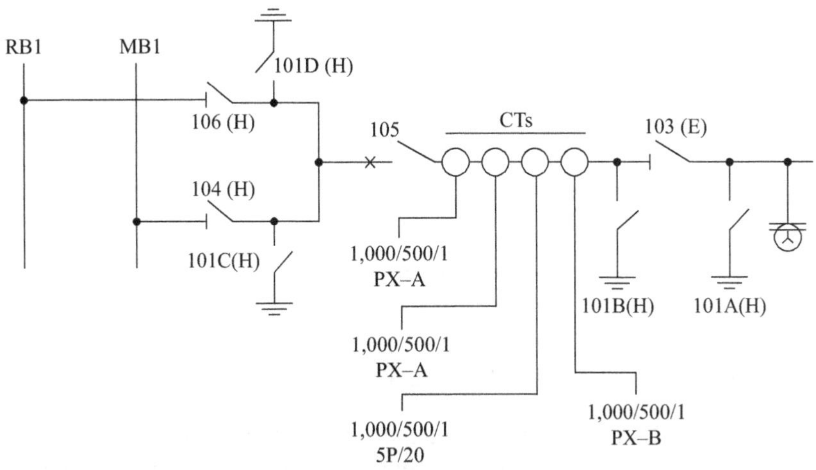

Figure 14.1 Extract from a single-line diagram

(iii) Transformer ratio, rating and vector group
(iv) Equipment mode of operation [i.e. motor (E) or hand (H)]
(v) Individual current transformer locations and types
(vi) Insulation boundaries (e.g. GIS/AIS)
(vii) Work (project) boundaries, if on an existing site.

The SLD forms the basis of the operations diagram(s), see Section 20.7. Figure 14.1 illustrates a typical SLD for a 132-kV feeder, although in practice the whole substation would usually be shown.

2. **Stages of work**
Diagrams showing the existing and final arrangements and stages in-between need to be prepared. This may be a mix of SLDs and block diagrams as judged appropriate. This forms the basis of the project stage-by-stage diagrams – see Chapter 24. Coordination with civil engineering design will need to be undertaken.
3. **Technical specifications schedule**
A final schedule of all required (generic) technical specifications needs to be prepared with identification of the equipment to which each applies. Importantly, if equipment is to be installed for which there is no existing technical specification, then a specification needs to be specifically prepared and agreed.
4. **Equipment ratings**
A schedule comprising the continuous, short time, short-circuit and cyclic current rating of all equipment to be installed needs to be prepared. This will eventually form the basis of the thermal and short-circuit rating schedules, see Chapter 20.

5. **Calculations**
   A schedule of calculations (that need to be undertaken) needs to be prepared, and the calculations subsequently undertaken. This will typically include:
   (i) Busbar forces (for AIS substations)
   (ii) Earthing
   (iii) Over-voltage studies (if not already finalised)
   (iv) Gantry design
   (v) LV AC supplies loadings
   (vi) Earthing system design and ROEP.
6. **Site supplies**
   The existing and final LV AC supply arrangements, including back-up generation, need to be specified together with the stages of work.
7. **Installation**
   The method, and staging of, equipment installation needs to be defined, identifying specialist installation equipment (e.g. cranes), erection techniques, the arrangements for site access and positioning during erection.
8. **Technical interfaces**
   The interfaces with other technologies involved in the project, or with adjacent projects, or existing equipment need to be defined together with the method of interface working.
9. **Earthing design**
   The substation earth mat (earth electrode system) and overall earthing system design need to be prepared.
10. **Commissioning**
    An overview of the commissioning methodology needs to be defined together with any specialist commissioning requirements – for later transfer to the commissioning panel.
11. **Type tests**
    New types of equipment that will be subject to type tests need to be identified together with a schedule of test requirements and dates. See Chapter 21.
12. **Impressed voltages (IV)**
    All site work where IV may be a hazard should be identified together with the method of control.
13. **Schedule of drawings**
    The range of drawings to be produced should include
    (i) Overall substation layout (showing all three phases)
    (ii) Substation elevation drawings
    (iii) Circuit general arrangement drawings (plan and elevation) – clearly showing clearance dimensions
    (iv) Site phasing diagram
    (v) MEWP access drawing
    (vi) Connections and fittings drawings
    (vii) Mechanical and electrical interlocking drawings
    (viii) Site AC supplies

(ix) Equipment labelling details
(x) Below and above ground earthing layout
(xi) Site laydown area
(xii) CDM hazard drawing (showing location of services, etc.).

14.2.3 Protection and control

With reference to Chapter 10, the main technical requirements at the time of detail design relating to protection and control include the following:

1. **Key diagram**
   A 'key diagram', sometimes termed a key line diagram, needs to be prepared. This comprises a SLD to which the following is added:
   (i) Relays shown in block diagram form, including the functionality of multifunction relays.
   (ii) Connections from CTs and VTS to the relays – shown as a single-line connection.

   The purpose of the key diagram is to provide an overview of protection requirements, as a reference document from which all other protection design flows. Figure 14.2 illustrates a relatively simple key diagram.

   NB: Historically, the output from the protection relays to trip relays and the associated circuit breakers to be tripped was also shown on key diagrams, but this tends to be omitted for the more complex higher voltage network schemes – but may be included in a separate tripping diagram shown in block diagram form.
2. **SCADA inputs/outputs schedule**
   The purpose of this schedule is to define all the alarms, indications metering and control inputs and outputs required by the substation control system (SCS) together with the correct nomenclature – after which the schedule is frozen. This may be termed a 'parameterisation' document.
3. **Stages**
   Diagrams, showing the existing and final arrangements and the intervening stages, need to be prepared (and aligned with the HV equipment stages). The stages may have to be coordinated with the key diagram and panel/cubicle layout arrangements – it may also involve temporary protection for emergency return to service purposes.
4. **Technical specifications schedule**
   A final schedule of all required (generic) technical specifications needs to be prepared with identification of the equipment to which each applies. Importantly, if equipment is to be installed for which there is no existing technical specification, then a specification needs to be specifically prepared and agreed.
5. **Equipment ratings**
   A schedule comprising the continuous, short time and short-circuit current rating of all equipment to be installed needs to be prepared for eventual transfer to the thermal rating schedule– See Chapter 20.

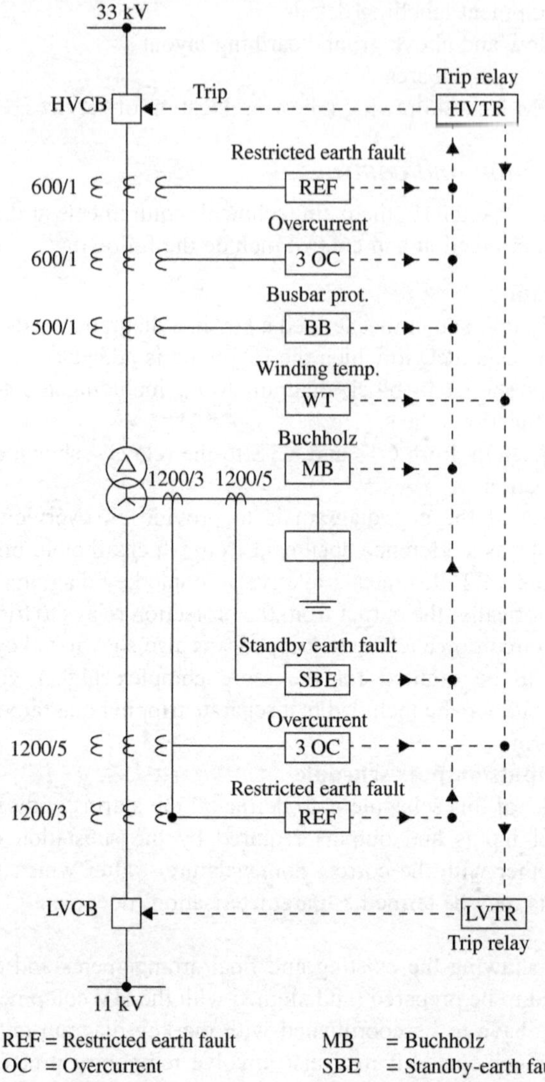

REF = Restricted earth fault
OC = Overcurrent
BB = Busbar protection
WT = Winding temperature
MB = Buchholz
SBE = Standby-earth fault
HVTR = HV trip relay
LVTR = LV trip relay

Figure 14.2 Example key diagram

6. **Calculations**
   A schedule of calculations needs to be prepared for each item of protection and control equipment that requires a setting. The correspondence settings guidance document for undertaking the setting should also be referenced. Calculations must also be undertaken to evaluate whether circuit breaker trip coil voltages, when subject to trip current, are within voltage limits, see Section 10.25.1.4.

7. **Fault-level review**
   A review of network fault levels should be undertaken to ensure that relay application is suitable for the range of fault levels that may be experienced.
8. **Site installation**
   The method and staging of equipment installation must be defined and aligned with that of the HV equipment. Consideration must be given to the risk of disturbing in-service circuits.
9. **Relay schedule**
   A schedule of all relays relating to the project should be prepared which should identify
   (i) The application, type, rating and range
   (ii) The corresponding CT type and required tapping range
   (iii) Relays that are remaining in service
   (iv) All relays to be set, including those remaining in service, and any outside the boundary of the project whose setting needs to be changed as a result of the project.
10. **Panel/Cubicle layout**
    Preparation of an overview drawing showing the location and purpose of all panels/cubicles.
11. **110 V/50 V battery supplies**
    Location and sizing of batteries needs to be undertaken, identifying loads/circuits that the batteries will supply.
12. **SCS database changes schedule**
    A schedule of all SCS database changes must be prepared and aligned with the stages of the programme.
13. **Communications facilities**
    The location of all required communications and telephony facilities should be prepared, identifying boundaries of responsibility.
14. **Drawings**
    Protection and control drawings fall into four different categories which are as follows:
    (i) **Circuit diagrams**
    Circuit diagrams were formerly termed schematic diagrams and comprise electrical circuitry drawn as simply as possible, with a minimum of crossovers to provide an explanation of how the circuitry functions (i.e. it is not concerned with physical positions). Prior to the advent of digital relays, it was possible to deduce the whole performance of the protection, or control, system with reference to the circuit diagram. However, with modern numeric relays, much is contained in software, and circuit diagrams are mostly concerned with relay input and output connections. Commissioning is undertaken by verifying the circuitry shown in circuit diagrams.
    (ii) **Wiring diagrams**
    Wiring diagrams show the physical position of relays, wiring terminations and the routing of wiring in a panel/cubicle/kiosk, including the

termination of multi-core cables. They are usually prepared on a per panel/cubicle/kiosk basis. Some circuits are also arranged with a marshalling panel for the marshalling of incoming multi-core cables – which again would be shown on a wiring diagram. Site installation work is undertaken in accordance with wiring diagrams.

(iii) **Multi-core cable schedules**
This comprises a schedule of all multi-core cables comprising the cable identity (marker number), cable end locations, the number of cores in the cable and their allocated usage (traditionally specifying ferrule numbers). NB: Each of the cores is inscribed with a unique number on the core sheath – repeated throughout the length of the sheath. Cables are laid and terminated in accordance with the schedule.

(iv) **Panel/Cubicle layout diagrams**
Panel/cubicle layout diagrams show the physical position of a panel/cubicle within the suite of panel/cubicles, together with the physical position of each relay on the panel/cubicle (drawn to scale) and the position of fuses, links, test blocks, labels, etc.

Prior to the production of drawings commencing a schedule of each of the categories of drawings identified above needs to be prepared. The order in which the diagrams need to be produced is usually as follows:

(a) Panel/cubicle layout diagram
(b) Circuit diagrams
(c) Wiring diagrams and multi-core cable schedules, simultaneously.

If standard bay solutions are employed, the panel layout, circuit and wiring diagrams are off-the-shelf solutions and largely complete requiring only title changes. This usually only requires the multi-core schedule to be prepared. In other instances, the drawings may be unique to the circuit in question.

By way of example, the following list defines typical circuit diagrams for a 400-kV feeder using numeric multifunction relays:

(a) Feeder CT and VT connections
(b) First main protection relay (including back-up protection, first inter-tripping and fault recorder)
(c) Second main protection relay (including back-up protection and second inter-tripping)
(d) Delayed auto-reclose relay
(e) Circuit breaker control (including control of disconnectors and synchronising relay)
(f) Busbar protection and circuit breaker fail (part of a site common diagram)
(g) Synchronising diagram (part of a site common diagram)
(h) SCS connections (part of SCS suite of diagrams).

The number of protection and control drawings for a multi-technology project usually greatly outnumbers those of any other technology.

15. **Technical interfaces**
    The interface boundaries with other technologies involved in the project or with adjacent projects or existing equipment need to be defined together with the method of interface working.
16. **Commissioning**
    An overview of the commissioning methodology needs to be defined together with any specialist commissioning requirements for later delivery to the commissioning panel, see Chapter 23.
17. **Type tests**
    New types of equipment that will be subject to type tests need to be identified together with a schedule of test requirements and dates. See Chapter 21.
18. **Factory acceptance tests**
    Protection and control numerical/digital relays are usually subject to factory acceptance tests to verify the circuit (or substation)-specific functionality. As such a schedule of required tests needs to be prepared, see Chapter 21.

14.2.4 Civil, structural and building engineering

Civil, structural and building engineering comprises a wide variety of specialist disciplines and therefore needs to be coordinated by an engineer who possesses a significant understanding of all disciplines. With reference to Chapter 13, a summary of the range of CSBE tasks includes:

- Geotechnical
- Earthworks (stabilisation, levelling and preparation of site)
- Drainage
- Landscaping
- Foundations
- Equipment support structures
- Buildings
- Building services
- Roadways and footpaths
- Fences and gates
- Substation lighting
- Fire protection
- Flood defences
- Oil containment works
- Gantries
- Temporary works.

Salient technical requirements at the time of detailed design include the following:

1. **Stages**
   Diagrams showing the existing and final arrangements, and stages in between, should be prepared.

2. **Technical specification**
   A final schedule of technical specifications will need to be prepared with identification of the equipment to which it will apply. To a greater extent than power engineering, instances will arise with CSBE of equipment/materials being proposed that will either not being covered by a technical specification nor subject to type tests. A judgement will therefore need to be made on suitability, taking into account factors such as the manufacturer's track record, standard of manufacturing, successful use of the equipment elsewhere, cost benefit, etc.
3. **Calculations**
   A schedule of calculations (and schedules of quantities) needs to be prepared, and subsequently undertaken, which will include
   (i) Foundations and supports
   (ii) Fences
   (iii) Gantry design
   (iv) Buildings, roads, etc.
   (v) Site lighting
   (vi) Building lighting and heating.
4. **Installation**
   The method and staging of equipment installation needs to be specified, identifying specialist installation equipment, positioning of equipment and method of site access – with suitable precautions if the substation is already operational.
5. **Technical interfaces**
   Technical interfaces, both within CSBE and between CSBE and other technologies (e.g. between busbar support foundations and structures and busbars), need to be defined together with the method of interface working.
6. **Commissioning**
   An overview of commissioning requirements needs to be defined, recognising that much in the CSBE domain is not subject to commissioning but QMS inspections and checks – and later transferred to the commissioning panel, see Chapter 23.
7. **Drawings**
   A schedule of all drawings that are required to be produced needs to be prepared, with reference to the range of tasks defined in the opening paragraph of this section.
8. **Temporary works**
   CSBE more than any other technology employs temporary works. Examples include access scaffolds, excavation and excavation supports, falsework (support of vertical loads), formwork (concrete shaping structures), temporary roads, fencing, etc. A schedule of all temporary works needs to be prepared together with the design solution.

14.2.5 OHL

With reference to Chapter 6, salient technical considerations at the time of detailed design for OHL include the following:

1. **Stages**
   Diagrams showing the existing and final arrangements, and stages in between, should be prepared. These may be a combination of SLDs and block diagrams, as the basis for the project stage-by-stage diagrams, see Chapter 24.
2. **Technical specifications**
   A final schedule of technical specifications needs to be prepared – with identification of the equipment to which each applies.
3. **OHL rating**
   A schedule defining the continuous, short-term and short-circuit current ratings of the OHL conductor(s) needs to be prepared. This will eventually feed into the thermal rating schedule, see Chapter 20.
4. **Calculations**
   A schedule of all calculations to be undertaken needs to be prepared, and subsequently undertaken. This will essentially cover tower/pole and foundation design, and conductor tension/sag design.
5. **Installation**
   The method and staging of OHL installation needs to be defined, identifying specialist equipment (e.g. cranes), and the method of site access.
6. **Impressed voltages**
   Site work where IV may be a hazard needs to be identified with method of control.
7. **Temporary works**
   OHL, like CBSE, contains significant temporary works involving scaffolding (for road/rail crossings, etc. and for temporary OHL) and including access roads, drainage, site offices, etc. These need to be specified together with the design solution.
8. **Technical interfaces**
   Technical interface boundaries with other technologies involved with the project, and with adjacent projects need to be identified, together with the method of interface working (e.g. terminal tower to substation gantry interface or crossing a lower voltage network OHL).
9. **Drawings**
   A schedule of all drawings needs to be prepared and typically would include:
   (i) OHL route drawings
   (ii) Tower/pole details/dimensions
   (iii) Tower/pole position drawings
   (iv) Foundation details

(v) Ground line profile (i.e. sag clearance)
(vi) Clearance drawings
(vii) Road/rail, etc., crossings
(viii) Swing clearances
(ix) Down lead arrangements
(x) OHL fittings drawings.

10. **Commissioning**
Consideration needs to be given to OHL commissioning requirements, particularly the location and method of energisation, and the phasing-out tests – for eventual transfer to the commissioning panel, see Chapter 23.

14.2.6 HV cables

With reference to Chapter 7, the main technical requirements relating to HV cables at the detail design stage include the following:

1. **Stages**
Diagrams showing the initial and final arrangements should be prepared, including any stages in between.
2. **Technical specifications**
A final schedule of technical specifications needs to be prepared – with identification of the equipment to which each applies.
3. **Calculations**
A schedule of all calculations to be undertaken needs to be prepared, and subsequently undertaken. These include
(i) Heat-loss calculations
(ii) Sheath bonding (where a bonded system is to be installed)
(iii) Pressure diagrams (relating to route and section length) for pressurised cables.
4. **Cable ratings**
A schedule defining the continuous, short-term, cyclic and short-circuit ratings of the cable needs to be prepared. This will eventually be incorporated into the thermal ratings schedule, see Chapter 20.
5. **Installation**
The method and staging of the cable installation needs to be prepared, identifying the requirements for cable sealing ends, joint bays, and road, rail and river crossings, etc. Specialist equipment should be identified and the arrangements for site access.
6. **Cable accessories**
A schedule of all cable accessories should be prepared including link boxes, link pillars, SVLs, etc.
7. **Technical interfaces**
Technical interface boundaries with other technologies and other projects need to be prepared together with the method of interface working (e.g. interface between cable sealing ends and OHL down leads).

8. **Commissioning**
   An overview schedule of commissioning requirements needs to be prepared with arrangements identified for HV pressure tests, the method of energisation and phasing out – for eventual transfer to the commissioning panel.
9. **Temporary works**
   A schedule of all temporary works needs to be prepared, which will typically include
   (i) Removal and replacement of the cable trench spoil
   (ii) Shuttering and barriers
   (iii) Scaffolding (e.g. for cable sealing end erection)
   (iv) Traffic management arrangements.
10. **Impressed voltages (IV)**
    Cable projects are a specific source of concern with reference to IV. Consideration should be given to the hazards that may arise from IV and the means of control.
11. **Drawings**
    A schedule of all drawings needs to be prepared. A typical list is provided in Section 7.9.2.

14.3 Manufacture/procure

14.3.1 Manufacture/procure specification

The narrative running through this publication is that a power network company awards a contract to a contractor who arranges for the manufacture or procurement of the equipment/materials required together with detail design, and subsequent site installation and commissioning. Within this context, the 'manufacture' task is often concerned with the contractor awarding another contract (between contractor and manufacturer) for the manufacture of equipment (or materials) – whereas the 'procurement' task is more usually associated with a purchase order for off-the-shelf items (of a smaller nature). It is recognised that instances arise of the contractor and manufacturer being in the one company – but none the less a formal interface is likely to exist between the two parts of the company. Instances also arise of a power network company contracting directly with a manufacturer (e.g. bulk purchase of equipment).

For the instance of the contractor interfacing with the manufacturer, then immediately prior to the manufacture/procure stage commencing the contractor will be in possession of three integral categories of specification (see Figure 1.3) namely:

- The project-specific design specification, which exists at time of contract award, i.e. the SDS
- The project-generic design specification, which exists at time of contract award, i.e. the technical specifications
- The detailed project-specific design specification prepared after contract award, i.e. the DDS.

At this stage, the contractor needs to prepare a further contract specification for the manufacture (or for procurement) of the equipment, which is based upon the above – but may require additional information and narrative together with the contractor-specific terms and conditions.

When acquiring the equipment/materials, a number of requirements need to be taken into account, which are as follows:

- The preparation of a schedule(s) which identifies all required equipment/materials down to the last nut and bolt. NB: For civil engineering works, 'bills of quantities' (itemising materials, parts and labour) may need to be prepared – although they may be available from the main contract tendering process.
- That an unambiguous technical specification exists for each item of equipment/material to be manufactured/procured.
- That the requirement for type tests (including the content of the tests) on new types of manufactured equipment is specified.
- That it is clear to where the equipment/material is to be delivered (i.e. factory or site) and when.
- That arrangements are in place for declaring manufactured items as acceptable before delivery (i.e. manufacturing surveillance).
- That all manufactured/procured items are accounted for as having been delivered and cross-referenced to the schedule in bullet point one above.

14.3.2 Manufacturing surveillance

Chapter 21 reviews the procedural requirements relating to the manufacture stage and manufacturing surveillance by the contractor/power network company and examines

- Expediting (manufacturing assurance)
- Type tests
- Routine tests
- Factory acceptance tests.

14.3.3 Manufacture/procure – detail design interface

The production of drawings during detail design, and the manufacture/procure stage, usually takes place simultaneously. Given that, that manufactured/procured equipment (e.g. a circuit breaker) will be included on the contractor's drawings (e.g. interface connections), it is essential that there is a two-way flow of technical information and requirements at the interface boundary (between drawings and equipment), to ensure that a mismatch does not arise.

14.4 Site installation

14.4.1 Site installation – overview

The following sections are concerned with the technical requirements and considerations relating to site installation and should be read in conjunction with the procedural requirements examined in Chapter 22. Within this context, site installation seeks to achieve two primary objectives which are as follows:

- The installation of power network equipment/infrastructure in accordance with the construction design specification (see Figure 1.3) – without errors (i.e. no rework)
- The undertaking of all site work to ensure the health and safety of all personnel on site and the general public, together with the implementation of environmental management measures.

To achieve the above objectives, site installation is undertaken in accordance with the following:

- Drawings and schedules (prepared during detail design) defining the requirements of both the permanent and temporary works
- Equipment (and infrastructure) installation instruction manual/handbooks, etc. (approved for the purpose)
- A CDM plan and environmental plan, together with all other related requirements detailed in Figure 22.6.

14.4.2 Site installation – content

In summary, the engineering tasks associated with site installation are as follows:

1. **Civil, structural and building engineering**
   - (i) Civil engineering work preparing and stabilising the ground
   - (ii) Construction of foundations and support structures
   - (iii) Construction of site infrastructure involving buildings, roads, fencing, etc.
   - (iv) Installation of building services, such as power, heat, lighting, water, etc.
   - (v) Environmental works such as oil containment and flood defences, etc.
2. **Substation HV equipment**
   - (i) The positioning, erection and interconnection of HV equipment
   - (ii) Installation of the earth mat (which in part is also a civil engineering task).
3. **Protection and control equipment**
   - (i) The positioning of panels/cubicles and mounting of relays
   - (ii) The laying and terminating of multicore cables and other communications mediums, together with the interconnection of the relays and cables, etc.
   - (iii) The installation of DC supplies and connecting cables.
4. **OHL**
   - (i) The positioning and construction of tower/pole foundations
   - (ii) The erection of towers/poles
   - (iii) The installation of OHL fittings (e.g. insulators) and the stringing/sagging of conductors.
5. **HV cables**
   - (i) The excavation and laying of cables and subsequent backfill
   - (ii) The construction of cable terminations
   - (iii) The jointing and terminating of cables.

The following sections will examine a wide range of site installation engineering requirements which are common to the above, many of which come under the banner of temporary works which are as follows:

(i) Site establishment
(ii) Demolition
(iii) Excavation
(iv) Piling operations
(v) Lifting operations
(vi) Erection of structures
(vii) Scaffolding
(viii) Electricity on site
(ix) Impressed voltages
(x) Roadworks
(xi) Overhead and underground services
(xii) Contamination on site
(xiii) OHL considerations
(xiv) HV cable considerations.

14.4.3 Site establishment

Site establishment comprises the physical establishment of the site accommodation for the workforce, together with laying down areas for equipment and materials. The site may be purposely built or within the confines of an existing operational substation – space permitting. Salient requirements are as follows:

1. **Access arrangements**
   This comprises civil engineering work to create suitable access (i.e. roadways) for the volume and weight of traffic required to execute the project on site. Some roads may be temporary.
2. **Ground work**
   This is civil engineering work associated with ensuring the ground level and weight-bearing capacity is suitable for site requirements, i.e. foundations roadways, paths, accommodation, car parks and laydown areas, etc. It also includes drainage, flood defence considerations and areas for storing waste until it can be removed from site.
3. **Site accommodation**
   Site accommodation includes the positioning and erection of office accommodation usually portable, light weight buildings delivered by low loader. Facilities within the buildings include:
   (i) Light, heat, power
   (ii) Communications
   (iii) Toilet and washing facilities requiring water and sewage connections
   (iv) Rest and eating facilities
   (v) Drawings storage
   (vi) Workshops
   (vii) General storage, etc.

4. **Site security**
   This usually requires the erection of a fence, not less than 2 m in height around the site, supported where necessary by an electronic security system.

14.4.4 Demolition/removal

In many instances, construction work commences with the demolition and removal of existing equipment or infrastructure. Within this context, the following may need to be undertaken:

- Creation of a demolition exclusion zone to provide controlled access to the zone
- The protection of operational or other facilities that may be impacted on by the demolition work
- Removal of flammable, toxic or dangerous materials e.g. oil, asbestos, PCBs, before the demolition work commences
- Implementation of the demolition through use of risk assessments and method statements (RAMS), which must stipulate the positioning of personnel, mobile equipment and static equipment to avoid danger
- The demolition workforce must be competent in the work to be undertaken
- Demolition may be carried out either piecemeal or via a single deliberate controlled collapse to the ground
- Remove the demolished material to a pre-defined location
- Clean and make good the ground following demolition.

14.4.5 Excavation

Excavation is an essential element in construction work and is required to facilitate the following:

- Drains
- Earthing
- Foundations
- Oil containment
- HV cable trenches
- Multi-core cable trenches
- Diversion of waterways, etc.

The work involved in excavation is dependent upon the ground type which may be rock, gravel, sand, clay, peat, loam, etc. Salient points associated with excavation work include the following:

- Excavation may take place at a variety of depths. In most instances, excavation is undertaken by mechanical excavators but may need to be undertaken by hand, either for small excavations or to avoid services such as cables and gas pipes, etc.
- Although each case needs to be evaluated on its own merits, excavations at a depth in excess of 1.2 m usually require the excavation walls to be supported to

prevent collapse (and danger). For trench excavations, this may comprise timber walls or steel trench sheeting as the support material, held in position by either timber or adjustable steel separation struts.

- Generally, excavated soil should be kept a minimum distance of 1.5 m from the excavation edge to avoid unintentional backfill – and heavy vehicles kept as far away as practical from the edge to avoid potential excavation collapse.
- HV cable trenches may be internal or external to the site. If external, they may be in roads or fields, etc. and enclosed by barriers (or covers) to ensure public safety.
- Consideration should be given to the impact of prolonged rain on the excavation.

14.4.6 Piling operations

With reference to Chapter 13, piling may be required for the strengthening and stabilisation of the ground, either within a substation or to support an OHL foundation. It usually comprises driven metallic rods. Piles are usually driven into the ground through the employment of specialist mobile equipment such as a 'pile hammer' (drop hammer). The installation of piles can be noisy – and as such the hours of work may be restricted (to avoid being a nuisance).

14.4.7 Lifting operations

Lifting operations are an important area of work on a construction site. They are covered by the LOLER Regulations, see Chapter 18. Lifting operations typically involve the loading/unloading of material, the lifting of materials into position, the excavation (lifting) of material, the repositioning of material and the transportation of materials. In summary, it includes the following equipment:

- Cranes
- MEWPS
- Excavators
- Winches
- Beam hoists
- Wire ropes
- Transportation vehicles
- Metallic slings and hooks
- Gin wheels and ropes.

All lifting operations should be carried out in accordance with a lifting operations plan. The plan should typically include the following:

- Location of the lift
- Specification of the weight of the load to be lifted
- The height, radius and path of the lift
- Hazards to be controlled (including the general public)
- The ground-bearing capability
- Use of an exclusion zone
- Communications arrangements
- Capacity of the equipment to undertake the lift

- Responsibilities for the lift, including competency requirements
- Confirmation that RAMS have been prepared.

The access route of all mobile equipment entering and leaving the site should be defined. Consideration should be given to the location of all stored materials that will be subject to lifting to ensure that it is accessible at the time of the lift.

14.4.8 Erection of structures

The erection of structures is a common installation activity particularly in AIS substations. Structures are usually metallic, pre-cast concrete or re-enforced concrete constructed on site. Installation requirements include the following:

- Confirming the position of the structure and fixing arrangements
- Determining the access route for transportation of the structure to site
- Determining the space requirements to lift the structure into position and the position of any lifting equipment
- Determining the method of lifting/erection
- Ensuring the structure is erected in the correct order (as identified in the programme).

14.4.9 Scaffolding

Scaffolding is widely used on all construction sites. In power network construction, usage includes the following:

- In the erection of, and to gain access to, substation equipment
- To protect road, rail crossings, etc. when stringing an OHL. NB: Some of these scaffolds are physically very large, e.g. providing protection over motorways
- In the installation of cable sealing ends.

The components which comprise scaffolding are generally standardised to aid rapid erection. Equally most scaffold designs (i.e. shape of the structure) follow a standard pattern, although some are purposely designed. All those with responsibilities for the design and erection of scaffolding must be competent and be in possession of an appropriate qualification.

Scaffolds must be erected on ground of suitable bearing capacity. If a hard surface such as asphalt or concrete is not available, the ground should be levelled and rammed firm. To spread the ground loading, scaffolding base plates are often laid on long timber 'sole plates'. The erection of scaffolding should include 'toe-boards' located at the edge of the working platform (to avoid the hazard of tools, etc. falling out of the scaffold). BS EN 12811 provides guidance on scaffolding performance and design.

14.4.10 Electricity on site

Construction sites require a 230-V (and sometimes 400 V) supply for a wide variety of reasons, e.g. heating and lighting, cooking, site lighting, pumps, electrically operated equipment, general site supplies, etc. It is a potential source of danger and must be responsibly managed. Considerations comprise the following:

- Installation of both the electricity distribution boards and distribution network– must accord with BS7671, IET Wiring Regulations. On completion, the installation must be subject to an inspection by a competent inspector and issue of an inspection certificate.
- The location of the main earthing terminal and earth-bonding arrangements must be considered.
- The interface boundary arrangements between the power network company who will run/operate the permanent site and contractor who usually only requires a temporary supply must be agreed.
- It is usual to power portable tools, etc. via a 230/110-V centre-tapped and earthed transformer to facilitate safe working.

14.4.11 Impressed voltages

Site work comprises that stage within the construction process where the workforce is subject to the hazards associated with IV. The impact of IV must therefore be considered in advance of any work activity commencing. It is a particular hazard when working either on de-energised OHL, cables or busbars or in the proximity of live OHL, cables or busbars. IV is examined in detail in Chapter 11.

14.4.12 Road works

Roadworks comprise the installation of roads within substations, the creation of temporary access roads or work involving public roads, which are as follows:

1. **Roads within a substation**
   Roads within a substation will usually be constructed early in the site installation stage to provide access for subsequent site work. Entrances to roadways (in AIS substations) may be via 'goal posts', to limit the height of vehicle access when the substation is operational – to ensure clearances to live equipment are maintained. Traffic rules may need to be put in place (such as one-way traffic) to facilitate either the volume of traffic or the staging of work – during site installation. Consideration should be given to the roadway arrangements for vehicle turn around.
2. **Temporary roads**
   Temporary roads may be required for access to either of the substations, HV cable or OHL work sites. They typically comprise either portable trackway or gravel – which is removed on work completion.
3. **Public roads**
   HV cables frequently require excavation in public roads. In doing so, the excavation process will need to avoid other services in the road. Disruption to public roads frequently requires the installation of temporary road traffic management system, including traffic cones, warning signs and a traffic light system. The arrangements must ensure both the safety of the workforce, and the general public, and the safe and efficient movement of traffic.

14.4.13 Dangers arising from OHL and underground services

Construction work invariably involves activities in close proximity to an OHL, often requiring the regular passage of vehicles under the OHL. In such instances, arrangements need to be put in place to ensure that statutory electrical clearances are not infringed. This is usually accomplished via the goal post and barrier system illustrated in Figure 14.3. The arrangement shown in the figure is positioned on the either side of the OHL and not less than 6 m from it. The barriers (often barrels) are surmounted by coloured bunting to form additional warning.

Underground services (i.e. electricity cables and gas pipes) also provide a hazard during site installation, and inadvertent contact with underground service over the years has resulted in major injuries and fatalities. It is therefore essential that adequate precautions are taken. HSG 47 'Avoiding Danger from Underground Services' provides guidance on how to manage the hazards arising from cables – salient points of which are as follows:

- Obtain copies of drawings from the utilities concerned to identify the location of the cables/pipes.
- Employ the use of detection devices (of which there are several available) to pin point the location of the cable/pipe.
- Hand dig trial holes to correctly identify the route and type of the service.
- Once the cable/pipe has been correctly identified, an exclusion zone should be marked out –with mechanical diggers permitted to operate outside the exclusion zone.

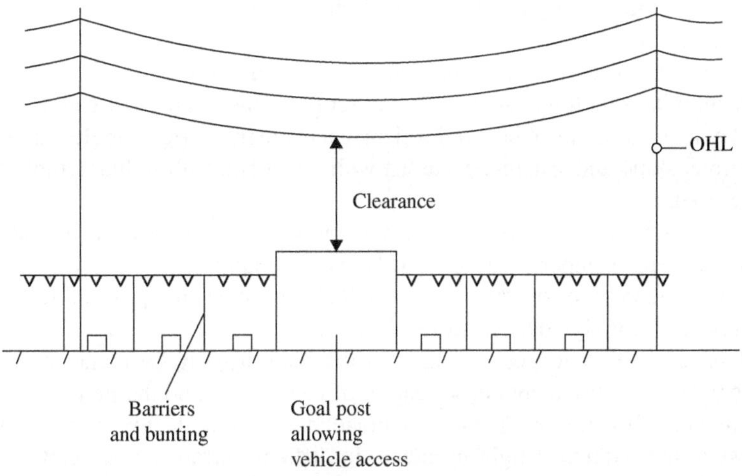

Figure 14.3 Goal posts and barriers for controlling vehicle access

14.4.14 Contaminated sites

It is occasionally the case that some construction sites are subject to contamination. The contamination may be within the land itself, or associated with buildings or equipment. Contamination may include

- Chemicals (e.g. PCBs)
- Petrochemicals (including asbestos)
- Metals
- Power station residues.

At an early stage, in the construction process, checks should be carried out to confirm whether contamination exists – and if so, determination of the measures to remove or control the contamination. This may require the following:

- Creation of an exclusion zone
- Control of vehicle and pedestrian access
- Contamination storage arrangements
- Protective clothing
- Washing and changing facilities
- Site hazard warning signage
- Food and water consumption restrictions.

RAMS will be required for the contamination extraction and removal from site.

14.4.15 OHL considerations

Installation of an OHL would usually include the following:

- Pegging out of the tower/pole position to determine the precise location. For a tower, this will involve the positioning of the tower centre and four legs, although the below ground configuration of the new 'T' tower is similar to that of a pole (see Chapter 6).
- Excavation of either a pole hole or tower foundations.
- Positioning of a tower stubs, i.e. stub setting. This requires care and precision. The four stubs must be set level to each other, at right angles, and at the correct slope and spacing. To assist with this work, stub setting templates may be used.
- Transportation of materials to site, i.e. poles or tower steelwork, conductor and fittings, etc. Temporary access roads may be required.
- Tower foundation concreting – which locks the stubs into position followed by backfill of the foundations (after the concrete has set).
- Raising a pole and positioning it in the hole, usually by crane or specialist vehicle and subsequent consolidation of backfill around the pole.
- Erection of the tower. The modern practice is to build the tower in sections and rise to the required height by crane. In difficult terrains, a helicopter may be employed.
- Installing the insulators and fittings.

- Arranging the positioning of the stringing equipment e.g. winding, pulling and tensioning machinery and pully blocks – a section at a time. OHL are usually strung under tension so the conductor is not grazed by the ground.
- Sagging the conductor. This is usually accomplished by the use of a sagging board placed on an adjacent tower at the correct height for the sag, and via the use of a sagging instrument (telescope) adjusting the sag until it is aligned with the sagging board as seen through the instrument.
- Conductors are then finally clipped (clamped) into position and as such supported by the insulators.

14.4.16 HV cable considerations

Installation of a new HV cable would usually include the following:

- Prior to site installation commencing, the necessary drawings should be available showing the proposed route, the position of joints, the position of obstacles and other services, etc.
- The trench should be excavated to the correct depth and width and the spoil placed appropriately away from the trench sides.
- Depending upon the depth of the trench and composition of the soil, it may be necessary to provide supports for the trench walls (see Section 14.4.5).
- The cable drums and pulling/winch equipment should be appropriately positioned.
- Cable rollers and skid plates should be located in the trench, both to keep the cable off the ground when laying, and for ease of pulling.
- A cable stocking should be attached to the cable, or a cable pulling eye plumbed onto the cable, to which the pulling wire should be attached.
- The cable should be pulled into position via the winch, taking account of the maximum tensile strength of the cable (i.e. not exceeded).
- On completion, the cable should be removed from the rollers/skid plates to rest on the trench floor.
- Cable jointing and terminations should be undertaken, which may require some form of weatherproof protection.
- In some instances, dewatering pumps may be required in either the trench or joint bay – to maintain dry conditions.
- The cable trench should be backfilled with an appropriate bedding material, such as sand or riddled soil to provide a compacted cover over the cable. Where the design dictates thermally conducting material must be used.
- On completion of the backfill, roadways that have been damaged must be resurfaced (e.g. tarmacadam).

14.4.17 Site installation – completion checks

Prior to completion of the site installation and handing over to the commissioning stage, final checks and inspections of the physical installation with reference to position, quality of work, labelling, pressure gauges, access restrictions and (importantly) electrical clearances should be undertaken. Minor works which are incomplete should be added to a 'snagging' list for remedial action.

Chapter 15

Equipment commissioning – technical

15.1 Commissioning – introduction

Commissioning is not only the final technical stage in power network construction but also one of the most critical. As one highly experienced commissioning engineer once put it, 'commissioning is the bucket and shovel stage of a project'. By this, the engineer meant that commissioning cleans up any errors or omissions that have taken place earlier in the project. Perhaps a slight overstatement, but commissioning is certainly that last point at which corrections may be made, thereby ensuring the equipment enters commercial service as a fully compliant design.

This chapter is concerned with the technical aspects of commissioning and should be read in conjunction with Chapter 23 which covers commissioning procedural requirements. As Chapter 23 points out, commissioning comprises equipment site tests and inspections which are sufficient to prove that the installation can be connected to the power network and enter service. Within this context, commissioning has historically been undertaken in accordance with the philosophy of:

assume everything is wrong until proven otherwise

Such an approach, albeit sound, dictates an all-encompassing test and inspection regime. However, modern day commissioning needs to take cognisance that an increasing number of tests are undertaken in the factory, under quality-assured tests and inspection arrangements, and as such many on-site tests/inspections do not need to be as exhaustive as was once the case. This applies particularly to equipment containing computer software where once it is confirmed that the software version on site is one and the same as that tested in the factory, then the site functionality testing of the equipment can be reduced. Within this context, equipment, etc., which is erected/connected/constructed on site and not subject to factory quality assurance arrangements must always be confirmed as correct via on-site commissioning. The essential point to keep in focus is that the commissioning engineers must be assured, by whatever means, that the equipment is fit for purpose and suitable for connecting to the live power network for the purpose of becoming operational.

This chapter will provide a broad overview of commissioning requirements and techniques, covering the most commonly installed types of equipment. The following will be included:

- Test procedures and documentation
- Substation earthing

- Insulation resistance
- Current transformers (CTs) (including primary injections)
- Protection relays (including secondary injections)
- VT supplies
- DC circuitry logic tests
- Protection and control common equipment
- Auto-switching equipment
- HV equipment
- Batteries and DC supplies
- Loadability
- Energisation and on-load tests
- Substation commissioning overview.

15.2 Test procedures and documentation

15.2.1 Test procedures and documentation – overview

Commissioning test procedures and documentation against which commissioning is undertaken may be categorised as follows (also see Section 23.9.1):

1. **Equipment test procedures**
   These may be divided into two types, which are as follows:
   (i) **Equipment specific**
   This type of procedure details the step-by-step requirements for testing an item of equipment, e.g. how to commission a specific type of protection relay.
   (ii) **Equipment generic**
   This type of procedure provides a framework of general requirements – but with a need to tailor the requirements to the site-specific situation, examples include:
   (a) Insulation resistance tests
   (b) Primary injections
   (c) Overcurrent relay testing (since there is more than one type of overcurrent relay).
2. **Drawings**
   Some equipment is commissioned by verifying the relevant drawings, examples include:
   (i) Protection and control circuit diagrams
   (ii) Equipment location drawings
   (iii) Air or oil systems drawings.
3. **Commissioning switching programme**
   A commissioning switching programme (or other similar title) is a circuit-specific, step-by-step procedure for the energisation of equipment from the HV network and subsequent on-load tests. It may include an equipment test procedure within it, see Section 23.7.

4. **Commissioning method statement**
   A commissioning method statement (or other similar title) is an equipment-specific, step-by-step procedure for all requirements which are not concerned with the energisation of the equipment from the HV network. Again, it may include an equipment test procedure within it. Examples include:
   (i) Primary injections
   (ii) Proving of VT circuitry
   (iii) Loadability test.
5. **Commissioning inspections**
   As described in Section 23.6.1, inspections are site specific and are undertaken to ensure both the safety of personnel, and the well-being of the equipment, during commissioning. Generally, three inspections are required, which are as follows:
   (i) Prior to commissioning commencing
   (ii) Prior to equipment energisation
   (iii) Following completion of the on-load tests.

15.2.2 Formal commissioning stages

It is pointed out in Section 23.4.1 that it is helpful to categorise commissioning into two formal stages: stage 1, which (in summary) comprises the off-load commissioning tests and inspections, and Stage 2, which comprises the method of energisation from the HV system and the subsequent on load tests. These stages will be cited in the remainder of this chapter.

15.3 Substation earthing

15.3.1 Substation earthing resistance

Commissioning tests associated with substation earthing are concerned with measuring the earth mat resistance and the subsequent determination of rise of earth potential (ROEP). Virtually, all methods for measuring earth mat resistance involves the passing of a test current through the earth electrodes, measuring the voltage drop and calculating the resistance through the application of Ohm's law. This type of test is generally called the 'AC fall of potential' (FOP) method. It should be carried out as one of the first commissioning tests and (ideally) prior to the substation being energised. The salient requirements of this test are as follows.

With reference to Figure 15.1(a), test current is injected into the earth mat with the return path being that of a current electrode inserted into the ground, some distance from the substation. The current electrode distance must be sufficient to ensure that it does not interfere with the flow and dissipation of the current flowing from the earth mat into the earth. It is generally required to be ten times the dimensions of one side of the earth mat – and therefore for large substations, a significant distance from the substation. A second electrode, the voltage electrode, is then used to measure the voltage to ground at increasing distances from the position of current injection; this is the distance *S–T* as shown on Figure 15.1.

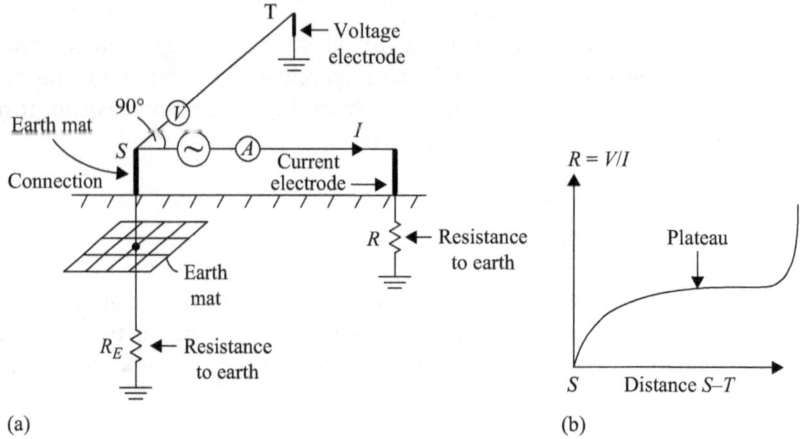

Figure 15.1 Determination of substation earth resistance: (a) test arrangement and (b) resistance profile

Consider in the first instance the voltage electrode being moved directly towards the current electrode (i.e. not the 90° angle shown in Figure 15.1(a)). Near to the earth mat, the measured voltage will be close to zero and will increase as point T is increasingly moved away from point S, in accordance with the lines of equipotential. When the lines of equipotential become constant, i.e. notionally at the point of true earth, the measured voltage then becomes constant, i.e. plateaus. As point T is again increasingly moved away from point S, the equipotential lines again become closer together, and the measured voltage increases, as the current electrode is approached. The resulting resistance profile is shown in Figure 10.1(b), and the resistance to true earth, R_E, is that given by the plateau region of the profile. The following points are worthy of note:

- Although the current electrode has a resistance, R, to true earth, which influences the voltage profile, a similar resistance associated with the voltage electrode can be ignored since it is relatively small compared with the voltmeter internal impedance.
- Situations arise where the earth mat may be subject to a standing voltage, if in the presence of energised equipment (i.e. adjacent OHL). Such voltages may be picked up by the test voltmeter, and therefore, the test voltage needs to be sufficiently large to render the standing voltage negligible. This may necessitate test currents in the range 20–200 A and may require specialist test equipment (of which there are numerous on the market).
- Large test currents can result in mutual inductance from the current electrode test lead to the voltage electrode test lead, so corrupting the test results. To minimise this interference, it is common practice to position the electrodes at 90° to each other, as illustrated in Figure 15.1(a). In this instance, the resistance profile shown in Figure 15.1(b) is likely to just plateau without a subsequent increase.
- Transmission substations may be designed with the earth mat arranged in a number of earth groups – with facilities for testing each group separately.

- If the test has to be undertaken with the substation energised, there is a risk of a system fault arising and the test equipment being subject to the ROEP. In such instances, necessary precautions against impressed voltages (see Chapter 11) should be undertaken, including insulated working and equipotential zone working.

15.3.2 Substation rise of earth potential

Once the substation earth mat resistance has been obtained as in Section 15.3.1, the ROEP can be derived by multiplying the resistance by the maximum earth-fault current. However, in those instances where the earthing system is extended via OHL earth wires and HV cable sheaths connected to the earth mat (see Section 12.6.1), the determination of ROEP must take this into account. This may be achieved by direct injection or calculation, which is as follows:

1. **Direct injection**
   This involves injecting current into the whole of the substation earthing system, including the earth connections of the OHL and HV cables. Again, the method used is that of FOP as explained in Section 15.3.1. Once the plateau voltage has been determined, the ROEP for the whole earthing system is obtained by dividing the substation maximum fault current by the injected current and multiplying this number by the plateau voltage (i.e. process of extrapolation). Again, when undertaking this test, the effects of impressed voltage must be protected against.
2. **Calculation**
   In practice, the direct injection method often proves impractical to undertake since it requires the simultaneous availability of all connected OHL and cables. In such circumstances, the derivation of ROEP must be via calculation (following determination of the earth mat resistance), as described in Chapter 12.

15.4 Insulation resistance

15.4.1 Insulation resistance – LV circuitry

Insulation resistance tests (commonly termed IR tests) are usually undertaken at the outset of commissioning and refer to the testing of the insulation of multicore cables and LV wiring in general. This comprises the application a voltage, usually 500-V DC (using a device such as a Megger insulation tester) between core and earth prior to any other tests being carried out on the cable/wiring; often numerous cores are connected together – to save time and effort. Generally, new wiring should have an IR of 100 MΩ and above, although in damp outdoor conditions, it could be as low as 5 MΩ and still be accepted as satisfactory. A single test will often pick up a significant amount of wiring. It is also common practice to test between CT, VT and DC circuits. Care should be taken to ensure that test voltages are not applied to electronic equipment (relays, etc.) that may be damaged. Such equipment may be protected by either disconnections or short-circuits.

15.4.2 Insulation resistance – transformers

IR tests are also usually applied to transformers using a 1,000-V DC test device, which are as follows:

- Between the transformer core (and clamping structure) and earth. A test bushing is usually available in the tank wall for this purpose which ensures the test can be undertaken without the need for oil handling. The test will require the permanent connection from the bushing to earth to be temporarily disconnected and the test wires placed across the disconnection.
- Between the winding terminals (both primary and secondary for a two-winding transformer) and earth.
- Across windings, when a two-winding transformer is installed.

Such tests are usually integral to the transformer equipment test procedure. Satisfactory test results are usually 100 MΩ and above.

15.4.3 Insulation resistance – cables

It is also usual to carry out an IR test on cable sheath insulation, once the cable has been laid, jointed and terminated. At the higher network voltages, this is usually carried out with a 10-kV DC test device for a period of 1 min between the cable sheath and earth. It is usual to monitor the leakage current which (usually) would not be expected to exceed 1 mA, i.e. equivalent to 10 MΩ. Again, this test would be integral to the cable equipment test procedure.

15.5 Current transformers

15.5.1 Current transformers – requirements

CTs are one of the most common items to be commissioned – the required tests are as follows:

- Flick test
- Primary injection
- Magnetising curve
- CT resistance

These will be briefly examined.

15.5.1.1 Flick test

The 'flick test' is arguably the most well known of all commissioning tests. It is undertaken to confirm the absolute polarity of a CT, i.e. confirmation of the instantaneous direction of primary and secondary current flow, relative to each other.

With reference to Figure 15.2, CTs which accord with BS7626, specification for CTs, have their terminals labelled to enable polarity to be determined. The primary terminals are labelled *P*1 and *P*2 and the secondary *S*1 and *S*2. The convention is that if the instantaneous primary current is flowing into terminal *P*1 and

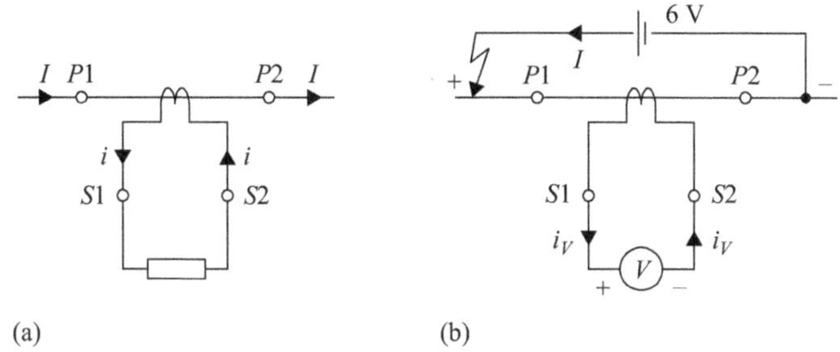

Figure 15.2 CT polarity and flick test: (a) CT polarity convention and (b) flick test

out of terminal *P*2, then the corresponding instantaneous secondary current flows out of terminal *S*1 and (via the external circuit) into terminal *S*2.

Figure 15.2(b) illustrates the circuitry for undertaking the flick test. When the battery positive pole is applied to terminal *P*1, correct polarity is indicated when (at the instant of application) the DC voltmeter exhibits a 'flick' in the positive direction. The voltmeter should ideally be an analogue device for observing the flick. Many CTs have tapped windings (e.g. 1,200/600/1), and all tappings should be flick tested. CTs located in transformer bushings may have to be flick tested prior to the bushing being installed in the transformer, since to undertake the test after installation may result in the high impedance of the transformer winding limiting the current flow from the battery to such a low value that the flick cannot be observed.

15.5.1.2 Primary injections

Whereas the flick test is concerned with the absolute polarity of a CT, primary injections are concerned with confirming the following:

- The ratio of each CT (normally a group of CTs requiring identical ratios)
- The relative polarity of a group of three (or more) CTs
- The matching ratio and polarity of two groups of connected CTs (e.g. unit protection)
- Stability of the relay when subject to through fault current (i.e. applicable to unit protections)
- Operation of the relay when subject to an internal fault
- The integrity of all connected wiring between the CT and the relay

The above requirements will be briefly reviewed.

1. **Group of three CTs**
   Figure 15.3(a) shows the arrangements for the primary injection of a group of three CTs. The following salient points apply:
   (i) The arrangement shows the primary injection of the *B* phase CT, and a *B–E* fault is simulated. The ratio of the CT is obtained by dividing the current in the primary ammeter by the current in the secondary ammeter.

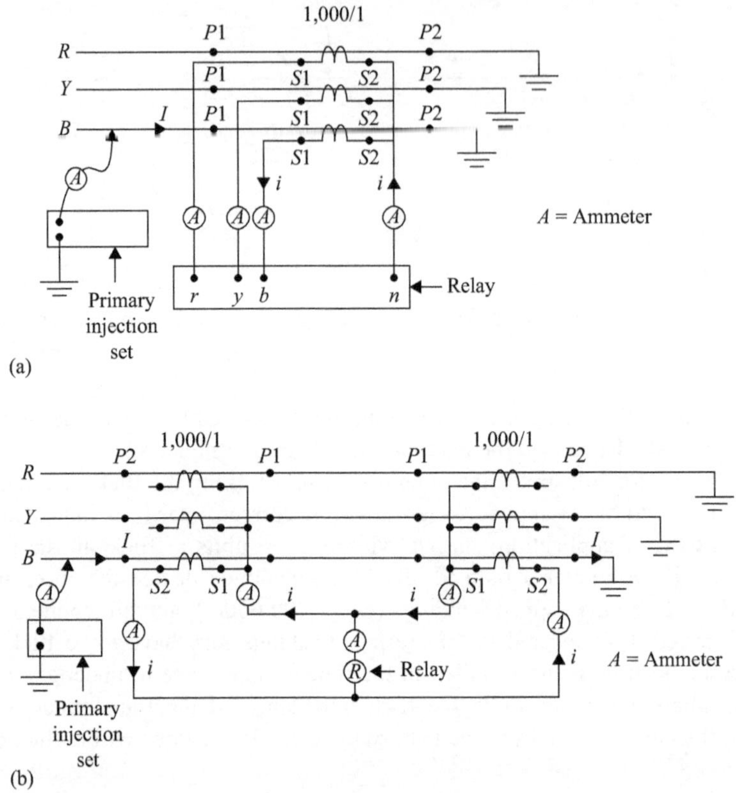

Figure 15.3 Primary injection – examples: (a) group of three CTs and (b) two groups of connected CTs

(ii) The relative polarity of two CTs (i.e. both facing in the same direction) is obtained by observing the current flow arising from a simulated phase-to-phase fault. For example, a *Y–B* phase fault is simulated by connecting the primary injection set across the *Y* and *B* phase conductors. As a result, the current flows through the CTs and around the earth loop. Correct (relative) CT polarity is when current is only observed in the *y* and *b* secondary winding ammeters.

(iii) The integrity of the wiring is verified through the above two tests.

(iv) It is usual to carry out six injections which cover ratio, relative polarity and integrity of wiring, i.e. *R–E*, *Y–E*, *B–E*, *R–Y*, *Y–B* and *B–R*. Theoretically, the *B–R* injection is superfluous but is often undertaken as a double check.

(v) A typical primary injection set can usually deliver a few hundred amps, but at a low output voltage, requiring the test connections to be very clean. A typical primary injection current of 200 A passing through the 1,000/1 CT ratio, as shown, would result in a secondary current of 200 mA.

2. **Two groups of CTs**
Figure 15.3(b) shows the instance of two groups of connected CTs (although in practice, it could be more, e.g. mesh corner protection). For reasons of diagram clarity, only the secondary of one phase is illustrated, the other two phases being similarly connected. Salient points are as follows:
    (i) Verification of the ratio and relative polarity of each group of CTs should be carried out as in (1). This would require the earth, as shown in Figure 15.3(b), to be repositioned in between the two of CTs. These tests would also verify relay operation for an internal fault. NB: The current may not be high enough to operate the relay.
    (ii) The injection as shown in Figure 15.3(b) would verify the matching ratio and polarity of the two groups of CTs. It would also verify relay stability on through fault, since for correct matching of the CTs, the current in the relay, as indicated on the ammeter, would be virtually zero. This injection should be repeated on the other two phases.

General observations on primary injections are as follows:

- The above describes general principles; there may be other methods of achieving the same result.
- Other sets of CTs may well exist in the same CT stack, and these should be short circuited to avoid damage.
- With relatively high-impedance equipment such as transformers, it may be necessary to prove the CT and associated circuitry by injecting a 400-V three-phase supply into the HV terminals to obtain sufficient magnitude of test current.

15.5.1.3 CT magnetisation curve

CT application is highly dependent upon the susceptibility to magnetic core saturation, i.e. magnetising impedance saturation, see Figure 3.18. Saturation is mostly caused by DC transient components of fault current as described in Chapter 3.

A CT magnetisation curve (generally termed a 'mag curve') test is used to determine the secondary voltage at which the CT commences to saturate. This voltage is compared with the manufacturers test data to confirm that the correct CT has been installed.

Figure 15.4(a) shows the test arrangement for a CT magnetisation curve. Prior to the test commencing, the test voltage is increased to drive the CT into saturation, as indicated by rapid increase of current (beyond the knee point), and then slowly reduced to zero – this process removes any remnant magnetism from the core which may corrupt the test results.

The test commences by increasing the test voltage (from zero) in increments, to provide a graphical plot of voltage against the corresponding current. This continues until the 'knee point' of the CT is reached. This is defined as that point on the curve of the graph where a 10% increase in voltage causes a 50% increase in current, as shown in Figure 15.4(b). The knee point indicates significant onset of saturation.

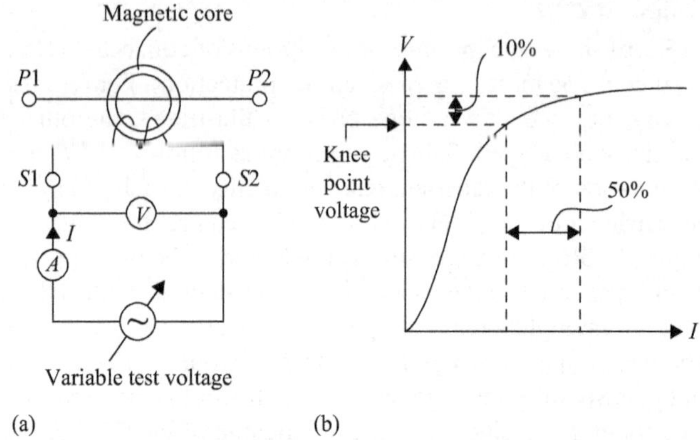

Figure 15.4 CT magnetisation curve: (a) test connections and (b) knee point

In practice, knee points range from typically 200 V, for distribution network CTs, to 3 kV for transmission networks. For CTs, with very high knee points, the test equipment may not be capable of delivering such a high voltage, and therefore only a partial plot for comparison with the manufactures test data will be taken – which will usually suffice. Furthermore, the injection of such high voltages risks damaging the secondary wiring.

15.5.1.4 CT resistance

The CT secondary winding resistance should be taken with an accurate ohm-metre. This is required for certain protection calculations (e.g. high-impedance schemes), and as a comparison with the manufacturers test results.

15.6 Protection and control equipment tests

15.6.1 Protection and control equipment tests – general

Tests on protection and control equipment generally fall under the heading of 'secondary injection'. That is the injection of current and/or voltage, usually into a relay, from the secondary side of the CT or VT, and usually from a test block positioned adjacent to the relay. Given that, the protection and control equipment is now subject to a greater degree of factory testing (with the advent of digital/numerical multifunctional relays) than was previously the case, the extent of secondary injection, on site, has generally been reduced. Within this context, secondary injection seeks to confirm the following:

- That the relay is the correct relay, with the correct version of software.
- That the settings in use, as specified on the relay settings record, see Section 20.3.3, are correct.

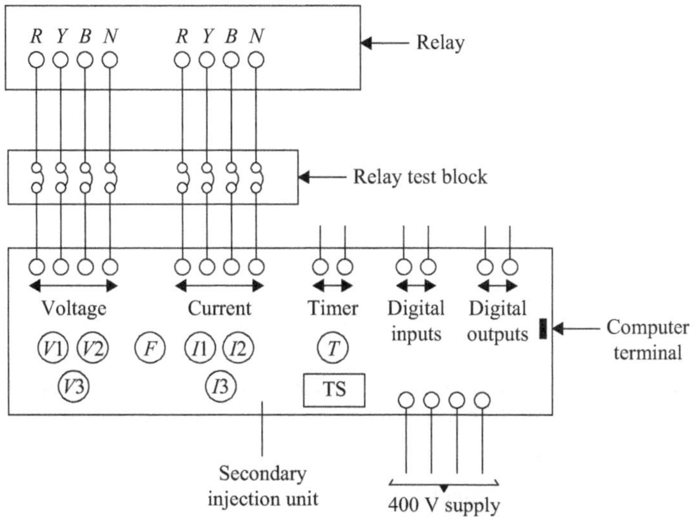

*V*1 = Voltage adjust, *V*2 = Phase(s) selected, *V*3 = Voltmeter
*I*1 = Current adjust, *I*2 = Phase(s) selected, *I*3 = Ammeter
F = Fault application, *T* = Timer, TS = Touch screen

Figure 15.5 Secondary injection unit and test arrangements

Historically, relays were injected with AC current and voltage via a manually controlled secondary injection test unit. This had the advantage of the operator understanding the performance of the relay. However, the modern practice is to inject relays via an auto-test test unit, which provides a statement of the test results on completion, which may be compared with the settings defined on the settings record.

15.6.2 Secondary injection test unit

The historical secondary injection equipment typically comprised a 400-V three-phase supply input and a variable voltage and current output which could be selected to either phase-earth or phase-phase. Fault simulation could be arranged via selector switches for either collapsing a specific phase voltage to a predefined level or suddenly applying current, typically up to 10 A. Fault application also resulted in the closure of contacts to start a timer, which could be stopped by contacts associated with relay operation. All of the preceding was manually controlled for each and every part of the overall tests.

Modern secondary injection units using microprocessor-based technology usually contain all of the above features but additionally have the facility to upload a test programme which drives the outputs of the unit to automatically test the relay. Many have touch screen control. Figure 15.5 illustrates typical features of a secondary injection unit. A disadvantage of auto-testing is that it does not require a specialist operator, as does manual testing, and as a consequence understanding of relay performance may be reduced.

By way of example, an overview of secondary injection requirements for a range of common relays will be provided, which are as follows:

- Trip and auxiliary relays
- IDMT overcurrent relay
- Feeder distance protection
- Biased differential feeder protection

15.6.2.1 Trip and auxiliary relays – secondary injection

Trip relays and other auxiliary relays such as DC supply supervision and circuit breaker trip/close interpose relays are generally powered by 110-V DC nominal battery supplies, although some may be from 48-V DC nominal batteries. Virtually, the only commissioning test required is to inject a DC voltage to find the pick-up and drop-off voltage. Such relays must be capable of operating with the following minimum applied voltage:

- 87.5 V – for 110-V nominal battery
- 32.5 V – for 48-V nominal battery

Some relays, including many trip relays are 'latched' and once operated require a second (reset) coil to be energised to reset the relay. These coils form part of any trip relay reset scheme.

NB: Some commissioning procedures also require the magnitude of the operating current to be measured.

15.6.2.2 IDMT overcurrent relay – secondary injection

These relays range from single-function devices to being one function in a multi-function relay. With reference to Section 10.3.2, secondary injection tests generally comprise the following (NB: The tests are usually on the setting in use and TM = 1):

1. **Minimum pick-up current**
   This is obtained by incrementally increasing the current until the relay commences to operate. With some modern relays, the extent of travel of the relay can be observed on the display unit.
2. **Minimum operating current**
   This is obtained by incrementally increasing the injected current until the relay is fully operated.
3. **Maximum reset current**
   This is obtained by first causing the relay to operate, and then reducing the current until the relay fully resets.
4. **Timing tests**
   A current of typically four times the setting should be injected into the relay (i.e. PSM = 4) and the corresponding operate time compared to the curve in Figure 10.4. For a 3/10 relay the operating time should be 5s. A second test may also be undertaken with the TM set to the setting in use, although it may be sufficient to note that the relay operates at the setting in use, when subject to significant injected current.

Historically, a greater number of tests were undertaken, however, with modern digital or software-controlled relays, the case for more tests is weak.

15.6.2.3 Distance protection – secondary injection

The following comprises the secondary injection tests generally undertaken on distance protection and should be read in conjunction with the sections on distance protection in Chapter10. All tests are at the settings in use.

1. **Reach setting**
   This requires secondary injection to measure the setting impedance for all relay comparators as follows:
   - (i) Zone 1, inject all six comparators (i.e. *R–E, Y–E, B–E, R–Y, Y–B, B–R*)
   - (ii) Zone 2, inject all six comparators as above
   - (iii) Zone 3, inject all six comparators as above, both in the forward and reverse directions
   - (iv) Zone 4, inject all six comparators as above – applies only to blocking schemes.

   The injections usually comprises an injection of current, in to the faulted phase(s), with full voltage applied, and subsequent incremental reduction of the voltage, of the faulted phase(s) in question, until the comparator operates. The injection would usually be carried out at both line angle (typically 75°) and load angle (typically 30°). For quadrilateral characteristics, a measurement of impedance on the resistive axis (i.e. zero degrees) may also be carried out.
2. **Operating time – phase and earth faults**
   A measurement of the operating time for each of the comparators identified above should be undertaken. This would typically be with injections equivalent to an impedance of 80% of the reach setting at the line angle, with a timer being started at the instance of the fault being applied and stopped on comparator operation. The zones 2 and 3 comparators are routed through time delays, see Section 10.10.1.2, and it is the time delays that are measured.
3. **Operating time – close-up three-phase fault**
   This comprises starting a timer coincidental with both removing all voltages from the relay and the injection of current. Ideally, this should be a three-phase current – but the test usually works equally well with single-phase current. The objective of this test is to prove the relay memory feature.
4. **Operating time – switch on to fault**
   This injection comprises the switching on of current only (i.e. with zero voltage) coincident with starting a timer. The timer should be stopped by operation of the switch onto fault feature.
5. **Blocking scheme circuitry**
   Where a blocking scheme is included (or other assisted tripping scheme), secondary injection should be carried out to prove the blocking circuitry.

Distance protection – manual vs automatic testing

With reference to Section 15.6.2, manual injection is usually undertaken by injecting a specific current (e.g. one amp) and manually, incrementally reducing the voltage until the injected comparator operates at the boundary of the characteristic – and comparing this voltage with a calculated voltage, which is derived from the setting impedance.

The calculated voltage is obtained from the expressions in Section 10.10.1.3, where the PPS impedance Z_1 is given by the following:

1. **Phase–earth fault**

$$Z_1 = \frac{V_R}{I_R + [I_R + I_Y + I_B]\{\text{RCF}\}} \tag{15.1}$$

or, on rearranging

$$V_R = Z_1(I_R + [I_R + I_Y + I_B]\{\text{RCF}\}) \tag{15.2}$$

where V_R is the faulted phase voltage (say red phase).

I_R, I_Y and I_B are red, yellow and blue phase currents, respectively, and RCF is the residual compensation factor.

If with reference to expression (15.2), the reach setting, $Z_1 = 10\ \Omega$; the injected (faulty phase) current, $I_R = 1$ A; the RCF = 0.7 (typical value) and $I_Y = I_B = 0$ A (healthy phases),

$$\text{then } V_R = 10[1 + 1(0.7)] = 17.0 \text{ V}$$

Thus, if 1 A is injected into red phase (to neutral) of the relay and the red phase to neutral voltage is incrementally reduced, the comparator should operate at 17.2 V. NB: It is assumed that the angle of the injected voltage, with reference to the injected current, is the same as the line angle of the setting impedance (typically 75°).

2. **Phase–phase fault**

$$Z_1 = \frac{V_Y - V_B}{I_Y - I_B}$$

but for a phase–phase fault $I_Y = -I_B$

so $Z_1 = \dfrac{V_Y - V_B}{2I_Y}$

Again, if $Z_1 = 10\ \Omega$ and $I_Y = 1$ A,

Then $V_Y - V_B = 10 \times 2 = 20$ V

Therefore, with an injected current of 1 A into yellow and blue phases, and the yellow-to-blue voltage incrementally reduced to 20 V, the comparator should operate. Again, the angle between voltage and current should match the line angle of the setting impedance.

Clearly, manual testing of a distance protection is a lengthy process, and in practice, a few hours to undertake. Secondary injection via an auto-test unit is much quicker with a more comprehensive set of tests obtained. For example, a 360° polar plot of the characteristics of each comparator is possible. It is usually necessary to input the value of residual compensation factor into the auto-test unit.

15.6.2.4 Biased differential feeder protection – secondary injection

The principles of biased differential feeder protection are discussed in Section 10.6.1 and illustrated in Figure 10.13. Early relays used electromechanical technology, whereas modern relays are numerical with the functionality derived from algorithms – although both operate on the same principles. By way of explanation, Figure 15.6 illustrates a simplified modern numerical relay but interpreted through a combination of analogue (the biased differential element itself) and digital technology. The biased differential element shows that for one phase only – where in practice there will be three, i.e. one per phase. The digital circuitry must ensure that the comparison of currents from both ends of the feeder, in the biased differential element, takes account of the time delay in the telecommunications (i.e. comms) medium and multiplexor (i.e. Mux), etc. This is arranged by time synchronisation techniques as explained in Section 10.9.1. The relay secondary injection requirements will vary with the equipment manufacturer but generally should encompass the following test regime.

1. **Minimum operating current**
   With reference to Figure 15.6, a red-to-neutral current should be incrementally injected into the relay (in effect injecting between points *X* and *Y* of the biased differential element) until the relay operates. This is the minimum operating current of the biased differential element and should accord with

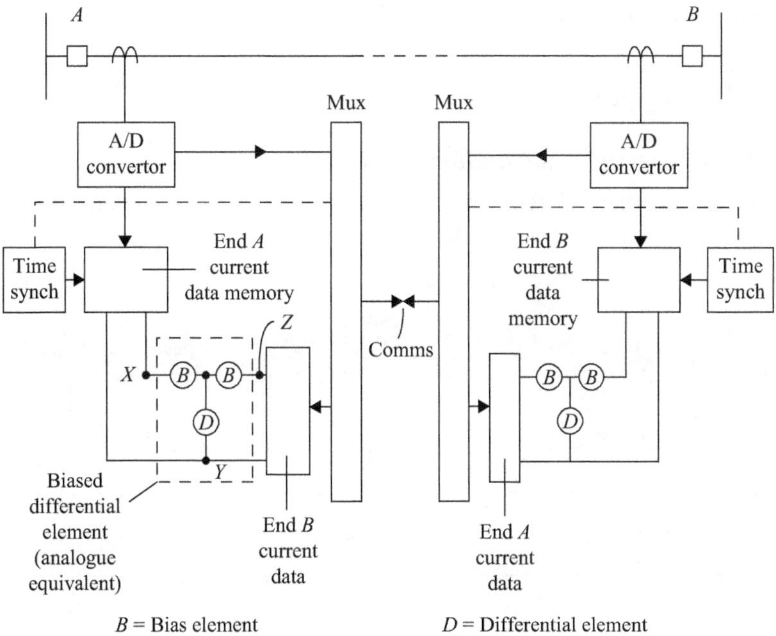

Figure 15.6 Numeric feeder protection working on biased differential principle – simplified

expression (10.7). The test should be repeated for the other two phases. The test also needs to be repeated for current flow from points *Z* to *Y*, and the means by which this is achieved is dependent upon the relay manufacturer.

2. **Bias slope**
   The bias slope is the gradient of the operating characteristic as shown in Figure 10.13(b). Its value is obtained by measuring at two points on the slope, e.g. points *M* and *N* in Figure 10.13(b). The slope *K* is determined from:

   $$K = \frac{I_{D(N)} - I_{D(M)}}{I_{B(N)} - I_{B(M)}}$$

   Where the subscripts in the above expression represent the currents at points *M* and *N*.

   The means by which this test is undertaken will vary with the design of the relay and from manufacturer to manufacturer but requires an injection of both through bias current between points *X* and *Z* in Figure 15.6 coincident with an injection of differential current between points *X* (or *Z*) and *Y*. Should the bias comprise two stages as shown in Figure 10.13, then the point at which the change of slope takes place together with the value of the second bias slope will need to be determined.
3. **Time synchronisation tests**
   The means by which time synchronisation proving tests between the two feeder ends are carried out will vary from manufacturer to manufacturer – some relays have arrangements integral to the relay for this purpose.

15.7 VT supplies

15.7.1 VT wiring – secondary injection

VT wiring from the VT terminal chamber to each item of equipment must be proven prior to energisation. With reference to Figure 15.7, this can be readily

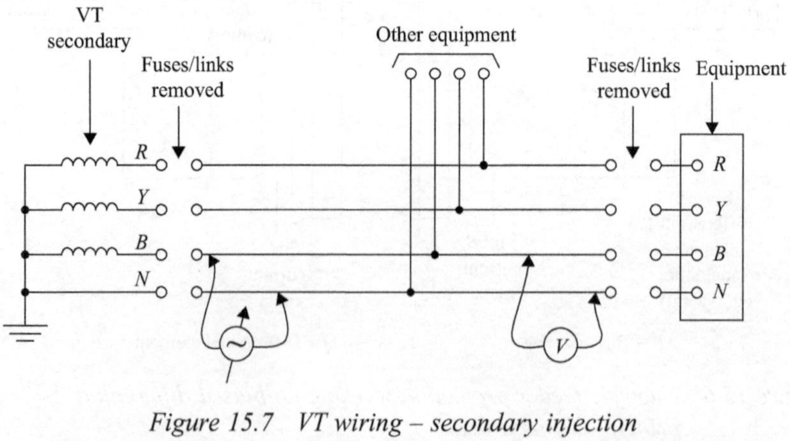

Figure 15.7 VT wiring – secondary injection

achieved by secondary injecting each phase and neutral wire with 63.5 V at the VT, and measuring the voltage on the correct phase only at the equipment fuse/links.

15.8 DC circuitry logic tests

15.8.1 DC logic tests – requirements

DC circuitry logic tests (alternatively termed DC operation tests) usually comprise the proving of DC circuitry as shown on circuit diagrams. Figure 15.8 provides two examples, which are as follows:

1. **DC logic chain**
   With reference to Figure 15.8(a), the task is to confirm that the installed circuitry accords with that shown on the drawing. This is accomplished by exhaustively checking the correctness of all contacts, fuses, links and connecting wires. For example, with reference to the tripping route through contact *A*, both positive and negative checks need to be undertaken as follows:
   (i) **Positive checks**
   Closing of contact *A* operates the trip relay, TR.

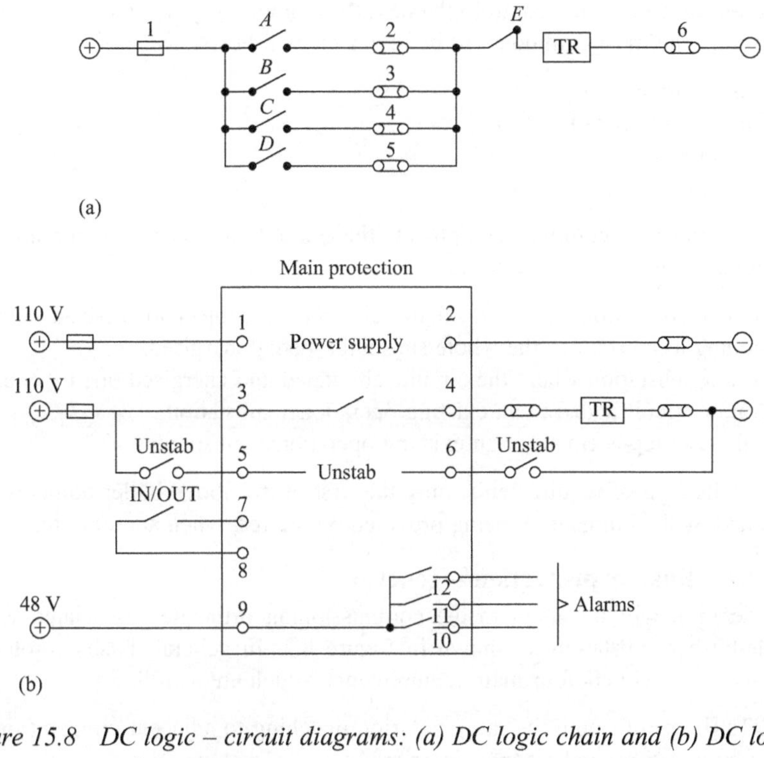

Figure 15.8 DC logic – circuit diagrams: (a) DC logic chain and (b) DC logic – main protection

(ii) **Negative checks**
The trip relay is operated when contact *A* is closed, but reset when:
(a) Fuse 1 or link 2 or link 6 is withdrawn.
(b) Contact *E* is opened.

The above is repeated for the circuitry routes via contacts *B*, *C* and *D*.

2. **DC logic – main protection connections**
With reference to Figure 15.8(b) (in which the relay CT and VT inputs are not shown), this test comprises the proving of the inputs and outputs to/from a main protection relay. This usually requires the test to be carried out in conjunction with relay secondary injection, which is as follows:
(i) Operation of the inputs to confirm correct operation/functionality of the relay
(ii) Operating the relay to drive the appropriate outputs.

These tests usually require the contacts or terminals on the drawing to be ticked (or the wiring coloured) as they are proven as correct, and the signing off of the drawings by the commissioning engineer on completion.

15.9 Protection and control – common equipment

15.9.1 Equipment under consideration

Numerous protection and control schemes are common to many or all circuits in the substation. The most common will be reviewed as follows:

- Busbar protection
- Circuit breaker fail
- Synchronising
- Substation control system

With reference to common equipment, there are four types of commissioning requirements, which are as follows:

- A new substation in which all the circuits are subject to off-load commissioning tests, prior to the whole substation being energised.
- A new substation where the circuits are added and energised one by one.
- A new circuit added to an existing operational substation.
- Full asset replacement of an existing operational substation.

As an indication of requirements, only the first of the above bullet points will be examined, with commentary being provided on the rest when appropriate.

15.9.1.1 Busbar protection – general

Consideration will be given to the commissioning principles associated with a double-busbar substation as shown in Figure 8.2. In general, busbar protection commissioning covers four main requirements, which are as follows:

- CTs [i.e. confirming the ratio, polarity, matching to adjacent circuits (via primary injection) and magnetising curves]

- Relay (i.e. confirming the settings and operating characteristics)
- Disconnectors (i.e. confirming the wiring between the disconnectors and the relay)
- Trip circuits (i.e. confirming the correct alignment of the busbar protection trip outputs with the busbar configuration).

With reference to Chapter 10 and the sections on busbar protection, two main designs of busbar protection will briefly be considered, which are as follows:

- Low-impedance biased differential busbar protection
- High-impedance busbar protection.

Low-impedance busbar protection

With reference to Figure 15.9, the salient features of a typical numeric low-impedance busbar protection are as follows:

- Each bay is equipped with a busbar protection 'bay unit' into which CT wiring and disconnector contacts are connected, together with a trip output to the circuit breaker.
- A 'central unit' is connected to each bay unit by a communications medium (often fibre). The central unit mirrors the substation busbar arrangement and current flows through the busbars. The biased differential relays (one per phase per busbar) are contained in the central unit (as an algorithm) – which when operated (i.e. busbar fault) sends trip outputs to the relevant bay units.
- The central unit usually comprises two fault detecting algorithms – both of which must operate to obtain a trip output.
- The central unit also usually contains a man/machine interface which allows both the bias current and differential current to be viewed, together with the layout of the substation busbars.
- Both the central unit and bay unit usually include integral test facilities.

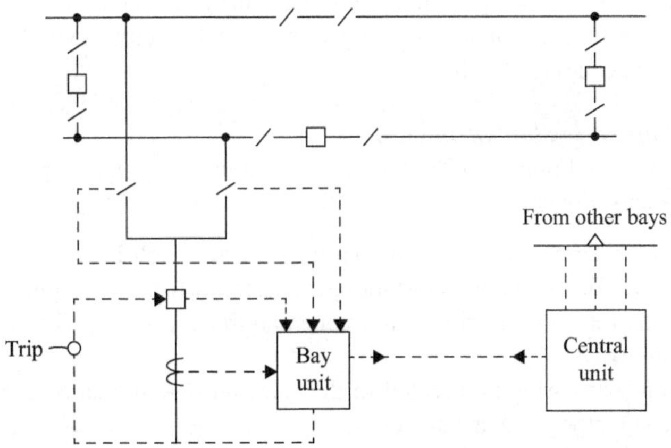

Figure 15.9 Busbar protection – bay and central units

The following commissioning tests would typically be undertaken.

NB: It is assumed that comprehensive factory acceptance tests (FAT) will have been undertaken to confirm the scheme accords with the specification, prior to dispatch to site.

1. **Off-load tests**
   - (i) Disconnector open/close position, observed both at the bay unit and central unit. Where practical this should include observation, at the central unit, of all disconnectors that can be selected to the same busbar.
   - (ii) CT magnetisation curves of all CTs on site associated with busbar protection.
   - (iii) Primary injection of all CTs for each circuit on site associated with busbar protection to prove ratio, polarity and matching. The ammeter measuring the secondary current should be placed at the bay unit.
   - (iv) Primary injection of each circuit against an adjacent circuit to prove CT matching between circuits. This should include a test of busbar protection operation and stability, with readings observed in both the bias and differential elements at the central unit.
   - (v) Secondary injection from each bay unit to confirm the minimum relay operating current and bias slope, similar to that described in Section 15.6.2.4. NB: Manufacturer-specific requirements will need to be adhered to undertake this test.
   - (vi) Operation of each relay differential element (within the central unit) to confirm trip outputs to the appropriate circuits, for each busbar selection.
2. **On-load tests**
   - (i) The on-load tests should prove that the relay is in a stable condition when subject to normal load flows through the busbars. This can be achieved by observing the current in both the bias elements (showing appropriate load current as determined by the algorithm), and the discrimination elements (showing virtually zero current) in the central unit.
   - (ii) An on-load disconnector changeover of each circuit should be undertaken to prove relay stability.

High-impedance busbar protection

With reference to Figure 10.28, the salient features of high-impedance busbar schemes are as follows:

- The scheme comprises a discrimination zone for each busbar and a common check zone for the whole substation, with circuit tripping requiring operation of both the relevant discrimination zone and the check zone (i.e. a two-out-of-two system).
- CT secondary wiring is directed to the relevant discrimination zone buswires via disconnector auxiliary switches.
- Trip signals from the relays are directed to the relevant circuit breakers by disconnector auxiliary switches, so reflecting the busbar running arrangements.

The following comprises typical commissioning tests:

1. **Off-load tests**
   (i) Disconnector auxiliary contact alignment tests to ensure the auxiliary contacts make prior to the main contacts reaching the pre-arcing distance.
   (ii) CT magnetisation curves of all CTs on site associated with busbar protection.
   (iii) Primary injection of the CTs associated with each circuit to confirm ratio, polarity and matching with ammeters placed across the high-impedance relays.
   (iv) Primary injection of each circuit against an adjacent circuit to prove CT matching between circuits. This should include a test of busbar protection operation and stability, as indicated by ammeter readings across the high-impedance relays.
   (v) Secondary injection of each high-impedance relay to confirm the setting. Secondary injection of the each setting resistor and metrosil should also be undertaken to confirm the setting and characteristic, respectively.
   (vi) A schedule of tests should be prepared and undertaken that confirm that the operation of each high-impedance relay trips the appropriate circuits, across the range of busbar selection permutations.
2. **On-load tests**
   (i) The on-load tests should prove that the relay is in a stable condition when subject to normal load flows through the busbars. This can be observed by placing voltmeters across each high-impedance relay (and stability resistor where applicable). For stability, the measured voltage should be negligible.
   (ii) An on-load disconnector changeover of each circuit should be undertaken to prove relay stability, i.e. negligible reading on a voltmeter connected across each high-impedance relay (and stability resistor where applicable) during operation of the disconnectors.

Busbar protection asset replacement

Asset replacement of busbar protection frequently requires the replacement of a high-impedance scheme with a low-impedance scheme (numeric relay). One possible way of achieving this, whilst maintaining the busbar protection operational is to transfer the high-impedance check CT to the low-impedance scheme – as the latter only requires one CT input – and short out the check zone in the high-impedance scheme. Thus, the substation is protected by the high-impedance discrimination zones. The low-impedance scheme can then be commissioned, and the tripping circuitry subsequently changed over from high-impedance scheme to low-impedance scheme, circuit by circuit.

15.9.1.2 Circuit breaker fail

Circuit breaker fail protection is applied to the 400 and 275-kV networks with limited application at 132 kV. Earlier installations were hard wired with discrete relays as shown in Figure 10.29, whereas new installations are invariably multifunction programmable numeric relays, and often incorporated into the same relay as busbar protection – as much of the circuitry is common to both. With reference to Figure 10.29(a), the current check relay, *CCK*, and the timer, *TD*, are usually positioned with the bay equipment with the remaining circuitry incorporated into the central relay. The following are typical of the commissioning tests to be undertaken.

1. **Off-load tests**
   (i) As with busbar protection, circuit breaker fail protection (i.e. the common relay) will have been subject to comprehensive FAT, prior to installation on site.
   (ii) Secondary injection of each current check relay, *CCK*, and timer, *TD*, should be undertaken to confirm the settings. The outputs should be observed in the central relay.
   (iii) Disconnector auxiliary contact inputs to the central relay need to be confirmed.
   (iv) A schedule of tests should be prepared and undertaken across the range of busbar selection permutations that confirm that when the circuit-specific *CCK* and *TD* relays are operated, the correct circuits are back tripped.
2. **On-load tests**
   (i) With the circuit on load, the current in the *CCK* relays should be measured and confirmed as correct.
   (ii) If the off-load schedule of tests identified above is comprehensive, the case for a repeat with the circuit on load is weak. However, to provide assurance, and if operational conditions permit, a limited number of tests should be undertaken. This may require injection of the *CCK* relay to cause operation of the scheme if the load current is not sufficiently high.

15.9.1.3 Synchronising scheme

The fundamentals of a synchronising scheme are described in Figure 10.37. It comprises a synchronising relay and synchronising selection circuitry per circuit, and common voltage selection circuitry. The latter is fed by all the VTs in the substation and operates at 63.5-V AC (i.e. VT secondary voltage). It is arranged to mirror the HV busbars with reference to the circuits connected to each busbar. Commissioning tests typically comprise the following:

1. Secondary injection of each synchronisation relay to confirm the check synchronise and power system synchronise settings.
2. The secondary injection of 63.5-V AC at each VT secondary terminal (with the fuses removed to avoid back-energisation of the VT) and confirmation with a voltmeter that the correct voltage appears at all relevant terminals. This will require functioning of the synchronise select switch.
3. DC logic tests to prove the synchronise select switch circuitry.
   (i) Following on-load phasing out tests, the circuit breaker should be closed both in the dead-line (over-ride) and check synchronise modes.

15.9.1.4 Substation control system

With reference to Chapter 10 and Figure 10.40, a substation control systems (SCS) is an integral part of a SCADA system. Prior to dispatch to site, the SCS will usually have been subject to comprehensive FAT to prove the inputs and outputs including the MMI, logger, communications link and the correct nomenclature/title of all data as seen on the screen of both the MMI and the logger.

The site commissioning tests will require verification of all inputs and outputs between the SCS and the equipment, with salient points as follows:

- Connections between the SCS and equipment may be either copper or fibre sometimes employing a local area network (LAN).
- Alarms should be initiated at source.
- Metering should be subject to secondary injection of the metering transducers.
- Indications will require confirmation of circuit breakers and disconnectors, etc. both in the open and closed positions. Motorised disconnectors also usually have a transitional stage between open and closed, often termed DBI (don't believe it). Tap change position indicators also need to be proven.
- Control signals emanating from the SCS will require the opening and closing of all equipment (in the de-energised state) together with the proving of in/out functions (e.g. blocking channel).

A decision needs to be taken on the extent of proving the alarms/metering/indications/control back to the control centre, taking into account the rigorous quality-assurance procedures in place, for confirming the compatibility of the SCS and control centre interfacing software. With a large substation, confirmation of all facilities can be both time consuming and resource intensive. It is usually the case that all controls, indications and metering are confirmed at the control centre but only a representative sample of alarms. In some instances, substation equipment simulators may be employed. Prior to circuit energisation, the status of all SCS inputs and outputs (including any standing alarms) will need to be confirmed both at the HMI and the control centre.

15.10 Auto-switching

15.10.1 Auto-switching – general

Auto-switching includes auto-reclose (AR), auto-opening, auto-isolation and auto-close. By way of example, a scheme comprising delayed auto-reclose (DAR) and auto-isolation will be considered. Prior to delivery to site, most auto-switching relays (certainly the more complex) are usually subject to FAT. Following site installation and with reference to the sections in Chapter 10, there are three main areas relevant to the site commissioning of auto-switching schemes, which are as follows:

- Relay inputs and outputs
- Off-load simulation tests
- On-load tests.

These will be briefly reviewed.

15.10.1.1 Relay inputs/outputs

With reference to Figure 15.10(a), verification of all the connections shown needs to be undertaken. With numeric relays, the inputs can usually be read on an MMI interface. Outputs can usually be provided either by injecting inputs, or in some instances, integral test facilities are provided.

15.10.1.2 Off-load simulation tests

These tests are required to prove both relay functionality and equipment interconnection. They require the availability of all connected equipment and are therefore undertaken as some of the final commissioning tests. The tests comprise

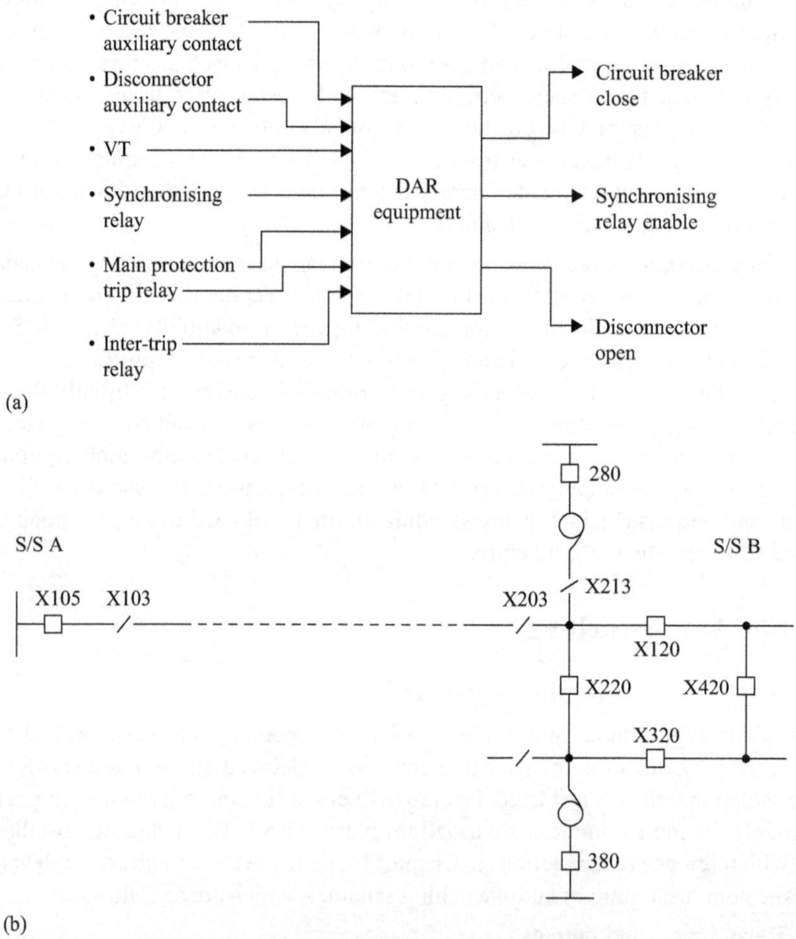

Figure 15.10 DAR and auto-switching considerations: (a) DAR inputs/outputs at S/S A and (b) network under consideration

the simulation of in-service conditions by closing the de-energised plant and simulating VT inputs by the use of switchable test voltages (63.5-V AC).

The test method comprises the operation of trip relays coincident with switching the VT supplies, so simulating a network fault, and observing and timing the tripping of the circuit breaker and subsequent DAR or auto-isolation, etc. All permutations of fault conditions should be simulated in accordance with a formally prepared schedule of tests. By this means, a high level of confidence can be obtained (and deficiencies identified and corrected) prior to the equipment entering service, thereby reducing the number of on-load tests.

15.10.1.3 On-load tests

The on-load tests are carried out with the equipment energised and in service. They are usually the final commissioning tests undertaken. They comprise manually operating the protection system (often involving synchronisation of activities with the remote end of the circuit by telephone) to cause the circuit breakers to trip and allow the auto-reclose or auto-isolation equipment to demonstrate their specified functionality. These tests should be undertaken in accordance with a schedule of tests. Figure 15.10(b) shows a typical part of a network with DAR and auto-isolation equipment installed at both circuit ends – and Figure 15.11 shows an example of the schedule format for one specific test, namely a line transient fault.

A more complex test would be that of a line permanent fault, where upon the DAR of *X105* at *S/S A*, the line fault is still present. In this instance, *X105* would trip again, and disconnector *X103* would open, and a persistent inter-trip signal sent from *S/S A* to *S/S B* (typically 60-s duration) to indicate to *S/S B* that *S/S A* had detected a persistent fault. *S/S B* would then instruct disconnector *X203* to open, followed by DAR of circuit breakers *X120* (deadline), *X220* (check synchronise) and *280* (check synchronise) to restore all healthy equipment to operational service.

		On-load test schedule **Line transient fault**
Stage	**Time**	**Event**
1	0	S/S A–S/S B feeder fault inception
2	40 ms	S/S A and S/S B protection operation and inter-trip send
3	80 ms	S/SA X105 trip and inter-trip receive S/S B X120, X220 and 280 trip and inter-trip receive
4	10 s	S/S A and S/S B trip relays reset
5	15 s	S/S A DAR (deadline)
6	17 s	S/S B X120 DAR (check synchronise)
7	19 s	S/S B X220 DAR (check synchronise)
8	21 s	S/S B 280 DAR (check synchronise)

Figure 15.11 Auto-switching (DAR) test schedule – example

15.11 HV equipment tests

15.11.1 HV equipment commissioning – general

The range of HV equipment is considerable and therefore the commissioning requirements of only four main and commonly used but diverse types of equipment will be examined as indicative of requirements. These are as follows:

- Circuit breakers
- Power transformers
- HV cables
- OHL

NB: Essentially, only off-load commissioning tests will be covered. On-load commissioning tests usually comprise energisation and operation tests and, in some instances, a period of observation of (apparent) satisfactory performance.

15.11.1.1 Circuit breakers

Commissioning tests associated with circuit breakers generally include the following:

1. Interrupter contact resistance
2. Alignment of auxiliary switches
3. Trip and close timing tests
4. Interruption medium tests
5. Pre-energisation operating tests.

1. **Interrupter contact resistance**
   This comprises a measure of resistance across each set of interrupter contacts, when closed, using a micro-ohmmeter. The measured values should be compared with the manufacturers quoted values and typically would be a few tens of micro ohms.
2. **Auxiliary switches**
   Auxiliary switches driven by the circuit breaker mechanism are usually provided in a variety of types which may make or break either coincident with, or in advance of, or after, the main contacts make or break. They are usually proven with the assistance of a 'slow close jack' which has to be connected to the circuit breaker mechanism and enables the latter to be incrementally moved. Where it is difficult to carry out a direct comparison with the interrupter position, the comparison is made to a defined point in the mechanism movement. Identification of auxiliary contact make/break is by means of observing the resistance across the contacts (e.g. the transition from open-circuit to short-circuit), usually with a multi-metre.
3. **Trip and close timing tests**
   This is an important commissioning test as it proves the whole electromechanical process from trip/close initiation to interrupter separation/closure respectively. The means by which the test is carried out varies from manufacturer to

manufacturer and usually involves a specialist timing-measuring device. For a modern SF6 circuit breaker, typical times would be:

(i) **Trip test**
From trip initiation to main contacts part = 30 ± 3 ms – with a contact spread of 5 ms.

(ii) **Close test**
From close initiation to main contacts close = 100 ± 5 ms – with a contact spread of 5 ms.
The test results should be compared with the manufactures specification.

4. **Interruption medium tests**
The four main common circuit breaker interrupting (and insulating) mediums are oil, air, vacuum and SF6 – with SF6 and vacuum currently being the most widely used. Tests are not usually carried out on interrupters that are factory sealed which mostly applies to vacuum and some SF6. The following tests apply to SF6 gas used in either GIS or dead tank SF6 circuit breakers:

(i) **SF6 leakage detection**
This test takes place following filling of the circuit breaker containment vessel with SF6 gas and generally consists of a hand-held probe which is moved around the seals.

(ii) **SF6 dew-point**
The dew-point test is usually carried out by passing the SF6 gas through a dew-point cell or moisture-measuring device. The results must be compared against the manufacturer's specification.

(iii) **Impurity levels**
The SF6 gas must be tested for impurity level limits as defined by the manufacturer, which are as follows:

(a) **Oxygen content**
The oxygen content may be assessed by passing the SF6 through an oxygen paramagnetic portable analyser.

(b) **Acidity content**
The acidity content can be determined by passing the gas through a chemical analysis unit manufactured for this purpose.

(c) **Oil content**
Oil impurities may arise during gas handling. The level may be determined by passing the gas through a chemical detector tube unit.

It is worthy of note that SF6 gas is subject to impurities when in operational service arising from interrupter arcing products.

(iv) **Pressure gauges**
SF6-filled equipment is usually fitted with pressure gauges, which must be confirmed as satisfactorily calibrated. The gauges provide both alarms (low SF6 gas), and at critically low levels of SF6, circuit breaker trip and close lockout interlocks.

(v) **SF6 density**
The correct filing of SF6 equipment requires a density level related to the temperature at which it is filled. Gas density is proportional to gas pressure, and therefore to determine whether the equipment is adequately filled, it is important to check the requisite pressure for the ambient filling temperature. This is usually achieved with reference to charts.

5. **Pre-energisation operating tests**
Following completion of all other tests and prior to energisation, the circuit breaker should be subject to a number of open/close operations, typically 20 or as specified by the manufacturer. This is to check that there are no apparent difficulties or deterioration with the circuit breaker.

15.11.1.2 Power transformer

The following comprises commissioning tests common to most power transformers:

1. Winding resistance
2. Vector group
3. Ratio, magnetising current and tap changer continuity
4. Buchholz
5. Winding temperature
6. Oil tests

The above will be briefly reviewed:

1. **Winding resistance**
A measure of the resistance of each winding should be undertaken, using a measuring device such as a Kelvin bridge. Measurements are usually taken on the transformer lowest, nominal and highest tap positions.
2. **Vector group test**
This classical transformer commissioning test is used to determine the phase shift between HV and LV winding to obtain the vector group. With reference to Figure 15.12, the HV terminal *A*2 is connected to the LV terminal *a*2, to provide a common point of potential. A three-phase 400-V supply is then connected to the transformer HV terminals. The test requires the measurement of the voltages between all terminals from which a table of results is populated (Figure 15.12(b)) and a vector diagram constructed (with compasses and ruler), as illustrated in Figure 15.12(c).
3. **Ratio, magnetising current and tap change continuity**
This test comprises injecting a 400-V three-phase supply into the transformer HV winding and measuring the current in each phase. Voltmeters are connected across the secondary windings, phase to neutral for a star-connected winding, and phase to phase for a delta-connected winding. The objective of the test is to measure the transformer ratio on all taps and, by observing the HV magnetising current and LV voltages, ensure there is continuity during the full range of the tap changer, i.e. the selector and diverter switches are causing neither an open-circuit nor a short-circuit. For each tap position, the HV and LV voltage and LV magnetising current should be recorded.

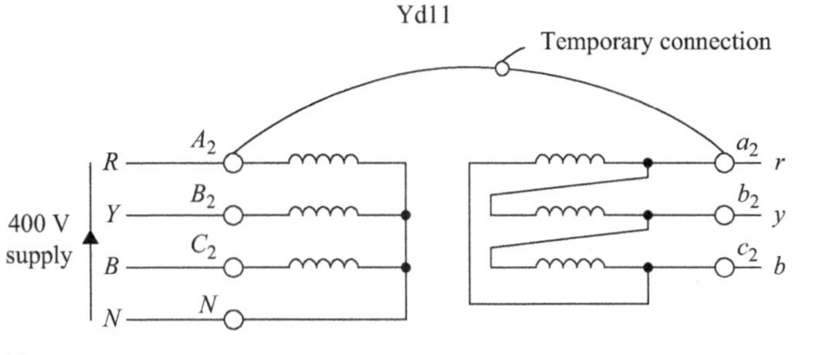

(a)

	R	Y	B	N	r	y	b
R	×	×	×	×	×	×	×
Y		×	×	×	×	×	×
B			×	×	×	×	×
N				×	×	×	×
r					×	×	×
y						×	×
b							×

× = Duplicated result

(b)

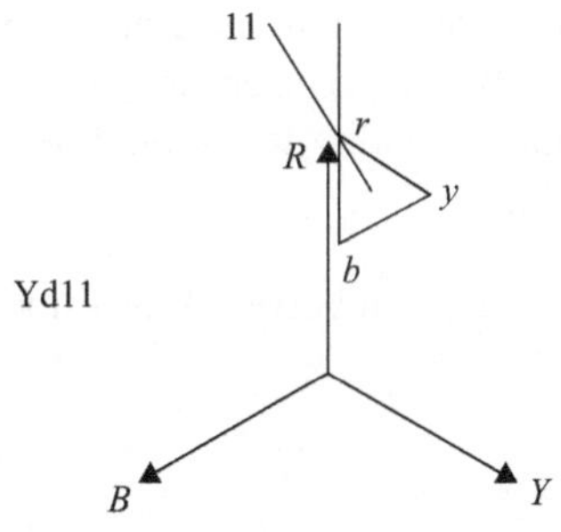

(c)

Figure 15.12 Vector group test: (a) test arrangements, (b) table of results and (c) graphical results

4. **Buchholz relay**
   The Buchholz relay is positioned in the pipework joining the transformer tank to the conservator and is described in Section 9.11.1.3. The following inspection and tests are usually undertaken:
   (i) The angle of inclination of the pipework is critical and must be checked. It is generally required to be at 3°–10° to the horizontal as defined by the manufacturer.
   (ii) Slowly inject air from a pressure vessel into the Buchholz chamber via the test connection until the Buchholz relay operates, so closing the

alarm contacts. The volume of air in the Buchholz inspection window should be noted.

(iii) Refill the Buchholz chamber with oil, and inject the chamber, via the test connection, with bursts of air, at increasing pressure until the Buchholz trip (surge) relay operates. Record the pressure at which this occurs.

5. **Winding temperature relay**

With reference to Section 9.11.1.3, the traditional method for testing a winding temperature relay is as follows:

(i) Withdraw the winding temperature capillary tube from the top of the transformer tank and immerse it in a small reservoir of oil. Heat the oil reservoir and monitor the temperature rise with a thermometer and check that the same temperature is indicated on the winding temperature instrument. The discrepancy ideally should not exceed 1°C.

(ii) With the capillary tube replaced back in the transformer tank inject current, via the winding temperature CT circuit, equal to transformer full load current, into the winding temperature heater. The temperature indicated on the winding temperature instrument should be recorded both at the outset of the injection and typically 30 min later (as the manufacturer recommends) and compared with the manufacturer's test results.

6. **Oil tests**

The integrity of the transformer oil is obviously essential to the well-being and successful operation of the transformer. A sample of oil should be taken from each of the main tank, the selector tank and the diverter tank and sent to a laboratory for analysis. If tests have to be carried on site, then both an electrical strength breakdown test and an acidity test should be undertaken.

15.11.1.3 HV cables

HV cable commissioning tests usually involve all, or some, of the following:

1. HV pressure test
2. Partial discharge (PD) test
3. Sheath insulation test
4. DC resistance test
5. Sequence impedance test (cross-bonded cables)
6. Cross-bonding test

1. **HV pressure test**

It is usual to carry out a HV pressure test when HV equipment is subject to significant on-site installation work. Therefore, instances may arise when it is deemed essential to carry out HV pressure tests on transformers and switchgear, however, in all instances, it is usual to carry out such a test on HV cables.

Historically, HV pressure tests were undertaken with the application of a DC voltage, since AC would result in a relatively large current flow into the cable capacitance, making the size of the test equipment impractical. However, XLPE cables may suffer degradation problems as a result of DC (and trapped charge), and as a result, AC test equipment using resonant principles has been developed. Using this technique, the current required by the test device is small but parallel-connected reactors, which take an equivalent current to the cable at resonant frequency, are employed. In some instances, series-connected reactors may also be employed, and frequencies to create resonance other than 50 Hz may be used (typically 70–300 Hz). This test is an all or nothing test since unlike the DC test, leakage currents are not measured. The applied test voltage is usually chosen to be just above system nominal voltage (depending upon the technical standard that is applied) and usually lasts for 1 h.

2. **Partial discharge test**
   PD is an electrical discharge that occurs in an area of insulation. It can be caused by discontinuities or imperfections in the insulation system. PD testing provides an indication of the deterioration of the insulation and incipient faults. PD can arise at any point on a cable installation particularly in cable sealing ends and joints. The test is usually carried out on XLPE installations, which are generally more prone to PD.

   PD can be observed by connecting a capacitor divider (e.g. capacitor voltage transformer) to the cable termination and observing bursts of PD on an oscilloscope. An alternative is to connect a CT to the earth connection of the cable and again monitor current flow to earth (passing through the CT) arising from PD on an oscilloscope. Some cable installations have PD-monitoring devices permanently connected.

   The PD test is usually undertaken at a voltage between 1.0 and 1.5 rms nominal. The test observes the number of pulses of PD on a monitoring device which is compared against manufacturer's data. Care must be taken to avoid background interference being inadvertently picked up by the test equipment. The test may be combined with the HV pressure test.
3. **Sheath insulation withstand**
   This test should be applied between the cable sheath and earth and is a measure of cable sheath insulation. The test voltage is usually 10-kV DC for 1 min. Maximum leakage current is not usually expected to exceed 100 mA.
4. **DC resistance test**
   The DC conductor resistance should be taken, using a suitable resistance bridge device. The temperature at which the test is carried out should be recorded.
5. **Sequence impedance measurement**
   This test is usually undertaken on cross-bonded cables. It comprises an injection of three-phase current, and by using the techniques described in Section 4.2.1, the PPS and ZPS impedance is determined. A wattmeter will be required in the circuitry to determine the resistive and reactive components of impedance. A current of 20–100 A is usually injected.
6. **Cross-bonding test**
   Proving the cross-bonding arrangements can be accomplished by injecting a three-phase balanced current (typically 100 A) down the cable conductors and measuring the sheath voltages and currents to earth at the end of each sheath section – these should be negligible. Some test regimes require the bonding links to be repositioned to prove incorrect bonding – which are then corrected.

15.11.1.4 OHL

OHL require relatively little with reference to commissioning and generally consist of the following requirements:

- Measurement of joint resistances using a micro-ohmmeter. Satisfactory joint resistances are usually in the range 10–100 μΩ.
- Measurement of tower footing resistance. This is generally undertaken with an earth-testing device similar to that described in Section 15.3.1.
- OHL inspection to confirm that all colour plates, flag brackets, danger plates, anti-climbing guards, etc. are fitted that all portable earths have been removed and that there is no apparent infringement of safety clearances.

OHL require phasing out tests to be undertaken at one of the connected substations.

15.12 Substation battery systems and DC supplies

15.12.1 Battery system and DC supplies

With reference to Chapter 10 and the sections on batteries and DC supplies, commissioning requirements usually comprise the following:

1. **Battery discharge tests**
   There are two types of battery discharge tests: low current and heavy current. The low current test comprises of switching off the battery charger and discharging the battery into a resistance network at either the 6-h rate or the 10-h rate, with the terminal voltage periodically measured such that a voltage discharge profile can be obtained. The results should be compared with the manufacturers test data. The high current test is more applicable to batteries that carry out a tripping function thereby requiring a high current for a short duration. This comprises of discharging at, say, the 1-h discharge rate (e.g. typically 120 A for a 200-A-h battery) into a resistor network for approximately 90 s, and comparing against the manufacturers test data.
2. **Alarm relays**
   This comprises tests to check the calibration of the high/low-volt alarm relays and the battery earth fault relay.
3. **Trip coil voltage**
   As described in Chapter 10, it is essential to ensure that under the most onerous circuit-breaker-tripping conditions (usually a busbar-fault tripping multiple circuit breakers) that the voltage measured at the trip coil does not fall below the limit of 87.5 V. This can be confirmed by the positioning of resistor banks, both in place of, and of equal value to, the trip coil in question, and at the battery terminals to simulate the other circuit breakers to be tripped. At the end of the battery 6-h discharge period with the standing load connected, the resistor banks are simultaneously switched on to simulate circuit breaker tripping (for a busbar fault), and the resultant voltage monitored on an oscilloscope. The voltage should be observed for at least 100 ms which represents typical circuit-breaker-tripping time.

15.13 Loadability

15.13.1 Loadability test

The 'loadability' test is virtually the last test to be undertaken prior to energisation. The necessity for this test increased with the introduction of multifunction relays – where due to the very large number of settings, some relays inadvertently entered service without disabling all of the settings functions that were required to be disabled – and this was not detected by traditional commissioning tests. The error was usually not revealed until the circuit load increased to a magnitude which was sufficient to cause operation of the function in question, with resulting (and unwanted) tripping of the circuit – sometimes causing loss of supply. The loadability test is undertaken to detect unwanted functionality, and to be assured that all relay settings are set above circuit maximum load current.

The test usually comprises the injection of three-phase balanced currents into the CTs from the CT secondary terminals and therefore into every connected relay – although it is usual to exclude high-impedance circulating current relays. A current equal to circuit maximum load current (as declared on the rating schedule) should be injected. With distance protections, three-phase balanced voltage should also be applied leading the injected current by, typically, 30°. A successful test is one whereby all relays remain stable during the test. Consideration should be given to ensuring the test duration does not exceed any equipment current rating.

15.14 Energisation and on-load tests

15.14.1 Circuit energisation and on-load tests – overview

The circuit energisation and on-load tests consist of not only the final stage of commissioning but also the final technical stage of the project. This comprises the energisation of the newly constructed equipment from the in-service power network itself, and the subsequent carrying out of on-load commissioning using network voltages and currents to confirm satisfactory performance of the equipment. Salient considerations comprise the following:

- Method of energisation
- Commissioning protections
- Energisation and soak tests
- Phasing out and synchronising
- Metering
- On-load tests

The above will be briefly examined.

15.14.1.1 Method of energisation

The energisation strategy should consider the following:

- The volume of equipment to be energised simultaneously. Within this context and with reference to a new substation, the merits of circuit by circuit

energisation compared to multiple circuit energisation or even energising the whole substation at one go should be considered.
- It is usual to energise new equipment from a circuit whose circuit breaker and protection is already proven, often utilising a bus section (or bus coupler) circuit breaker selected to an isolated busbar. Advance consideration should be given to this requirement.
- Consideration should be given to the location of sources of generation when energising a circuit. It is preferable not to place generation at risk should a fault occur on energisation.
- Consideration should be given to the source of VT supplies for the purpose of phasing out tests.

15.14.1.2 Commissioning protections

As noted above, the energisation of new equipment for the first time must be from a proven circuit breaker and proven protection. The following considerations are relevant:

- It is common practice to energise new equipment from a bus section or bus coupler circuit breaker selected to the same busbar as the new equipment. Protection is often provided by a high-set overcurrent (HSOC) relay installed on the section/coupler circuit for commissioning purposes. Mesh substation circuit breakers may alternatively be used or other circuit breakers as appropriate.
- When energising transformers for the first time, the commissioning protection must be both immune to magnetising inrush current, whilst at the same time, being able to detect a fault on the transformer LV side.
- Instances may arise where it is convenient to use a proven distance protection as the commissioning protection with the zones 2 and 3 time delays reduced to zero to provide instantaneous tripping should a fault occur.
- Calculations should be undertaken to ensure that the commissioning protection can detect all possible faults under minimum plant conditions that may arise at the time of commissioning – and a suitable setting determined.
- Commissioning protection relays should be secondary injected at the setting in use prior to circuit energisation.

15.14.1.3 Energisation and soak test

With reference to the example arrangement illustrated in Figure 15.13, in which circuit *A* (and connected circuit breakers) is a new circuit to be commissioned, energisation arrangements would typically be as follows:

- With circuit breaker *A*2, *C*1 and *D*1 open, close circuit breaker *A*1
- Close circuit breaker *C*1 (already proven) to energise circuit *A* and carry out a soak test (see below)
- Open circuit breaker *A*1 to de-energise circuit *A*
- Close circuit breaker *A*2

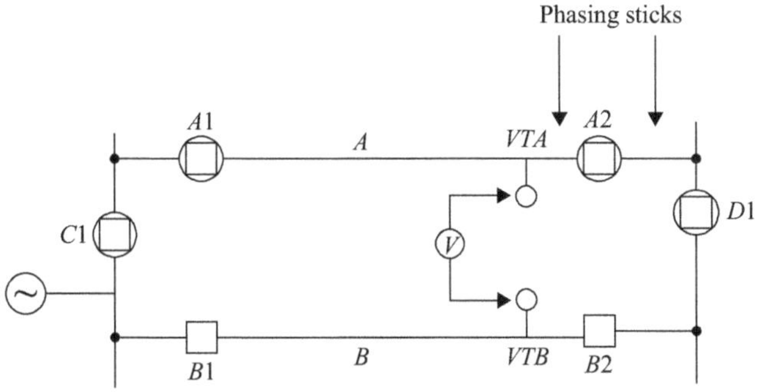

Figure 15.13 Energisation and phasing out test

- Close circuit breaker *D*1 (already proven) to energise circuit *A* and carry out a soak test
- Open circuit breaker *A*2 to de-energise the circuit.

Initial energisation of a circuit is usually followed by a soak test of HV equipment, prior to the phasing out tests and placing on load. The soak test is a visual observation to ensure substation-based equipment does not appear to be in a state of distress when subject to network voltage. It typically lasts up to 1 h. For transformers, the tests may last longer (e.g. 2 h) to observe that no gas accumulates in the Buchholz chamber.

15.14.1.4 Phasing out and synchronising

Phasing out is the process of verifying that the phase connections of the new equipment are correct, and in-phase with the existing power network, prior to closing a circuit breaker to parallel the new equipment with the existing network.

With reference to Figure 15.13, consider circuit breakers *C*1, *A*1 and *D*1 closed and circuit breaker *A*2 open. At 33 kV (and below), it is often acceptable, and practical, to use phasing sticks for phasing out. These comprise two long-insulated probes connected at the base by a conducting bond and fitted with an indicating voltmeter. Each phase across the circuit breaker is tested against each other phase. In-phase connections would indicate 0 V, and out-of-phase connections, 33 kV. The test must be undertaken with considerable care with reference to safety and rigorously accord with the manufacturer's instructions for the phasing sticks.

Above 33 kV, it is necessary to undertake phasing out tests via the use of VTs. With reference to Figure 15.13, the requirements are as follows:

1. With circuit breaker *A*1 open, close circuit breakers *D*1 and *A*2. This energises *VTA* (a new VT) from a proven source of supply.
2. Check phase rotation of *VTA* at the secondary terminals, followed by the voltage of each phase to earth (63.5 V) and between phases (110 V).
3. Using a voltmeter (as shown), carry out a phasing out test by measuring the voltages between *VTA* (new VT) and *VTB* (already proven VT). For correct phasing, the corresponding phases, e.g. red phase to red phase should indicate

0 V and between phases, e.g. red phase of *VTA* to yellow phase of *VTB* should indicate 110 V. This test proves the phasing of *VTA* by using an already proven source of supply.

4. Open circuit breaker *A*2 and close circuit breakers *C*1 and *A*1 to energise *VTA* from the new circuit.
5. Carry out phasing out tests between *VTA* and *VTB* as stated in 3 above. Again for correct polarity, 0 V should be indicated between corresponding phases and 110 V across phases. This test proves the correct phasing of the circuit *A* conductors.

The phasing out tests are now complete.

With the VT secondary fuses/links now replaced, a check must be made to ensure that the correct voltages appear at each and every item of equipment connected to the VT, including the synchronising circuitry.

The synchronising scheme can now be employed to check synchronise close circuit breaker *A*2 – which upon closing places the circuit breaker on load.

15.14.1.5 Metering

Following the placing of the circuit on load, polarity sensitive metering (i.e. MW, MVAr) should be proven – and when proven, assist in the proving of polarity conscious protection.

Figure 15.14 illustrates the metering convention for the current flow considered as flowing out of the busbars. The new unproven metering should be proven against proven metering on another circuit, i.e. arranging the current to flow though both circuits is series.

NB: One circuit current will be into the busbars and the other out of the busbars. If this cannot be achieved then wattmeters, voltmeters and ammeters will be required to determine the phase angle of the current.

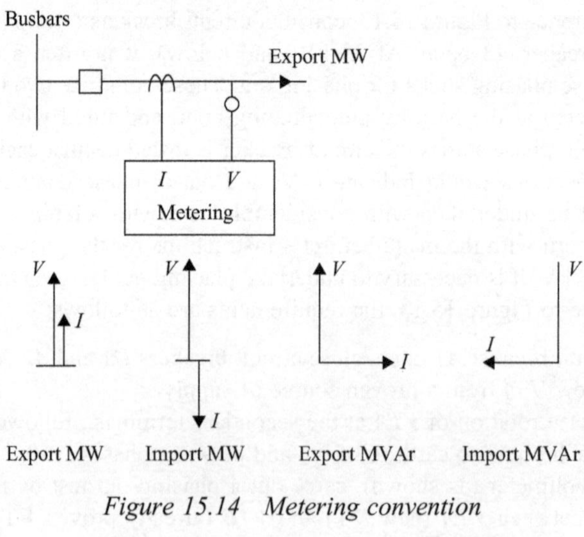

Figure 15.14 Metering convention

15.14.1.6 On-load tests

When carrying out on-load tests, it is advantageous to have as much load current as possible. In some instances, the power network may need to be reconfigured to maximise the load current. On-load tests comprise the following:

- Confirming the correct functioning and direction of transformer fans, pumps and tap changers from the various sources of AC supply.
- As noted above, a check of all VT supplies at all fuses and links, a check of correct phase rotation, correct phase connection and correct voltage magnitude. It is worthy of note that some VT do not have the neutral earthed but rather the yellow phase.
- Confirmation of CT ratios, and that each item of equipment is supplied from the correct ratio of CT. This is accomplished via comparing the current in the secondary wiring, measured either in the relay or via a clip-on ammeter, with the current in the HV circuit as indicated by the metering.
- Check of directional and distance protection directionality (i.e. polarity), in accordance with manufacturer's instructions.
- Check of unit protection stability as per manufacturer's instructions. It is worthy of note that feeder capacitance current will appear in the relay as operating current and must be taken into account when evaluating correct CT ratio and polarity.
- Auto-switching equipment tests and ATCC equipment tests.

  Adequate time should be allocated for these tests, and depending upon network operational availability, they may have to be undertaken at a slightly later date.

Upon completion of the on-load tests, the newly constructed and commissioned assets are available for operational and commercial service.

15.15 Substation commissioning overview

15.15.1 Substation commissioning tests and inspections

Figure 23.4 provides the procedural framework for commissioning tests and inspections. Within this context, it is instructive to summarise the commissioning requirements for (say) a double busbar substation, similar to that shown in Figure 20.16, to provide an insight into the scale and range of requirements. Within this context, the following will be considered:

1. Feeder circuit
2. Transformer circuit
3. Bus section/bus coupler circuit
4. Substation common equipment
5. Civil, structural and building engineering (CSBE)

15.15.1.1 Feeder circuit

For each of the feeder circuits shown in Figure 20.16, the commissioning requirements will include the following (NB: It is assumed that the protection relays are programmable multifunction numeric relays incorporating main protection, backup protection, fault recorder, circuit breaker fail current check relay, inter-tripping, etc.):

1. **Stage 1**
   - (i) Cabling and wiring insulation resistance tests
   - (ii) CT flick tests and magnetising curves
   - (iii) Primary injections
   - (iv) VT wiring injections
   - (v) Circuit breaker equipment tests
   - (vi) Busbar joint resistances
   - (vii) Disconnector equipment tests (*x* number of disconnectors)
   - (viii) Earth switch equipment tests (*x* number of earth switches)
   - (ix) First main protection relay equipment test (secondary injection)
   - (x) Second main protection relay equipment test (secondary injection)
   - (xi) Synchronising relay equipment test
   - (xii) DC logic circuit diagrams comprising:
     - (a) Circuit breaker trip and close
     - (b) First main protection relay
     - (c) Second main protection relay
     - (d) Auto-reclose
     - (e) Trip relay reset
   - (xiii) Loadability test
   - (xiv) Stage 1 inspections
2. **Stage 2**
   - (i) Switching programme for energisation and on-load tests
   - (ii) Stage 2 inspection.

15.15.1.2 Transformer circuit

For each of the transformers shown in Figure 20.16, the commissioning tests will include the following:

1. **Stage 1**
   - (i) Cabling and wiring insulation resistance tests
   - (ii) CT flick tests and magnetising curves
   - (iii) Primary injections
   - (iv) Transformer tests
     - (a) Vector group
     - (b) Ratio, magnetising current and tap changer continuity
     - (c) Winding temperature and Buchholz
     - (d) Cooler fans and pumps
   - (v) HV circuit breaker equipment test
   - (vi) LV circuit breaker equipment test

(vii) Busbar joint resistances
(viii) Disconnector equipment tests (*x* number of disconnectors)
(ix) Earth switch equipment tests (*x* number of earth switches)
(x) VT wiring injections
(xi) Transformer overall protection equipment test (secondary injection)
(xii) Transformer backup protection equipment tests (secondary injection)
  (a) HSOC
  (b) Two-stage overcurrent
  (c) LV backup protection
  (d) Circuit breaker fail
(xiii) HV connections equipment test (secondary injection)
(xiv) LV connections equipment test (secondary injection)
(xv) Synchronising relay test
(xvi) ATCC equipment test (secondary injection)
(xvii) DC logic circuit diagrams comprising
  (a) HV circuit breaker trip and close
  (b) LV circuit breaker trip and close
  (c) Transformer overall protection
  (d) Transformer protection (first supply)
  (e) Transformer protection (second supply)
  (f) Transformer tap-change control
  (g) Transformer cooler fans and pumps
  (h) Trip relay reset
(xviii) Loadability test
(xix) Stage 1 inspections.

2. **Stage 2**
  (i) Switching programme for energisation and on-load tests
  (ii) Stage 2 inspection.

15.15.1.3 Bus section/bus coupler

For each of the bus section/bus couplers shown in Figure 20.17, the commissioning tests will include the following:

1. **Stage 1**
  (i) Cabling and wiring insulation resistance tests
  (ii) CT flick tests and magnetising curves
  (iii) Primary injections
  (iv) Circuit breaker equipment tests
  (v) Busbar joint resistances
  (vi) Disconnector equipment tests (*x* number of disconnectors)
  (vii) Earth switch equipment tests (*x* number of earth switches)
  (viii) Backup protection equipment test (secondary injection)
  (ix) Circuit breaker fail equipment test (secondary injection)

   (x) DC logic circuit diagrams comprising
      (a) Circuit breaker trip and close
      (b) Backup protection
      (c) Trip relay reset
   (xi) Loadability test
   (xii) Stage 1 inspections.
2. **Stage 2**
   (i) Switching programme for energisation and on-load tests
   (ii) Stage 2 inspection.

15.15.1.4 Common equipment

With reference to Figure 20.17, the common equipment tests will comprise the following:

1. **Stage 1**
   (i) Busbar joint resistance
   (ii) Reserve section disconnector equipment test
   (iii) Busbar earth switches resistance test
   (iv) Mechanical interlocking (some will be circuit specific)
   (v) Electrical interlocking (some will be circuit specific)
   (vi) Busbar protection equipment test (secondary injection)
   (vii) SCS equipment test (secondary injection)
   (viii) ARS equipment test – if fitted (secondary injection)
   (ix) Synchronising scheme functionality test (some will be circuit specific)
   (x) DC logic circuit diagrams comprising:
      (a) SCS inputs/outputs (control/alarms/indications)
      (b) Busbar protection
      (c) Circuit breaker fail
   (xi) Battery system tests
   (xii) Diesel generator and LV AC supplies tests
   (xiii) Stage 1 inspections.
2. **Stage 2**
   (i) Switching programme for the energisation of the busbars and on-load tests of common equipment. NB: This may be undertaken on a circuit-by-circuit basis as the circuit is energised.
   (ii) Stage 2 inspection.

15.15.1.5 Civil structural and building engineering

Commissioning of CSBE relates to those items that require on-site tests. This includes

- Oil containment
- Drainage (pumps)
- Site lighting
- Building services – heating, lighting, etc.
- Site security systems.

Part 2

Construction QMS procedures

Construction QMS procedures – overview

1. **Quality management system – requirements**
   Most quality management systems (QMS) accord with the international standard on quality management: ISO 9001 (see Sections 1.5.1 and 2.8.2). Whilst it is not the intention of this publication to examine in detail the requirements of a QMS, but rather the procedures which form an integral part of a QMS, a summary of QMS requirements which satisfy ISO 9001 is described below to provide the necessary background and context.
2. **ISO 9001 – purpose and structure**
   The purpose of a QMS based upon ISO 9001 is to help organisations, of all shapes and sizes, achieve the following:
   (i) Organise processes
   (ii) Improve the efficiency of processes
   (iii) Continual improvement

   ISO 9001 is based upon a methodology of 'plan–do–check–act' and provides a process-orientated approach to documenting and reviewing the structure, responsibilities and procedures required to achieve effective quality management in an organisation. Specific sections of the standard specify requirements relating to the following:
   (i) The requirement for a QMS, including the preparation of a quality manual, procedural document control and procedural interactions
   (ii) Management responsibilities for a QMS
   (iii) Resource management, including human resources encompassing work environment and competence requirements
   (iv) Product realisation, encompassing the steps from initial design to final delivery
   (v) Measurement, analysis and improvement encompassing periodic audits, customer satisfaction evaluation and the incorporation of preventative and corrective measures.

   All of the above points contribute to the preparation and implementation of an effective and efficient suite of QMS procedures.
3. **Quality management system – procedures**
   The following chapters will examine the range of typical QMS procedures required to undertake power network construction. The procedures are essentially based upon the instance of a scheme being developed by a power network company and then an Engineer, Procure and Construct (EPC) contract being awarded to a power engineering contractor (via a competitive tender

process). This is typical of many power network construction schemes/projects both in the United Kingdom and worldwide.

The procedures examined will mainly be those relevant to a power network company, since it is the power network companies that are involved in the complete end-to-end construction task. However, reference will also be made to the contractor's procedures which, in many instances, will be similar to, and must dovetail with, those of the power network company. In practice, the number of procedures produced and the title and the content may differ from company to company – but will broadly follow those described in the following chapters. Thus, both the range of procedures discussed, and the procedural content, may be taken as being 'typical' and one 'best practice' way of doing it.

Chapter 16
Construction delivery models and contracts

16.1 Introduction

This chapter will examine construction delivery models and contract arrangements that have relevance to power network construction. The chosen delivery models and contracts must be incorporated into the QMS procedures of each of the parties involved with the construction work. In summary, this chapter will encompass the following:

- Construction delivery risk considerations
- Construction delivery models
- Contract price options
- Contract process and terms and conditions.

16.2 Construction delivery risk considerations

16.2.1 Construction delivery risks

The construction delivery risk is usually shared between the client (power network company) and the contractor – and in some instances a consultant. It is to be noted that in contract documentation, the client is often referred to as the 'employer'. The decision on how and where the risk is shared is usually determined by the client – since the client makes the decision on the type and timing of the contract to be awarded. Generally, the risk may be categorised as follows:

- Risk that the contract is not delivered to the required time quality or cost
- Risk that the power network is disrupted either as a result of the construction work, or as a result of inadequacies arising after construction completion.

The two types of risk will be evaluated below:

1. **Contract risk**
   With reference to Figure 16.1, the client may award a contract to a contractor at any stage following completion of the Need Case (a client task) – and should do so where the risk to the client is a minimum/acceptable. Generally, the earlier the contractor is involved the greater the loss of control by the client and usually the greater the risk for the contractor. The decision on at what stage of a

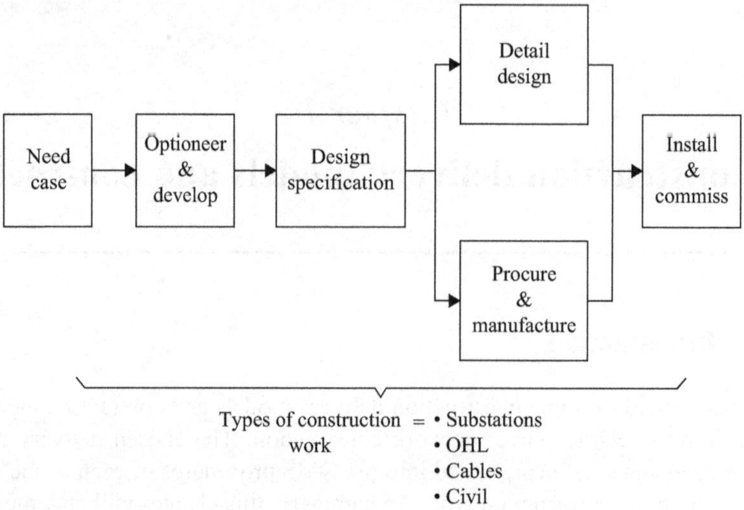

Figure 16.1 Construction scheme stages

construction scheme to place the contract is influenced by one or more of the following factors:

(i) In-house resource capability and size of both the client and the contractor organisation. The term 'informed client' is often used for a client who carries sufficient capability to fully understand and significantly participate in the work to be undertaken. Most clients are informed – but some for strategic considerations are not, and, therefore, outsource as much as possible to contractors – usually via a long-term relationship.
(ii) The extent to which the client wishes to influence the specification (i.e. at a functional level or more detailed level).
(iii) The nature of the work. A novel, complex or research and development (R&D) project may benefit from the contractor being involved earlier, e.g. at the development stage.
(iv) Optimal contract arrangements for either driving down costs or time, or increasing quality.
(v) Size of the client's construction programme, i.e. too large or too small for the in-house resource.

Contract risk also arises from contract specification deficiencies by the client, and contract delivery deficiencies by the contractor, e.g.

(a) Failure to apply the right levels of competent resource
(b) Failure to adhere to, or correctly interpret, the specification
(c) Failure to adhere to QMS procedural requirements.

2. **Power network construction risk**

The risk of the power network being disrupted as a result of construction work is essentially a client side risk, although a contractor may suffer loss of reputation. In many instances, the impact of power network disruption (e.g. loss of electricity supply or disconnection of a generator) has greater consequences and cost than those associated with a construction scheme/contract alone, and therefore, every endeavour must be made to ensure that this risk is minimised/removed.

Both of the above risks may be minimised/managed in the following ways:

- Through the client letting the contract at a later stage (see Figure 16.1). Generally, the greater the involvement of an informed client, the less the risk to the power network. This arises from the greater familiarity and knowledge of the operational power network by the client's engineers. Clearly, if the client does not have this resource, this is not an option.
- Through the client ensuring the preparation of a watertight and well-defined specification.
- Through watertight QMS systems which are strictly adhered to by both client and contractor.
- Through both the client and contractor employing appropriate numbers of experienced people who are competent in the requirements of the work.
- Through the client employing an 'engineering assurance' resource. This comprises engineering surveillance and acceptance of the contractor's deliverables at various key stages in the work sequence. Most power network companies (clients) employ an in-house engineering resource at the design, installation and commissioning stages, for this very purpose. Engineering assurance is a risk-management process with the depth of surveillance depending upon.
  - Confidence in the contractor, i.e. the contractor has a good track record and the assurance team has successfully worked with the contractor before.
  - Complexity/novelty of the construction work to be undertaken.
  - Whether errors are found, thereby requiring greater surveillance.
  - Availability of assurance resource.
- By ensuring the equipment is correctly commissioned. Proficient commissioning has the potential to detect and correct errors not previously identified.

16.2.2 Construction design specification risk

The content of a construction design specification is always a source of risk. The dilemma is how much detail to include. Generally, the greater the detail, the more watertight the specification.

Section 1.4.1 explained that there are two categories of construction design specification:

1. Project-generic design specifications (originated by each client and applicable to most of that client's projects – NB: this suite of documents is omitted from Figure 16.1)
2. Project-specific design specification (applicable only to the project in question – NB: this is shown in Figure 16.1).

These will be discussed below:

1. **Project-generic design specifications**
   It is again worthy of note that project-generic design specifications are sometimes termed 'Technical Specifications' (TSs). Prior to *ca.* 1990, the practice in the United Kingdom relating to project-generic design specifications (for power equipment) resulted in very detailed specifications contained in standard diagrams that defined the design almost down to the wire, nut and bolt (sometimes known as 'plant standards'). Thus, the UK network was built to a very standardised and consistent design, and usually delivered by UK-based contractors. Circa 1990, European Union directives, in an endeavour to open up and stimulate the market, required that all generic design specifications be defined in terms of equipment and system performance requirements, i.e. a functional specification. This requirement remains to the present day. However, by not defining the detail, functional specifications embody greater risk. This risk is mitigated by appropriate factory and site testing of the equipment. Functional specifications have, however, resulted in a greater variety of equipment and designs than was previously the case, with concomitant implications for familiarity, training and spares.
2. **Project-specific design specification**
   The greater potential for risk probably resides in project-specific design specifications since they are unique to the project in question, whereas the project-generic design specifications, which apply to most projects, become refined and perfected over time. The risk is how much detail to specify and at what stage. The greater the content of the specification, the less the risk for errors/omissions/ambiguities, but the greater the resource required and the longer it takes the client to prepare. There are probably three levels at which this specification may be prepared by a client:
   - (i) **High-level functional specification.** Such a specification, for example, would simply state the requirement for a substation connecting two generators and four outgoing circuits, with defined loadings and fault levels, with the substation occupying a defined land area. The project-generic design specifications would probably apply. The detailed requirements within this specification would be left to the contractor to define.
   - (ii) **Low-level functional specification.** This level of specification would add to the above by typically defining the substation layout (e.g. double busbar with two bus couplers and a bus section), single-line diagram, the protection and control functional arrangements together with a key diagram, and buildings, roadways and drainage system outline requirements, etc. The project-generic design specifications would apply.
   - (iii) **Detailed specification.** A detailed specification would comprise all of the above but would additionally specify the exact type of equipment to be used. It would typically also include the staging of the work, and the installation and commissioning requirements. This specification would be comprehensive enough for the production of the detailed design/drawings to commence immediately.

Which level of project-specific design specification to be used will again depend upon a range of factors such as the resource levels and capability of both the client's and contractor's organisations, complexity and novelty of the work and the client's preferred delivery model. Generally, the above two functional specifications result in a greater risk of the client not getting what is wanted – but minimises the resource input from the client. If functional specifications are used, a detailed specification will need to be prepared at a later stage, i.e. a two-stage specification. With both functional specifications, the second-stage detailed specification may be left either solely within the domain of the contractor or may require the agreement of the client. The most common type of specification in use probably approximates to a low-level functional specification with the second-stage detailed specification agreed to by the client, through the engineering assurance process.

16.2.3 Single or multiple contract risk

Figure 16.1 illustrates the stages for a construction scheme that may include a number of categories of construction work, i.e. substations, OHL, cables and civil. The question arises that whether it is preferable, from a risk perspective, to issue separate contracts for each category of work with the client coordinating the interfaces – or whether to award the contract to a single contractor (say the substation contractor) and that contractor subcontracts and coordinates the work of the OHL, cables and civil contractor. The risk to the client is usually minimised if a single contract can be awarded, but this is dependent upon whether the single contractor has the capability to manage the subcontractors or whether the single contractor even wants this task or the risk.

16.3 Construction delivery models

16.3.1 Range of construction delivery models

The term 'construction delivery model' essentially relates to the division of tasks between client and contractor in delivering the stages described in Figure 16.1.

Figure 16.2(a) illustrates a traditional client/contractor contract arrangement where via contract 1 the client contracts with a consultant to discharge most of the

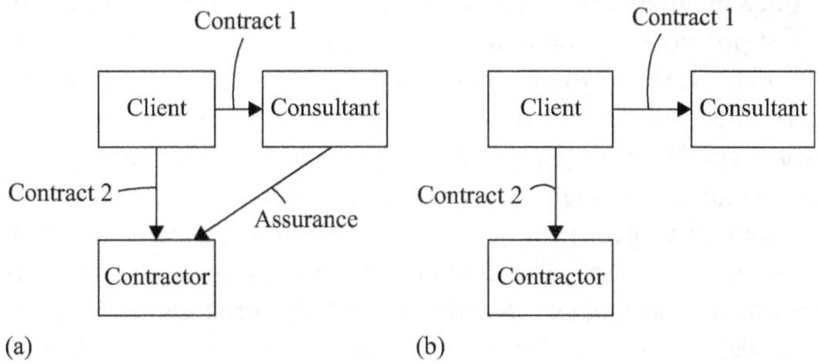

Figure 16.2 Client/consultant/contractor contractual relationships

technical and engineering assurance tasks associated with the scheme. In this model, the consultant may be required to develop the design, prepare the design specification and engineering assure the contractor's work. The level of interaction between contractor and consultant is defined in contract 2. This model has been used when the client either has a limited in-house engineering resource, or where the client has a very large capital works programme requiring delegation of many of the client-side tasks to an engineering consultant organisation.

Figure 16.2(b) illustrates the instance of a consultant being contracted to provide engineering resources to the client, whereby the resources merely augment the client's own resources. In this instance, the consultants interact with the contractor as if they were the client. This model has been frequently used in more recent times.

With reference to construction delivery models, the range of options employed in power network construction align to the type of contract and broadly encompass the following:

- Turnkey
- Functional
- Alliancing
- Engineer, procure and construct (EPC)
  - Single contract
  - Multiple contract
  - Free issue of equipment
- Framework agreement
- Build only
- Client in-house construction

Each of the above will be briefly examined.

16.3.1.1 Turnkey

A turnkey delivery model is one whereby following determination of the Need Case by the client, the contractor executes and delivers the whole construction scheme and hands it back to the client once the project is completed and operational. The analogy is that of a contractor handing to the client the key to a completed house. In this model, the client is essentially 'hands off', thereby minimising the client's resource input. In practice, the model is difficult to execute for all, but the simplest of schemes, since invariably there are some tasks that the client (power network company), is best positioned to undertake e.g. obtaining planning permission and connecting the new works to the existing power system (stage 2 commissioning). In addition, the client usually requires some degree of confirmation, to mitigate risk, that the works are fit for purpose, which requires a level of engineering assurance.

Where the turnkey model has been used, it is usual for the contractor to work to the client's suite of project-generic TSs. This model suits the instance of the client employing minimum in-house resource but at the expense of loss of control of, and ability to influence, design, site installation and commissioning – with concomitant risk. This model is best suited to a long-term relationship in which the contractor becomes very familiar with the client's requirements.

16.3.1.2 Functional specification

A functional specification delivery model is a more practical alternative to a pure turnkey. In this model, a high-level functional specification, as outlined in Section 16.2.2, is provided by the client, with the contractor required to adhere to the client's suite of project-generic TSs, and subject to engineering assurance through all stages of the scheme. The client would also usually direct the connection of the new works on to the operational power system (i.e. stage 2 commissioning). This model has the benefit of allowing the contractor to innovate with reference to the project-specific design specification, but with some loss of control to influence and shape the design by the client.

16.3.1.3 Alliancing

Alliancing comprises the integration of client and contractor personnel into a separate work unit usually tasked with undertaking a significant portion of the clients capital works programme. The impetus behind the alliancing model is to forge collaborative and efficient working between the client and contractor, i.e. to remove the more traditional adversarial working relationship which at times has been problematic for efficient project delivery. The alliance usually takes over the work following completion of the Need Case by the client and work is usually given to the alliance rather than compete for it. Contracts are often arranged on a cost reimbursable, target cost, open book basis including pain/gain incentives/penalties. Merits of alliance working are summarised as follows:

1. **Advantages**
   - (i) Secures the contractors resource for large capital works programmes which span long periods of time.
   - (ii) Reduces the adversarial relationship between client and contractor.
   - (iii) Synergies arising from a more widely skilled workforce jointly undertaking the work and jointly resolving issues arising.
   - (iv) Has been known to work well on high-volume work where each job is relatively small in content and low on complexity.
2. **Risks**
   - (i) No natural incentive for driving down costs as with competitive models
   - (ii) Cost control is facilitated by open book continual benchmarking, target setting and assessment and visibility of costs – increases bureaucracy.
   - (iii) The client may still have to fund engineering assurance from outside the alliance to be reassured of work quality and standard.
   - (iv) The culture of the alliance partners may not successfully integrate.
   - (v) Cost-control drivers difficult to achieve on complex multi-faceted power network schemes.

16.3.1.4 Engineer, procure and construct

EPC is a generic term in which the contractor is appointed to design (i.e. usually detailed design), procure the equipment and install and commission. Generally, a project-specific specification (usually low-level functional specification as described in Section 16.2.2) is prepared by the client and a contract is awarded to a contractor following a competitive tender process, usually on a fixed price lump sum basis. The contract may be a single contract where a lead contractor employs sub-contractors and coordinates their work (sometimes termed an EPCM contract where the M stands for management) or multiple contracts as outlined in Section 16.2.3. A variant of EPC is where the client purchases (often in bulk) the equipment and 'free issues' to the contractor, i.e. design and construct only. EPC is one of the most common delivery models for power network construction.

16.3.1.5 Framework contract

A framework contract model comprises competent and interested contractors subject to pre-qualification evaluation, following which the best of those that qualify are nominated to be part of the framework. Projects are then offered to those contractors who are on the framework – who may still be required to compete for the work. This arrangement has the advantage that the contract documentation can be simpler and briefer by virtue that those on the framework have pre-qualified.

16.3.1.6 Build only

With reference to Figure 16.1 in this model, the client does all the work up to and including both detailed design and procure/manufacture. A contract is then let to install and commission only. This arrangement was popular when clients had their own in-house drawing offices (or were prepared to coordinate the work of an external drawing office) – particularly for protection and control, or civil orientated projects. A variation of this is to also let the contractor procure/manufacture the equipment. This model is less prevalent than was once the case.

16.3.1.7 Client in-house construction

In this model, the client does all the work, and there is no contractor involvement. There are few instances in the United Kingdom where this model is now used. With in-house construction, the client takes all the risk including the requirement to maintain significant expertise and expand/shrink the workforce as the construction capital programme varies. There are instances of where a client may maintain a small workforce for this very purpose to deliver small scale, complex work, often concerned with protection and control, and usually involving a complex operational site or multiple sites.

16.4 Contract price

16.4.1 Types of contract price

Integral to the terms and conditions of any contract is the criterion for the price, i.e. the criterion for determining the sum of money that the employer pays to the contractor. Some of the most common found in power network construction are as follows:

- Lump sum, fixed price
- Target price

- Cost reimbursable with a fee
- Measured contract

These will be briefly examined below.

16.4.1.1 Lump sum, fixed price

This usually comprises a competitive tender to determine the lowest cost contractor – whose accepted bid forms the fixed price. Payments are made at the end of the work or staged as the work proceeds. Sums of money may be withheld at the end of the work until certain deliverables are complete e.g. 'snagging' (outstanding relatively small amounts of work e.g. missing labels) or return of final marked up drawings into the clients filing system. This type of contract places maximum risk with the contractor since failure to fully understand the contract specification, or unforeseen circumstances such as problems with a manufacturer, must be resolved and funded by the contractor. However, if the employer's specification is not watertight, any change or addition to the work is a variation to the contract funded by the employer. This tends to be one of the most commonly used types of contract.

16.4.1.2 Target price

In this arrangement, the employer and contractor agree a target price for the works. If the contractor completes the work under the target price, then a bonus is paid. In some instances, if the cost of work exceeds the target price, the employer funds a percentage of the cost. This is sometimes termed a pain/gain incentive. Target pricing has been used in alliance delivery models – but it usually requires substantial visibility of contractor's costs.

16.4.1.3 Cost reimbursable with a fee

This arrangement requires the employer to pay the contractor the actual cost of carrying out the work – whatever it may be. The employer then pays an agreed fee (either a percentage or lump sum) which forms the contractors' profit. This places the client at high risk since there is little incentive for the contractor to keep costs under tight control. This type of contract may be suitable for an R&D or equipment trial project, where it is difficult to evaluate precise costs in advance. Usually, employer and contractor would scope out the costs at the outset to minimise risk.

16.4.1.4 Measured contract

This is sometimes termed a 'bill of quantities'. In this arrangement, each item of work is comprehensively specified in advance (by virtue of identical work being previously undertaken) and a rate per unit of work attached. The contractor is paid in accordance with the number of units of work performed. This arrangement simplifies the contract documentation. Measured work may apply to say, the laying of cables or repetitive equipment replacement – usually on a large scale.

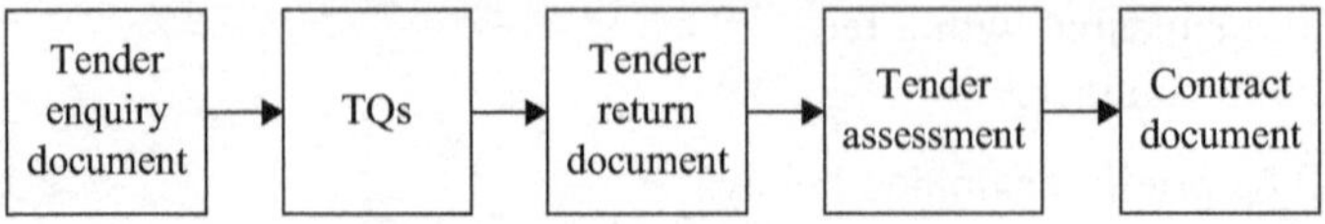

Figure 16.3 Contract process

16.5 Contract process and terms and conditions

16.5.1 Contract process

The contract process is summarised in Figure 16.3. The process is similar no matter when in the construction scheme process, as shown Figure 16.1, the contract is released, although the detail may vary from stage to stage. The text below will assume that following the optioneer/develop stage that an EPC contract process is undertaken. The stages of Figure 16.3 are summarised below:

1. **Tender enquiry document (TED)**
   The employer (client) prepares a TED incorporating the specification (including the contract design specification) and issues a copy to each contractor inviting them to tender for the contract.
2. **Technical questions (TQs)**
   On receiving the TED, each contractor assesses the requirements and, where the information appears insufficient in the TED, formally requests information/clarification from the employer. This process is frequently referred to as TQs.
3. **Tender return document (TRD)**
   Each contractor prepares a TRD which specifies the contractor's proposals to satisfy the TED, i.e. a design solution, responses to questions asked in the TED (e.g. an organogram, safety management arrangements, installation programme, etc.) and a price for delivering the works.
4. **Tender assessment**
   On receiving the TRD from each contractor, the employer assesses the responses against defined criterion to determine the best proposal for delivering the works, and hence the successful contractor.
5. **Contract document**
   The final stage of the process is the preparation by the employer of the contract document, the format of which is almost identical to the TRD, with the addition of relevant TQs, and legally binding documentation for employer and contractor to sign. The contract document is then issued to the successful contractor (i.e. contract award).

Each of the above stages of the contract process will be examined in more detail in the following sections.

16.5.1.1 Tender enquiry document

A TED comprises the information and instructions provided by the employer to a number of suitably qualified contractors [NB: EU directives (if applicable) define how/which contractors are notified] inviting each contractor to submit their proposals/solutions/prices for delivering the works specified in the TED. The TED is generally divided into following three categories:

- **Level 1: conditions of contract**
  (Applies to the projects of many employers)
- **Level 2: project-generic tender requirements**
  (Applies to most projects of one employer only)
- **Level 3: project-specific tender requirements**
  (Applies to one specific project of one employer only).

The above-mentioned levels will be briefly examined.

Level 1 conditions of contract

The conditions of contract essentially define and provide the legal framework for common contract requirements which would be applicable over a large range of contracts (i.e. able to be used by many employers). Typical contents include: formal documentation to be used under the contract (e.g. variation instructions), arrangements for payment, liquidated damages, insurance arrangements, resolution of disputes, etc. Most employers used model forms of conditions of contract (occasionally amended by the employer to reflect employer-specific requirements). Typical model conditions of contract include:

- IET/IMechE
- FIDIC
- NEC

These will be briefly examined below:

1. **IET/IMechE**
   The Institution of Engineering and Technology and the Institution of Mechanical Engineers have jointly formulated model forms of contract. There are three variants of the contract contained in forms MF/1, MF/2 and MF/3, respectively. Form MF/1 provides for the supply and erection of equipment and also includes minor civil works. Form MF/2 provides for the supply of equipment but not for erection by the contractor. Form MF/3 provides for the supply of equipment but excludes both design work requiring approval by the employer (purchaser) and erection. These forms are generally more applicable to minor construction work or procurement only.
2. **FIDIC**
   FIDIC stands for Federation Internationale Des Ingenieurs – Conseils; it is an international standards organisation for the construction industry and has played a significant role in specifying model contract terms and conditions. The FIDIC model contract relevant to power network construction provides for the design, procurement, installation (including commissioning) of the equipment and is suitable for the inclusion of major civil engineering works. FIDIC is widely used and is suitable for international procurement.

3. **New engineering contract (NEC)**
   NEC is a suite of contracts originated by the Institution of Civil Engineers. It is equally suitable for power network contracts and like FIDIC provides for the design, procurement, installation (including commissioning) of equipment, with the inclusion of major civil engineering works. NEC differs from FIDIC; in that it allows a more collaborative approach to managing the contract such as that required by an alliance delivery model. NEC also allows more options for the client to contractor payment arrangements e.g. cost reimbursable or target price. NEC is also suitable for international procurement.

Role of the engineer
Most Level 1 conditions of contract define formal roles, the most notable of which is probably the role of the 'engineer'. Traditionally, under English law, the engineer named in the contract discharges the following tasks:

- Employers agent in overseeing the progress of the work
- Certifier under the contract, e.g. signing off certificates of payment, completion, defects, etc.
- Resolving disputes between the parties. This requires the engineer to act honestly, fairly and reasonably.

The engineer is usually both of high standing and reputation in the employer organisation. Responsibilities are frequently delegated to the engineer's representative, which is usually the employer's project manager. It is an occasional point of contention that the engineer drawn from the employer organisation arbitrates disputes between the employer and the contractor; however, this is the usual arrangement. In some instances, the engineer may be from a consultant engineering company to bring greater impartiality – but even so the employer would normally fund the cost of the engineer.

Overview of FIDIC terms and conditions
To provide an insight into model conditions of contract, Figure 16.4 provides an overview and summary of those specified in FIDIC. A full and comprehensive understanding should be via reference to the actual document.

Level 2 project-generic tender documentation
Project-generic tender documentation is essentially employer (client)-specific requirements which are common to most of that employer's contracts. Typically, this documentation will include the following:

- Arrangements for returning the TRD
- Employer's tender assessment criteria
- Requirement to comply with the employer's project-generic design specifications (see Chapter 1)
- Applicable standards to be used if requirements are not defined in the project-generic design specifications e.g. BSI, IEC, ENATS, etc.
- Arrangements for planning permission, consents and wayleaves (usually acquired by the employer).

FIDIC terms and conditions – content summary

1. **Definitions**
   Covers a range of definitions such as those of contractor, defects liability period and taking over certificate
2. **Engineer and engineer's representative**
   Specifies the role and duties of the engineer and engineer's representative
3. **Contract documentation**
   Specifies requirements for contract drawings, erection information, O&M manuals, manufacturing drawings, etc.
4. **Obligations of the contractor**
   Covers both obligations and rights, e.g. requirement to prepare a programme
5. **Obligations of the employer**
   Defines employer obligations such as requirement to acquire consents and wayleaves and issue PTW, etc.
6. **Labour**
   Covers contractor requirement for managing his labour; working hours and working hour restrictions, etc.
7. **Workmanship and materials**
   Covers the requirement for the contractor to allow the employer the opportunity to inspect the works
8. **Supervision of work**
   Specifies the right of the employer to suspend the work and the implications of doing so
9. **Completion**
   Specifies the arrangements associated with completion and late completion of the works
10. **Tests on completion**
    Covers the commissioning tests, failure of either contractor or employer to attend the tests, disagreement over the test results, failure of the equipment to pass the tests, etc.
11. **Taking over**
    Defines the formalities associated with the employer taking over the works e.g. issue of a taking over certificate
12. **Defects after taking over**
    Specifies the arrangements for resolving defects within the defects liability period
13. **Variations**
    Covers the arrangements associated of the employer varying the works before take over and use of the variation instruction
14. **Ownership of plant**
    Defines the timing of the employer taking over the equipment
15. **Certificates and payments**
    Specifies the formalities relating to the contractor being paid for the works via certificate of payment
16. **Claims**
    Covers the timing and requirements for the contractor making a claim for additional payment
17. **Provisional sums**
    This covers the requirements relating to provisional sums. NB: provisional sums cover possible additional work (usually of a routine nature) not specified in the contract e.g. laying of extra multicore cables

Figure 16.4 Continued

18. **Risk and responsibility**
Specifies the division of responsibilities relating to risks such as wars, errors in design, consents and wayleaves, damage, unforeseen events, etc.
19. **Damage to property and personal injury**
This refers to employer and contractor liabilities with reference to injury/death of contractor's personnel, damage to property other than the works and defects in design/workmanship, etc.
20. **Insurance**
Covers arrangements for insuring the works
21. **Custom duties**
Specifies responsibilities for paying custom and import duties for the importation of plant
22. **Disputes and arbitration**
Covers arrangements for resolving disputes between employer and contractor
23. **Law**
Defines which legal code applies to the contract
24. **Formal documentation**
Specifies model documentation for both the tender return and contract agreement

Figure 16.4 Overview and summary of FIDIC terms and condition

- Normal site working hours
- Use of public highways, etc. and the requirement for a traffic management plan
- Emergency contact arrangements (e.g. fire, local authority, etc.)
- Information security arrangements
- Arrangements for employer's personnel to be trained on new equipment
- Health, safety and environmental arrangements e.g. requirements for implementing CDM regulations, working at height, etc.
- Application of employer's safety rules
- Requirement for contractor to deliver work in accordance with a QMS system including the provision of a quality plan and a site quality plan.
- Requirement for contractor to seamlessly interface with employers QMS system
- Project audit requirements
- Design submission audit requirements (engineering assurance) by the employer
- Quality control of equipment manufacture arrangements
- Monthly reporting requirements
- Equipment testing requirements
- Site accommodation facilities and site security arrangements
- Format of the design documentation and arrangements for confirming that detailed design production may commence
- Requirement to interface with the employers drawings management system
- Commissioning management and procedural arrangements (e.g. formation of a commissioning panel).

Level 3 project-specific tender documentation

Project-specific tender documentation relates to information and requirements that are unique to the project in question. At its core is the project-specific design specification as described in Chapter 1. The employer will usually form a team to provide input to, and prepare, the document – usually headed by the employer's project manager. Typical contents comprise the following:

- Summary and scope of the work
  - Purpose of the project
  - Description of the site
  - Scope and extent of the works
  - Sections of the work
  - Interfacing work/contracts
- Health, safety and environmental information relevant to the work
- Outage programme
- Drawings, records, schedules, audits, reports, etc. relevant to the works
- Traffic management plan requirements
- Planning permission, consents and wayleaves status/requirements
- Subcontractor appointment proposals
- Supply of site services, i.e. water, electricity, telephones, etc.
- Training requirements
- Tender assessment criterion
- Payment milestones and contractual key dates
- Contractor resource profile for the project
- Project-specific design specification (as specified in the scheme design specification – see Chapter 17). This comprises the specification of the employers engineering requirements as outlined in Section 1.4.3. In summary, it comprises the following (as required):
  - Power system electrical design
  - Substation spatial layout design
  - Protection and control accommodation design
  - OHL route and electromechanical design
  - Cable route and electromechanical design
  - Civil structural and building engineering design
  - Temporary works design
- Emergency return to service requirements
- Schedule of equipment being offered – to be completed by the contractor
- Schedule of drawings and reports – provided by the contractor
- Specific proposals to be returned by the contractor:
  - High level stage by stage (installation and commissioning)
  - Build sequence programme
  - MEWP/maintenance access drawing
  - High-level programme
  - Procurement/manufacture programme
  - Key issues risk programme
  - Design/drawing production programme

- QMS arrangements
- Price summary and equipment price schedules – to be completed by the contractor
- Prices for cancellation of the works by the employer – to be provided by the contractor

Completion of the three levels of tender documentation is a very significant piece of work – particularly Level 3. It is essential that the completed document contains a clear and unambiguous specification of requirements and watertight contract terms and conditions.

16.5.1.2 Technical questions

The TQ stage follows the contractor's examination and evaluation of the TED. Any perceived omissions or ambiguities are clarified with the employer through a formal question and answer process. When responding to a question from a specific contractor, a response may be sent to all contractors, if the issue has relevance for all. This begs the question of whether a contractor wishes to raise the TQ since it alerts all other contractors to the issue – possibly loosing competitive advantage. With some projects, the TQs are very significant in number. TQs are also advantageous to the employer since they assist in making the contract watertight.

16.5.1.3 Tender return document

The TRD will be based on the TED and will contain the contractors proposed design solution, and responses/proposals to requirements contained in the TED. It will also contain the completed price schedules and the proposed contract price. Preparation of the TRD is a resource intensive piece of work for the contractor and usually comprises the following four stages:

1. **Stage 1 – management arrangements**
   The contractor forms a structured team charged with developing and preparing the TRD.
2. **Stage 2 – TED validation**
   Examination of the TED by the contractor's team to understand the employer's requirements and to determine whether the TED is watertight or there are omissions/ambiguities. Where the latter exists, the TQ process with the employer will be enacted.
3. **Stage 3 – design proposal**
   The formulation and pricing of an optimum design solution
4. **Stage 3 – TRD**
   The preparation of the TRD and collation of all supporting documents and drawings, etc. to meet the employer's requirements. Completion of the TRD will usually require an officer of the contractor organisation of appropriate standing to sign off the TRD.

Contract award criterion	Weighting
1. Contract price	60%
2. Design solution and compliance	15%
3. Project delivery programme	10%
4. Health, safety and environmental management	10% (NB: must achieve at least 8% to qualify)
5. Contractor-proposed organisational structure and capability	5%

Figure 16.5 Typical contract award criterion

16.5.1.4 Tender assessment

The tender assessment stage is undertaken by a team selected for this purpose. It would usually be chaired by either the employer's project manager or the procurement officer for the project and would typically include the lead design engineer and health and safety representative. Contract award is rarely based solely on the contractor offering the lowest price but usually takes account of a range of weighted factors. A typical contract award criterion is given in Figure 16.5.

16.5.1.5 Contract document

The contract document is prepared by the employer and forwarded to the successful contractor with the award of contract. It will closely resemble the TRD together with any relevant TQs. The final content will be greatly influenced by the employer's project manager and procurement officer. The contract document will contain a contract agreement form to be signed by both the employer's representative and the contractor's representative – thereby making the contract legally binding on both parties. On receipt, the successful contractor may then commence work on the project.

16.5.2 QMS considerations

The requirements addressed in this chapter must be incorporated in the QMSs of the parties involved, i.e. usually the employer (client) and the contractor. From the range of options examined, each client will choose construction delivery models and contract arrangements that best suit the delivery of their capital works programme – influenced by strategic considerations such as what proportion of the scheme they want to deliver with in-house resources (the modern tendency has been to increasingly outsource). Depending upon the prevailing workload, clients may elect to use a mix of delivery models, some of which may be a hybrid of those examined. Whatever the delivery model and contracts used, successful project delivery is dependent upon the sound working relationship between client and contractor, often forged over time.

Chapter 17
Scheme investment procedure

17.1 Scheme procedures

The range of QMS procedures required to deliver a scheme typically fall into three levels. At the highest level (tier 1) is the scheme investment procedure (typical title). At the level below (tier 2), there are a number of other important procedures which define the key stages/requirements of the scheme process, and which are common to most schemes. They are as follows:

- Contracts and procurement (covered in Chapter 16)
- Health and safety – including CDM (covered in Chapter 18)
- Environmental (covered in Chapters 2 and 22)
- Project management (covered in Chapter 19)
- Financial management (covered in Chapter 19)
- Scheme design management (covered in Chapter 20)
- Manufacturing assurance (covered in Chapter 21)
- Site installation (covered in Chapter 22)
- Equipment commissioning (covered in Chapter 23)
- Project stage by stage (covered in Chapter 24)
- Competency (covered in Chapter 25).

The lowest level (tier 3) is usually a raft of subordinate procedures, mostly covering specialist requirements, but which are of no less importance.

17.2 Construction QMS procedures – structure and subject-matter

Figure 17.1 illustrates a suite of typical procedures required by a power network company for delivery of a construction scheme, incorporating the procedures identified above The list may vary from power network company to power network company – and not all are required from each and every scheme. Part of the scheme process is to select those procedures that are relevant to the scheme in question, as part of a scheme quality plan.

As previously stated, all parties participating in the construction process will have their own QMS which must dovetail at the procedural interfaces, especially those of power network companies (clients) and contractors. This chapter will

1. Scheme Investment Procedure

Project management

2.1 Project management*
2.2 Programme management
2.3 Resource management
2.4 Outage management
2.5 Risk management
2.6 Consents and wayleaves
2.7 Outstanding work
2.8 Lessons learned
2.9 Project audit
2.10 Project filing

Financial management*

3.1 Scheme investment management
3.2 Scheme financial management
3.3 Customer financial management

Manufacture, install, commission

4.1 Manufacturing assurance*
4.2 Site installation*
4.3 Equipment commissioning*
4.4 Project stage-by-stage*
4.5 Customer connections

Environmental management*

5.1 Waste management
5.2 Water management
5.3 Air management
5.4 Land management
5.5 Biodiversity
5.6 Nuisance, etc.

Design and technical

6.1 Scheme design management*
6.2 Protection settings
6.3 Drawings management
6.4 Earthing management
6.5 HV equipment nomenclature
6.6 Operations diagrams
6.7 SCADA systems
6.8 Metering
6.9 Factory testing
6.10 Operational communications
6.11 Thermal rating schedules
6.12 Fault level schedules
6.13 Protection and DAR schedules
6.14 Technical data
6.15 Asset data management

Health and safety

7.1 CDM regulations*
7.2 Temporary works
7.3 COSHH regulations
7.4 Work at height regulations
7.5 LOLER regulations
7.6 RIDDOR regulations
7.7 Fire safety
7.8 PPE regulations
7.9 Danger from underground services
7.10 First aid at work
7.11 Impressed voltages, etc.

Procurement*

8.1 Contract management
8.2 Purchasing management

Competency*

9.1 Engineering competency
9.2 Workforce competency

*Refers to Tier 2 procedures or suite of procedures

Figure 17.1 Suite of typical power network construction procedures

essentially focus on the scheme investment procedure (tier 1 procedure) as required by a power network company, since it encompass the end-to-end construction process. Contractor's procedures will largely mirror and align with those of the power network company – but commence later in the scheme process.

17.3 Scheme investment procedure

17.3.1 The scheme investment procedure – application

The scheme investment procedure (the title may differ from company to company) is the highest level procedure in the suite of QMS procedures and is effectively an enabling document linking to all other documents. For purposes of illustration, it will be assumed in the following sections of this chapter that the construction delivery model will be that of engineer, procure and construct (EPC) with a lump sum, fixed-price contract (as described in Section 16.3.1.4). It is assumed that the contract is awarded after a competitive tender process. The scheme investment process will be that illustrated in Figure 1.4. These arrangements are commonly used in practice, but other alternative arrangements will broadly follow the requirements outlined.

17.3.2 Scheme investment procedure – content

The scheme investment procedure usually comprises the stages shown in Figure 1.4, but only up to and including sanction, i.e. the decision to commit to the project. It additionally includes review and closure, i.e. an evaluation of whether the commissioned equipment accords with the specification and sanctioned cost together with formal close down of the scheme. The scheme investment procedure is, therefore, only concerned with the purely capital investment aspects on the scheme. The remaining stages of Figure 1.4 are covered in separate procedures. In summary, the scheme investment procedure includes:

- Need case
- Optioneer
- Develop
- Sanction
- Review/closure

The above stages will be examined in Sections 17.3.5–17.3.10.

17.3.3 Scheme team

The work involved in progressing the scheme from the need case to the point of sanction is usually very significant, requiring consultation, agreement and input from a wide range of departments in the power network company. This usually results in the formation of a 'scheme team'. The decision to form a scheme team usually resides with the scheme 'sponsor' – who is usually a senior figure in the power network company with the authority to commence the scheme process. The scheme team leader (chairman) is usually drawn from a team devoted to the investment process from need case to sanction, and later review/closure. A separate team under the project manager usually takes responsibility for the works following sanction. In summary, the scheme team typically comprises:

- Scheme team leader
- Power system design representative

- Customer facing representative (if a customer scheme)
- Procurement representative
- Legal representative
- Project manager
- Health and safety representative
- Environmental representative
- System operations representative (covering system outages and system integrity)
- Consents and wayleaves representative
- Engineering assurance representative
- Site engineering representative (covering site installation and commissioning).

The above represents the range of disciplines that are required to deliver a scheme.

17.3.4 Scheme timescales

Scheme timescales typically range from 6 months at one extremity to 8 years at the other. However, the majority probably fall in a range between about 1 year minimum, and about 4.5 years maximum. The example shown in Figure 17.2 shows typical minimum and maximum times for an averaged size new substation with connecting OHL and cables (NB: Most OHL and cable schemes would also fall into these timescales). The longer time durations are generally associated with the higher voltages; the obtaining of consents, wayleaves and planning permission; and significant earth/ground works.

17.3.5 Need case

The need case is based upon the drivers for the investment. These drivers generally fall into the categories of schemes summarised in Figure 17.3.

In the United Kingdom, customer connections are subject to certain regulatory requirements. For example, connections to the transmission system allow the transmission system owner (i.e. National Grid) a 90-day period to make a connection proposal to a customer, from the date of the customer making a valid application. In turn,

	Typical minimum time (months)	**Typical maximum time (months)**
1. Need case	1.0	3.0
2. Optioneering	1.5	4.0
3. Development	3.0	8.0
4. Sanction	0.5	2.0
5. Contract tender/release	2.0	6.0
6. Detail design/manufacture	3.5	16.0
7. Install/commission	3.0	14.0
8. Review/closure	0.5	2.0
Total time	**15.0**	**55.0**

Figure 17.2 Range of typical scheme times for a new substation

Load-related schemes

- New generation connection (customer driven)
- New transformer connection (customer driven)
- Strategic network reinforcement (non-customer driven)

Non-load-related schemes

- Asset replacement (like for like)
- Asset replacement (upgrade/modify)

Figure 17.3 Scheme categories

the customer has 90 days to accept the proposal. Connections to the UK transmission system must adhere to the 'connection and use of system code' (CUSC).

In summary, typical need case content will encompass the following:

- Type of scheme as outlined in Figure 17.3
- Name of customer if customer related
- System design report
- Projected completion date and stages
- Likelihood of meeting the completion date with risks
- Likelihood of achieving planning permission, consents and wayleaves
- Resource availability to deliver the scheme
- Scheme design specification (SDS) – skeleton
- Whether the scheme can be delivered in accordance with current policy and process
- Constraints and method of mitigation
- Cost estimate and error margin
- Approval to proceed by nominated officer of the company.

The last bullet point applies to each and every stage of the process and is sometimes referred to as a 'decision gate'. With connection schemes the 90-day offer period will often only allow time for the need case to be prepared and approved. In this instance, the chosen option would be subject to a proportionately greater degree of design than a non-load related scheme and would involve evaluating more options. It may be necessary to undertake system performance studies (i.e. load flows, voltage levels, etc.), and these would be recorded in the system design report.

17.3.6 Optioneer

The optioneer stage involves evaluating a number of possible options for the scheme with the objective of determining an optimum solution for development and taking forward to sanction. The preferred option may be determined by use of a table, which summarises the considerations, see Figure 17.4. The table can be scored and the scores weighted to obtain a numeric total for each; however, it is usually preferable for the scheme team to collectively evaluate the criterion based

Criterion	Option 1	Option 2	Option 3	Option 4
1. Design/technical performance				
2. Equipment availability				
3. Health and safety				
4. Environmental				
5. Flood risk				
6. Delivery model				
7. Programme				
8. Resource levels				
9. Sustainability				
10. Business risks				
11. Planning, consents, wayleaves				
12. System connection risk				
13. System complexity				
14. Buildability				
15. Maintainability				
16. Extendibility				
17. Operational communications				
18. Safety rules strategy				
19. ERTS				
20. Commissioning				
21. Spares holdings				
22. Training				
23. Cost-and-risk-margin estimate	£	£	£	£

Figure 17.4 Typical optioneering report scoring matrix

on a simple scoring system, e.g. 3 = good; 2 = satisfactory; 1 = concern – or other similar system, and the preferred option discussed and agreed.

The core activity of design and technical performance would be specified in a skeleton SDS – i.e. an upgraded version from that created at time of the need case. This is one and the same as the project-specific design specification – see Section 17.3.7. If not already undertaken during the need case, it may be necessary to undertake system performance studies during the optioneer stage.

17.3.7 Develop

The develop stage comprises the development of the preferred option to a level of completeness and accuracy to achieve sanction. All of the criterion shown in Figure 17.4 would be further developed plus additional technical requirements summarised in Figure 17.5. Surveys may need to be undertaken both to be assured that

Scheme Design Specification

1. Scheme title and reference number
2. Scheme approval names/titles
3. Scheme team names/communications
4. Scope and summary of the work
5. Key work interfaces
6. Summary of design to be resolved
7. Related documents and drawings – status
   (i) System design report
   (ii) Single-line diagram
   (iii) Layout and elevation drawings
   (iv) OHL and cable route drawings
   (v) Phasing diagram
   (vi) High-level project stage-by-stage diagram
   (vii) Survey reports
8. Circuit current rating

Circuit end points and kV	Circuit pre-fault, post-fault continuous – and short-term ratings

9. Substation rating

Substation name and kV	Substation continuous and fault level ratings

10. Fault levels

Substation name and kV	3 Phase/single-phase winter subtransient and summer minimum transient for infeeding circuits

11. Earth return current

Substation name and kV	Earth return current for each substation

12. Substation layout

Substation name and kV	Busbar layout type	New/existing	Number of additional bays	Insulation requirements	SLD reference

13. Substation earthing

Substation name and kV	Earth mat requirements – and linkage with other earth mats

14. Circuit breakers

Substation name and kV	CB nomenclature	Continuous current rating	Fault current rating

Figure 17.5 Continued

15. Busbars

Substation name and kV	Busbar type and current rating	Busbar pollution withstand levels

16. Disconnectors and earth switches

Substation name and kV	Disconnector/earth switch nomenclature	Disconnector current rating and requirement	Earth switch current rating and requirement

17. CTs and VTs

Substation name and kV	Circuit name	CT location and requirements	VT location and requirements

18. Substation auxiliary supplies

Substation name and kV	LVAC supplies	110-V DC supplies	48-V DC supplies

19. Transformers

Substation name and kV	Circuit name	Transformer vector group	Nominal rating	Cyclic rating	% Impedance	Phase connections

Tap changer type	Range and no. of taps	Earthing arrangements	Tertiary requirements

20. Surge arrestors

Substation name and kV	Surge arrestor location	Requirements

21. Feeder main protection

Substation name and kV	Feeder name	First main or second main	Protection type and comms link	Prot. signal type and comms link	Intertrip type and comms link

22. Feeder back-up protection

Substation name and kV	Feeder name	Back-up protection requirements

23. Transformer/other protection

Substation name and kV	Circuit name	Requirements

24. Busbar protection and circuit breaker fail

Substation name and kV	Busbar protection new/existing requirements	Circuit breaker fail new/existing requirements

Figure 17.5 Continued

25. Substation automatic control

Substation name and kV	Control system • Delayed auto reclose • Auto switching • Synchronising • Interlocking • Tap change control	Requirements

26. Metering

Substation name and kV	Circuit name	CT and VT location	Requirements

27. SCADA system

Substation name and kV	SCADA type and location	Remote control point location	Facilities to be provided

28. Recording and monitoring equipment

Substation name and kV	Circuit/location	Requirements

29. Telephony requirements

Substation name and kV	Requirements

30. Communication links

Substation name and kV	Requirements

31. Protection and control accommodation requirements

Substation name and kV	P&C cubicle/panel/Kiosk location and requirements

32. Substation civil, structural and building engineering

Substation name and kV	Task: • Buildings • Earth works • Environment • Oil containment • Roads/paths • Fences/gates/security • Landscaping • Line termination structures • Building engineering, etc.	Requirements

Figure 17.5 Continued

33. OHL

Circuit end points and kV	Length (km)	Conductor type and operating temp	Continuous current rating	Tower/pole type	Circuit phasing

34. OHL considerations

Location	Task:	Requirements
	Task: • Scaffolding and road/rail crossings • Duck under arrangements • Temporary OHL requirements • Access considerations • Substation structure terminations • Tower earthing arrangements • Tower earthworks • Other temporary works, etc.	

35. HV cable

Circuit end points and kV	Length (km)	Summer/winter current ratings	Cable type	Earthing/bonding arrangements

36. HV cable considerations

Location	Task:	Requirements
	• Cable route crossings • Oil considerations • Cable tunnel accessories/services • Cable system design calculations • Cable trenching • Civil engineering • Route traffic management • Temporary works, etc.	

37. Site installation

Location	Requirements (including stages)

38. Equipment commissioning

Location and kV	Requirements

39. Surplus equipment

Location	Requirements

Figure 17.5 Scheme design specification – typical content

ambiguities are removed, and to facilitate accurate costing. Sums of money may need to be allocated to undertake the surveys – sometimes requiring a mini scheme in its own right (typically termed a 'Preliminary Works Scheme'). Typical surveys may include

- Environmental
- Environmental impact assessment
- Asbestos
- Ecological
- Geotechnical/contamination survey
- Flood assessment
- Topographical
- Historical/archaeological
- Equipment condition assessment
- Fencing
- Earthing
- LVAC
- Ground penetrating radar (for underground infrastructure)
- Noise
- Transport routes
- OHL/cable route survey.

The completed design would be defined in a SDS (i.e. project-specific design) or a document of similar name serving the same purpose. A single SDS would usually be prepared for the whole scheme covering all substations, OHL, cables, etc. relevant to the scheme. The SDS content may be presented in a variety of ways, but many follow a tabular format similar to that illustrated in Figure 17.5. The scheme team leader would usually populate the SDS.

Stand-alone substation, OHL or HV cable schemes or permutations of the same, require only the relevant portion of the SDS to be populated. The format of Figure 17.5 may not be ideally suited to asset replacement schemes – and in such instances, a schedule of requirements may need to be appended to the SDS.

17.3.8 Need case, optioneer, develop – range of documentation

Figure 17.6 gives the typical range of documentation that usually needs to be prepared across the three stages of need case, optioneer and develop.

The purpose of the documentation listed in Figure 17.6 is summarised as follows:

1. **System design report (SDR)**
   The SDR provides the case for the construction work from a power system need perspective. It will identify the drivers for the scheme as summarised in Figure 17.3, and proposed solutions by which the drivers are satisfied, e.g.:
   (i) Establish a substation at a certain point in the power network for reasons of strategic reinforcement.

Pre-Sanction Documentation – Typical				
Document title	**Need case document status**	**Optioneer document status**	**Develop document status**	**Comment**
1. System design report	F	–	–	–
2. SDS	SD	FDA	F	R
3. Single-line diagram	SD	FDA	F	R
4. OHL/cable route diagram	SD	FDA	F	R
5. Outage programme	SD	FDA	F	R
6. Resource programme	SD	FDA	F	R
7. Delivery model	–	SD	F	–
8. Project programme	–	SD	F	R
9. Risk register	SD	FDA	F	R
10. Buildability	–	–	F	R
11. H&S/CDM	–	SD	FDA	Ongoing work
12. Environmental	–	SD	FDA	Ongoing work
13. Project stage-by-stage	–	SD	F	R
14. Panning permission	–	SD	FDA	Ongoing work
15. Consents/wayleaves	–	SD	FDA	Ongoing work
16. Cost estimate model	SD	FDA	F	–
17. Asset data workbook	–	–	SD	Ongoing work
18. Tech specs availability	–	–	Status confirm	R

SD, skeleton document; FDA, further detail added; F, finalised; R, refined after contract release.

Figure 17.6 Range and status of typical pre-sanction documentation

(ii) Replace all circuit breakers of a certain type due to reasons of obsolescence and failure.

The SDR will be supported by evidence from power system studies and accord with network planning standards or alternatively from asset condition reports. Power system studies generally examine the following (see Chapter 5)

(i) Load flows to evaluate
   (a) Voltage levels
   (b) Thermal ratings
(ii) Fault levels
(iii) Harmonics (power system quality)
(iv) Standing NPS and ZPS levels

2. **Scheme design specification (SDS)**
   The SDS contains the project-specific design specification (as outlined in Chapter 1) with typical model content shown in Figure 17.5. It is effectively the technical specification for sanction (i.e. that which is going to be constructed) and later forms the (project specific) technical specification for the Tender Enquiry Document.
3. **Single-line diagram (SLD)**
   A SLD is a single-phase representation of the electrical equipment to be constructed using standard symbols for equipment types (e.g. transformers, circuit breakers, CTs, etc.), see Section 14.2.2. The SLD is integral to the SDS but is frequently referred to in its own right.
4. **OHL and cable route diagrams**
   These comprise single-phase line diagrams of the OHL and cable routes usually drawn on an ordnance survey map or similar to specify the route.
5. **Outage programme**
   The outage programme comprises the dates and durations when parts of the existing power network are taken out of service both to facilitate the construction work, and the subsequent entry of the newly constructed equipment into operational service. This requires careful assessment and planning by the power network company's system operations department (i.e. those responsible for the control of the network) to ensure satisfactory performance of the in-service power system and continuity of electricity supply whilst construction work is undertaken.
6. **Resource programme**
   The resource programme comprises the profiling of the different resources and skills to deliver the project – with subsequent confirmation that the resources will be available when required.
7. **Delivery model**
   The delivery model is the chosen client/contractor working arrangements and interfaces for delivering the project (as defined in the contract) as outlined in Chapter 16.
8. **Project programme**
   The programme comprises a schedule of key tasks and dates for ensuring the project can be completed to the scheduled time. It usually specifies the degree of float (i.e. slack) associated with each date in the programme. Typical high-level programme dates would include:
   - Tender enquiry release
   - Tender return
   - Contract award
   - Detail design complete
   - Site access
   - Commissioning commences
   - Equipment enters commercial service
   - Scheme closure

9. **Risk register**
   A risk register consists of a list of hazards that may place project delivery at risk – together with measures to mitigate the risk. Many of the risks are common from project to project. Typical risks may include the following:
   - Failure of new equipment to pass factory tests by the due date. Mitigation may be by building slack in the programme and progressing the manufacture programme.
   - Failure to achieve consents by the due date. Mitigation may be by not releasing the contract until consents are obtained.
10. **Buildability**
    Buildability is confirming that there is an acceptable and logical sequence of activities for building the installation – and that it is achievable by the planned date. It includes access arrangements, site accommodation, clearances and the correct sequence of build.
11. **Health and safety/CDM**
    This comprises the requirement that health and safety considerations are accounted for in the construction works and that the CDM information pack will be ready for handover to the contractor after contract release.
12. **Environment plan**
    The environment plan must take account of environmental considerations such as waste management, biodiversity, air and water regulations, etc. – and the means of managing these issues. The environment plan will be continued by the successful contractor.
13. **Project stage-by-stage**
    This requirement comprises a single-line block diagram of the stages required to build the equipment, and at the appropriate time bringing the equipment under power network company safety rules, and subsequent commissioning on to the existing power network. This document will be progressed in greater detail by the successful contractor.
14. **Consents/wayleaves and planning permission**
    This comprises the evidence trail that provides assurance that the required consents/wayleaves and planning permission will be obtained by the due date required by the project programme. Work will usually continue after the development stage.
15. **Asset data workbook**
    The asset data workbook comprises a draft workbook listing all equipment that will form part of the scheme, with detailed requirements usually populated later by the successful contractor. The finalised data will be included in the company technical asset register – which produces data for a number of purposes including future asset replacement, identification of equipment types which may, for example, be subject to failure and to facilitate maintenance planning (i.e. to identify the equipment and its location).
16. **Cost estimate model**
    This facility comprises a model for estimating the price of the work (i.e. for sanction purposes) together with an estimated risk margin. Input to the model

will be from price schedules for previous similar work (e.g. cost of an eight-bay GIS substation). Contractors may also be requested to provide contract prices for generic types of construction work i.e. an indicative cost.

17. **Technical specification availability**
This is essentially a question of whether the client has sufficient technical specifications (i.e. project generic specifications) to cover all anticipated equipment types that will be required by the project. If this is not the case (perhaps because it is a new type of equipment) then a plan to remedy the situation must be prepared.

17.3.9 Sanction

With reference to Figure 1.4, the sanction stage consists of confirmation by appointed senior officers of the power network company (sanction committee) that the scheme may proceed – or alternatively either the scheme is rejected or additional work is required. A sanction paper is usually prepared and presented to the sanction committee by the scheme team leader. The sanction paper is invariably required to be brief, concise and to the point – with typical content as shown in Figure 17.7.

17.3.10 Scheme review/closure

Review and closure are usually two distinct stages at the end of the project when the work has been completed and final costs determined.

Sanction Paper – Typical Content

1. Title of the scheme and scheme no.
2. Why the work is required, i.e. a summary of the need case
3. Brief description of the chosen option
4. Other options considered and why the chosen option was selected
5. Outline of the project delivery programme with confirmation that it can be delivered
6. Outline outage programme with confirmation that circuit outages can be obtained
7. Confirmation that planning/consents/wayleaves can be obtained
8. Confirmation that resources are available to deliver the work
9. Health, safety and environmental concerns and method of mitigation
10. Major risks and method of mitigation
11. Confirmation that a SDS and associated documentation has been prepared, that it specifies an appropriate and optimum technical solution, and has been technically approved
12. Confirmation of customer acceptance of the proposal – where applicable
13. Financial appraisal. This comprises an evaluation of the cost of the project (usually at outturn prices), that it can be delivered for this cost, and that it represents value for money. The sanction risk margin is typically ±5% but may vary with the company and the scheme. The appraisal additionally comprises a commercial evaluation relating to predicted future revenues from the completed works, and return on capital
14. Confirmation of authority to proceed

Figure 17.7 Scheme sanction paper – typical content

1. **Scheme review**
   This comprises a lesson's learned review, i.e. what were the learning points from the whole scheme that may be factored into future schemes to improve performance, including changes to procedures.
2. **Scheme closure**
   This usually follows the scheme review and comprises a financial review of the investment and would typically consider:
   (i) Whether the scheme was delivered for the sanctioned value and reasons for deviation
   (ii) Whether the scheme has met its objectives and will provide predicted future revenues
   (iii) Financial lessons learned
   (iv) Whether the scheme should be subject to internal audit
   (v) Formal sign off, to confirm the scheme is completed, by an authorised officer of the company

17.4 Major infrastructure projects

17.4.1 Major infrastructure projects – outline

Major infrastructure projects may generally be defined as those that have a significant impact on the general public requiring public consultation, as required by the UK Planning Acts – see Chapter 2. They invariably involve significant addition/change to substations, OHL, cables, convertor stations, etc., with OHL being the most difficult to progress because of public concerns and frequent opposition.

As described in Chapter 2, all significant infrastructure projects are required to accord with the UK Planning Acts and cannot proceed until a Development Consent Order (DCO), sanctioned by the Secretary of State, has been obtained. With reference to OHL at 132 kV and above, the National Policy Statement for Electricity Network Infrastructure (EN-5) must be satisfied to achieve a DCO. EN-5 requires that a developer takes account of the following with reference to good OHL design practice:

- Biodiversity and geological conservation
- Landscaping and visual impact – including a requirement to accord with the Holford rules for OHL
- Noise and vibration (i.e. OHL crackle and hum and vibration from associated infrastructure such as transformers)
- Electric and magnetic fields (i.e. according with the levels specified in Chapter 11).

17.4.2 Scheme alignment with the planning act 2008

The scheme investment process must align with the requirements of the Planning Act 2008 and EN-5. Within this context, National Grid, the transmission system owner, has published a six-stage process to accord with the planning act – and by way of example, this model will be aligned with the scheme investment process outlined above. The six stages contained in National Grid's publically accessible document on the design and routing of new OHL are summarised as follows:

1. **Stage 1: strategic options**
   This comprises an evaluation of whether new infrastructure is needed, identifying and appraising options, evaluating the benefits, consultation with stakeholders and the determination of preferred options. This stage aligns with the need case.
2. **Stage 2 OHL routing and siting**
   OHL routing corridor studies are undertaken and evaluated. The Holford Rules underpin good OHL practice and are applied to each routing corridor considered to obtain an optimum solution. In such instances, recourse to (expensive) HV cables may have to be considered. After the end of Stage 2, full public consultation with stakeholders and communities is undertaken – following which a preferred route corridor is established. This aligns with the optioneer stage.
3. **Stage 3: detailed routing and siting**
   This comprises the development of the preferred option. Considerable attention is given to design that minimises impacts on people, settlements and environmentally sensitive areas. Consideration is again given to undergrounding if the Holford guidelines are difficult to satisfy. Ongoing consultation with stakeholders and communities take place together with an environmental impact assessment.

   NB: An environmental impact assessment is called for by the Town and Country Planning Acts 2011 (linked to EU directives). It is used to assist local planning authorities to assess whether planning permission should be awarded to projects, which have a significant impact on the environment. It may also be undertaken independently of planning permission requirements as good practice to assist evaluate and underpin development proposals. An environmental impact assessment usually follows a structured format and typically evaluates impact on humans and buildings; effect on flora, fauna and geology; effect on land, water and air; and other impacts such as traffic management and volume. Mitigation measures are specified.
4. **Stage 4: proposed application**
   This comprises a full public consultation with a consultation report – following which refinements may be made to the proposed development solution. The DCO would then be prepared.
5. **Stage 5: application for development consent**
   At this stage, the DCO is submitted to the planning authority; it may also include 'additional development' such as associated substations.
6. **Stage 6: consideration and hearing**
   This stage comprises the hearing and decision process by the secretary of state, resulting in the DCO being granted or refused.

Stages 4–6 align with the develop stage of the scheme investment procedure. Following granting of the DCO and finalising all relevant documentation such as the SDS, a sanction paper will then be prepared for the purposes of obtaining sanction to proceed with the work. For very major infrastructure schemes, it may be that a number of sanctions are undertaken to divide the scheme into manageable and logical stages.

Major infrastructure schemes are usually both challenging and demanding for those involved. Arriving at an optimum proposal requires the balancing of a number of key considerations – which are summarised as follows:

1. **Environmental**
   This includes visual, landscape, ecology, historical sites, air noise, water, waste, ground, geology, energy, etc.
2. **Socio-economic**
   This category includes people and communities, aviation, traffic, transport and local economic impact, etc.
3. **Technical**
   This is the construction of works that result in optimised power network performance taking into account factors such as risk of flooding, exposure to climate and security from third-party interference.
4. **Health and safety**
   This requires a proposal that results in good health and safety performance both during construction of the works, and later when in operational service.
5. **Cost**
   This relates to the capital cost of the project, the lifetime costs and the financial return.

17.4.3 Holford and Horlock Rules

The Holford and Horlock Rules greatly influence scheme design, particularly for major construction work. The UK Electricity Act 1989 requires that power network companies must consider the impact of projects on communities, landscape, cultural heritage, ecology and visual amenity. The Holford Rules (for OHL) and Horlock Rules (for substations) were both developed for the UK Central Electricity Generating Board *ca.* 1960 as a code of practice for good design. They are essentially concerned with landscaping and visual impact, and underpin modern-day design guideline documents such as EN-5 and are specifically referenced by National Grid in their design documentation. The following summarises the requirements:

1. **Holford rules (for OHL)**
   There are seven rules as outlined below:
   (i) **Rule 1**: Avoid altogether, if possible, the major areas of highest amenity value, even if route mileage is increased.
   (ii) **Rule 2**: Avoid smaller areas of high amenity value by deviating OHLs, providing that not too many angle towers are required.

- (iii) **Rule 3**: Choose the most direct route with no sharp changes in direction.
- (iv) **Rule 4**: Wherever possible, choose tree and hill backgrounds in preference to sky backgrounds.
- (v) **Rule 5**: Preferably route the OHL through open valleys with woods where apparent tower heights are reduced.
- (vi) **Rule 6**: Wherever possible, arrange that parallel and closely related OHL routes are planned with tower types, spans and conductors forming a coherent appearance.
- (vii) **Rule 7**: Where an OHL needs to pass through a development area, route it where possible to minimise impact on development.
- (viii) **Supporting notes**: Avoid routing close to residential areas; choose routes which minimise the effect on special landscapes; evaluate the use of alternate tower designs.

2. **Horlock Rules (for substations)**
   The guidelines contained in the Horlock Rules for substation design are summarised as follows:
   - (i) **Overall system options and site selection:** From the outset, environmental issues must be balanced against technical benefits and capital cost considerations.
   - (ii) **Amenity, cultural or scientific value:** Siting of new substations, sealing end compounds and line entries should as far as reasonably practical avoid international/national areas of highest amenity, cultural or scientific value, through the planning of substation connections.
   - (iii) **Local context, land use and site planning:** Siting of substations and extensions should take advantage of screening provided by land and existing features, together with the use of site layout levels to keep intrusion to a practical minimum.
   - (iv) **Design:** Early consideration should be given to the use and location of terminal towers, buildings and equipment, etc. to minimise visual impact.
   - (v) **Line entries:** In an open landscape line entries should be kept as far as possible visually separate from lower voltage OHL so as to avoid confusing appearance.

An environmental report is usually prepared with mitigation measures.

Both the Holford and Horlock Rules form the basis of much of OHL and substation spatial design requirements and are referenced by numerous power network companies.

17.5 Scheme investment procedure – complexity

This chapter has examined the scheme investment procedure, which is the tier 1 and highest level (umbrella) document in a power network construction QMS suite of procedures. The objective of the procedure is to obtain an optimum design and cost effective solution that satisfies the need case. This in turn requires a relatively complex process that must take into account a very significant number of considerations. Those who lead, and have responsibilities within, this process must be well versed and competent in the requirements, since the decisions made have a profound impact on the integrity of the power system and the financial wellbeing of the companies involved.

Chapter 18
Construction health and safety management

18.1 Safety management system

A safety management system (SMS) is an integral part of a quality management system (QMS), although it may also considered as being separate and sitting along-side the QMS. A SMS is a structured approach to health and safety (H&S) management. It is not however a legal requirement but is recommended by the HSE as an appropriate means for ensuring satisfactory management of H&S. In doing so, it satisfies the protective and preventative measures for managing H&S risk, as required by the management of H&S at work regulations 1995. It also satisfies the guidance provided in HSG65, successful H&S management, by following the specified approach to H&S based upon criterion of plan/do/check/act.

There are numerous definitions of an SMS, one example of which is as follows:

> An SMS is systematic approach to managing health and safety, including the necessary structure, accountabilities, policies and procedures

In essence, an SMS defines the organisation's structure (from CEO down), responsibilities, procedural arrangements and competency arrangements for controlling H&S risk and improving H&S performance. The procedural arrangements must specify how the HASWA and supporting regulations are enacted. All parties contributing to power network construction, i.e. power network companies' consultants, contractors and manufacturers, should have their own SMS. Figure 18.1 provides an overview of a typical SMS and how the construction design and management (CDM) regulations in particular are incorporated within it.

18.2 Health and Safety at Work etc. Act 1974

18.2.1 HASWA etc. 1974

Chapter 3 has overviewed the HASAWA including the main requirements of the act, relevant supporting documents and the duties of the HSE. Figure 18.2 summarises the key requirements of the act relevant to power network construction. These place significant responsibilities on manufacturers, contractors, power network companies and consultants.

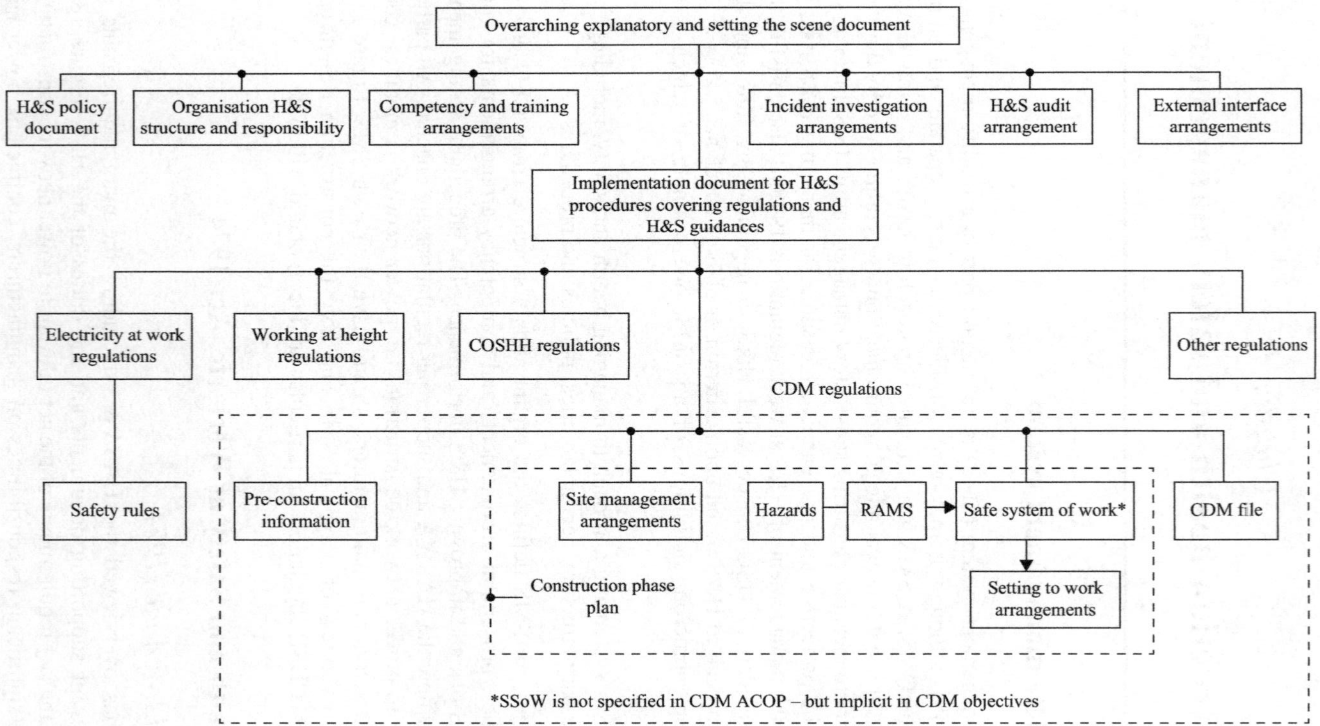

Figure 18.1 Structure and component parts of a typical SMS

• People with control of premises	Must ensure	• People other than their employees are safe without risk to health
Any person who • Designs • Manufacturers • Imports • Supplies Any article for use at work	Must ensure	The article is • Designed and constructed to be safe and without risk to health • Tested and examined to confirm safe and without risk to health • Provided with adequate information to ensure usage is safe and without risk to health
• Any person who erects or installs any article for use at work	Must ensure	• Erection or installation of article is safe and without risk to health

Figure 18.2 Summary of key requirements of the HASAWA relevant to power network construction

18.3 Key regulations and guidance documents

18.3.1 Key regulations and guidance documents

Key regulations and H&S guidance documents relevant to power network construction will be examined and the main points summarised. The documentation will need to be transformed into operational procedures, especially by power network companies and contractors – to ensure the requirements are enacted on construction sites. The following will be examined (NB: Reference should always be made to the documents themselves for a complete and comprehensive understanding):

- Management of H&S at work regulations 1999
- Control of substances hazardous to health (COSHH) regulations 2002
- Pressure system regulations 2000
- Work at height regulations 2005
- Lifting operations and lifting equipment regulations 1998 (LOLER)
- Reporting of injuries, diseases and dangerous occurrences regulations 2013 (RIDDOR)
- HSG 168 fire safety in construction
- Personal protective equipment at work regulations 1992
- Provision and use of work equipment regulations 1998
- HSG 47 avoiding danger from underground services
- Occupiers Liability Act 1957/84
- Electricity at work regulations 1989
- Safety rules
- The addition/removal of equipment to/from the system

- Temporary works
- Construction (design and management) regulation 2015.

The above list is not exhaustive but in the authors experience gives an insight into many of the most significant activities relevant to a construction site. NB: See also OHSAS 18001, see Section 2.10.1.

18.3.2 Management of Health and Safety at Work Regulations 1999

These regulations are essentially concerned with methods for reducing and eliminating H&S incidents. Some of the key requirements are summarised as follows:

1. **Risk assessment (regulation 3)**
   This regulation requires every employer to make an assessment of the risks to H&S both of employees at work, and non-employees who may be impacted upon by the work. Where the employer employs more than five employees, he shall record the findings of the risk assessment (RA).

   The regulations state that a 'risk' exists if there is a chance or possibility of danger or injury arising. The extent of the risk is dependent upon the product of (1) likelihood of harm arising and (2) severity of that harm. The regulations also define a 'hazard' as being something with the potential to cause harm. Therefore, a RA establishes the likelihood of potential harm arising from a hazard being realised.

   Once the RA has been undertaken, the actions to manage the risks need to be specified. This is usually referred to as the 'method statement' (MS). In practice, the RA and MS are collectively known as the RAMS – and organisations should have procedures in place defining the format of RAMS – which will be applicable to most work undertaken.
2. **Principles of prevention (regulation 4)**
   Regulation 4 specifies the principles to be adopted in deciding on the protective and preventative measures to manage the risk, these are specified as:
   (i) Avoiding situations with risk
   (ii) Evaluating risks that cannot be avoided
   (iii) Combating the risk at source
   (iv) Adapting the work to the individual
   (v) Adapting to technical progress
   (vi) Removing danger from a situation
   (vii) Developing a coherent overall protection policy
   (viii) Giving collective protective measures priority over individual protective measures
   (ix) Giving appropriate instructions to employees.

   In practice, the acronym ERIC, or more recently ERICPD, is used to summarise the above and is termed the hierarchy of risk control.
   (i) E – Eliminate
   (ii) R – Reduce
   (iii) I – Isolate

(iv) C – Control
(v) P – Personal protective equipment
(vi) D – Discipline

3. **Health and safety assistance (regulation 7)**
This regulation requires every employer to employ one or more competent persons to assist in delivering the measures to comply with legal requirements, i.e. the employment of a competent H&S professional. In practice, larger construction sites usually have an H&S specialist attached to the site.
4. **Contact with external services (regulation 9)**
This requires every employer to ensure that the necessary contacts with external services are arranged, particularly as regards first-aid, emergency medical care and rescue work. This would be required for major construction sites.
5. **Information for employees (regulation 10)**
This regulation requires every employer to provide his employees with comprehensive and relevant information on the following:
   (i) Risks to H&S
   (ii) Preventative and protective measures
   (iii) Procedures dealing with serious dangers, dangerous areas and fire fighting
   (iv) Identification of persons nominated to implement evacuation and fire-fighting measures
   (v) Risks notified by another employer in a shared workplace.
6. **Capabilities and training (regulation 13)**
This requires employers to provide employees with adequate H&S training:
   (i) When recruited
   (ii) When exposed to new and increased risks, new responsibilities, new equipment, new technology and new systems of work.

   The employer should ensure the demands of the job do not exceed the employees' ability to carry out the work. This re-enforces the requirement to ensure engineering competency.

18.3.3 Control of substances hazardous to health regulations (COSHH) 2002

Although construction sites are not the most extreme places to work, in interacting with hazardous materials, this is none the less a significant H&S management issue.

COSHH refers to any substance harmful to health and includes substances that are toxic, corrosive, biological agents, dust and a substance which because of its chemical properties creates a risk to health. Many everyday products such as paint, glue, lubricants, detergents and inks are included. Activities such as prolonged contact with wet cement in construction leading to chemical burns or dermatitis are covered by COSHH. Workers might be exposed to hazardous substances in the following way:

- Breathing in gasses/fumes/dust
- Contact with skin

- Contact with eyes
- Skin puncture

The COSHH regulations require employers to not carry out work where substances hazardous to health arc involved, without first undertaking a RA followed by the implementation of exposure control measures. The control measures should follow the principles of prevention outlined in Section 18.3.2.

Manufacturers of hazardous substances are required to label their products with symbols (which are common throughout the European Union). The symbols are accompanied by descriptions such as toxic, harmful, irritant, highly flammable, corrosive, etc. In addition, a safety data sheet must be provided with the substance – this will stipulate the dangers, the handling and storage requirements and the emergency response measures.

The regulations also require that where employees may be exposed to hazardous substances, that they are monitored and, in certain circumstances, placed under health surveillance. Employers are also required to provide sufficient information, instruction and training to employees. Construction sites require a procedure for implementing COSHH which must consider the following:

- Identification of hazardous substances entering site
- Handling and storage arrangements for hazardous substances
- Obtaining and storage of data sheets/information of the risks and the means of control and use of hazardous substances
- Those nominated and trained to access and use hazardous substances
- Means of confirming that proposed use of substance will not cause H&S danger
- Audit requirements.

18.3.4 Pressure system safety regulations 2000 (PSSR)

The pressure system safety regulations (PSSR) apply to all equipment that operate at a pressure greater than 0.5 bar (or at any pressure in the case of steam). With reference to power networks, PSSR apply for example to circuit breakers (SF6 and air) and air-storage vessels.

The regulations require employers to ensure that pressure systems are:

- Installed correctly
- Operated within safe limits
- Adequately maintained
- Accompanied by written schemes of examination (WSE)
- Examined by a competent person in accordance with WSE
- Accompanied by securely held records on construction, repair, modification and examination.

The WSE is a document that identifies the parts of a pressure system to be examined by a competent person. It includes all safety valves, pressure vessels and certain pipework.

A competent person is someone with the qualifications and skill to perform the examination identified in the WSE. Competent persons are usually accredited under the UK accreditation service. Relevant equipment cannot enter service without a WSE.

18.3.5 Work at height regulations 2005

Falls from height are one of the biggest causes of fatal injuries on construction sites. Work at height means any place either above or below ground level where a person could fall a distance to cause an injury. Thus, work associated with a cable trench is working at height. Previous issues of the regulations considered any height above 2 m as working at height, but this has been replaced by any height liable to cause injury.

The regulations require access equipment such as ladders, scaffolding, cradles, etc., to be of sufficient strength and stability, and to rest on a sufficiently stable and strong surface. On construction sites, the top of any guard rail is required to be 950 mm above the edge of the platform being worked from.

As much work as possible should be done from ground level and designers should design accordingly. Employers are required to ensure that all work at height is planned, supervised and carried out by competent people. HSE guidance states that a common-sense approach should be taken to working at height and the following should be adhered to:

- Do as much as possible from the ground
- Ensure that workers can get safely to/from their work
- Ensure equipment is suitable, stable, strong enough, maintained and checked regularly
- Do not overload or overreach
- Take precautions on or near fragile surfaces
- Provide protection from falling objects
- Consider emergency evacuation and rescue requirements.

18.3.6 Lifting Operations and Lifting Equipment Regulations 1998 (LOLER)

These regulations also have great relevance to construction sites. They apply to all lifting equipment including hoists, cranes, ropes, winches and loading/unloading of machinery, etc. In summary, the regulations require the following:

1. Lifting equipment shall be of adequate strength and capability.
2. Lifting operations and equipment must ensure that a person cannot be crushed, trapped or fall.
3. Lifting equipment shall be installed and positioned in such a way as to reduce the risk of striking a person or the load drifting or falling freely or being released unintentionally.
4. Machinery for lifting equipment should be clearly marked to indicate safe working loads.

5. Before lifting equipment is put into service for the first time, it must be thoroughly examined for defects. There must be a thorough examination after installation and, depending on conditions of use, examined every 6 months.
6. Examinations must be carried out by a competent person who is independent and impartial, and records must be kept.

18.3.7 Reporting of Injuries, Diseases and Dangerous Occurrences Regulations 2013 (RIDDOR)

The RIDDOR regulations require employers and other people in control of premises to report and keep records of:

- Work-related accidents that cause death
- Work-related accidents that cause specified serious injury
- Certain industrial diseases
- Certain dangerous occurrences

Specified serious injuries include (list is not exhaustive):

- Fracture other than for fingers, thumbs or toes
- Amputation of arm, hand, finger, thumb, leg, toe, foot
- Permanent loss or reduction in eyesight
- Crush injuries leading to internal organ damage
- Serious burns
- Unconsciousness
- Where a worker is away from work and is unable to perform normal duties for more than seven consecutive days
- Work-related accident involving a member of the public if that person is taken to hospital.

Examples of dangerous occurrences that are reportable include collapse, overturning or failure of load bearing parts of lifts and lifting equipment; plant or equipment coming into contact with an overhead power line; explosions causing work to be stopped for more than 24 h; electrical short-circuits or load which result in stoppage of plant for more than 24 h; failure of pressure systems.

Reporting must be done by a 'responsible person', usually the employer, to the HSE (for some occupational diseases reporting may be to the local authority). Notification in the first instance shall be without delay and the quickest means (e.g. telephone, email) and subsequently a report of the incident in an approved manner within 10 days of the incident. Forms are available on the HSE website.

18.3.8 HSG168 fire safety in construction

This guidance document is linked to a number of regulations concerned with fire safety. The following will summarise main requirements of HSG65; there are key five steps.

Step 1 is to identify the hazards, which are as follows:

- Source of ignition (flammable liquids, gasses, combustible solids)
- Fuel (naked flames, smoking, electrical equipment)
- Oxygen (air, oxygen cylinders, chemicals).

Step 2 is to identify the people most at risk. This would include the workforce, security staff, disabled people, general public, etc.

Step 3 is to evaluate, remove, reduce and protect from the risk. This may be achieved by:

- Evaluating the circumstances that give rise to the fire
- Removing/reducing sources of ignition/fuel/oxygen
- Reduce risk to people by use of escape routes, fire-fighting equipment, emergency plans, etc.

Step 4 is to record, plan, inform, instruct, train, i.e.:

- Record significant findings and the action to be taken
- Inform and instruct the workforce and visitors
- Train both those with designated fire-fighting duties and the workforce in general
- Prepare emergency plans. Consult the fire and rescue authority and incorporate into plans.

Step 5 is to review by monitoring the risk to fire on site; monitoring changes in site circumstances; review the emergency plans on a periodic basis.

18.3.9 HSG 47 avoiding danger from underground services

HSG 47 is concerned with construction work involving penetrating the ground at or below the surface, and in doing so avoiding damage to underground services – which if damaged may become a source of danger to H&S. The term services means all underground pipes, cables and equipment associated with electricity, gas, water, sewage, telecommunications and petro-chemical.

In summary, HSG 47 describes the following process:

- Locate buried services on plans and records
- Use locating devices to trace services
- Confirm services location by hand dug trial holes
- If the work is close to services carry out excavation work by hand digging – if not use tools of choice, e.g. excavator.

Problems encountered comprise the following:

- The plans/records are either not accurate or do not exist
- The locating device proves inadequate, or those using them do not possess requisite skills
- Little or no training on HSG 47, so no awareness of the risk.

In the author's experience, striking underground services remains an ongoing problem. As a result, some electrical network companies require that all those with responsibilities on electrical construction sites are trained and assessed against HSG 47.

18.3.10 Provision and use of work equipment regulations 1998 (PUWER)

These regulations require that equipment, including that used for construction, is maintained in good repair. Employers should ensure that employees who use work equipment or who manage/supervise its use have received adequate training in the risks involved, the methods of use and the precautions to be adopted.

Where equipment for ensuring safety at work depends upon installation conditions, it must be inspected before being put into service. Where work equipment is exposed to conditions causing deterioration, it must be inspected at suitable intervals and records of such inspections kept.

18.3.11 Person protective equipment (PPE) at work regulations 1992

Person protective equipment (PPE) refers to protective clothing, helmets, goggles, gloves, boots, high visibility clothing, respiratory equipment, ear protection, etc. The hazards addressed by PPE comprise physical (falling objects), electrical, heat, noise, chemicals, biohazards and airborne particulates. The regulation state that work should be arranged so that PPE is only used as a final resort. Where PPE is used, the following applies:

- It should be provide free of charge and be personally allocated to each employee.
- In no circumstances, should exemptions from PPE be allowed – even for a few minutes.
- Employers should ensure that PPE is maintained in sound condition, and repair and replaced and cleaned as appropriate.

It is worthy of note that the supporting guidance notes state that when providing PPE, there is need to avoid 'overkill'. An employee who wears every single item of PPE resembles someone from outer space and is likely to be a hazard to himself and others!

18.4 The occupier

18.4.1 Occupiers liability

Two acts address the occupier's liability, they are:

- Occupiers Liability Act 1957/1984 emanating from common law
- HASWA 1974 emanating from statute law.

Under both acts, an occupier of premises (i.e. land, buildings, offices, etc.) means that person who has 'control' of the premises. Both acts define in similar words the duty of the occupier to others who may enter the premises. The relevance of the above acts to power network construction is that they regulate the interfaces and responsibilities of the occupier, usually a power network company, with reference to a contractor undertaking work on an operational site.

18.4.2 Occupiers Liability Act 1957/1984

Section 2 (3) (b) stipulates that when a trade person, such as a painter or window cleaner, is exercising his trade, the person will not only appreciate the specialist skills associated with his trade, but will also guard against them. This means that when the trade is exercised on an occupier's premises, the occupier will be free to leave the tradesman to execute their trade – and there is not a higher duty of care placed on occupier, i.e. the occupier will not be liable for the trade person working in an unsafe manner.

The situation is somewhat different for a contractor working on an occupiers premise. Section 2 (4) (b) stipulates that where work is being done on the premises by a contractor, the occupier is not liable for the contractors safe working if the occupier:

- Took care to select a competent contractor and
- Satisfied themselves that the work is being properly done by the contractor.

Therefore, in the instance of a power network company (the occupier), employing a competent contractor all that is required by a power network company is that they monitor the work of the contractor to be assured it is being done properly. This monitoring is usually termed 'sensible' monitoring.

18.4.3 HASWA 1974 and control of premises

Section 4 of the HASWA covers the general duties of people concerned with premises to people other than their employees, who use the premises as a place of work. This would apply to a contractor working on the premises. The act goes on to say that every person who has control of a premises (i.e. the occupier) must ensure that the premises and all means of access/egress, and any plant/substance on the premises, or provided for use, is safe without risk to health.

The implication of Section 4 of the act is that an occupier, i.e. a power network company must take all reasonable practical measures, i.e. sensible monitoring, to ensure that a contractor does not breach any statutory requirement. It is to be noted that sensible monitoring (see also Section 18.9.12) serves two purposes:

- That site construction work is carried out to the required standard to safeguard both the contractors' workforce and any others who may be affected by the work, e.g. adjacent contractors or the general public.
- To ensure that equipment is installed and commissioned to the required standard, so that when it enters operational service, it is not a hazard to others, e.g. to ensure that a circuit breaker does not explode, or that a protection system works correctly, or that cable trench coverings do not collapse.

It is worthy of note that the concept of sensible monitoring by a client (to use CDM terminology) of a contractor also applies to the design phase of a project – since the client (in conjunction with the contractor) has a duty of care to ensure that equipment (including civil engineering installations) on entering operational service does not create a hazard as a result of design deficiencies.

It is acceptable for a contractor to assume some occupier responsibilities on an operational site when occupying a CDM zone within the site. This will be examined in more detail under the section on CDM.

18.5 Electricity at work regulations 1989

18.5.1 Guidance on regulations (2007)

Whereas the ESQC regulations, see Section 2.4.1, are focused on the power network, the electricity at work regulations are broader and more wide ranging, and apply and encompass everything from a 400-kV overhead line to a battery powered hand lamp. To assist in the interpretation, the electricity at work regulations are supported by a comprehensive guidance document, which will form the basis of this summary.

The regulations are very influential and underpin the following:

- BS7671 requirements for electrical installations (i.e. the IET wiring regulations)
- Power network company safety rules
- The design of power system equipment from a H&S perspective.

As with most other regulations, these regulations place duties on employers to comply with the provisions and employees to cooperate with their employer. The following summarises the main requirements of the regulations.

1. **Systems, work activities and protective systems (regulation 4)**
   This regulation requires that all electrical systems shall be constructed, maintained and operated to avoid danger. The regulations also require safe systems of work incorporating safety isolation procedures with a preference for conductors to be dead before work commences.
2. **Strength and capability of electrical equipment (regulation 5)**
   This regulation requires that no electrical equipment shall be put into use where its strength and capability may be exceeded in such a way as to give rise to danger. Therefore, a power network company, or consultant, when preparing a technical specification for equipment to be procured, or as part of a contract, must be assured that the specification satisfies this regulation. This has resulted in some power network companies subjecting equipment to rigorous test and inspection regimes – typically termed type approval, type registration or equipment certification. All of the latter would usually comprise very comprehensive tests and inspections on the first installation of an equipment, and once approved/registered/certified, subsequent installations would be subject to more limited tests and inspections and the equipment considered an 'off-the-shelf solution', see also Section 21.3 on type tests.
3. **Earthing or other suitable precautions (regulation 8)**
   Regulation 8 requires precautions to be taken by earthing or other suitable means, to prevent danger arising on a conductor, which is not a circuit conductor (i.e. not energised from the system), as a result of either the power

system being in normal use or subject to a fault. This refers to capacitive or inductive coupling from the system to the other conductor and the requirement for precautions to protect against such impressed voltages.

4. **Means for protecting from excess current (regulation 11)**
   This regulation recognises that faults and overloads may arise on electrical systems and the requirement that systems and parts of the system be protected against the effects of short-circuits and overloads, i.e. resulting in electrical currents that may give rise to danger. This regulation recognises that protection is likely to be in the form of fuses or circuit breakers controlled by relays. Due regard must be paid to the maximum short-circuit current with which the protective device may have to deal.
5. **Means for cutting off supply and for isolation (regulation 12)**
   The objective of this regulation is that to ensure that where necessary to prevent ganger, suitable means are available by which the electricity supply to any piece of equipment can be switched off. This also requires provision to enable the prevention of unauthorised, improper or unintentional energisation, e.g. locking off facilities. The requirement for locking has in some instances been problematic in matching UK equipment specifications to manufacturers' standard equipment designs.
6. **Precautions for work on equipment made dead (regulation 13)**
   This regulation relates to situations in which electrical equipment has been made dead in order that work either on it or near it may be carried out without danger. In particular, the regulation requires adequate precautions to be taken to prevent the electrical equipment upon which work is being carried out becoming electrically charged. It is recognised that safety isolation procedures formalised in written procedures or rules such as permits to work may form part of the written procedures. This regulation underpins the production of a suite of electrical safety rules which are prepared by virtually all power network companies.
7. **Work on or near live conductors (regulation 14)**
   Regulation 14 recognises that it may be necessary to undertake work on conductors that are live. But this is only acceptable if:
   (i) It is unreasonable for the conductors to be made dead
   (ii) It is reasonable for the work to be undertaken live
   (iii) Suitable precautions are taken to avoid injury.

18.6 Safety rules

18.6.1 Purpose of the safety rules

All power network companies and most consultants and contractors usually prepare their own safety rules. For electricity power systems, safety rules are provided to achieve the following:

- Safety from the system
- General safety.

General safety relates to all safety requirements which are not safety from the system. This would include safe methods of work (e.g. working at height, lifting operations, working with hazardous substances, etc.), a safe place of work, a safe working environment, safe access to and from the place of work, the correct use of PPE, etc. Some organisations may specify these requirements in a suite of procedures rather than a book of safety rules.

Safety from the system relates to safeguarding persons who are working on or near the electrical power system from dangers which are inherent in the power system. Such dangers include:

- Safety from the power system. This would typically include safety precautions when working on dead equipment (e.g. to prevent inadvertent re-energisation) or maintaining safe distance from live equipment or safety precautions when working on live equipment
- Safety from impressed voltages
- Safety from the mechanical or chemical, etc. dangers which are inherent within the power system – for example compressed air or sulphur hexafluoride.

The purpose of safety rules is to ensure that the requirements of the HASWA and subordinate regulations are enacted, and in doing so, people are working safely and without risk to health. The safety rules associated with safety from the system (i.e. the electrical safety rules) are closely aligned with the requirements of the electricity at work regulations. They are often issued in book format and in effect comprise a 'bible' of safe working practice on an electricity network. The remainder of this section on safety rules will examine electrical safety rules. It is to be noted that whereas all power network companies will have a suite of electricity safety rules, contractors will also require their own suite when constructing equipment prior to the equipment becoming under the jurisdiction of the power network company (e.g. when the equipment under the control of a contractor is within a CDM zone).

18.6.2 Safety rule certificates

Safety rules usually contain the following certificates:

1. Permit to work (PTW)
2. Sanction for test
3. Limited work/access certificate
4. Certificate for live LV work
5. Earthing schedule

These will be briefly examined below.

1. **Permit to work**
   A longstanding slogan for carrying out work on the HV power system is 'dead, isolated, earthed, cautioned and Permit to Work issued', and this greatly sums up the requirements of the safety rules for working on the HV system, see Figure 18.3. A typical PTW is shown in Figure 18.4 – the format may vary

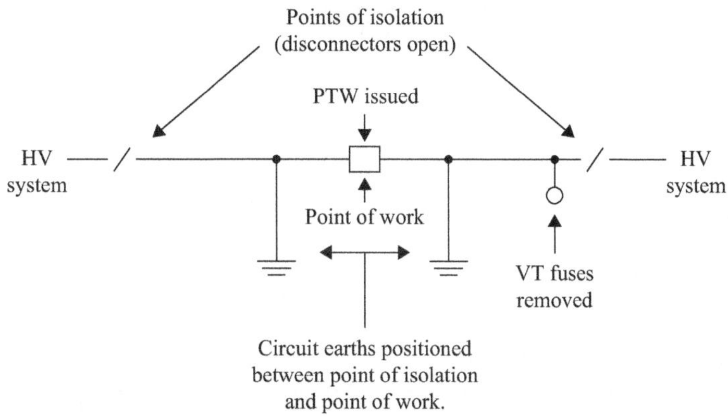

Figure 18.3 Safety rule requirements for working on the HV system

slightly from organisation to organisation. Some organisations term this a permit for work.

2. **Sanction for test**

   Activities such as primary injection of current transformers or high-voltage testing of cables require circuit earths to be removed to undertake the work. This type of work comes under the heading of testing, requiring the issue of a sanction for test (SFT). The SFT certificate has a very similar format to a PTW – but allows the removal of circuit earths for the work to proceed.

   Some organisations alternatively term a SFT a sanction for work, thereby allowing earths to be removed when undertaking work and not just testing.

3. **Limited work/access certificate**

   This certificate is used by many organisations to enable work to be carried out, under the safety rules, in close proximity to the power system – but outside the safety distances. It defines the limits within which work may be carried out and specifies the necessary precautions. The format of this certificate is similar to a PTW except:

   (i) It does not define 'points of isolation' and 'circuit earths' (which is required when working on the HV system).
   (ii) But instead it does specify the 'limits of work' and the 'precautions to avoid system derived hazards'.

4. **Certificate for live LV work**

   The certificate for live LV work is issued to enable work to be carried out on live LV equipment. The format of the certificate typically comprises:

   (i) Unique certificate number
   (ii) Location of work
   (iii) Work to be undertaken
   (iv) Hazards (to be avoided)
   (v) Precautions to be taken for a safe system of work
   (vi) Issue, receipt, cancellation and clearance similar to that for a PTW.

PERMIT TO WORK

No: ___________

1. Location:

   Equipment identification:

   Work to be done:

2. Points of isolation:

   Circuit earth positions:

   Action taken to avoid danger:

3. Issue and receipt:

   Issued by ____________________ Time _______ Date ________

   (Senior authorised person)

   Consent of ____________________ Time _______ Date _________

   (Safety control person)

   Received by ____________________ Time ______ Date _________

   (Competent person)

   Circuit Identification: ____________ Flags no _____ Wristlets no ______

   Earthing schedule no: __________ No of IV earths ________

   Safety keys: ______________

4. Clearance and cancellation

   Cleared by ____________________ Time ________ Date _________

   (Competent person)

   Consent of ____________________ Time ________ Date _________

   (Safety control person)

   Cancelled by ____________________ Time ________ Date _________

   (Senior authorised person)

Figure 18.4 Simplified example of a permit to work

5. **Earthing schedule**
   An earthing schedule stipulates earthing requirements to safeguard against impressed voltages, for each stage of the work. It typically comprises:
   (i) Unique schedule number
   (ii) Location of work
   (iii) Associated safety document number (e.g. PTW)
   (iv) Schedule comprising: stages of work; work description; location and number of IV earths

(v) Sketch showing IV earth position
(vi) Issue, receipt, cancellation and clearance similar to PTW.

18.6.3 Safety distance

Safety distance, with reference to the safety rules, is that distance from the nearest HV exposed conductor, which is not earthed, or from an insulator supporting a HV conductor, which must be maintained to avoid danger. Figure 18.5 gives the safety distances applicable to the United Kingdom, and Figure 18.6 illustrates the safety distances. Any infringement of safety distance must only be undertaken with the HV conductors dead, isolated, earthed and a PTW or SFT issued.

18.6.4 Safety rule duty holder responsibilities

Most safety rules in the United Kingdom would usually define the following duty holders:

- Person
- Competent person

Rated system voltage (kV)	Safety distance (m)
400	3.1
275	2.4
132	1.4
66	1.0
Up to 33	0.8

NB: Nearest distance of approach to any insulator supporting an HV conductor – but which is outside safety distance is 300 mm

Figure 18.5 Safety distances

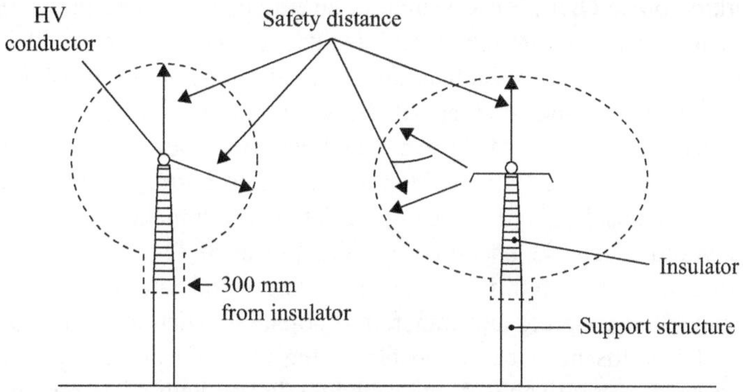

Figure 18.6 Illustration of safety distance

- Authorised person
- Senior authorised person (SAP)
- Safety control person (SCP).

The responsibilities of the above will be briefly examined:

- **Person**
  This is an individual with sufficient technical knowledge to avoid danger. Person status usually confers unsupervised mobility to move around a substation.
- **Competent person**
  This is a person who has been appointed to receive and clear safety documents such as PTW, SFT, limited work/access certificate. This is usually the person both undertaking the work and supervising any working party assisting with the work.
- **Authorised person (AP)**
  An authorised person is usually a person who has been appointed to carry out defined duties in writing which usually includes switching of substation equipment, e.g. circuit breakers, disconnectors, earth switches, etc.
- **Senior authorised person (SAP)**
  This is usually the most senior appointment under the safety rules. The SAP is appointed in writing to carry out specified duties such as the preparation, issue and cancellation of safety documentation such as PTW, SFT, limited work/access certificate, certificate for live LV work or earthing schedule. The SAP has a very influential role on any operational construction site.
- **Safety control person (SCP)**
  This is a person who has been appointed to be responsible for operational control and coordination of the power system within and across safety rule boundaries. The SCP would issue the unique certificate number and give consent for the issue of PTW or SFT, etc. by having an overview of the status of the whole power system under his/her control. They would normally operate from a central control room.

18.6.5 Safety rule practice when working on a OHL

When working on an OHL, it is essential to be sure that it is the right OHL. To this end, OHLs usually have a unique identifiers attached. Double circuit OHLs usually have a unique colour code for each circuit (e.g. yellow/white/black coloured bands) attached to the tower, and wristlets of the same colour coding are issued to the competent person in charge of the work and each member of the working party. In addition, a colour-coded circuit identification flag corresponding to the tower and the wristlets must be fixed to the appropriate socket/bracket on the tower. OHL identifiers would usually be specified on the PTW or SFT.

OHL are subject to significant levels of impressed voltage – i.e. capacitive coupling, inductive coupling and transferred potential. Therefore, before work can commence, IV earths must be connected to the OHL, in proximity to the work, usually in accordance with an earthing schedule. Power network companies usually have standard requirements for the number and location of IV earths. If, for example, work is being carried out on specific tower on an OHL, it is usual to fit IV earths not only on that tower but also to the towers on either side, and to earth all three phases. For a twin conductor circuit, this could involve as many as 27 IV earths, the fitting of which is not an insignificant task.

In the United Kingdom, all work on OHL is carried out in conjunction with a fall arrest system – to prevent injury from falls.

18.6.6 Safety rule practice when working on HV cables

As with overhead lines, it is essential to identify the cable to be worked upon. In the first instance, cable plans, etc., should be consulted followed by use of a cable locating device. Where a de-energised cable is to be worked upon and the cable cannot be conclusively identified, it must be spiked before a PTW or SFT is issued. This comprises a projectile fired in to the cable from a spiking gun to cause a short-circuit on the cable. If the cable had been live, it would be de-energised via protection systems. A disadvantage of spiking a cable is that further damage is done to the cable – which must subsequently be repaired.

HV cables may also be subject to considerable impressed voltages. Prior to work commencing the optimum method of protecting against IV must be determined. This may be by the application of IV earths or insulated working. This is discussed in more detail in Chapter 11 covering the subject of impressed voltages.

18.7 The addition/removal of equipment to/from the power system

18.7.1 Addition/removal of equipment to/from the system

When equipment is brought on to an operational site subject to the safety rules of a power network company, it is initially not part of the power system. Although on entering the site, the equipment is subject to safety rules such as those concerned with the movement of objects around the site and safety clearances to HV equipment, at some stage, it needs to be both formally declared as part of the power system and subject to those specific rules that apply to equipment which is part of the power system. A similar but reverse argument follows for equipment being removed from the system. The addition/removal of equipment to/from the system therefore covers the following activities:

1. Adding or removing HV equipment to/from the HV power system – and in doing so bringing the equipment under or removing it from, the safety rules that apply to HV equipment. This is usually undertaken via a formal certificate.
2. Adding or removing LV equipment, mechanical equipment and earthing to/from the power system – and in doing so bringing the equipment under or removing it from, the safety rules applicable to such equipment.
3. Changing the name of a circuit or the nomenclature of equipment. Again, this is usually undertaken via a formal certificate.

18.7.2 Addition/removal of HV equipment to/from the system

Figure 18.7 shows the typical format of a safety rules clearance certificate for formalising the addition/removal of HV equipment to/from the system. The usual

SAFETY RULES CLEARANCE CERTIFICATE

Part 1: Notice

To: Contractor/Department ____________________________

____

The equipment described below will be:

- Added to the HV System* (* delete as appropriate)
- Removed from the HV System*
- Subject to circuit/nomenclature change*

Effective from: Date: _______ Time: _________

No work shall be undertaken on or near the equipment after the above date/time unless sanction by a Senior Authorised Person in accordance with power network company safety rules.

Description of equipment

Issued by: ____________________________ Date: ________ Time: ________

(Power network company site representative)

Confirmed by: ____________________________ Date: ________ Time: ________

(Safety Control Person)

Part 2: Acknowledgement

I/we acknowledge the notice served in this Safety Rules Clearance Certificate and that all persons in my/our charge have been duly warned of the above.

Acknowledged by: ________________________ Date: ________ Time: ________

(Contractor/department)

________________________ Date: ________ Time: ________

(Contractor/department)

Part 3: Effective

I confirm that all contractors and/or power network company departments have acknowledged receipt of this Safety Rules Clearance Certificate which is effective at the Date /Time below.

Declared effective by: ____________________ Date: ________ Time: ________

(Power network company site representative)

Confirmed by: ________________________ Date: ________ Time: ________

(Safety Control Person)

Figure 18.7 Simplified example of a safety rule clearance certificate

method by which equipment is added and physically connected to the system is as follows:

- Issue a safety rules clearance certificate to declare both the new equipment and the connections between the new equipment and the system, as part of the system
- Issue a PTW to erect the connections between the new equipment and the system.

An alternative method, used by some companies, is as follows:

- Issue a PTW to erect the connections between the new equipment and the system
- Issue a safety rules clearance certificate to declare both the connections and the new equipment, as being part of the system, simultaneous with cancelling the PTW.

Removing HV equipment from the system uses the same process as the above with the wording on the safety rules clearance certificate being that of removal instead of addition. It is worthy of note that simultaneous with issue of a safety rules clearance certificate, any associated operations diagrams, and control and indications systems for the equipment (both locally and remotely) must become operational.

Careful consideration needs to be given, in a project, to the timing of making equipment subject to the safety rules. With reference to the example given in Figure 18.8, of the instance of a new substation connecting to a new cable and OHL, the following options need to be evaluated:

- Option 1: bring the new substation, cable and OHL under the HV safety rules all at the same time and on the one certificate – once site installation of all is complete.
- Option 2: delay the substation being subject to the HV safety rules until the off-load stage 1 commissioning tests are complete. Once the equipment becomes subject to the safety rules control moves from the contractor to the power network company, and of necessity much greater formality is required to work on the equipment, possibly slowing progress – so delay may be advantageous.
- Option 3: the cable and OHL contractors may complete their work in advance of the substation, and it may be beneficial to bring the cable and OHL under HV safety rules before the contractors depart site.

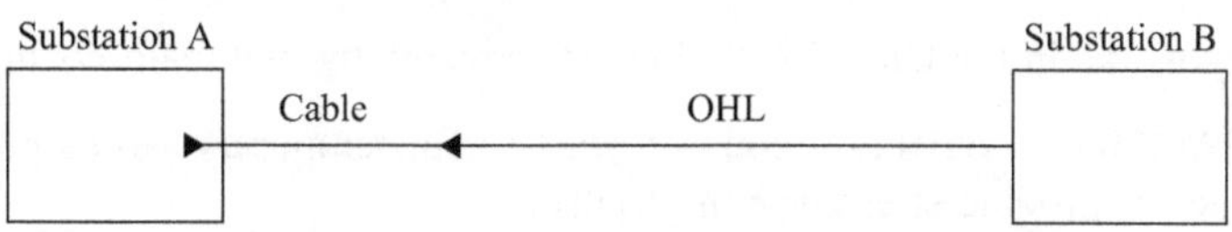

Figure 18.8 Safety rule clearance certificate – issue considerations

18.7.3 Addition/removal of LV equipment, mechanical equipment and earthing to/from the system

The requirements for adding/removing LV equipment, mechanical equipment and earthing to/from the system is usually a less formal process than that for the HV system. This is because the dangers are usually less, and therefore, the requirements for safety management are less extensive than those for the HV system. Methods by which this is achieved are as follows:

- Option 1: the equipment is considered to be part of (or removed from) the system immediately before (or after) it is connected (or disconnected) to (or from) the existing system, respectively. At this stage, all drawings and records must be correct and all who need to know have been notified.
- Option 2: the equipment is connected to the system whilst in a CDM zone and considered to be part of the system when the CDM zone is removed. The initial connection from the system to equipment within the CDM zone may require a documented agreement.
- Option 3: some companies still use the safety rules clearance certificate to add/remove LV equipment, mechanical equipment and earthing to/from the rules – but in some instances, this can become very bureaucratic.

18.7.4 Change of circuit name/equipment nomenclature

Change of circuit name and/or equipment nomenclature is usually formalised through issue of a safety rules clearance certificate. At the time of issue of the certificate, any operations diagrams, and control and indications systems must simultaneously be changed.

18.8 Temporary works

18.8.1 Temporary works – composition

Temporary works may be defined as those parts of the on-site works that allow or enable either the construction of, the protection of, or provide support for, or access to, the permanent works. Temporary works are usually removed after use. Examples include:

1. **Earth works:** e.g. trenches, structures, falsework, excavations, shoring, edge protection, temporary bridges, site fences and coffer dams, etc.
2. **Equipment/plant foundations:** e.g. tower crane bases, supports, anchors and ties for hoists, ground works for plant erection (such as mobile cranes and piling rigs) and scaffolding, etc.

Temporary works in the United Kingdom are covered by two main documents:

- BS5795 (2008 + A2011) – code of practice for temporary works procedures and the permissible stress design of falsework
- HSE publication SIM 02/2010/04.

Much of the site construction activity is associated with temporary works, and its design and execution is fundamental to the success of a project. As such a procedure defining the requirements for temporary works should be integral to the quality plan for the project.

18.8.2 Temporary works duty holders

Duty holders associated with temporary works comprise the following:

- Temporary works designer (TWD)
- Temporary works design checkers
- Temporary works coordinators (TWC)
- Temporary works supervisors (TWS)

These roles will be examined below:

- **Temporary works designer**
  TWDs are usually distinct and separate from the main works designers, largely because the nature of the design and skill set are different. They must be engaged early in the project to determine the requirements of the temporary design and therefore be assured that the permanent works can be constructed.
- **Temporary works design checker**
  Temporary works design checkers should be independent of the TWD. The greater the scale, complexity and novelty of the temporary works, the greater the required independence of the checker (e.g. the use of a separate company). Work should not proceed until confirmed by the checker.
- **Temporary works coordinator**
  A TWC is responsible for ensuring that a contractor's procedures for the control of temporary works on site are implemented. The TWC is preferably not the designer (particularly on larger scale works) and is responsible for ensuring both that a design specification is prepared and that the detail design is prepared, checked and implemented correctly on site, including the carrying out of on-site inspections and audits. The TWC would also ensure that the TWSs are briefed and their work coordinated.
- **Temporary works supervisors**
  On larger sites where a number of contractors are involved, it may be helpful to appoint one or more TWSs. A TWS would usually be responsible to the TWC and assist the TWC in the supervision and coordination of temporary works.

18.8.3 Temporary works documentation

Temporary works documentation usually comprises the following:

1. Temporary works design brief
2. Temporary works design
3. Design check certificate (where appropriate)
4. Permit to load certificate
5. Temporary works register

These will be briefly examined below:

1. **Temporary works design brief**
   A design brief should be prepared for each item of temporary works. The TWC should prepare the design brief and forward to both the TWD and temporary works design checker.
2. **Temporary works design**
   The temporary works design should accord with the design brief. The preparation of design calculations, drawings and specifications must be undertaken with the same rigour as that of the permanent works.
3. **Design check certificate**
   A design check certificate should be issued by the design checker for more complex temporary works design, to certify that the strength, structural adequacy and associated calculations are fit for purpose. Where empirical solutions are used, evidence should be available to demonstrate that they are based upon sound engineering practice.
4. **Permit to load certificate**
   A permit to load is recommended by the HSE for load bearing temporary works, especially for falsework. It would normally be provided and signed by the TWC to give permission for loading to commence. In some instances, a permit to dismantle is issued to formally signify that the temporary works can be dispensed with.
5. **Temporary works register**
   A temporary works register should be prepared for each project by the TWC. It should contain a list of all temporary works associated with the project. The format of the register should typically include:
   (i) Unique number and date of issue
   (ii) Description of temporary works
   (iii) Required completion dates
   (iv) Category of temporary works (as defined in BS5975)
   (v) TWD and temporary design checker names
   (vi) Date temporary works design complete
   (vii) Date temporary works design checked as satisfactory and approved
   (viii) Date permit to load was issued
   (ix) Date dismantling commenced or permit to dismantle was issued.

18.9 Construction (design and management) regulations 2015

18.9.1 CDM regulations – scope

The CDM regulations define legal duties for the safe operation of construction sites. They place specific duties on clients, designers and contractors to plan and manage H&S. They apply throughout a construction project from inception to final demolition and removal. The CDM regulations are the most significant regulations relating to a construction project and may be considered as umbrella regulations, since successful execution of CDM is dependent upon the successful execution of many other regulations, including all of the other requirements that are covered in this chapter. The CDM regulations implement the European Union Directive 92/57/EEC relating to construction sites and therefore contain requirements common to other EU countries.

The CDM regulations were first introduced in 1994 and were revised in 2007. The 2015 regulations contain the following main changes from those of 2007:

- Introduction of the principle designer (PD) role and withdrawal of the CDM coordinator (CDMC) role. The duties of the PD and CDMC are different.
- Introduction of a duty of information, instruction, training and supervision to replace the duty to assess competence.
- Replacing the previous ACOP with tailored guidance.

An ongoing objective of the updated CDM regulations is to reduce bureaucracy, with a focus on removing unproductive paperwork which can cause inefficiency and the obscuring of tangible H&S risks (it is argued that earlier versions of the regulations unintentionally created bureaucracy).

The CDM regulations apply to building, civil engineering or engineering construction work and includes the following (the list is not exhaustive):

- Construction, alteration, conversion, fitting out, commissioning, renovation, repair, upkeep, relocation or other maintenance, decommissioning, demolition or dismantling, of a structure.
- The preparation for an intended structure – including site clearance, investigation (but not site survey) and excavation (excluding pre-construction archaeological investigations), etc.
- Installation, commissioning, maintenance, repair, removal, of mechanical, electrical, gas, compressed air, hydraulic, telecommunications or computer services which are associated with a structure.

NB: The term structure is wide in its meaning and includes buildings, metal or reinforced concrete structures, railway lines, tunnels, bridges, pipes, cables, roads, river works, earthworks, dams, caissons, pylons, towers, formwork, falsework, scaffolding and any structure similar to the aforementioned. It also includes any formwork, falsework, scaffold or other structure designed to be used to provide support or means of access during construction work. The list is not exhaustive.

The CDM regulations do not apply to routine maintenance work.

18.9.2 CDM construction phases

The CDM regulations specify two phases of construction as follows:

1. **Pre-construction phase**
   The pre-construction phase means that period during which design and preparatory work is carried out – up to the point of site access commencing for site construction work to begin.

2. **Construction phase**
   The construction phase means that period from the point of site access commencing, for site construction work to begin – to the point of site construction work ending and the construction site closed down. The construction phase commences when the pre-construction phase ends – however, in practice, there may be some overlap.

18.9.3 CDM duty-holders

The CDM regulations stipulate the following duty holders:

- Clients
- Designers
- Principal designers
- Principle contractors
- Contractors
- Workers

The main duties of each will be briefly summarised below:

18.9.4 Clients

Clients are organisations or individuals for whom a construction project is carried out, e.g. an power network company. The main duties comprise:

- Appointment of the other duty holders
- Ensure sufficient time and resources are allocated to the project
- Make sure relevant information is prepared and provided to the other duty holders
- Make sure the PD and principle contractor (PC) carry out their duties
- Make sure that welfare facilities are provide for all on the construction site.

18.9.5 Designers

Designers are those who prepare or modify designs. They include architects, consultants, quantity surveyors, chartered surveyors or anyone who specifies a design. The main duties are as follows:

- Using the principles of prevention, eliminate, reduce or control foreseeable risks that may arise both during construction, and subsequent maintenance or use of the structure or workplace.
- Provide information especially relating to risks, to all other duty holders, to enable them to discharge their duties.
- NB: There may be a number of designers on a project including both the power network company or consultant's designers who prepare the contract technical specification, and contractor's designers who undertake the detail design following contract release.

18.9.6 Principal designers

A PD is appointed by the client for projects involving more than one contractor. The principal designer can be an organisation or individual. The main duties comprise:

- Planning, managing and monitoring the pre-construction phase of the project
- Identifying, eliminating or controlling foreseeable risks – using the principles of prevention
- Ensuring coordination and cooperation of those individuals involved in the pre-construction phase including ensuring the designers comply with their duties
- Providing pre-construction information to every designer and contractor appointed on the project
- Liaise with the PC and share information relevant to the planning, management and monitoring of the construction phase, including coordination of H&S matters
- NB: Instances may arise where the PD role may need to be split between the pre-contract release stage at which time a power network company or consultant may be appointed, and the post-contract release stage when a contractor representative will be appointed.

18.9.7 Principal contractors

A PC is a contractor appointed by the client where a project involves more than one contractor. The role of the PC is to coordinate the construction phase of the project; the main duties comprise the following:

- Plan, manage, monitor and coordinate H&S during the construction phase of a project. This includes:
  - Liaising with the client and PD
  - Preparation of the construction phase plan
  - Organising cooperation between contractors and coordinating the work.
- Ensuring suitable site induction.
- Ensuring that steps are taken to prevent unauthorised access.
- Consulting workers about H&S.
- Ensuring welfare facilities are provided, i.e. sanitary conveniences, washing facilities, drinking water, changing rooms, lockers and facilities for rest.
- NB: for the instance of only one contractor being on the project then the contractor must discharge some of the PCs responsibilities, e.g. prepare a construction phase plan.

18.9.8 Contractors

Contractors are those who do the actual construction work and can be an individual or a company. Main duties include:

- Plan, manage and monitor construction work under their control so they are carried out without risk to H&S.
- Comply with directives given by the PD and PC.

18.9.9 Workers

These are people under the control of a contractor. Their main duties include:

- The need to be consulted on matters which affect their health, safety and welfare.
- Taking care of their own H&S and that of others.
- Reporting anything which is likely to endanger their own and others' H&S.
- Cooperating with their employers, fellow worker, contractors and other duty holders.

18.9.10 CDM documentation

The CDM regulations require the following documentation to be prepared.

- F10 notification form.
- Pre-construction information.
- Construction phase plan.
- Health and safety file.

These will be summarised briefly below.

18.9.10.1 F10 notification form

The F10 notification form is required for a notifiable project. It is completed by the client. A notifiable project must be notified to the relevant enforcing authority, usually the HSE as soon as practicable before construction phase begins. The criterion for a project being notifiable is that the construction work on a construction site is scheduled to either:

- Last longer than 30 working days and have more than 20 workers working simultaneously at any point of the project.
- Exceed 500 person-days.

The F10 form can be obtained from the HSE website. Typical information to be entered onto the F10 include:

- Address of the construction site
- Brief description on the project
- Name, address and contact details of the client
- Details (names, addresses, etc.) of both the principal designer and principle contractor
- Date planned for start of the construction phase
- Planned number of contractors and estimated number of people on site
- Details of any designers and contractors already appointed.

The F10 in addition to being a source of information to the HSE also prompts the HSE to consider inspecting the site. When completing the F10 many organisations state the duty holder as being the company but adding a named individual as the point of contact.

18.9.10.2 Pre-construction information

Pre-construction information is information relevant to the project which is already in the clients' possession (e.g. existing H&S file, structural drawings, etc.). The client has the main duty of providing pre-construction Information to the PD/designers and the PC/contractors. It is expected that the PD will assist the client in assembling the information.

With reference to power network construction the client, usually the power network company, will be required to provide pre-construction information to the PD and PC in the contractor organisation. Where the project takes place on an already operational site (e.g. an operational substation), the interface arrangements with the occupier, i.e. the representative of the power network company, often the SAP, must form part of the information flow. This often results in a structured approach to providing pre-construction information – often necessitating a formal meeting. Figure 18.9 provides a typical agenda for such a meeting where information is discussed, communicated and handed over.

18.9.10.3 Construction phase plan

The construction phase plan must be prepared by the PC prior to setting up the construction site. The plan is sometimes referred to as the 'Construction Phase H&S Plan' or even the 'Construction Phase SHE plan' where the 'E' in SHE refers to environmental plan considerations, see Section 22.4.1 (and 'SH' comprises safety and health). The CDM regulations require the construction phase plan to address the following:

- The site rules (i.e. site management arrangements) for managing the construction site).
- The arrangements for identifying and controlling the H&S hazards and risk.

A typical list of contents for a construction phase plan for a large power network construction project is given in Figure 18.10.

NB: The list is not exhaustive.

18.9.10.4 Health and safety file

The H&S file is only required for projects involving more than one contractor. It is prepared by the PD. At the end of the project, the PD or PC (if the PD has completed his work earlier) must transfer the file to the client for the client to retain. The file should be complied as the project progresses, and typical contents are summarised as follows:

- Brief description of the work
- Hazards not eliminated and how they have been addressed
- Key structural principles
- Hazardous materials used
- Information for the future removal or dismantling of installed plant
- Location and marking of significant services, e.g. cables and pipes
- The as-built drawings

Pre-construction Information Meeting – Agenda

- Project description
- Details of duty holders
- Health and safety goals
- Project programme and key dates
- Status of design, site installation and commissioning documentation
- Schedule of known H&S hazards and risks – and method of control (possible inclusion in construction phase H&S plan)
- Workplace specific requirements
- Status of key drawings [e.g. site hazard plan; services search records; stage-by-stage drawings; survey results (geotechnical, contamination, asbestos, etc.)]
- Procedural arrangements (e.g. pre-site access meetings; site installation meetings; relevant clients procedures)
- Welfare provision
- Site vehicle movement arrangements
- Safety rules application arrangements (clients and contractors)
- CDM zone arrangements
- Fire and emergency procedures
- Arrangements for coordinating on going design
- Site security arrangements
- Environmental health hazards
  - Asbestos
  - Storage of materials
  - Contaminated land
  - Structures containing hazardous materials
  - Waste management
  - Waste/effluent discharge
  - Air emissions
  - Impact on adjacent land use

Figure 18.9 Typical pre-construction information meeting agenda

The purpose of the file is to enable risks arising from the present project to be identified – particularly if they should impact on future projects. The H&S file should be categorised into civil engineering, HV substations, OHL and cables. Careful thought should be given to where it is located and where that location is documented, so that those requiring to access the file in future are able to find it. Locations may be in the company documentation office of a major substation or an office location.

18.9.11 Principal contractor – occupier interface

It will often be the case that construction and the application of the CDM regulations take place on an operational site, such as a substation, under the control of the occupier (usually the representative of the power network company). In such an instance, the work could be undertaken under the occupiers' safety rules. However,

Construction Phase Plan

Contents

Site management

- Description of the project
- Programme and key dates
- Details and communications with duty holders
- Management structure and responsibilities
- Arrangements for liaison, consultation and exchange of information between workers
- Design change process
- Site induction
- Site security
- Site training
- Welfare and first aid
- Site audit
- Fire and emergency plans
- Reporting of investigations and accidents
- Clients site-management requirements
- Contractor site-management arrangements
- Site filing

Hazard and risk control

- Hazard and risks register
- RAMS preparation
- Safety rules
  - Contractors
  - Clients
- Procedures
  - COSHH
  - Working at height
  - LOLER
  - Pressure systems
  - Control of IV, etc.
- Environmental procedures
  - Contaminated land
  - Air emissions
  - Waste effluent discharge, etc.
- Third party impacts

Figure 18.10 Typical construction phase plan – contents

where the scale of the construction work is large, it is usually more convenient and efficient to create a separate, secure and delineated area in the operational site where occupier responsibility for that area is transferred from the power network company to the contractor, i.e. from the occupier for the substation, usually the

SAP, to the PC. The transferred area is frequently termed the CDM zone and access to the zone must be on roads allocated exclusively for that purpose. The PC becomes responsible for safe working in the zone, including the general duty of care, access/egress arrangements, welfare facilities and the application of the PCs own safety rules and safety procedures. In this instance, the substation occupier (e.g. the SAP) will sign into this zone as a visitor. The creation and removal of a CDM zone should be via formal transfer, and the zone and all access roads clearly marked on a drawing and notified to all working on site. CDM zones should be enclosed by a fence (e.g. heras), barriers or walls of a building, proportionate to the risk of unauthorised entry. It should be noted that the substation occupier is still required to undertake a level of sensible monitoring in a CDM zone. CDM zones can also apply to cable or OHL projects.

A CDM zone may apply equally to a green field site but again requires formal transfer of occupier responsibility from power network company to PC.

18.9.12 Sensible monitoring

The earlier section examining the role of the occupier explained why sensible monitoring by the occupier of a contractor is required. In summary, sensible monitoring is required to provide assurance for the following:

- That satisfactory H&S requirements are being carried out by the contractor whilst site construction work is being carried out and
- That the quality of work being undertaken by the contractor during site construction is satisfactory and does not result in H&S incidents arising from the equipment after entering operational service.

Sensible monitoring is essentially a visual inspection of site construction work proportionate to the scale and complexity of the work. It typically should include the following:

- Site access arrangements
- Safety from the system
- Scaffolding suitability
- Traffic and plant management
- Use of tools and machinery
- Fire/emergency arrangements
- Welfare facilities
- Protecting the public
- Contractor working to own procedures
- Filing and records

Sensible monitoring is about getting the balance of monitoring correct. Where the occupier has confidence in the contractor, say through working together previously, then the scale of sensible monitoring can be reduced accordingly. Sensible monitoring is carried out by representatives of the occupier, i.e. the power network company. As such it could be carried out by the SAP or a site construction engineer. Sensible monitoring involving safety from the system will usually involve the SAP.

18.9.13 Interfaces between CDM zones

Often situations arise involving interfaces between CDM zones and the PCs or contractors with responsibility for those zones. Figure 18.11(a) shows the instance of two PCs each with a CDM zone and working for different clients. This is allowable but access arrangements must not cause the PC for one project to pass through the CDM zone of the other PC. Should this be the case the PCs must cooperate at the interface to find a solution that ensures satisfactory H&S.

Figure 18.11(b) illustrates the case of an OHL PC/contractor requiring to connect a down-lead to a landing structure which is in the CDM zone of another PC/contractor. Two options exist for managing the interface:

- PCs/contractors cooperate at the interface, or
- One PC/contractor takes overall control.

In most instances, the work is most easily accomplished by cooperation at the interface – with the OHL PC/contractor being inducted into the CDM zone of the substation PC/contractor.

Figure 18.11(c) shows the case of a cable under the control of PC/contractor 1 passing through the CDM zones of PC contractors 2 and 3. Options for managing the interface are:

- PC/contractors 2 and 3 transfer some of their CDM zone to PC/contractor 1 so that PC/contractor 1 has a CDM zone covering the whole of the cable, or

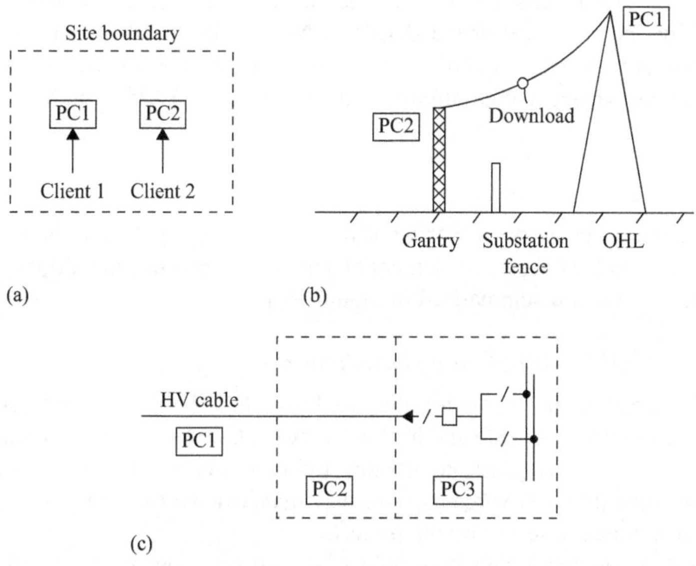

Figure 18.11 Examples of interface arrangements between CDM zones: (a) two PCs on same site, (b) PC interface on different sites and (c) the cable passing through two CDM zones

Safe System of Work

1. Description of work to be undertaken
2. List the step-by-step sequence of work in a 'schedule of work'
3. Specify the requirements to make the work area safe
4. Specify the requirements to make the environment safe (e.g. lighting, ventilation, heating, etc.)
5. Specify PPE requirements
6. Subject the work to RAMS
7. Incorporate 3, 4, 5, 6 into 2 above
8. Enact setting to work process prior to commencing work
9. Incorporate lessons learned into future SSoW

Figure 18.12 Typical contents of a safe system of work

- PC/contractors 2 and 3, respectively, become the sole PC/contractor for all work in their CDM zone including that of cable. In effect PC/contractor 1 assumes contractor status to PC/contractors 2 and 3, respectively, when in their CDM zones.

Advance consideration and planning should be given to CDM interface arrangements, and which safety rules will apply. This will require the agreement of all relevant parties – usually requiring a meeting specifically for this purpose.

18.9.14 Safe system of work

All work must be undertaken in accordance with a safe system of work (SSoW) as required by Section 2.2(a) of the HASWA. Neither the HASWA, any supporting regulations nor the HSE actually specify the content of a SSoW, but numerous specialist texts suggest the requirements shown in Figure 18.12 form the basis of a SSoW.

18.9.15 Setting to work

The term setting to work (StW) is not defined by any regulations or supporting documents – but StW forms an element of sound H&S practice in a SSoW. Typical StW requirements are summarised in Figure 18.13.

18.9.16 CDM competency requirements

The CDM regulations recognise the need for duty holder competence as an essential requirement for safe and healthy working. CDM regulation 8 requires that designers, principal designers, contractors and PCs have the skills, knowledge and experience (and if they are organisations the organisational capability) to secure the H&S of any person affected by the project.

Clients when appointing duty holders should carry out sensible and proportionate enquiries about their capability to undertake work – and designers and contractors must be able to demonstrate that they possess the requisite H&S skills, knowledge and experience. It is worthy of note that the competencies required are

Setting to Work

1. Supervisor and worker both confirm
   (i) That they have been inducted onto the site.
   (ii) That they have the necessary proof of competence for the work to be undertaken.
2. With reference to a schedule of work, the supervisor explains to the worker
   (i) The work to be undertaken
   (ii) The RAMS to be followed
   (iii) Interfaces with the work
   (iv) Action to be undertaken if there is a problem with the work (usually notify the supervisor)
3. The worker demonstrates to the supervisor that the requirements of 2 above are understood
4. The worker discharges the work
5. On completion of the work, the worker demonstrates to the supervisor that the work has been discharged as specified
6. The supervisor assures himself that the work has been discharged as specified and signs off the schedule of work as complete
7. NB: In some instances, a safety document such as a PTW may be required

Figure 18.13 Typical contents of a setting to work procedure

not only those relating to H&S when undertaking site construction work, but the engineering skills to design, install and commission the equipment so that equipment is not a hazard during operational service.

The CDM guidance notes recognise that membership of appropriate professional institutions is a good indication of competence but in addition recommends that questions should be asked of individuals to ensure they have sufficient skills, knowledge and experience relevant to the work to be undertaken. The guidance notes also make reference to the use of PAS 91 (2013) and the standard competency-based questions relating to construction pre-qualification tendering, see Section 2.9.1.

18.9.17 Execution of CDM within a safety management system

Figure 18.1 illustrates how the CDM and other regulations are nested within a SMS, and in turn how both SSoW and StW requirements are nested within the CDM regulations. As stated previously, the CDM regulations are the umbrella regulations for site construction, and to successfully deliver site construction in accordance with CDM, all the other requirements in this chapter must be satisfactorily undertaken. To achieve this objective, all of the documents examined must be transformed in to operational procedures and safety rules, with the concomitant requirement that the network construction workforce is competent in working to the requirements of those procedures/safety rules. Sound H&S practice invariably results in an effective, efficient and professional construction site. It is worthy of note that although this chapter has been written from a UK perspective, many of the requirements are common to the rest of Europe and increasingly the rest of the world through harmonisation of standards and best practice.

Chapter 19
Project management procedures

19.1 Project management – overview

Project management is the overarching task in the delivery of a project. Many authoritative texts have been written about the theory and practice of project management, and therefore, this text will focus narrowly on the requirements of power network construction. The only exception to the previous statement may be to recognise the three classical dimensions of a project, i.e. all projects involve the three interlinking factors of time, cost and quality. These are illustrated in Figure 19.1. The relative importance of these factors will vary from project to project. For example, a new substation and associated circuits for feeding, say, an Olympic stadium may have to be completed by a certain date, and therefore time becomes the dominant factor. In practice, of course, all three factors have to be kept under control for a project to be deemed successful.

This chapter will focus on both the role of the project manager (PM) and the key tasks that comprise project management (NB: health and safety (H&S) is covered in Chapter 18 and environmental management in Chapters 2 and 22). In summary, the following will be examined:

- The role of the PM
  - Power network company PM
  - Contractor PM
- Project programme
- Resource management

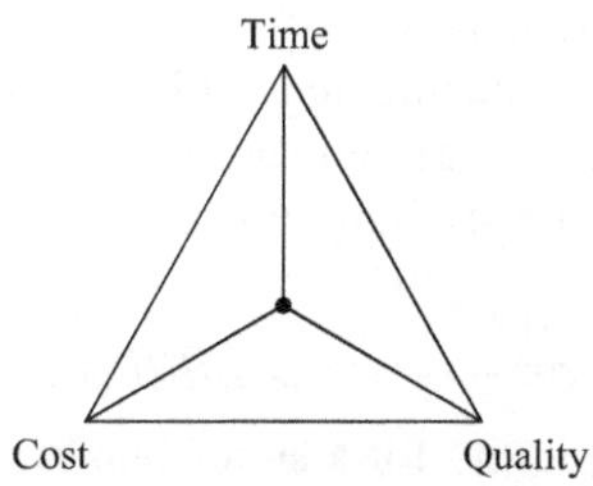

Figure 19.1 Time/cost/quality triangle

- Outage management
- Risk management
- Financial management
- Contract management
- Consents and wayleaves
- Project filing and project audit
- Project quality management
- Outstanding works and post-project review.

As implied above, a project would usually have two PMs: that of the power network company; and that of the contractor. This text will focus mostly on the role of the power network company PM – as this role tends to run for a longer period of time, although the two roles are similar in their tasks and the responsibilities that they are required to manage.

It is worthy of note that the term 'programme management' is frequently used for the management of a portfolio of projects.

19.2 Project manager (power network company)

19.2.1 Project manager (power network company) – accountabilities

The power network company, PM, is usually accountable for delivery of the sanctioned scheme in accordance with the sanctioned specification, the sanctioned programme and the sanctioned sum (i.e. quality, time and cost). It has famously been said that the means by which a PM delivers a project is about two key competencies: planning and managing interfaces. However, a more comprehensive list is summarised as follows:

- Preparing project plans (programmes)
- Reviewing progress against the plans and taking corrective action
- Building and motivating the project team
- Ensuring the smooth passage of information throughout the project to all concerned
- Preparation of progress reports, especially for key stakeholders (e.g. the scheme sponsor)
- Conducting lessons learned reviews
- Constant focus on the big picture and not becoming distracted by detail
- Managing the project to ensure 'no surprises', i.e. to identify the mire in the distance and not to end up standing in it.

19.2.2 Project manager – responsibilities to be managed

The PM has overall responsibility for a significant range of tasks as illustrated in Figure 19.2. The content of each will be examined in greater detail in both this and later chapters.

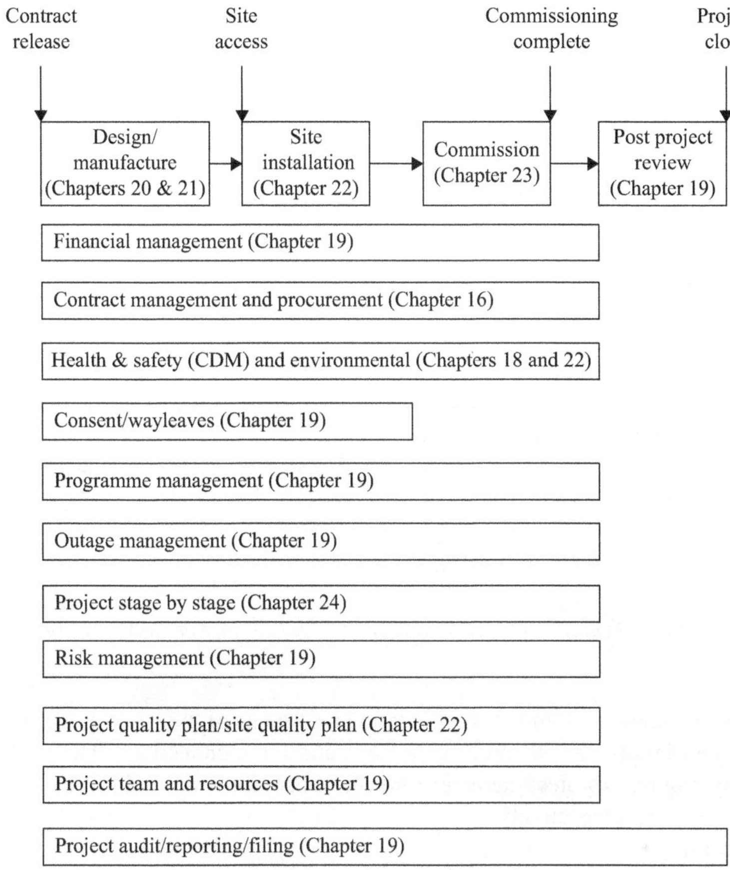

Figure 19.2 Project manager responsibilities – typical

19.2.3 The project team

Figure 19.3 illustrates a typical project team (sometimes termed as multi-disciplinary team). Not all projects require each and every team member, and not all are required at each and every stage of the project. The main duties of each team member are as follows:

1. **Project manager**
   The PM has overall client side accountability for the delivery of a project, and in accordance with any contractual terms and conditions. Some PMs specialise in substations, OHL, HV cables or civil projects, respectively. On very large projects, the term 'project director' may be used.
2. **Project engineer**
   On larger projects, the scale of the work may exceed the capacity of the PM to deliver the whole project, and it would be usual to appoint one or more

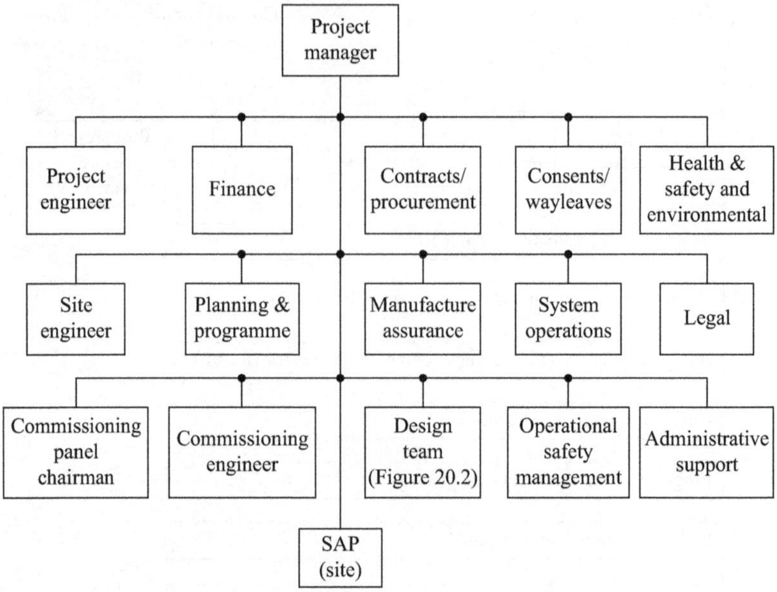

Figure 19.3 Typical project management team – power network company

'project engineers' to assist with defined elements of the work. For example, a project involving say three substations and interconnecting OHL and cables may require a project engineer for each substation, and each of the OHL and cable work, respectively.

3. **Finance**
   The project finance officer for the project (usually an accountant) would be responsible for assisting the PM with all financial aspects of the project – particularly project accounting.
4. **Contracts/procurement**
   The officer representing contracts/procurement would usually be a specialist in this discipline and would provide the formal interface with the contractor on the terms and conditions of the contract – and on any items of equipment, etc., to be procured. For example, an item of equipment may be procured outside of the contract and free issued to the contractor.
5. **Consents/wayleaves**
   The consents/wayleaves officer would again be a specialist in this field and would provide the project interface on all matters associated with planning consents and wayleaves, etc.
6. **Health and safety and environmental**
   This may be combined or separate functions. The H&S team member would advise the PM on all matters relating to H&S particularly relating to the CDM regulations and on H&S audits. The environmental team member would

advise on all matters relating to the environment, e.g. flora, fauna, historical sites, waste management, etc.

7. **Manufacturing assurance**
   The manufacturing assurance engineer is responsible for engineering assurance during the manufacturing stage – in particular, expediting the manufacturing programme and quality standard of the manufactured equipment.
8. **System operations**
   System operations (typical term) is responsible for the instantaneous control of the power system to ensure continuity of electricity supply. The system operations representative will both advise on, and grant, circuit outages to facilitate the project, in addition, to agreeing and participating in stage 2 commissioning requirements.
9. **Legal**
   The company legal department, if required, will provide legal advice to the PM, e.g. contract law, H&S law, third-party interfaces, etc.
10. **Operational safety**
    Operational safety refers to the department who consent to the issue of safety documents on-site, and undertake the role of safety control person as described in Chapter 18. In some instances, this role may be one and the same as that of system operations.
11. **Planning and programming**
    The planning/programming officer is responsible for resource planning, outage coordination and project programme.
12. **Design team**
    This is the team described in Chapter 20 and illustrated in Figure 20.2.
13. **Site engineer**
    The site engineer is in effect a project engineer with site-based responsibilities. The role is essentially that of contract management and monitoring of progress on-site, including engineering assurance of the equipment being installed, see Chapter 22.
14. **Commissioning panel chairman (CPC)**
    The CPC is responsible for managing the commissioning panel, see Chapter 23. The PM, project engineer or site engineer may in some instances also be the CPC.
15. **Commissioning engineer**
    The power network company commissioning engineer will shape the commissioning programme and testing regime. Responsibilities also include engineering assurance during stage 1 commissioning and directing stage 2 commissioning, see Chapter 23.
16. **Senior authorised person (SAP)**
    The SAP is responsible for all work on-site that fall under the power network company safety rules. This is a very key and significant role with reference to planning the site works and monitoring progress. In some instances, the SAP may be one and the same as the commissioning engineer and/or the site engineer.

17. **Team administrator**
    Administrative support is the lubricant that keeps the project management machine in motion and moving efficiently. Team administrator will provide the PM (and the project team) with the following typical support:
    (i) Project filing
    (ii) Meeting/event organising and scheduling
    (iii) Typing facilities
    (iv) Documentation collation
    (v) Presentation preparation and management
    (vi) Diary management

19.2.4 Progressing a project

Progressing the tasks and responsibilities shown in Figure 19.2 requires two categories of activity:

1. **Documentation**
   Documentation is required for a number of purposes: to formalise plans/programmes; to formalise certain stages of the project, e.g. certification and to record minutes of meetings and project progress. It is good practice for the PM to list the documentation required and its timing. Figure 19.4 provides a list of typical documentation.
2. **Meetings**
   Meetings are essential for a number of reasons: to plan, evaluate progress and take corrective action; to resolve emergency and unexpected occurrences; to resolve differences of opinion and to maximise team working. Again, it is good practice for the PM to create a schedule of meetings. The following is a list of typical meetings.
   (i) Project review meetings (Chapter 19)
       (a) Internal
       (b) External (with contractor)
   (ii) Design review meeting (Chapter 20)
       (a) Internal
       (b) External (with contractor)
   (iii) CDM and safety rule coordination meeting (Chapter 18)
   (iv) CDM pre-construction information handover meeting (Chapter 18)
   (v) Inaugural H&S and environmental meeting (Chapter 18) (including CDM construction phase requirements)
   (vi) Project stage-by-stage meetings (see Chapter 21)
   (vii) Site meetings (see Chapter 21)
       (a) H&S and environmental meetings (including CDM construction phase requirements)
       (b) Contract progress meetings
   (viii) Commissioning panel meetings (see Chapter 21)

Project Documentation

- Project review meeting minutes
- Monthly project progress report
- Financial (see Chapter 19)
  - Monthly project finance report
  - Financial asset register documentation
- Design (see Chapter 20)
  - Detail design specification
  - Design review meeting minutes
  - Design acceptance form
- Contract/purchasing (see Chapter 17)
  - Contract staging payment certificate
  - Contract variation instructions
  - Procurement payment certificates
- Health & safety and environmental (see Chapter 18)
  - CDM form F10
  - CDM pre-construction information pack
  - Environmental pre-construction plan
  - CDM health and safety file
- Project programme – monthly update (see Chapter 19)
- Project risk register – monthly update (see Chapter19)
- Project outage programme – monthly update (see Chapter 19)
- Manufacturing assurance reports
- Planning consents and wayleaves documentation (see Chapter 19)
- Project stage by stage plan (see Chapter 24)
- Site installation
  - Health and safety meeting minutes
  - CDM construction phase health and safety plan
  - Contract progress meetings minutes
  - Site quality plan
- Commissioning panel documentation (see Chapter 23)
- Technical asset register data

Figure 19.4 Typical project documentation

19.2.5 Project team performance

Project teams usually only exist for the duration of the project – and frequently comprise matrix managed team members. The PM must therefore, actively and positively manage the team to ensure maximum team working and team performance. In doing so, the PM must ensure that each team member is fully aware of their deliverables, the time-scales for delivery and the interface working requirements to achieve the deliverables. Good communications are essential. In particular, all members of the team must be aware of the requirement for early notification of potential problems. To achieve this, the PM must foster a spirit of approachability and accessibility. Team bonding events may well be worth considering.

Of equal importance to project team working is the working relationship between the two project teams, i.e. that of the power network company and that of the contractor. Both have a joint objective, namely the successful delivery of the project in accordance with contract terms and conditions. From the outset, the two PMs must create a joint team culture. The team qualities listed in the above paragraph apply equally to both the internal teams and to the joint team.

19.2.6 Project review meeting

Project review meetings, chaired by the PM and attended by relevant members of the project team (as determined by the PM and dependent upon the stage of the project), would typically be held once per month (or less frequently for some projects). Typical agenda items of a project review meeting are given in Figure 19.5. The meeting should primarily review progress against programme and determine remedial action where necessary.

Project Review Meeting Agenda

1. Project overview – status summary
   - (i) Detail design
   - (ii) Manufacture/procurement
   - (iii) Site establishment
   - (iv) Site installation
   - (v) Commissioning
   - (vi) Post-project review
2. Health & safety and environmental
3. Financial status
4. Contract status
5. Design status
6. Manufacturing assurance status
7. Consents and wayleaves
8. Project programme status
9. Outage programme status
10. Risk register status
11. Project stage by stage status
12. Site installation status
13. Commissioning status
14. Project audits
15. Lessons learned
16. Resources and competency
17. Project team performance
18. Project filing
19. Schedule of actions
20. Any other business (AOB)

Figure 19.5 Project review meeting agenda – typical

19.2.7 Project manager – competency requirements

Project management is a skill set in its own right and extends far beyond power network construction. The question thus arises of what competencies in addition to that of PM are required, if any, to discharge the role of PM on a power network project. Within this context, the twin dimensions of qualifications and experience need to be considered as follows:

1. **Qualifications**
   - (i) Power engineering qualification (usually a minimum of degree level)
     or
   - (ii) Other engineering (or engineering science) qualification (e.g. civil or mechanical)
     or
   - (iii) Non-engineering qualification
2. **Experience**
   - (i) Power network construction experience (typically 3–4 years minimum including site experience)
     or
   - (ii) Other construction experience (e.g. civil)
     or
   - (iii) No construction experience

At one end of the competency spectrum, PMs who have both a power engineering qualification, and power engineering construction experience are generally the most competent. Equally, PMs with a civil or mechanical engineering qualification, in addition to power network construction experience prove very competent. At the other end of the competency spectrum, PMs with no engineering qualification and no construction experience would clearly struggle with the necessary competency. As a generalisation, the weaker a PM is in power engineering qualifications and experience, the stronger and more proactive the project team must be – particularly the design team supporting the PM. Some companies produce an experience/qualifications scoring matrix to assist with the selection of PMs. Some just employ judgement against the backdrop of the above criterion.

It is worthy of note that where power network projects comprise a significant element of civil engineering work, then PMs with a civil engineering qualification and background may prove the most competent. In some instances, a 'civil' PM may commence the project and hand over to an 'electrical' PM at a suitable juncture.

With major infrastructure projects, the dominant competency is usually that of project management itself (particularly skills of leadership, communications, stakeholder management and staying focused upon and managing, the big picture). It is however greatly advantageous when managing stakeholders and third parties for the PM too have a strong grasp of, and be able to articulate, the power network/engineering considerations underpinning of the project.

NB: the above considerations apply equally to the contractor's PM.

19.2.8 Equipment transfer – stages and formalities

As leader of the project team, it is important that the PM is cognisant of the stages and formal arrangements by which equipment is transferred from the control of the contractor to the control of the power network company (which ideally ought to be defined in the contract). In general, this is a three stage transfer, as follows:

1. **Safety rules**
   At some planned point in the site installation and commissioning process, the equipment will become subject to the power network company safety rules. HV equipment will usually be formally transferred via a safety rules clearance certificate (or similar title), see Section 18.7.2. From the instant of being subject to the power network company safety rules no work can be undertaken on (or near to) the equipment without the agreement of the power network company.
2. **Operational control**
   Prior to equipment being subject to stage 2 (on load) commissioning, it will be formally transferred to the operational control of the power network company, usually via a commissioning acceptance certificate (or similar), see Section 23.8.1.2. This allows the power network company to open and close circuit breakers, etc., for the purposes of energising the equipment and placing on load.
3. **Contract taking over certificate**
   Full commercial taking over of the equipment (i.e. transfer of ownership and complete control) usually follows successful stage 2 commissioning. This is usually formalised through issue of a contract taking over certificate (or similar) as defined in the contract, the completion of which usually commences the equipment defects liability period.

19.3 Project manager (contractor)

19.3.1 Project manager (contractor) – accountabilities

The contractor's PM has very similar tasks to those of the power network company PM, albeit on the delivery side of the contract, see Figure 19.6. Within this context, the accountability is usually to deliver the project in accordance with the contract terms and conditions, but preferably for less than the contract sum to maximise the contractor's profit. The responsibilities are very similar to those illustrated in Figure 19.2 excepting that consents and wayleaves, and outage planning, is usually solely in the domain of the power network company. The structure in Figure 19.6 also differs from that in Figure 19.3 in the following ways:

1. **Contract/procurement**
   This is generally a much more significant task for the contractor than the power network company. In addition, to progressing the contract with the power

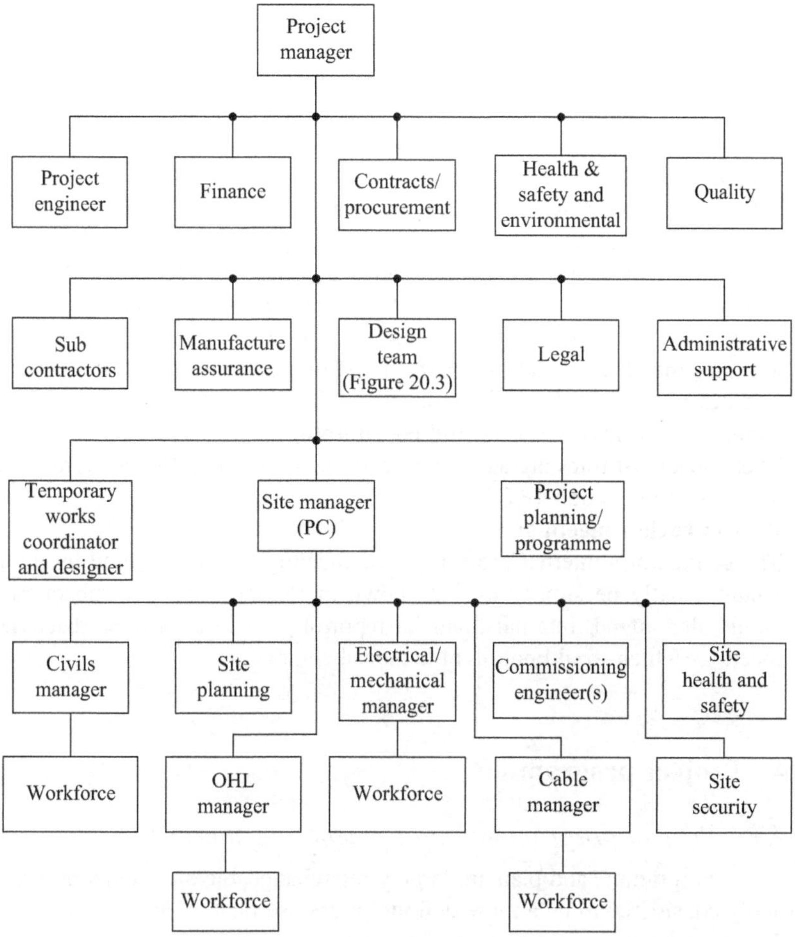

Figure 19.6 Typical project management team – contractor

network company, the contractor's PM will need to progress contracts and orders with equipment manufacturers, suppliers and subcontractors.

2. **Quality**
   The contractor's PM will be responsible for providing a quality plan (as required by ISO 9001) defining the QMS procedures to be used. These may need to be adapted or refined to suite the interface requirements of the power network company's QMS procedures. There may also be a requirement to train the project team in the procedures.
3. **Subcontractors**
   It would be usual to define subcontractors as an integral part of the project team. They would usually attend project team meetings and project review meetings.

4. **Project planning/programme**
   Again, this would be a much greater task for the contractor's PM than that of the power network company – since the programme would need to cover the detailed requirements of design, manufacture, site installation and commissioning.
5. **Site manager**
   The site manager would usually be one and the same as the principal contractor's (PC) representative on-site, as required by the CDM regulations. The contractor would usually be named on the form F10 as the PC. This is covered in more detail in Chapter 18. The site work is the major task for the PC requiring a significant site presence. The model shown in Figure 19.6 assumes that the substation, OHL and cable work are under the control of one PC.
6. **Temporary works designer and coordinator**
   These specialist roles are associated with the delivery of the temporary works as outlined in Section 18.8.1.
7. **Project review meetings**
   The contractor's internal project review meeting agenda (chaired by the PM) would usually be similar to that shown in Figure 19.5. The subcontractors would also attend. Internal financial reporting would usually be discussed at meetings which would not involve the subcontractors.

19.4 Project programme

19.4.1 Project programme requirements – general

The terms programme and plan are largely interchangeable although a programme is usually considered to be a more definite or precise plan.

In power network construction, a project programme essentially covers three categories of event, as follows:

- A physical activity (e.g. site establishment commences)
- Documentation preparation/completion (e.g. safety rule clearance certificate issued)
- A meeting (e.g. a design review meeting).

The content of a project programme essentially comprises the following:

- Statement of the event
- Time duration of the event
- Event dependencies (e.g. one event must precede another)
- Resources required to achieve the event
- Duty holder responsibilities for the event.

Programmes may be presented in a variety of ways. At a high level, they may be presented in block diagram form, sometimes referred to as a 'Gantt' chart – see

Figure 19.7(a). However, at the detailed level, they are often depicted as shown in Figure 19.7(b), salient features of which are as follows:

- The term 'critical path' is usually given to the shortest time in which the programme can be completed. With reference to Figure 19.7(b), the critical path is determined by the transformer events.

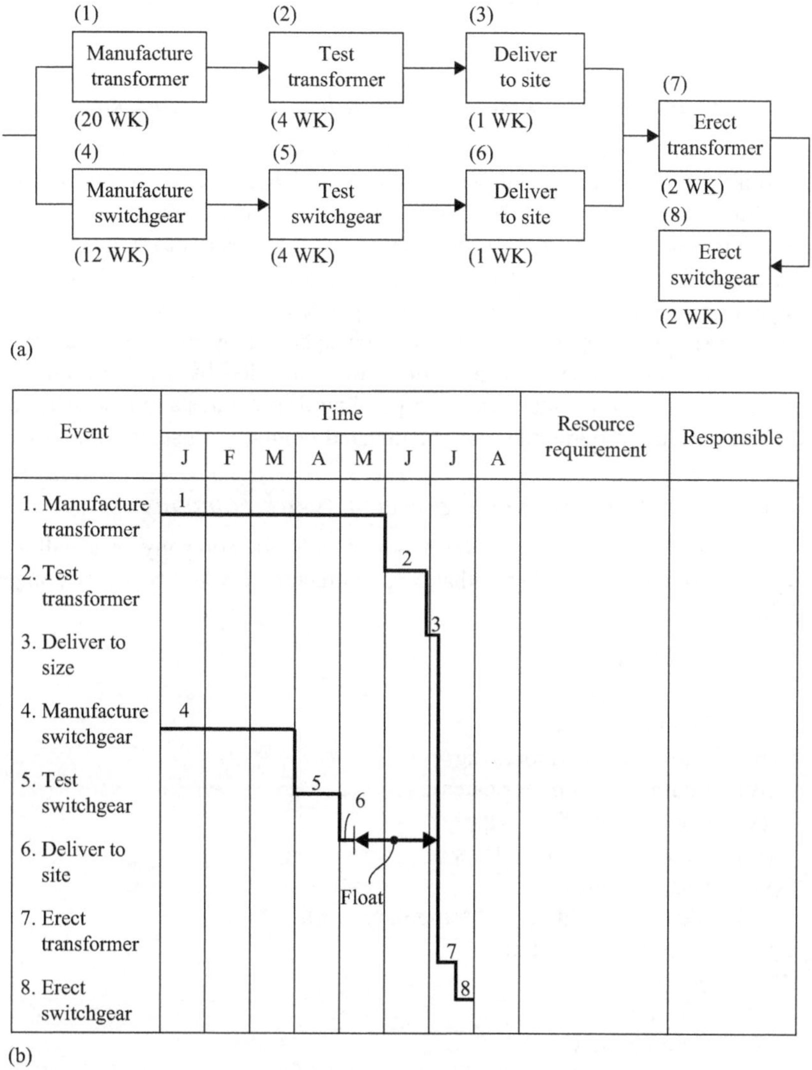

Figure 19.7 Project programme – resources: (a) block diagram and (b) detail programme

- The term 'float' depicts a time of no work being undertaken for a certain event, it represents how long an event can be delayed (i.e. the switchgear) without delaying the overall duration of the project.
- Event dependencies are connected by the vertical lines. Thus, the transformer cannot be tested until manufactured, and the switchgear cannot be erected in advance of the transformer.

The critical path is an important factor in any project since it determines the earliest time of project completion. The float time (essentially time in hand) is also important since it provides options of when events may take place i.e. to execute events differently or at a later stage.

19.4.2 Project programme commercially available packages

There are a range of computer-based packages available for formulating project programmes. 'Primavera' is one widely recognised package for controlling large and multidisciplinary projects. It requires specially trained planners to operate the system. With larger projects, the programme is often divided into stages to make it more manageable. Another widely recognised package is 'Microsoft Project'. This package is probably better suited to small-to-medium-sized projects. Again, it requires a skill-set to operate it – but can be operated by many PMs/engineers. Small and simpler projects can quite often be controlled by a programme comprising a simple list of events with corresponding times, dates and responsibilities. In all instances, the programme must be updated regularly – usually monthly.

19.4.3 Project programme – power network company

The project programme required by a power network company is usually at a higher and less detailed level than that of a contractor. It would typically comprise the following:

1. **Physical activities**
   (i) Circuit outages
   (ii) Temporary circuits
   (iii) Site access commencing
   (iv) Equipment delivery dates
   (v) Connection of LV supply
   (vi) Factory acceptance tests
   (vii) Road closures
   (viii) Major scaffolding and temporary works
   (ix) Circuit commissioning
2. **Meetings**
   (i) Project review
   (ii) Design review
   (iii) Project stage by stage
   (iv) Health and safety
   (v) Public consultations

   (vi) Third-party meetings
   (vii) Post-project review
3. **Documentation**
   (i) Detail design specification
   (ii) Pre-construction information
   (iii) Design acceptance
   (iv) Manufacturing assurance
   (v) Contractual documentation
   (vi) Consents and wayleaves
   (vii) P&C settings calculations
   (viii) Switching programme
   (ix) Technical workbooks
   (x) Financial asset register
   (xi) Third party agreements.

19.4.4 Project programme – contractor

The contractor's project programme will have similar content to that of the power network company, but in much greater detail. The two programmes must of course be in alignment. Design aspects may be split into a separate design programme. The programme will cover the detailed requirements of both power network equipment and that of the supporting infrastructure (e.g. buildings, roads, substation lighting, temporary works, etc.). It is advantageous to have a high-level programme in addition to the detailed programme – as there is a danger of losing sight of the big picture with the latter.

19.5 Resource management

19.5.1 Resource management – requirement

Resource management is that of ensuring there are sufficient resources to successfully complete the project. In general, the following needs to be considered:

- The range of tasks to be undertaken
- The resource levels and resource capability to deliver the tasks
- The start and end dates and the profiling of resources between those dates.

The task of resource management is rarely limited to the project being undertaken in isolation – but usually relates to the ability of a company or organisation to resource a total work programme comprising a variety of projects.

Resource management is greatly assisted by companies having their own generic resource profiles across a range of project types. Such projects are of value in assisting with the costing of a project, in the first instance.

A specific task of resource management is to identify the requirement for specialist types of resource (e.g. specialists in tunnelling or HV DC commissioning) and to ensure their availability at the required time.

Engineering resources – types		
	Company	**Contractor**
• Project manager/engineers	Y	Y
• Temporary works coordinator	–	Y
• Design engineers (all technologies)	Y	Y
• Temporary works designer	–	Y
• P&C settings engineers	Y	Y
• Drawing office (all technologies)	–	Y
• Factory testing engineers	–	Y
• Principal contractor/site manager	–	Y
• Quantity surveyor	–	Y
• Substation equipment erector	–	Y
• Substation wiremen	–	Y
• Crane/equipment operator	–	Y
• Civil constructor	–	Y
• Building services technician	–	Y
• OHL wireman	–	Y
• Tower/pole erectors	–	Y
• OHL winchman	–	Y
• Cable winchman	–	Y
• Cable jointer	–	Y
• Cable layer	–	Y
• Cable test engineer	–	Y
• Commissioning panel chairman	Y	–
• Commissioning engineer	Y	Y
• Site engineer	Y	–
• SAP	Y	–

Figure 19.8 Engineering resource types – typical

19.5.2 Resources to be considered

In practice, it is frequently the engineering resources that prove to be the most critical. Figure 19.8 summarises some of the main engineering resources to be considered. Resource profiles are usually best presented on a histogram, i.e. a bar chart.

19.6 Outage management

19.6.1 Outage management – requirements

Outages are required on power network company circuits to facilitate projects in the following ways:

- To enable construction work to take place on an existing circuit.
- To connect newly constructed equipment to the existing network – via stage 2 commissioning.
- To facilitate the connection of a temporary circuit (e.g. temporary OHL).
- As a 'proximity outage', i.e. an outage on a circuit not directly concerned with the project, which is required to be removed from service because there is a

risk of the construction work breaching safety distances (see Section 18.6.3) of the circuit in question.
- To facilitate commissioning e.g. busbar protection ratio and polarity tests against an adjacent circuit.

In addition, the system operations department (system operator), which controls the power network, needs to be informed of work that requires a 'risk of trip'. That is, work on or near to an in-service circuit which may unintentionally cause the circuit to trip. For example the drilling of one panel which may cause a relay on an adjacent, and in service, panel to operate and trip the circuit.

The PM therefore needs to identify well in advance all circumstances requiring circuit outages or a risk of trip – and subsequently, formulate and agree an outage programme with the system operations department that facilitates the project programme.

19.6.2 Outage management – planning

Provisional outages would in the first instance be agreed during the scheme development stage, prior to sanction as confirmation that that scheme is deliverable. However, in the project phase when detailed requirements are formulated, the outage programme needs to become firm and in some instances expanded (e.g. identification of proximity outages).

The outage programme needs to be based upon a three-way agreement between the power network company system operations department, the power network company PM and the contractor's PM. It should be subject to monthly review.

Occasionally, network circumstances beyond the control of the project will necessitate changes to the programme. Late delivery of the project can play havoc with an outage programme – impacting on other circuit outages and possibly other projects. If security of electricity supply is in jeopardy, the project may have to be delayed until the network can tolerate the outages required by the project. The outage programme may require an emergency return to service of the circuit being worked upon and contingency plans will need to be formulated for this eventuality.

The outage programme would usually be incorporated into the main project programme – but it is helpful to also have it as a stand-alone programme to provide a clear and concise focus on required outages.

19.7 Risk management

19.7.1 Risk management – requirements

All project plans should take account of the possibility that things may go wrong, and set up procedures to deal with this. Assessing and managing risk is therefore an important responsibility for any PM. The risks involved in a project may arise from a number of sources, and the PM needs to consider their impact on the three project parameters of time, cost and quality – and very importantly H&S of personnel. Figure 19.9 summarises the generic risks to be considered.

Risks

- Financial (control of costs)
- Contractual/legal (delivery confidence/liabilities incurred)
- Technical (specification/equipment reliability)
- People (resource levels/capability)
- Environmental (weather/geology/access/habitation, etc.)
- Political (stakeholder commitment)
- Health and safety (risks to people)
- Third party (local communities/pressure groups, etc.)

Figure 19.9 Generic risks – typical

19.7.2 Risk register

The standard method of recording risk is via a 'risk register'. The scheme team leader will usually commence a risk register at the outset of a scheme, and this will be later transferred to, and developed, by the PM. Similarly, the contractor will have commenced a risk register at the tender stage and again this will be later transferred to, and developed by, the contractor's PM.

A risk register evaluation works by examining two key factors as follows:

- The likelihood of the risk arising.
- The likely severity of the risk.

Both factors are usually scored, typically, between 1 and 5 (with 5 being the most likely) and the risk is scored and evaluated as follows:

$$\text{Risk} = (\text{likelihood of risk arising}) \times (\text{severity of risk})$$

Using a scoring system based on 5, the maximum risk score would be 25 and the lowest would be 1. There are numerous ways of presenting and configuring a risk register. One typical example is given in Figure 19.10. This model re-evaluates the risk after applying mitigation measures. After scoring the risk, it may typically be categorised as, say, high (score of 15 or more), medium (score of 8–12) and low (score of 1–6) and the risks managed accordingly. It is important not to overpopulate the risk register with low risks, as there is a danger of not being able to clearly see the big issues.

19.7.3 Typical risk register risks

Although many risks are common from one project to another, it is helpful to undertake a brain storming session to identify risks that are peculiar to the project in question. A list of typical common risks is provided below:

- Lack of existing as-built drawings as the basis for the design of the new project
- As-built drawings do not agree with site installation
- Design/drawings not completed on time
- Errors found on drawings when subject to engineering assurance with subsequent late delivery to site

Risk description and cause	Risk impact	Severity score	Likelihood score	Risk score	Risk mitigate measure	Severity score after mitigate measure	Likelihood score after mitigate measure	Risk score after mitigate measure	Estimate cost of risk (£)	Risk owner

Figure 19.10 Typical risk register format

- Equipment fails type tests
- Abnormally heavy load delays equipment delivery to site
- Site access blocked due to impact of inclement weather
- Work delayed on-site due to impact of inclement weather
- Cable laying delayed due to flooded trenches
- Unforeseen ground conditions
- Contaminated waste found on-site
- Asbestos found on-site
- Equipment damaged on way to site
- Equipment found to be incomplete when unpacked on-site
- Equipment found to be incomplete when assembled on-site
- Equipment not delivered to site by the due date
- Theft of critical equipment/materials from site
- Invasive species such as Japanese knot weed encountered
- Protected species such as great crested newt encountered
- Public complaints due to noise levels on-site
- Route to site results in excessive traffic through villages/country lanes
- Route to site difficult for heavy vehicles such as cranes
- Price movement in metallic materials
- Circuit outages changed
- Unexpected sources of impressed voltages encountered
- Shortage of competent resources, e.g. commissioning engineers, SAPs, etc.
- Equipment fails commissioning tests
- Protest groups encountered.

The risk register would usually be reviewed and updated at each project review meeting.

19.8 Financial management

19.8.1 Financial management – overview

The management of project finances is a fundamental and major task for both the power network company's and contractor's PM, respectively. Suffice it to say that whilst it is not the intention of this text to stray into the realms of project finance and accounting, it would be remiss not to provide an overview of the key principles and requirements to be managed by the PM. In so doing, it will be assumed (in line with the other chapters) that a power network company enters into a fixed price, EPC type of contract, with a contractor. In brief, the following will be covered.

- The project financial 'S' curve
- Value of work done (VOWD)
- Tasks of the power network company PM
  - Project accounting
  - Financial asset register
- Task of the contractor PM

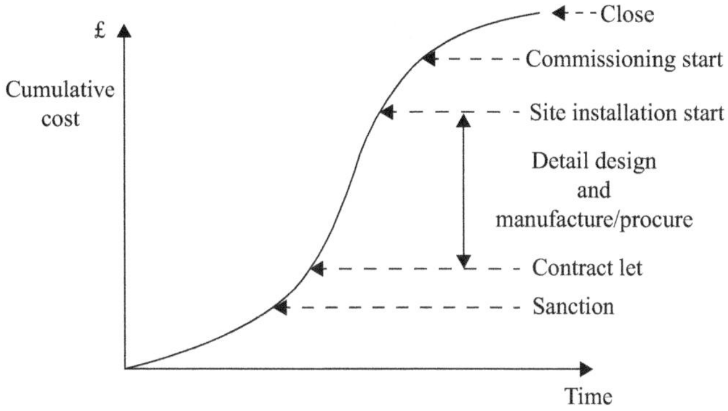

Figure 19.11 Typical project S curve

19.8.2 Project finance S curve

With reference to Figure 19.11, cumulative project costs follow a typical 'S' curve. As can be seen, the greatest percentage of the cost is usually incurred during the detail design and manufacture/procurement stage. This is perhaps not surprising since much of the cost of a scheme/project is incurred in the cost of the equipment.

19.8.3 Value of work done

VOWD is a project management technique for estimating and measuring project cost. Its purpose is to obtain an accurate as possible estimate for the cost of a project at any point in time regardless of payment. The VOWD is not measured against planned cost but against 'committed' cost. Within this context, the term committed is the confirmed (i.e. legally binding) price for either contracts or orders for the future provision of goods and services. For example, a fixed price contract, once let, would result in a committed cost equal to the value of the contract.

Figure 19.12 illustrates an example of a VOWD profile. Curve (a) shows the cumulative VOWD culminating in the committed cost. Curve (b) shows the cumulative payment (cash flow) up to time 't1'. The payment profile in this example lags behind the VOWD profile resulting in a cost variance, £V. Ideally, and reasonably the payment profile should match the VOWD profile and where this is not the case the PM should take corrective action.

NB: the amount by which curve (b) lags curve (a) i.e. the cost variance is often referred to as an 'accrual'.

Figure 19.13 illustrates two VOWD profile curves. Curve (a) shows the planned VOWD profile, and curve (b) the actual VOWD profile to point 't1'. As can be seen, in this example at time 't1', the actual VOWD lags behind the planned, and again the PM would need to understand the reasons why, and whether corrective action was required to align the two curves. In determining the actual VOWD profile, the power network company will frequently require assistance from the contractor who will be closer to the manufacturing progress. In addition,

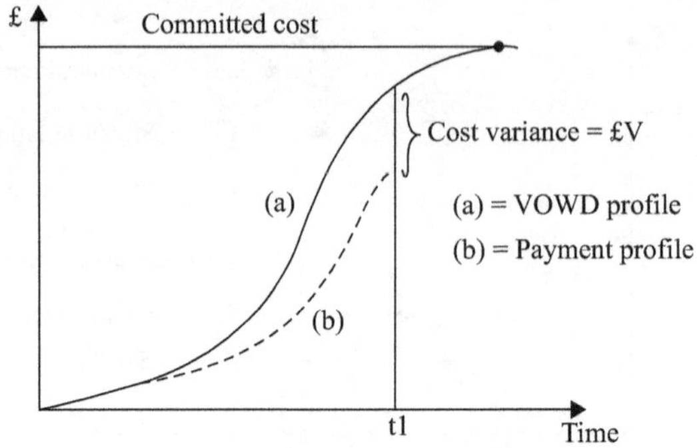

Figure 19.12 VOWD vs payment profile

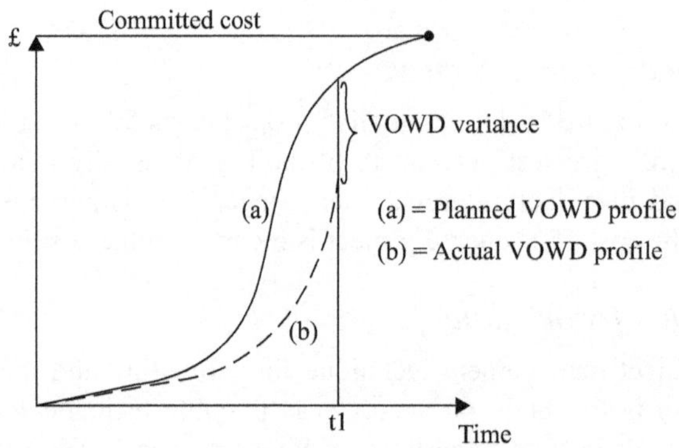

Figure 19.13 VOWD planned vs actual profile

confirmation of manufacturing progress can be obtained through manufacturing assurance, see Chapter 21.

It is worthy of note that an alternative project finance technique termed 'earned value' may be employed. This is very similar to VOWD but evaluates the cost profile against the approved budget rather than committed cost. The company finance department will specify which technique is to be adopted – usually as a company standard across all projects.

19.8.4 Project accounting – power network company

Project financial reporting is usually undertaken monthly. Financial data will usually be provided by the project finance officer (i.e. accountant) and the monthly report prepared by the PM. There will usually be an ongoing dialogue between the two relating to data accuracy, financial profiling and required corrective action. Figure 19.14 illustrates typical (and simplified) project financial data that may form part of a monthly financial return from the PM. The data format will vary from company to company, and therefore, Figure 19.14 is merely an example.

	VOWD prev years (£) (a)	**VOWD this year to date (£) (b)**	**Total VOWD to date (£) (c)**	**Total payments to date (£) (d)**	**Cost variance to date (£) (e)**	**Estimated total VOWD at year end (£) (f)**	**Estimated VOWD future years (£) (g)**	**ECTC (£) (h)**	**Commit. cost (£) (i)**	**Sanction cost (£) (j)**
Con 1										
Con 2										
Pur 1										
Pur 2										
Internal labour										
Pre-sanction										
Travel and sub										
Other										
Etc.										
Total										

Con=Contract; Pur=Purchase order.

Figure 19.14 Typical monthly project financial report

With reference to Figure 19.14:

1. **Column (a)**
This is the cumulative VOWD in all previous years up to the start of the current financial year.
2. **Column (b)**
This is the cumulative VOWD this financial year up to the present month.
3. **Column (c)**
This is the addition of columns (b) and (c).
4. **Column (d)**
This is the total cash payment to date.
5. **Column (e)**
This is the difference between columns (c) and (d).
6. **Column (f)**
This is the estimated cumulative VOWD at the end of this financial year. It comprises the addition of column (c) with an estimate of the additional VOWD incurred between this month and the end of the financial year.
7. **Column (g)**
This is the estimated VOWD that will be delivered in future years. It is often broken down on a year by year basis.
8. **Column (h)**
ECTC is the estimated cost to completion. It comprises the addition of columns (f) and (g).
9. **Column (i)**
This is the committed value of costs to date. At the end of the project columns (h) and (i) will be equal.
10. **Column (j)**
This is the sanctioned cost. Comparison of columns (h) and (j) shows whether the project is on track or there is an under/over spend.

As can be seen from the above, an extensive spreadsheet is required to manage the financial report. Costs are usually collected on an annual basis to reflect company annual budgets. This also gives a breakdown and measure of the construction activity in a specific year.

19.8.5 Financial asset register – power network company

At the end of a project, there will be a requirement to enter the asset financial values of newly constructed assets into the power network company financial asset register. This usually requires the PM to complete financial templates which categorise the assets from a financial value perspective (e.g. circuit breakers, transformers, protection relays, buildings, etc.). In doing so, the PM will have to conclude the cost of the work to bring the asset into operational service. The asset cost is not just that of purchasing the equipment (albeit via a contract) but must include all costs that contribute to the cost of completion of the project. This can be an exacting exercise. Guidance will usually be provided by the project finance officer. It is worthy of note that the contents of the financial asset register must align with those of the technical asset register, see Chapter 25.

19.8.6 Project accounting – contractor

The contractor's PM will also be accountable for the financial management of the contract – but from the contractor's perspective. The requirements will be very similar to those of the power network company. However, the contractor's PM will have the added complexity of not only managing the cost of the contract with the power network company – but also the payments for work undertaken by others (e.g. subcontractors) under separate contracts and for equipment procured via procurement orders, i.e. responsibility for both income and expenditure. The focus of the contractor PM will be that of maximising profit.

19.9 Contract management

19.9.1 Contract management – requirements

The task of contract management is to ensure that the contract is delivered as specified in the contract to achieve the interacting factors of specification, cost and time. More definitively, it relates to many of the activities that comprise project management, as discussed in this chapter, particularly the following:

1. **Specification**
   This requires the designed, installed and commissioned equipment, together with the associated infrastructure and temporary works to accord with both the contract specification, and the detail design specification. All stages should be subject to power network company engineering assurance.
2. **Finance/costs**
   Costs should be ideally managed to accord with the contract price, and certainly within the sanctioned risk margin (assuming contract variations are incurred). This is achieved by financial monitoring and management control, as outlined in Section 19.8.
3. **Time**
   The contract deliverables should be managed to accord with the times specified in the project/contract programme.
4. **Variations**
   Contract variations arise as a result of an imperfect contract specification. As far as possible they should be minimised and, where incurred, rigorously managed.
5. **Risk register**
   A risk register including risks to the contract should be prepared, kept under regular review and managed to eradicate/reduce the risk.
6. **Disputes management**
   The means by which disputes are managed should be specified in the contract. Dispute minimisation and subsequent management depends greatly on the strength of the working relationship between the respective PMs and project teams.

7. **Stakeholder management**
   Stakeholders (e.g. scheme sponsors, project boards or third parties) need to be kept informed and comfortable with the progress of the project, and managed accordingly.
8. **Records**
   Appropriate documents relating to the project need to be kept by both the power network company and the contractor. These provide a record of project progress, remove ambiguity in times of dispute and provide an audit trail of appropriate contract management.

The means by which contracts are efficiently and effectively managed comprise:

- Sound QMS procedures together with strict adherence
- Ongoing meetings (to confirm progress against programme, and resolve issues arising), particularly:
  – Project reviews
  – Design reviews
- Proactive communications (to ensure no surprises)
- Effective team and team-to-team working.

19.10 Consents and wayleaves

19.10.1 Consents and wayleaves – overview

Schemes/projects will, in many instances, require interaction with third parties landowners. Within this context, the terms 'consents and wayleaves' encompass a range of legal requirements relating to access to, and ownership of, land. The term wayleaves may alternatively be termed 'permissions'. In brief, the following will be examined:

- Wayleaves
- Lease
- Deed of grant of easement
- Transfer
- Street works
- Bodies providing statutory consents and permissions
- **Wayleaves**
  A wayleave is a terminable licence for which a power network company pays annual rent and compensation to a landowner. A wayleave gives a power network company the right to install an OHL or cable on, under or above, land, and subsequently to inspect, maintain, repair, replace or remove the OHL or cable. Compensation is usually payable to the land owner for interference with agricultural activities.
- **Lease**
  A lease is when a landowner gives the right (to a power network company) to occupy land for a set period of time, e.g. a temporary installation.

- **Deed of grant of easement**
  A deed of grant of easement is a legal right in perpetuity for the power network company to install and keep installed an OHL or cable on, over or under land and subsequently to inspect, maintain, repair, replace or remove the OHL or cable. It can be used to secure equipment access routes.
- **Transfer**
  Transfer is when the land owner sells the land to a power network company (in this instance).
- **Street works**
  A power network company like many other utilities (water gas, telephone, etc.) requires access to streets for the purpose of laying, maintaining and repairing cables, etc. Access arrangements are regulated by the traffic management Act 2004/2005 and other related acts. This requires the following typical requirements.
  - Notification to the appropriate authority of intended work
  - Confirmation by the authority that the work may proceed – sometimes requiring a permit
  - Ensuring that defined arrangements are in force during the work e.g. traffic management.
  - Provision of records of the cable location
  - Reinstatement works on completion.

19.10.2 Bodies providing statutory consents and permits

In addition, to legal requirements between a power network company and a third-party land owner – there will be occasions when further consents are required with reference to the land to be accessed and worked upon. Such additional consents are typically associated with protected sites, buildings, animals and plants, etc. and concern planning permission for new or modified OHL or substations. Bodies providing such consents include:

- Local planning authorities for substation permissions
- Department of energy and climate change for OHL below 132 kV
- Planning inspectorate for OHL at 132 kV and above – see Section 17.4
- Natural England – for work on or near protected sites (and other similar bodies for other parts of the United Kingdom)
- English heritage for work on or near heritage sites with archaeological importance (or similar bodies for rest of United Kingdom)
- Environment agency for work on or near rivers
- Highways authority for street and road works.

19.10.3 Project management considerations

Consents and wayleaves have significant potential to delay or even halt a scheme, or project. It is a specialist area and most power network companies employ a specialist, dedicated team for acquiring consents and wayleaves. A member of this team would usually sit on the scheme and project teams and would both advise the scheme team leader and PM and obtain the required consents and wayleaves.

Consents and wayleaves should ideally be obtained prior to scheme sanction, and certainly before any contracts are let. OHL in particular are usually heavily impacted by consents and wayleave requirements – including access to land holder's property to work on existing OHL.

19.11 Project filing and project audit

19.11.1 Project filing

Project filing, although arguably not the most stimulating of tasks, is none the less an essential project requirement. A comprehensive suite of project files facilitates the following:

- Reference source for key project documentation and certificates
- Record of project progress
- Data source for resolution of disagreements
- Data source for project audit
- Data source for lessons learned
- Historical record – for future reference

Project files generally contain three types of information: formal documentation and certificates (as required by QMS procedures and contracts); minutes of meetings; and correspondence. The filing system may be structured in a number of ways, a typical example is as follows:

- Project finance
- Contracts
- H&S and environmental
- Legal
- Consents and wayleaves
- Design and technical
- Site installation
- Commissioning
- Third party/general public
- Projects general
  - Risk register
  - Project programme
  - Outage programme
  - Project team structure, etc.

Most present-day filing systems are held electronically. This results in a more structured approach, saving of space and multi-location access. Some filing systems are common to both the power network company and the contractor – although separate financial files would usually be held. On completion of a project, the files are usually archived. There are certain legal requirements associated with the retention of project documentation, particularly the financial records, which may require retention for up to 6 years.

19.11.2 Project audit

The term 'audit' refers to systematic and independent examination of accounts, documentation, certificates, etc. It is good practice to subject a project to periodic audit to ensure that it is progressing (or has been completed) in accordance with stipulated requirements (e.g. QMS procedures). A typical range of audits are as follows:

1. **Financial audit**
   A financial audit would usually examine the project financial records to confirm that they accord with procedural and legal requirements.
2. **Health and safety audit**
   H&S legislation requires all employers to provide healthy and safe conditions for all employees at work. Within this context, a H&S audit will establish the H&S status of the project. In particular, it will both monitor whether employees are observing H&S practices and procedures, and whether the company is providing an appropriate healthy and safe work environment.
3. **Environmental audit**
   Environmental audit will examine whether environmental procedures for discharging legal obligations are being met. It is a requirement of ISO 14001.
4. **Quality audit**
   A quality audit is a requirement of quality system standard ISO 9001. It should be directed towards an examination of procedural adherence and the effectiveness of the QMS. Quality audits should report not only non-conformances and corrective actions but should highlight areas of good practice. An important element of a quality audit is that associated with the design process.

The PM must include a schedule of audits in the project programme. They must be planned to avoid the project being subject to an excessive number of audits. Sources of audit typically comprise the following:

1. **External audit**
   External audits are typically concerned with ISO 9001, ISO 14001, etc., where the audits are required to satisfy ongoing accreditation, and are carried out by an externally appointed auditor.
2. **Company internal audit**
   Most companies undertaking power engineering will have their own internal audit departments. They are most frequently concerned with financial adherence – but also encompass QMS and H&S and environmental adherence.
3. **Project generated audit**
   These are audits generated by the PM – often in accordance with project QMS procedural requirements. They usually have a significant focus on H&S, and design, adherence.

19.12 Project quality management

19.12.1 Project quality plan

A project quality plan (PQP), alternatively termed a project execution plan, is a vehicle for delivering project quality management. A PQP describes the activities, tools and processes (procedures) to achieve quality in the delivery of a project. A PQP typically comprises the following:

- Description of the project
- Quality standards to be applied, e.g. ISO 9001
- Activities and stages required to deliver the project
- Resources/responsibilities and project organisational structure
- Applicable QMS procedures
- Subcontractor QMS procedure interface
- Training requirements
- Tools to be used (e.g. computer-based systems).

An element of the PQP is the site quality plan which is applicable to site quality, see Section 22.4.1.

19.13 Outstanding work and post-project review

19.13.1 Outstanding work

It is frequently the case that at the point of circuit commissioning and entering commercial service that there is outstanding work on-site. This is often relatively small in nature and typically referred to as 'snagging'. Examples include:

- Outstanding civil and building engineering work.
- Replacing temporary labels on cubicles, kiosks, etc.
- Items of equipment yet to be commissioned (e.g. auto-reclose equipment)
- Cable trench reinstatement works
- Removal of temporary works (e.g. temporary roads)
- Outstanding cable markers
- Equipment damage repair (e.g. paintwork).

Some outstanding work may require circuit outages to complete the work.

It is essential that outstanding work is actioned as soon as possible – since once the contractor leaves site it may become difficult to reschedule the work at a mutually convenient time. To this end, the commissioning panel should arrange a schedule of outstanding work which specifies:

- The outstanding work
- The planned date for completing the outstanding work
- Outage requirements
- Responsible people (both in contractor and power network company organisations)
- Date the work was completed (with sign off by the PMs).

19.13.2 Post-project review

A post-project review should be convened following commissioning of the final circuit. The post-project review differs from the scheme review/closure in as much as the latter is concerned with the power network company's review of the investment, whereas the former is a joint review by the power network company and contractor of the both the good, and less good, aspects of the project.

Typical post-project review content would include:

- Review of the contract specification and whether it was delivered as specified
- Identification of errors and areas of poor practice with proposal for improvement
- Identification of what went well and how this would be factored into future projects
- Identification of deficiencies/weaknesses in the QMS procedures with proposal for improvement
- To reflect on how well the two project teams worked together with proposal for improvement
- To generally determine 'lessons learned' for incorporating into future best practice.

Finally, it is always good practice to acknowledge the success of a project through a joint project team celebratory lunch (or similar).

Chapter 20
Scheme design procedures

20.1 Scheme design procedures overview

This chapter will examine both the key scheme design management procedure and a number of other major design-related procedures. The scheme design management procedure will apply to virtually all schemes, where as many of the others will only be used as and when the scheme requires. In summary, the procedures examined will comprise the following (NB: The titles used are typical but may vary from company to company):

- Design management
- Protection and control settings management
- Thermal rating schedules
- Protection and automatic reclose/switching schedules
- Equipment nomenclature
- Operation diagrams
- SCADA systems management
- Drawings management

20.2 Design management

20.2.1 Scheme design process

Figure 20.1 provides an overview of a scheme design process commonly used in practice. The purpose of the scheme design management procedure is to define responsibilities and tasks which will ensure that the chosen design is holistic across all technologies, and that the end to end process is watertight, seamless and free from error. An important objective for the scheme design management procedure is to ensure 'no rework', i.e. get it right first time. As can be seen, the tasks are divided between power network company and contractor – and has been stated previously the QMS procedures of each must dovetail at the interface. Key considerations in a scheme design management procedure are outlined below.

20.2.2 Scheme design team

Figure 20.2 illustrates the composition of a typical power network company scheme design team. Depending upon the requirements of the scheme, some or all

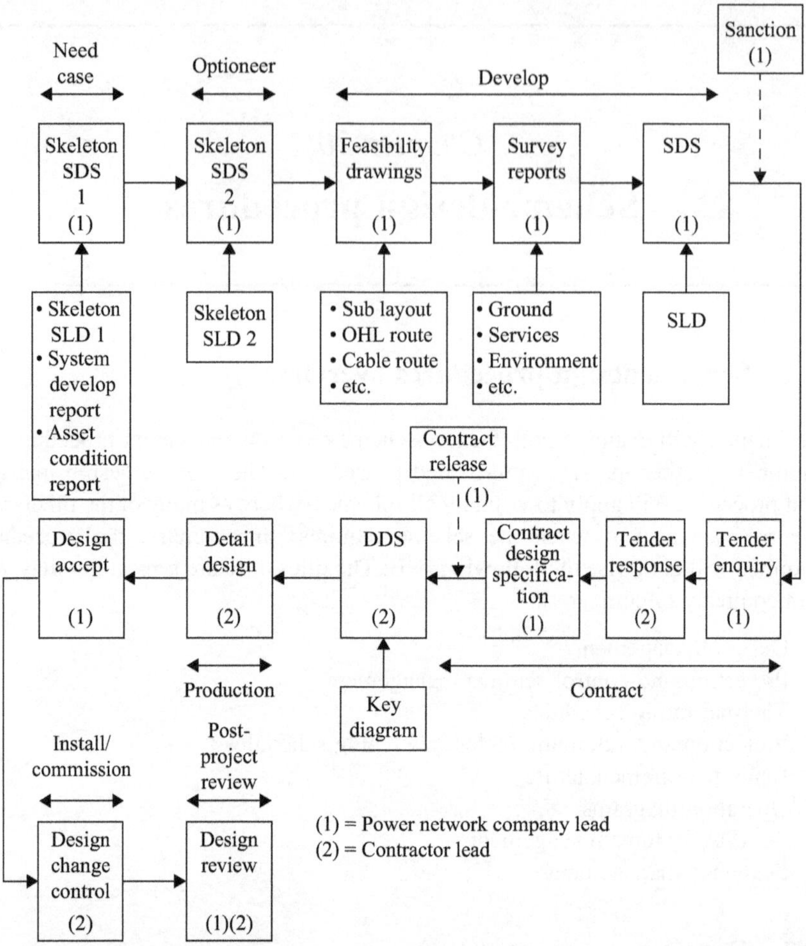

Figure 20.1 Scheme design process overview – typical

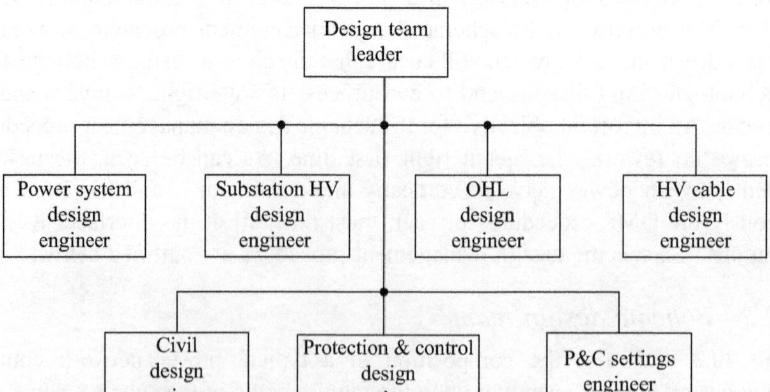

Figure 20.2 Typical scheme design team – power network company

of those shown will be involved. Power system design is concerned with network planning and system studies (i.e. load flows, voltage levels, fault levels, etc.), and it may be that this task is undertaken by a separate team which interfaces with the design team.

The design team leader is often one and the same as an engineer from one of the technology disciplines (e.g. Substation HV design). Each of the technology disciplines may have a team under them, depending upon the scale of the work. The design team would typically prepare and oversee the design to the point of contract release. Thereafter, they would undertake an engineering assurance role to confirm that that design deliverables provided by the contractor accord with the contract specification. Finally, the team would contribute towards the design review. In some instances, consultants may be utilised in the process. One school of thought suggests that those that prepare the design specification should not be one and the same as those that engineer assure the design, however, although this is sound process governance it is also resource intensive. It is critically important that the various design technology disciplines coordinate at the interface to achieve a holistic design (e.g. OHL down leads interfacing to a substation gantry). Duties of a scheme design team are typically as follows:

1. **Need case**
   Propose design considerations (NB: In some instances, there may be no requirement for design input other than that of power system design).
2. **Optioneer**
   Prepare and evaluate design options, and skeleton design documentation.
3. **Develop**
   Specify design requirements and prepare design documentation, most notably the SDS.
4. **Contract**
   Prepare design elements of the tender enquiry document, resolve tender TQs and prepare the design elements of the contract specification.
5. **Detailed design specification**
   Undertake engineering assurance and confirm that the production of the detailed design may commence.
6. **Detail design and design accept**
   Liaise with and advise the contractors design team, see Figure 20.3, on issues arising out of the specification, progress the design at design review meetings and engineer assure the contractors final design (drawings) to confirm that the design is compliant and that site installation may begin.
7. **Design change control**
   Engineering assure changes to design during site installation and commission.
8. **Design review**
   Lead role in the post-project design review to determine lessons learned for factoring back into the design procedures.
9. **Generally**
   Provide a source of expert advice on power network company requirements.

The structure and role of the scheme design team should be defined at an early stage in the scheme design management procedure.

20.2.3 Scheme design management procedure – purpose

A typical scheme design management procedure should serve the following purpose:

- Ensure that the design solution is technically compliant, cost effective and satisfies the need case.
- Ensure the design accords with health, safety and legal requirements.
- Ensure the design complies with power network company's network planning standards and technical specifications (i.e. project-generic design specifications).
- Ensure that the design solution is holistic and seamless across all technology disciplines.
- Ensure that the contractors' design submission is subject to engineering assurance to confirm that it is compliant and accords with the construction design specification.
- Ensure that the design interfaces between the power network company and contractor are defined and are unambiguous, and promote collaborative working.
- Ensure that the design process is to a standard that minimises the risk of inefficient design, i.e. no rework.
- Ensure that all design duty holders are competent for the work undertaken.

20.2.4 Scheme design management – stages

The stages of the design process are shown in Figure 20.1. In brief, the stages comprise the following:

1. **Need case**
   To support the need case stage the following is generally undertaken.
   (i) System development report (incorporating network planning proposals, system studies or asset condition evaluation) to define the requirements of the power system.
   (ii) Skeleton SDS, required network ratings, fault levels and other major requirements.
   (iii) Skeleton single line diagram (SLD).
2. **Optioneer**
   Following evaluation of a number of design options and selection of the chosen option, the skeleton SDS and SLD will be upgraded. This will usually be based on an optioneer report which justifies the selected design. The system development report may require updating.
3. **Development**
   Design requirements during development are typically as follows:
   (i) Preparation of feasibility drawings, particularly substation layout for the position of busbars and circuits, and the location of buildings, roads, etc.
   (ii) OHL and cable routing corridors.

(iii) Survey reports e.g. ground suitability for earthing, location of underground infrastructure, historical sites, etc., see Section 17.3.7.
(iv) Preparation of the final SDS (see Section 17.3.7) and SLD, which form the technical component of the sanction paper.

4. **Contract**
During the contract stage, the tender enquiry document is prepared (see Chapter 16) which will include the (project specific) technical specification as provided in the SDS. The contractor (tenderer) will then submit TQs to clarify with the scheme design team any ambiguities or omissions. Each contractor will then prepare a tender response document stipulating the contractor's proposed design solution. The scheme design team then undertake an evaluation of the contractors' tender response documents to determine the successful contractor, as outlined in Chapter 16. The contract document will then be prepared, including the contract design (technical) specification, and released to the successful contractor to commence work.
5. **Detail design specification**
The standard of design specification that exists at time of the contract stage will have been sufficient to ensure that contractor's tender price is within the risk margin of the sanctioned scheme (or thereabouts). However, at the instant of contract release, the design specification is usually not sufficiently well defined to enable the mass production of the detail design drawings to commence. It is therefore usually necessary for a lower and more detailed level of specification, i.e. detail design specification (DDS) (or other similar title) to be prepared to satisfy this requirement. The typical content of a DDS is provided in Section 20.2.5. Integral to a DDS is the preparation of a key diagram. This comprises a block diagram of protection relays superimposed on the SLD, showing CT and VT connections to the relays and the circuit breaker tripping arrangements associated with each relay.
6. **Detail design**
The detail design stage comprises the mass production of drawings and schedules which specify, in detail, the design of the works to be undertaken. It is a major task for the contractor and requires planning and coordination of design activities to ensure the drawings progresses smoothly, harmoniously and with no rework. Areas such as interfaces between substation HV plant and civil engineering must be resolved at an early stage (e.g. the size, strength and exact location of equipment bases and plinths). Detail design also needs to be harmonised with the simultaneous stage of equipment manufacture, since details from either impact upon the other (e.g. the weight and weight distribution of equipment, or the number and purpose of terminations in HV equipment cubicles to size the connecting and routing of multicore cables). Site installation and commissioning will be undertaken with reference to detail design drawings, which will eventually form the part of the CDM health and safety file.
7. **Design acceptance**
Design acceptance (or other similar title) comprises an examination of the contractor's submitted detail design, by (power network company) engineering

assurance (usually the scheme design team) – as confirmation that the design is compliant (or otherwise returned to correct errors). This is the formal acknowledgement (and notified via a design acceptance form – typical title) that the design may proceed to site to commence site installation. Design acceptance is invariably a risk managed strategy since it is usual to examine only selected aspects of the design (i.e. not a 100% check). The extent of the design assurance examination will depend upon the complexity, novelty and relative importance of the work, together with confidence in the contractor (e. g. a contractor with a proven track record). Example areas of important design acceptance comprise an examination of the following (list is not exhaustive and greatly abbreviated):

(i) Confirmation that all of the required drawings, surveys and reports, etc. have been provided.
(ii) Whether the quality of the drawings, etc. is to an acceptable standard, i.e. logically drawn and presented?
(iii) Confirmation that specific requirements defined in the DDS have been adhered to?
(iv) Confirmation that the equipment shown on the drawings is that which was specified?
(v) Confirmation that substation layouts, access and safety clearances are compliant?
(vi) Confirmation that trip circuitry from protection relays to circuit breakers is compliant. NB: It is usual to examine circuit diagrams (schematic diagrams) only.
(vii) Confirmation that cable design calculations are correct (e.g. conductor core and sheath temperatures do not exceed specific values, IV magnitude from core to cable sheath is within limits).
(viii) Confirmation that OHL critical conductor clearances are compliant.
(ix) Confirmation that the location of an oil dump tank is outside the fire damage zone.
(x) Confirmation that the interfaces between technology disciplines have been correctly designed and are compliant (e.g. between OHL terminal tower and substation gantry, between protection relay and communications cubicle).

It is good practice for the scheme design team leader and the contractor's design team leader to agree the extent of engineering assurance likely to be undertaken in advance – to factor into the design programme. It is also good practice to have this stage completed at least 4 weeks before site work commences, particularly if circuit outages are concerned.

8. **Design change control**

Design change control comprises the control process for changes to design after it arrives on-site, either due to error or alternative design requirements. The process must be clearly specified, tutored to all who need to know and rigorously adhered to, to avoid the design getting into a muddled state. It invariably requires the drawings to be marked up by the contractor's representative on-site, and subject to engineering assurance. It may require formal reissue of drawings if the scale of the change warrants it. This can be a bureaucratic, albeit important, process consuming time and resource, and therefore very minor changes (e.g. ferrule number change) may in some instances be left to the discretion of the site-based teams.

9. **Design review**
   Design review is an integral part of the post-project review, evaluating good/bad practice and lessons learned for incorporating into future projects. It usually requires the contribution of the contractor.

20.2.5 Detail design specification

The DDS is integral to the project-specific design (see Chapter 1) and forms the final specification prior to the commencement and mass production of drawings. It is usually prepared by the contractor and subject to (power network company) engineering assurance. The objective of the DDS is to ensure that sufficient information is provided to be assured that the detail design (mostly drawings), once commenced, is a compliant solution and produced correctly first time, without error. The DDS specifies the technical implementation of the project, including site installation and commissioning requirements. It will be based upon, and will reference, the SDS. However, whereas the SDS usually defines the requirements of a scheme, i.e. all substations, lines, cables, etc. encompassed by the scheme, the DDS is usually prepared on a per substation, OHL or cable basis. The DDS is usually structured as an aid memoir document. Typical content includes:

1. **Common**
   (i) Detailed scope of work
   (ii) Stages of the work
   (iii) Drawings to be prepared prior to detail design commencing and agreed
   (iv) Schedules of calculations to be undertaken
   (v) Technical specifications (project-generic specifications) to be used aligned to equipment
   (vi) Designs that involve hazard, complexity or risk and method of management
   (vii) Method of installation, risks to be managed and outages required
   (viii) Method of commissioning
   (ix) List of drawings to be prepared
   (x) Rating schedule data requirements
   (xi) Inter-technology interfaces, boundaries and risks to be managed
   (xii) Third-party interfaces, boundaries and risks to be managed
   (xiii) Handbooks and maintenance manuals required
   (xiv) IV risks and method of management
   (xv) Temporary works and emergency return to service arrangements
   (xvi) Deviation from standards
   (xvii) Outstanding surveys
   (xviii) Maintenance/operation/repair and replace requirements

(xix) Spares requirements
(xx) Design still to be resolved at the time of the DDS being signed off – and programme for resolution
(xxi) Redundant equipment disposal

2. **Substation HV**
   (i) Substation HV equipment design requirements
   (ii) Earthing design requirements
   (iii) Busbar design requirements
   (iv) Gantry design
3. **Protection and control**
   (i) Detailed protection relay specification, CT type and tap
   (ii) Protection panel and cubicle location arrangements
   (iii) Multicore termination requirements
   (iv) Protection relays requiring settings, and setting changes not covered by the scheme
   (v) Fault level changes and implication for settings
   (vi) Factory acceptance tests (FAT) requirements
4. **Substation civil, structural and building engineering**
   (i) Civil engineering design including geotechnical and environmental
   (ii) Buildings and building engineering
   (iii) Substation lighting
   (iv) Structure design
   (v) Fences, gates, etc.
   (vi) Landscaping
5. **HV cables**
   (i) Cable system design
   (ii) Cable accessories design
   (iii) Thermal monitoring
   (iv) Crossings design (road/rail, etc.)
   (v) Traffic management
6. **OHL**
   (i) Wire clearances
   (ii) Tower/pole design
   (iii) Foundation design
   (iv) Temporary OHL arrangements
   (v) Scaffolding design
   (vi) Non-standard arrangements (e.g. duck-unders)
   (vii) Fittings design.

20.2.6 Project design team – contractor

The contractor's project design team becomes active following receipt of the tender enquiry document. Wherever possible, the contractor's design team should mirror the power network company's design team to maximise team to team working relationships. Although the contractor's design team is involved in fewer stages of

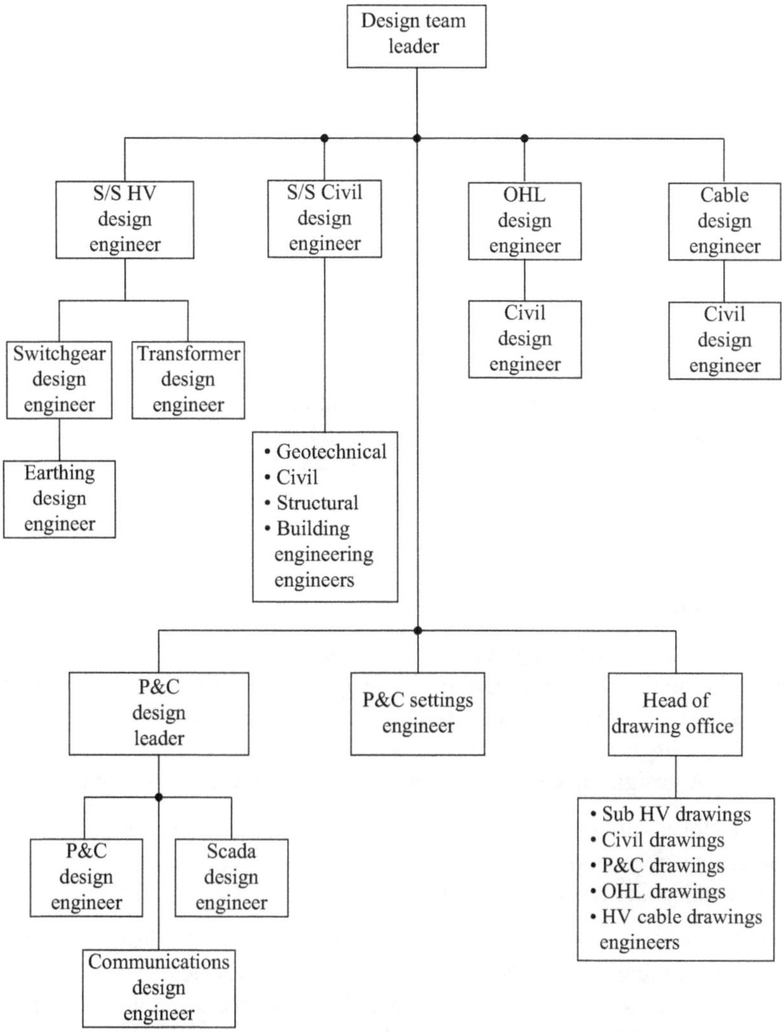

Figure 20.3 Typical project design team – contractor

the design process than the power network company's design team, the resource and input required from the contractor is usually much greater. Figure 20.3 illustrates the typical structure of the contractor design team. It is assumed that a single contractor (e.g. substation) is coordinating the work of subcontractors (e.g. OHL, HV cable).

In general, the contractor design team require a greater number of specialists since all aspects of detail must be understood and resolved. Civil engineering in particular comprises a number of subspecialist activities. The tender enquiry stage may require a smaller team and many contractors have a team dedicated solely to the tender stage.

A significant task for the contractor is that of drawings production – which can almost be a mini project in its own right. Many power engineering drawings will be based upon standard solutions and have precedent. However, equipment interface drawings are usually specific to the project, as are most OHL and HV cable drawings, and many civil and building engineering drawings. It is usual for drawings to have three signatures attached: drawn, checked and approved. Drawn is the person preparing the drawing. Checked is by an engineer with the ability to check for accuracy/mistakes/omissions. Approved is usually by an engineer with sufficient understanding of the drawings and inter-technology interfaces, e.g. head of the drawing office, to confirm that they may proceed to the design acceptance stage.

The contractors' design team must also interface with the equipment manufacturers. This will frequently require the contractor's design team to prepare a specification of requirements (based upon the SDS and DDS) for the manufacturer. Integral to this will be the requirement for the contractor's design team to interface with factory tests to prove that the manufactured equipment is compliant with the specification.

20.2.7 Design review meetings

An essential part of the design process is the requirement for design reviews, particularly during the DDS and detail design stages (see Figure 20.1). These comprise meetings between the power network company and contractor design teams. They may also comprise meetings internal to either the power network company or the contractor. A design review meeting has two prime objectives, as follows:

- To evaluate progress against the design programme and instigate remedial action where necessary.
- To identify and resolve design issues.

A joint design review meeting could be chaired by either of the design team leaders although the preference is probably for the contractor. A typical joint design review meeting agenda is summarised in Figure 20.4. As the meetings progress and requirements are finalised, the agenda will reduce in size.

Typical design review meeting timescale would probably be every 4 to 6 weeks but is dependent upon the requirements of the project. A not insignificant reason as to why project design veers off-course is the absence of design review meetings.

Section 6 of the agenda identifies a 'hazard review'. This usually comprises brain storming sessions to identify health and safety risks arising from the project and the measures required to mitigate the risk. The hazard review may comprise a separate meeting in its own right or alternatively included in the design review. An aid-memoir list of considerations usually assists the review, examples on the list may include: electrical shock; vibration; noise; access and egress; build sequence; excavation; working from height; third-party hazards, etc.

Design Review Meeting
Agenda

1. Design programme and documentation status
   - (i) DDS
   - (ii) Detail design (drawings/schedules/calculations)
   - (iii) Manufacture/procure
     - (a) Specification
     - (b) Quality assurance
     - (c) Factory tests/FAT
   - (iv) Build sequence programme
   - (v) Temporary works design
   - (vi) Project stage by stage
   - (vii) Site installation
   - (viii) Commissioning
   - (ix) Post project design review
2. Technical specification and standards compliance
3. Key drawings status
   - (i) SLD
   - (ii) Key diagram
   - (iii) Phasing diagram
   - (iv) Layout/elevations
   - (v) Buildings and rooms
   - (vi) MEWP access
   - (vii) OHL/cable routing
   - (viii) LV AC
   - (ix) Other
4. Physical position design
   - (i) Ground conditions finalised?
   - (ii) Location of below ground infrastructure finalised (e.g. earthing, drainage, etc.)?
   - (iii) Location of all foundations/plinths finalised?
   - (iv) Location of all support structures/gantries finalised?
   - (v) Location of all equipment finalised?
   - (vi) Location of all roads, walkways, fences, site lighting finalised?
   - (vii) Location/profile of all buildings/rooms finalised?
   - (viii) Building engineering heating, lighting finalised?
   - (ix) Location of all panels/cubicles finalised?
   - (x) Location of all towers/poles finalised?
   - (xi) OHL down-lead position finalised?
   - (xii) Precise cable route finalised?
   - (xiii) Cable sealing end location finalised?
   - (xiv) Access routes for all incoming equipments finalised?
   - (xv) Location of laydown area finalised?
   - (xvi) Location of scrapped equipment storage area finalised?
5. Interfaces' design and responsibilities (e.g. OHL to substation)
6. Hazard review
7. Technical data
8. Resources and competency
9. Training
10. Schedule of actions
11. AOB

Figure 20.4 Design review meeting agenda – typical

20.3 Protection and control settings management

20.3.1 P&C settings management procedure – relative importance

Many power network companies will have in the region of 200,000 to 1,000,000 settings on their network. A single incorrect relay setting has the potential to cause loss of electricity supply, or even break-up of the power system, and therefore with such a relatively large number of settings the potential for error is significant. A robust and watertight settings management procedure is therefore essential to the well-being of the power system, and within this context, it is one of the most important procedures in a power network company's suite of QMS procedures.

20.3.2 P&C settings management procedure – objectives

Generally, a settings management procedure should deliver the following objectives:

- That the relay application and type are correctly specified
- That the power system fault levels and relay settings are correctly calculated
- That the calculated settings are correctly applied to the relay on-site.
- That the relay is subject to commissioning tests to confirm the correct settings
- That a relay settings' database is established as a reference source for the following:
  - Confirmation of the settings applied to the relay to assist operational decisions, fault investigations and maintenance tests.
  - To enable searches to be made on relay types and location, e.g. for asset replacement.
- That a relay settings calculations' database is established as both a data source for future settings changes, and as an audit trail.

20.3.3 P&C settings management – single function relays

Prior to circa the year 2000, virtually all relays on the power system were 'single function' (e.g. overcurrent, distance, transformer overall protection) and the only variables on the relay were power system settings (to match individual circuits) i.e. volts, amps, ohms, time, etc. As a result, the calculation of settings associated with these relatively simple relays was mostly managed in-house by power network companies (or in some instances by consultants). Figure 20.5 illustrates a typical process that was undertaken with this type of relay. As can be seen, a 'relay settings record' (sometimes termed a relay settings sheet) was populated with settings for the relay in question, and this provided an instruction to site on how to set the relay. A copy of the relay settings record was mostly held both on-site, and in a centralised relay settings' database (either in paper or electronically or both). The maximum number of settings for the biggest relay on the system was usually of the order of 25, and this could usually be accommodated on single sheet of paper. Some of these relays may well be on the power system until around 2,030, and therefore, this process is still valid for setting changes to these relays.

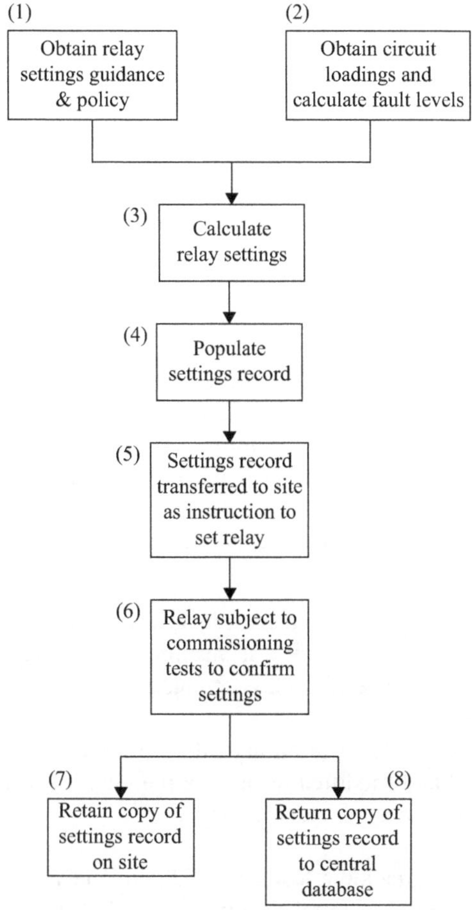

Figure 20.5 Typical P&C relay settings process – single function relays

20.3.4 P&C settings management – programmable multi-function numeric relays

Around the year 2000, computer technology commenced to have a significant impact on P&C relay design, rapidly replacing the installation of single function relays. The new computer-based relays were/are termed 'programmable multi-function numeric relays' or numeric relays for short. Numeric relays are considered to be more reliable since they are much less prone to component drift, and have self-monitoring features. In addition, there is a much greater range of options in the one relay case, such as:

- The ability to vary the number of functions (i.e. applications), e.g. main protection, back up protection, inter-tripping and fault recorder all simultaneously selectable in the one case.

- Varying the inputs to the relay (e.g. amps, volts, auxiliary contacts) to reflect the use of the relay.
- Varying the number and position of the output connections (e.g. via miniature relays) to reflect the use of the relay.
- Greater variety of setting options and each option being continuously variable within the settings' range.

The huge expansion in the options inbuilt into numeric relays greatly increased the number of settings, many of which are associated with the internal wiring of the relay (albeit in software). This in turn greatly enlarged the number of pages of settings as defined on settings' records (sheets). In extreme instances, there could be as many as 100 pages. Such a huge number of settings records made the settings management process, as shown in Figure 20.5, unmanageable. Furthermore, the detailed knowledge on how to configure and set a relay began to move from the power network company to the contractor/manufacturer. As a result, many power network companies moved to a settings management process typical of that shown in Figure 20.6.

The main stages of Figure 20.6 are as follows:

1. **Stages 1, 2 and 3**
   Based upon circuit loadings and fault level data (provided by the power network company), the contractor (or manufacturer) relay settings' engineer determines the circuit-specific relay settings.
2. **Stages 4 and 5**
   The master software file F1 (arbitrary title) prepared at time of the relay Type Tests is obtained and modified to include the circuit-specific settings, by the contractor/manufacturer, to become file F2.
3. **Stages 6, 7 and 8**
   The contractor/manufacturer loads file F2 into the relay in the factory/test laboratory and populates the relay settings record(s). Within this context, only the power system settings (the minority), as shown in Figure 20.7, are included on the settings record, with the fixed settings (majority and established time of type test) held as either a paper or electronic file. An example of the structure of a relay settings record, for recording power system settings only, is shown in Figure 20.8. The relay is then subject to FAT, to exhaustively confirm the relay application and settings, and in particular that the settings on the relay agree with those on the settings record. This is usually engineering assured by the power network company.
4. **Stages 9, 10 and 11**
   The relay with file F2 is transferred to site and installed. If file F2 is uploaded on-site, it is imperative that QA arrangements ensure that file F2 is one and the same as subject to FAT. The relay settings record is also transferred to the power network company settings' engineer who again confirms the settings, signs the settings record, retains a copy and forwards a copy to the power network company commissioning engineer on-site. The relay on-site is accessed by the contractor's commissioning engineer and the settings on the

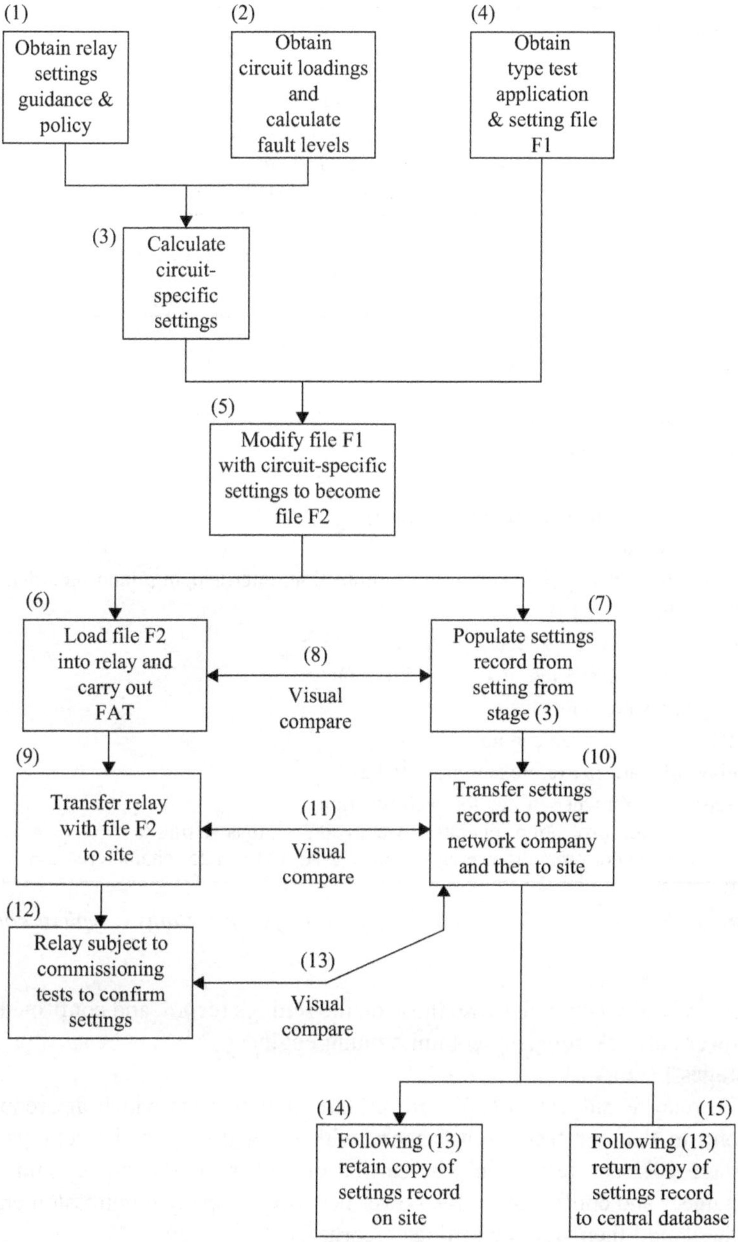

Figure 20.6 Typical relay settings process – programmable multi-function numeric relays

Power system settings (project specific)	• Relay functions (e.g. distance, O/C, fault recorder) • Power system measurements (i.e. V, I, Z, time) • Selectable logic sequence (e.g. auto-reclose direction)
Fixed settings (project generic)	• Input/output configuration • Fixed logic configuration (e.g. blocked distance scheme, auto-reclose logic)

Figure 20.7 Numeric relays setting categories

Relay Settings Record

- Company name
- Settings record revision no
- Substation name and voltage
- Circuit name
- Relay type (i.e. manufacturers reference)
- Relay serial no
- Relay functions (e.g. blocked distance protection, inter-tripping, fault recorder, etc.)
- Relay rating (e.g. 1 A)
- Relay range (e.g. 50%–200%)
- CT ratio (e.g. 600/1 tap of 1,200/600/1 CT)
- Relay hardware reference no
- Relay firmware reference no
- Relay software file reference (e.g. file F2)
- Relay measurement settings for each function
  - e.g. Main protection: operate current = 0.2 A; bias setting = 20%
  - Overcurrent protection; plug setting = 2 A; TM = 0.25; characteristic = SI etc.

Figure 20.8 Relay settings record for power system settings – typical contents

relay visually compared with those on the settings record, and confirmed by the power network company commissioning engineer.

5. **Stages 12 and 13**
   The relay is subject to (off load) commissioning tests which are required to confirm that the test results accord with the settings on the settings record (stage 10). The tests would be carried out by the contractor commissioning engineer and confirmed by the power network company commission engineer, who would then sign the settings record.
6. **Stages 14 and 15**
   The power network company commissioning engineer would then file one copy of the settings record on-site, and return one copy to the power network company settings' engineer to file in the relay settings' master database. NB: Master relay settings' databases (and even site records) may be in electronic form – but usually with paper copies held as back up.

Relay Settings Monitor Log

- Substation name
- Circuit name
- Relay function
- Settings record revision number
- Date settings record received from contractor/originated in-house
- Date settings record dispatched to site
- Date signed copy of settings record received from site
- Date relay settings record filed in master database

Figure 20.9 Relay settings record monitor log – typical

20.3.5 Relay settings record – administration

In large power networks with significant relay settings activity, the administration of the settings record process is not insignificant, often necessitating a monitor log (or similar) both to keep track on progress, and to provide an audit trail for future reference. With reference to the procedure shown in Figure 20.6, a typical monitor log would comprise the procedural requirements shown in Figure 20.9. This is particularly important in ensuring that a copy of the relay settings' record has been received from site and filed in the master relay settings' database, i.e. that the process is finalised.

20.3.6 Numeric relay settings process – electronic comparison

The process outlined in Section 20.3.4 assumed that the relay and file F2 that were subject to FAT were one and the same as that transferred to site. Instances may arise where this is not the case and the FAT is carried out on a representative relay. In this instance, it is imperative that the file loaded into the relay on-site is identical to that subject to FAT. If there is any doubt then the settings record should record all settings depicted in Figure 20.7 (i.e. including the fixed settings) – with on-site visual comparison between all settings on the relay and the settings record. This results in a dramatic increase in size of the settings record, and a tedious and resource intensive comparison process. As a consequence, some power network companies have developed an electronic comparison and transfer technique which enables the settings record to be retained solely to record power system settings. A typical example of how this may be achieved is outlined below.

With reference to Figure 20.10, stages (1)–(6) are identical to those in Figure 20.6. With reference to stage (7), a text file F4 is prepared containing the power system settings as shown in Figure 20.8. This is added to file F2 to create file F3.

An electronic comparison, stage (8) is then undertaken to confirm that file F2 in the relay is identical to file F2 contained in the settings' record.

In stage (9), file F3 is transferred to the power network company who load it into their electronic central relay settings' database. A copy of file F3 is then transferred to site, stage (10).

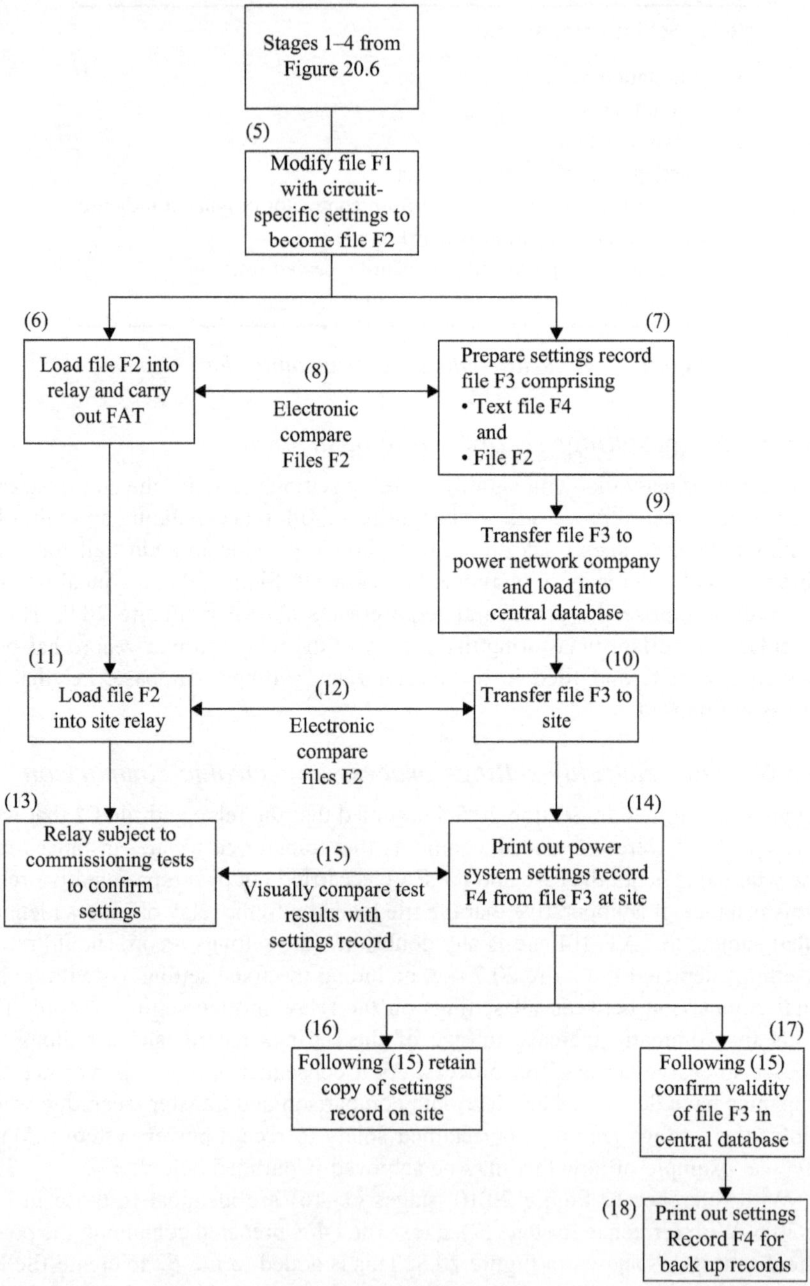

Figure 20.10 Typical numeric relay settings process – electronic comparison

In stage (11), file F2 is transferred to site, and in stage (12), an electronic comparison is made with file F3 to confirm that the files F2 (in the relay and on the settings record) are identical – i.e. a reconfirmation of stage (8).

In stage (13), the relay is subject to commissioning tests, the results of which are visually compared with the power system settings on the settings' record [which is obtained from a print out of file F4, which is nested in file F3 – see stage (14)].

A copy of the settings' record (i.e. print out of file F4) is retained on-site – stage (16).

Following successful on-site tests, confirmation of the validity of file F3 is made in the central database – stage (17), and a copy of file F4 is printed, i.e. the power system settings, settings' record and filed centrally as a backup – stage (18). Should absolute confidence in electronic records ever arise, stage (18) could be dispensed with.

20.3.7 Settings calculations' timescales

It is worthy of note that with the older single function types of relay that the settings record had to be forwarded to site prior to commissioning tests commencing. This meant that the settings calculations could be left as late as 6 to 8 weeks prior to the commissioned circuit being energised. With numeric relays, the settings' calculations must be undertaken prior to the FAT i.e. typically 20 weeks prior to circuit energisation, as such the whole process needs to commence much earlier.

20.3.8 Relay settings process – significance

As stated earlier, rigorous relay settings management is critical to the well-being of the power system – as such, all concerned must be competent in, and diligently adhere to, the relay settings process.

20.4 Thermal rating schedules

20.4.1 Equipment thermal ratings

The current carrying capability of electrical equipment is limited by the maximum equipment design temperature (otherwise the equipment would be damaged). This varies not only from equipment type to equipment type, but also with both the ambient temperature, and the temperature rise time constant of the surrounding medium (i.e. air, oil, SF6, earth, etc.)

With reference to Figure 20.11, each of the items of equipment that comprise the example circuit is subject to temperature rise with increasing load current.

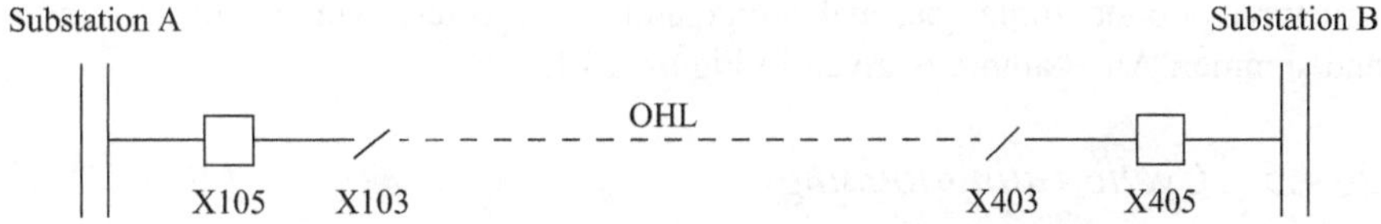

Figure 20.11 Example circuit – thermal rating schedule

The purpose of a thermal rating schedule is to identify the item of equipment which provides the load current limit for the circuit.

Each power network company will require a document equivalent to a thermal rating schedule for each circuit to determine circuit maximum loading. Many equipment thermal limitations can be obtained from IEC or similar documentation. However, in some instances, power network companies will determine their own thermal limitations from equipment modelling, and computer simulations.

20.4.2 Continuous current ratings

Some items of equipment possess two current ratings, as follows:

- Post-fault continuous current rating
- Pre-fault continuous current rating

The two ratings recognise that following a circuit fault, the current in a circuit remaining in service is likely to be subject to an increase in load current. Circuits are therefore loaded to the pre-fault continuous current rating – with further increase in load current allowed following fault clearance (on another circuit) up to the post-fault continuous current rating. The post-fault rating is typically about 15% greater than the pre-fault rating. The two ratings are generally associated with OHL and switchgear by virtue of the short thermal time constants (relatively rapid temperature rise) of the surrounding medium (air or SF6).

20.4.3 Short-term overloads – emergency ratings

If the current in a circuit is less than the continuous current rating, then short duration overloads are permissible until the equipment's maximum allowable temperature is reached. Short-term overloads are usually stated as 3, 5, 10 and 20 min. After short-term overload periods have elapsed, the current in the circuit must be reduced to the continuous current rating (or post-fault continuous current rating for equipments with dual continuous ratings) otherwise equipment damage would occur. Short-term overloads less than 10 min would usually require automatic action (as opposed to manual action) to reduce the current to the continuous current. Short-term overloads are illustrated in Figure 20.12.

20.4.4 Seasonal ratings

Equipment exposed to seasonal temperature variation, i.e. any equipment that is not in a temperature controlled environment is subject to ambient temperatures which vary with the seasons. As a result, equipment thermal rating tables are often structured around three seasonal temperature categories; winter; spring/summer and summer. An example is given in Figure 20.12.

20.4.5 Cyclic rating/loading

Cyclic rating (or loading) is a term used to describe daily load profiles which are imposed upon a circuit, e.g. the daily morning and evening peaks of a distribution

Thermal Ratings Table 2 × 700 mm all aluminium alloy conductor OHL				
		Winter (kA)	**Spring/Autumn (kA)**	**Summer (kA)**
Post-fault continuous		4.0	3.8	3.6
Pre-fault continuous		3.4	3.2	3.1
Overload			**Overload**	
Pre-overload*(%)	**Duration (min)**			
84	20	4.4	4.2	3.9
	10	5.0	4.7	4.4
	5	6.0	5.7	5.2
	3	7.2	6.8	6.3
60	20	4.8	4.5	4.2
	10	5.8	5.5	5.0
	5	7.5	7.1	6.5
	3	9.5	9.0	8.2
0	20	5.1	4.8	4.5
	10	6.6	6.2	5.7
	5	8.9	8.5	7.7
	3	11.7	11.0	9.9

*% of post-fault rating.

Figure 20.12 Typical thermal rating table

system. Both transformers and cables, in particular, are specified in terms of cyclic ratings since their insulating mediums have relatively long time constants, typically of a few hours or more (i.e. relatively slow heating). This means that these equipments can operate above their nominal rating for a period of time (e.g. 6 h at 125% nominal), on the basis that they operate below their nominal rating for longer periods (e.g. 18 h at 60% nominal) and with no deterioration, or ageing, of their insulation. As can be seen, this provides a much greater duration of overload compared to that shown in Figure 20.12.

20.4.6 Transformer loading

Transformer loadings are restricted by the winding temperature trip/alarm settings – which monitor the temperature of the transformer hot spot. Transformer loading is also dependent upon the operation of fans/pumps, see Chapter 9.

20.4.7 Protection relay loadings

As with HV equipment, protection relays and their CTs are subject to current thermal rating considerations. Relays are usually located in rooms with a controlled environment, and are therefore not subject to seasonal temperature variation considerations. They are however, still categorised with 3-, 5-, 10- and 20-min short-term overload ratings similar to that shown in Figure 20.12. Published relay thermal data tends not to be as comprehensive as that for HV equipment, and thus some interpolation may be required when determining the limiting equipment or relay. Modern relays with a nominal rating of 1 A are usually specified as having a continuous current rating of 2.4 A. When determining relay current, consideration needs to be given to the CT ratio.

20.4.8 Protection tripping considerations

None unit protections such as distance or back up overcurrent protection will trip as the load current rises above the setting. As a result, they are usually set around maximum credible load current situations and therefore, the trip current setting must also be factored into thermal rating schedules.

It is very advantageous if a power network company can deduce and declare a system maximum loading – against which all protection relays can be set. For example, the national grid transmission system at one time adopted a system maximum loading for all circuits of 7,600 A (for 10 min), at 400 kV. This was based upon maximum circuit loading following loss of connected generation, and redistribution of load current. Thus, the protection tripping limit, for thermal ratings schedules, can be standardised.

It is also worthy of note that back up protection earth fault relays can operate as a result of standing ZPS current arising from unbalanced load currents.

A factor of safety (e.g. 10%) should be factored into protection trip currents – for inclusion in thermal rating schedules.

20.4.9 Thermal rating schedules – procedural requirements

Procedural requirements for thermal rating schedules typically comprise the following:

- Preparation of generic templates for each type of equipment for collecting the data. NB: In some instances, the template will require data similar to Figure 20.12, and in other instances only the type of equipment and the nominal current rating (since the data as per Figure 20.12 is already held). In all instances, protection tripping currents will have to be provided. Equipment to be included will be identified from the SDS.
- Subjecting the completed templates to engineering assurance by the power network company.
- Transferring the data from the templates into a computer-based system for producing the thermal rating schedule. The date the schedule becomes operational will need to be included. NB: The schedules can be produced manually – albeit that it is a laborious task.
- Circulating the thermal rating schedules to the system operations department, etc.

The format of the rating schedule will be similar to that shown in Figure 20.12 with the limiting item of equipment and associated limiting current stated – for both continuous and short-term current ratings.

20.5 Protection and automatic reclose/switching schedules

20.5.1 Protection and automatic reclose/switching schedules – requirement

Protection and automatic reclose/switching schedules provide summary information for system operators on the protection, auto-reclose and auto-switching facilities on each and every circuit. They are an aide to operational decisions, particularly when the power network is subject to fault conditions, or to aide decision making when the equipment named in the schedule is either in failure mode (e.g. inter-trip channel fail alarm) or does not perform as expected.

20.5.2 Protection and automatic reclose/switching schedules – typical format

Figure 20.13 illustrates the requirements of a typical schedule. A SLD similar to that in Figure 20.11 would be included in the schedule.

The schedule would be populated to reflect the circuit facilities. Table 9 of the schedule on 'delayed auto reclose/switching sequence' would stipulate the sequence of events for the most common fault conditions e.g. line transient fault, line persistent fault.

20.5.3 Protection and automatic reclose/switching schedules – procedural requirements

This schedule is probably easier prepared by the power network company (although it could be equally prepared by the contractor). It needs to be finalised prior to stage 2 commissioning tests commencing.

Protection and Automatic Reclose/Switching Schedule

1. Circuit name and kV
2. Circuit diagram (see Figure 20.13)
3. Circuit protection type

	Substation A	Substation B
First main protection		
Second main protection		

4. Protection signalling equipment type

	Substation A	Substation B
First main protection		
Second main protection		

Figure 20.13 Continued

5. Inter-trip equipment type

	Substation A	Substation B
First inter-trip		
Second inter-trip		

6. Inter-trip initiation source

	Substation A	Substation B
Busbar protection		
Circuit breaker fail		

7. Synchronising facilities

	Substation A	Substation B
Check synchronise		
Power system synchronise		

8. Delayed auto-reclose

	Substation A	Substation B
Dead line close time		
Dead busbar close time		
Check synchronise close time		

9. Delayed auto reclose/switching sequence

Substation	Event	Time

10. Description of auto-switching scheme

Figure 20.13 Protection and automatic reclose/switching schedule – typical format

20.6 HV equipment nomenclature

20.6.1 HV Equipment nomenclature – requirement

To ensure the safe and effective operation of power networks and to reduce the risk of human error, it is common practice to adhere to a common system of identification (numbering) for each item of equipment. This largely concerns substation equipment – although unique identifiers also apply to OHL and cables as will be discussed in the next topic on operations diagrams.

By way of example, the nomenclature system in common use in the United Kingdom on the 400 and 132 kV systems is summarised in Figure 20.14. This nomenclature system is applied to the two substation layouts illustrated in Figures 20.15 and 20.16. Alternative numbering sequences are available for the 275 kV system and voltages less than 132 kV. It is worthy of note that metal clad switchgear circuits with withdrawable trucks (typically found at 33 kV and below) is usually only identified by the circuit name (e.g. London road circuit).

20.6.2 Equipment nomenclature – procedure

A procedure on equipment nomenclature should typically include the following:

- The standard numbering format that would apply at each and every voltage
- The timing and freezing of equipment numbering. This would normally be proposed by the power network company prior to sanction when preparing the SLD. It must be agreed with the contractor and frozen on finalisation of the DDS.

Characters			
1 **First character**	**2** **Circuit no**	**3** **Circuit type**	**4** **Equipment Type**
400 kV = X 132 kV = No first character	First OHL (or cable) = 1 Second OHL (or cable) = 2, etc. First BS/BC = 1 Second BS/BC = 2, etc. First transformer = 1 Second transformer = 2, etc.	0 = OHL 1 = Transformer HV 2 = Main BS (or Mesh CB) 3 = BC 6 = Reserve BS	0 = CB (Not OHL) 1 = Earth Switch 3 = Disconnector 4 = Main bar disconnector 5 = CB (OHL) 6 = Reserve bar disconnector 8 = Main Bar disconnector – Second choice

Figure 20.14 400 and 132 kV substation nomenclature – summary

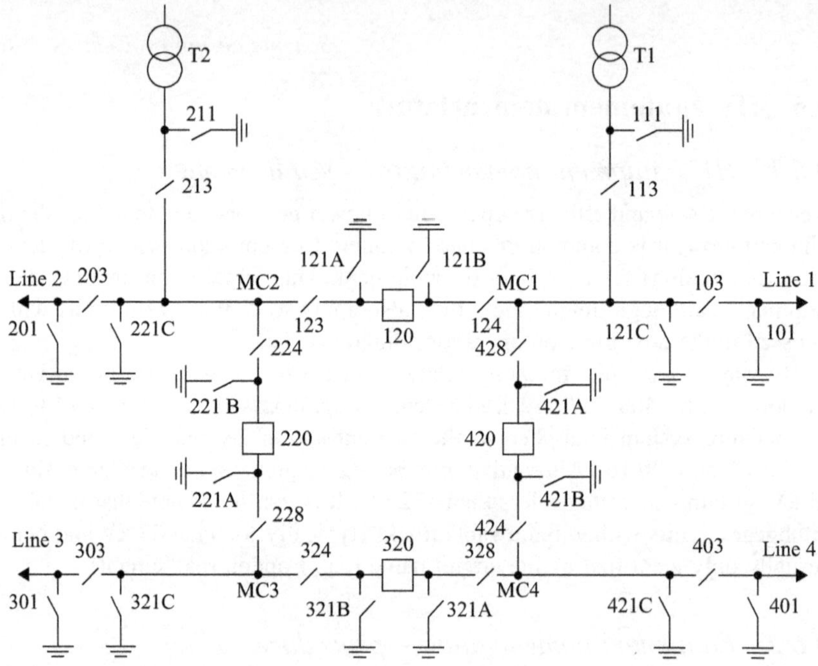

Figure 20.15 132-kV mesh substation nomenclature

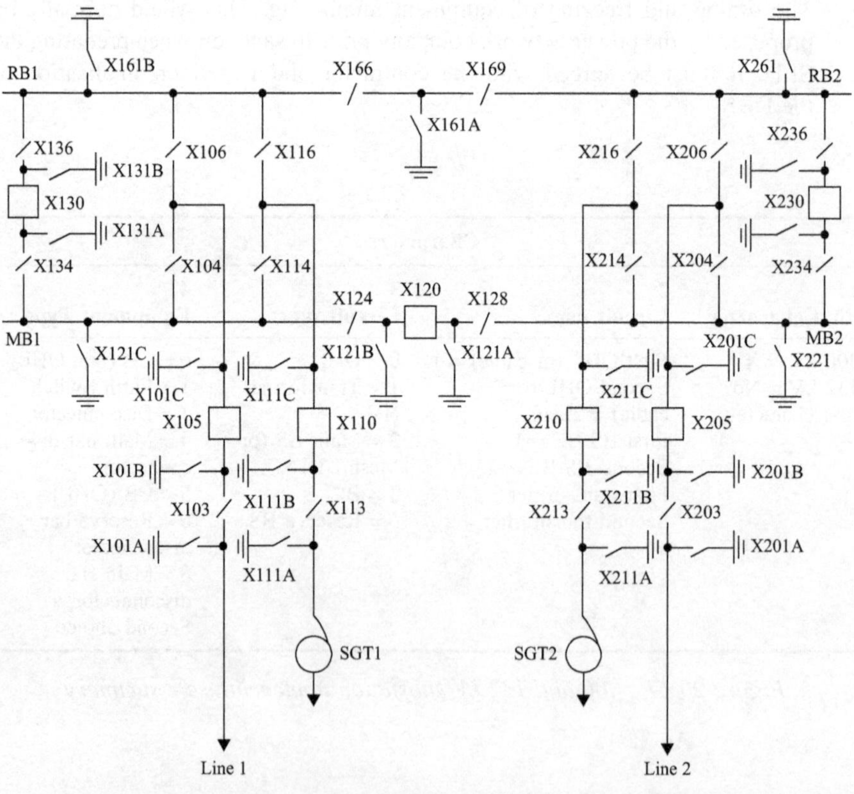

Figure 20.16 400-kV double busbar substation nomenclature

20.7 Operation diagrams

20.7.1 Operation diagrams – purpose

Operation diagrams (or other similar title) are prepared by most power network companies and provide a definitive reference document for facilitating the following:

- Operational and safety switching of the power system
- Defining the technical interface between a power network company and either another power network company, or a generation company, or a connected major customer.

Operation diagrams as a minimum comprise a substation operation diagram – but may consist of a suite of diagrams comprising:

- Substation operation diagram
- Substation gas zone diagram
- OHL/cable routes diagram

The above diagrams will be briefly examined in the following sections.

20.7.2 Substation operation diagram

A substation operation diagram shows the circuit layout and interconnections of a substation on a SLD. The diagram must accurately show the HV equipment layout and connection interfaces to any third part HV systems. The layout of the diagram should reflect the geographic layout of the substation. All equipment must be suitably numbered as per standard nomenclature requirements specified in Section 20.6.2. Figures 20.15 and 20.16 illustrate most of the content that typical operation diagrams should contain – but require the addition of CTs, VTs, line traps, surge diverters, i.e. any other HV equipment. For example Figure 20.17 illustrates the complete requirements of an operations diagram for line 1 of Figure 20.16.

20.7.3 Substation gas zone diagram

Gas zones are applicable to GIS substations. A gas zone diagram comprises the over-layering of gas zones onto an operation diagram that would be applicable to an AIS substation. Figure 20.18 provides an example for a single circuit of a substation.

20.7.4 OHL and cable route diagram

An OHL/cable route diagram would typically show the following (Figure 20.19):

- Substations at the ends of the routes shown as block diagrams
- OHL/cable route reference (e.g. YK tower route – where YK is a reference code usually provided by the power network company)
- OHL circuit colours (e.g. red/black/white)

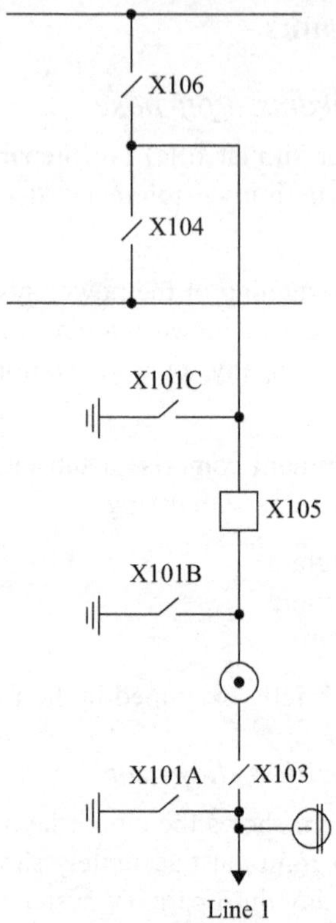

Figure 20.17 Substation operations diagram showing typical requirements for a circuit

- OHL conductor phasing colours (e.g. R, Y, B from the top conductor down)
- OHL tower (or pole) numbers showing:
  - Terminal tower number
  - Towers where the numbering is not continuous (otherwise it is assumed the numbering is sequential and continuous).

20.7.5 Procedural requirements

Operations diagrams are based upon the layout and numbering of SLDs (which are largely complete prior to scheme sanction). They are usually prepared by the power network company who interface with a single drawing office who become familiar with requirements. Their accuracy is usually checked by a SAP on-site who will diligently confirm the diagram against the constructed equipment.

When newly constructed equipment is declared as being subject to the Safety Rules for HV equipment (via an SRCC – see Section 18.7.2), a corresponding operation diagram must simultaneously become operational.

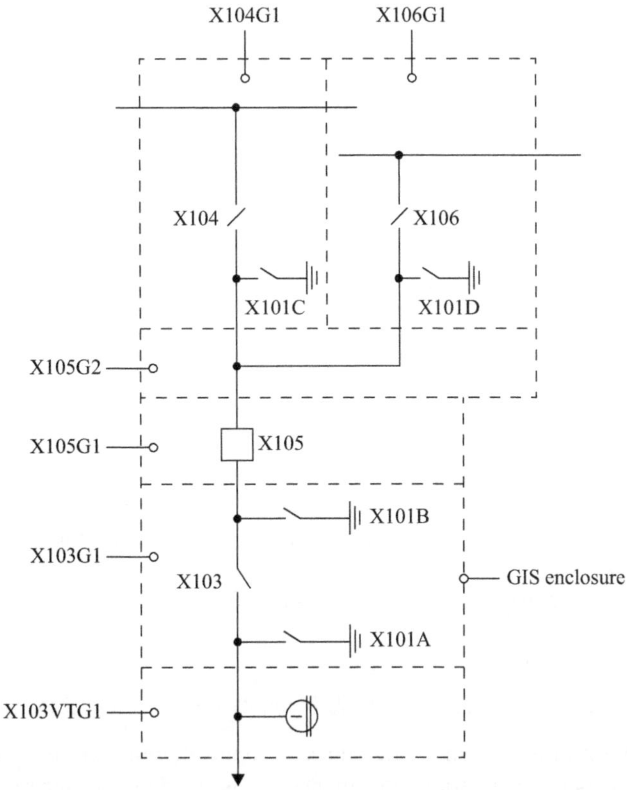

Figure 20.18 Typical gas zone diagram – circuit only

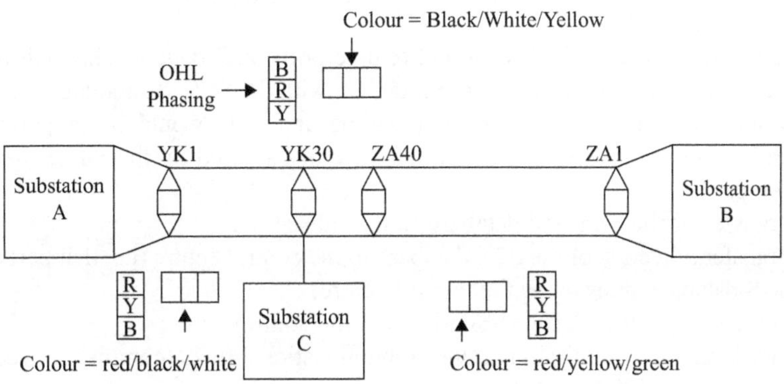

<u>Substation A – C – B circuit</u>

Route = YK, Towers = YK1 – YK30 (colour = red/black/white)

Route = ZA, Towers = ZA40 – ZA1 (colour = red/yellow/green)

<u>Substation A – B circuit</u>

Route = YK, Towers = YK1 – YK30 (colour = black/white/yellow)

Route = ZA, Towers = ZA40 – ZA1 (colour = black/white/yellow)

Figure 20.19 OHL route diagram – typical

20.8 SCADA systems management

20.8.1 SCADA system management – requirements

Figure 10.40 illustrates a typical SCADA system as employed by power network companies. SCADA systems management generally covers two types of activities, which are as follows:

- A new substation control system (SCS) for the whole substation
- A change to an existing SCS (e.g. addition of an existing circuit).

The first requirement in a SCADA management procedure is the specification of generic equipment models – usually prepared by and retained by the power network company. That is to say, the specification of alarms, metering, indications and controls for each type of installed equipment (i.e. specific type of circuit breaker, protection relay etc. and nomenclature). Assembling the model is a considerable task.

NB: It may be that some alarms are common.

20.8.2 New SCS

Procedural requirements for a new SCS typically comprise the following:

- Preparation of a substation-specific SCS inputs/outputs document (sometimes termed a parameterisation document) – based upon the generic equipment models. This would usually be prepared by the contractor then subject to engineering assurance by the power network company.
- Construction of the SCS software to reflect the parameterisation document – by the contractor.
- A full factory test of all local and remote ports and loggers. This will necessitate specialist test equipment to interface with the SCS to simulate inputs and observe outputs. Both the testing regime and tests would be prepared and undertaken by the contractor and engineering assured by the power network company.
- Transfer of the SCS and database to the substation.
- Transfer of a copy of the SCS database to the control centre (i.e. that part of the SCS database relevant to the control centre).
- Full site tests at both the substation and the control centre.
- Installing and commissioning the communications' links (comms) between the substation and the control centre.
- Undertake substation to control centre end to end tests. NB: This is a more onerous process for the control centre since the IM (see Figure 10.40) is likely to be in operational service scanning many substations.
- It would be usual for the new SCS to become operational, simultaneously, with both the substation operation diagrams, and any equipment being subject to the HV equipment safety rules – therefore careful planning and timing is required, usually by the commissioning panel.

20.8.3 Addition to an existing SCS

With changes to an existing SCS, the FAT for proving the new database is often carried out on representative hardware in the factory. Exacting QA arrangements must be in place and observed to ensure that the existing (and remaining) requirements are not changed. Full tests are usually carried out on the additional or changed inputs/outputs with random checks on those existing/remaining. The new database is then transferred to the substation and control centre and uploaded on to the SCS and IM, respectively (see Figure 10.40). Site commissioning tests similar to those of the FAT are then undertaken.

20.9 Drawings management

20.9.1 Drawings management – requirement

The detail design stage of a project is essentially that of the production of a suite of drawings by the contractor against which site installation and commissioning takes place. On completion of the project, the drawings are usually transferred into the stewardship and ownership of the power network company. The drawings subsequently form the as installed design, which are referenced when any future changes to that installation take place. Power network companies therefore hold a huge suite of drawings covering all their sites and circuits. Both the storage of the drawings, and the ebb and flow of drawings between contractor and power network company therefore needs careful management. Sections 20.9.2 and 20.9.3 will identify some of the key requirements of a drawings' management procedure.

20.9.2 Drawings manager procedure – outline

Figure 20.20 illustrates a typical process where a contractor undertakes a project requiring updating of existing drawings which are stored in a power network company's master drawing record system. The various stages are outlined below:

1. **Stage (1)**
   Drawings are held in the power network company's master drawings record system. These are usually held electronically with the drawings indexed on a per site and a per circuit basis, and categorised into the various technologies (e.g. P&C, civil engineering, etc.).
2. **Stage (2)**
   The contractor retrieves the index to determine the drawings required for the project and is given permission by the power network company to electronically withdraw the drawings.
3. **Stage (3)**
   The contractor updates the drawings in the contractor drawing office, and creates any additional drawings required. All drawings are subject to a quality assurance process of drawn/checked/approved with related signatures included on each drawing, see Section 20.2.6.

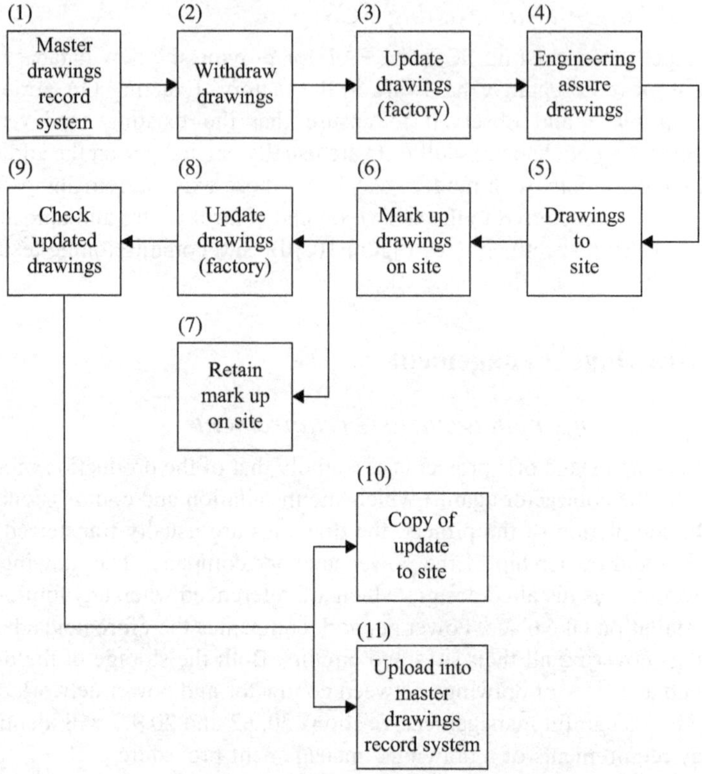

Figure 20.20 Typical drawings management process – overview

4. **Stage (4)**
The contractor forwards copies of the completed drawings (together with an index) usually in paper form, to the power network company who subject the drawings to engineering assurance (i.e. the Design Accept stage in Figure 20.1). The contractor revises the drawings against any comments received.
5. **Stage (5)**
The contractor forwards copies (usually three copies) of each drawing plus drawing index, to site to facilitate site installation and commissioning.
6. **Stage (6)**
Should errors or corrections be identified on-site, two copies of the drawings are marked up accordingly (usually with coloured pencils depicting additions and removals, respectively).
7. **Stages (7) and (8)**
Following circuit commissioning, one marked up copy of the drawings is retained on-site as the 'as installed record' and the other copy returned to the drawing office to be updated (i.e. revision).

8. **Stage (9)**
   The updated/revised drawing is compared against the markup drawing for an accuracy check. Sometimes, this is undertaken by the contractor and sometimes by the power network company.
9. **Stage (10)**
   Updated/revised paper drawings are filed on-site, with the site markup removed, and the site index updated accordingly. These drawings form part of the CDM health and safety file.
10. **Stage 11**
    The contractor electronically uploads the new suite of drawings arising from the project into the power network company's master drawings record system, and the index updated. The power network company confirms satisfactory completion of the upload to the contractor.

20.9.3 Drawing management procedure – requirements

Some of the key requirements of a drawing management procedure are as follows:

1. **Drawing numbering**
   Drawings often contain two sets of numbers, that of the contractor and that of the power network company. The numbering conventions will usually include:
   (i) Drawing number
   (ii) Sheet number (if more than one sheet)
   (iii) Revision number
   (iv) Electronic file reference
2. **One master principle**
   Each and every drawing shall have only one version recognised as the 'master' to avoid multiple markup (and revision numbers) of the same drawing – sometimes by different contractors.
3. **Substation master template**
   It is good practice for providing clarification of substation infrastructure to create a multilayered substation master template, typically comprising:
   (i) Overall site plan
   (ii) Oil containment and site drainage
   (iii) Site hazards plan
   (iv) Underground hazard plan
   (v) Earthing
   (vi) Plant layout
   (vii) Overhead structure supports

Chapter 21
Manufacture procedure

21.1 Introduction

This chapter will examine the procedural arrangements associated with the 'manufacture' stage of the scheme process as shown in Figure 1.4. This would usually involve QMS procedures from each of the manufacturers involved with the project, the contractor who places the contract/orders with the manufacturers and the power network company that places the contract with the contractor. This chapter will not examine the QMS procedures of the manufacturers, since they would be concerned with detailed factory processes, but will examine the procedural arrangements that would usually be required by the power network company or contractor at the interface with manufacturers. In brief, the following will be examined:

- Manufacturing assurance
- Equipment testing

21.2 Manufacturing assurance

21.2.1 Manufacturing assurance – requirements

Manufacturing assurance is, in effect, engineering assurance of the manufacture stage. The term 'expediting' may also sometimes be used for the task of manufacturing assurance. Expediting may be defined as securing the timely delivery of goods and components that comprise the manufactured article together with the timely delivery to site. The person discharging this role may be from the power network company, or the contractor, or a third-party specialising in this work.

Manufacturing assurance/expediting is concerned with deadlines, milestones and quality, and whether the manufactured equipment will be delivered on time. It additionally assists in the evaluation of VOWD associated with the equipment, see Section 19.8.3, and it may be categorised into the following:

1. **Production control**
   This comprises an inspection of the factory to ensure that the production arrangements are up to the necessary standard for manufacturing the

equipment. Within this context, audits/inspections are undertaken to ensure the following:

  (i) That the correct components are ordered, and are from approved suppliers.
  (ii) That the correct components are received.
  (iii) That the equipment is assembled in accordance with defined quality procedures.

2. **Quality control**
   Components are inspected (and sometimes tested) prior to assembly, and the assembled equipment is subject to inspections and tests to ensure that the equipment specification is met. See later sections for testing arrangements.
3. **Packaging/shipment**
   Packaging and shipment arrangements must be in accordance with specified quality standards – to ensure the equipment will withstand the adversities of transportation and delivered satisfactorily to site.

The level of assurance will depend upon the complexity, uniqueness and relative importance (i.e. risk to the project) of the manufactured item. It will also depend upon the level of confidence in the manufacturer (i.e. new or tried and tested). The manufacturer is usually required to provide a quality plan that is approved by those undertaking the assurance.

21.3 Equipment tests

21.3.1 Equipment tests – requirements

As part of the holistic manufacturing process, equipment manufacturers are required to undertake equipment tests to confirm that the equipment accords with the technical specification (usually the project generic specification). Technical specifications usually reference an IEC or similar standard document, and these documents usually define the testing regime to confirm the required performance. Testing may be categorised as follows:

- Type tests (factory based)
- Routine tests (factory based)
- Factory acceptance tests (factory based)
- Commissioning tests (site based)

These will be briefly examined below.

21.3.1.1 Type tests

Type tests (and inspections) are usually carried out on the very first item of a manufactured equipment, e.g. a type of circuit breaker, to comprehensively confirm that the item of equipment fully conforms with the technical specification. Generally, the tests confirm the following:

1. **Functionality**
   Confirmation of the functionality of the equipment, i.e. that it satisfies the range of performance requirements specified (e.g. that a distance protection comprises three zones, each with a defined setting range and each with an output signal per zone).
2. **Equipment strength and capability**
   This requires confirmation that the equipment can withstand the range of operational conditions to which it may be subject, including factors of safety, e.g. voltages and overvoltages, current ratings, thermal capacity, mechanical loadings, etc.
3. **Physical**
   This would include access arrangements, weather proofing, interface arrangements, labelling, etc.

Type tests can be expensive to undertake and some high-voltage tests require the facilities of specialist test laboratories. However, once the tests are complete, it is not usual to repeat them on subsequent manufacturing runs of that item of equipment unless one of the following occurs: a fundamental change is made to the design; or a change of manufacturer of one or more of the components; or a change of either place or method of manufacture. In effect, satisfactory completion of the type tests results in something similar to an off-the-shelf product.

Type tests are alternatively referred to by some power network companies as 'Type Registration' or 'Type Approval' or 'Type Certification'. Some power network companies require manufacturers to provide a 'bay solution' or a 'system solution', e.g. type test a complete bay (i.e. circuit) of protection as an integrated whole, or a complete OHL insulator string, including insulators, conductor types, dampers, spacers, clamps and terminations.

The awarding of type registration/type approval/type certification, etc. by a power network company often requires a package of supporting documentation as part of the type tests, typically comprising the following:

- Technical description of the equipment
- Schedule of type tests
- Equipment layout drawings
- Key assembly drawings
- Commissioning test and inspection schedule
- Protection equipment specific
  - Tripping circuit diagrams
  - Standard application diagrams
  - Numeric protection settings configuration file reference
  - Settings guidance document
- Installation, operation and maintenance manuals

21.3.1.2 Routine tests

Once type testing of an item of equipment has been achieved, each subsequent manufactured item of that equipment is usually subject to 'routine tests'. These are less onerous and comprehensive than type tests and are intended to demonstrate that the items of equipment are to the same standard as the original type-tested equipment.

21.3.1.3 Factory acceptance tests

Factory acceptance tests are a specific and additional type of routine test usually associated with protection and control equipment. Software-based multifunctional numeric protection and control relays are available with numerous settings options. For example, a distance protection relay will have a different reach setting applied to it dependent upon the length of the feeder – and depending upon the configuration of the circuit to be protected the relay may be set either as a plain or blocked distance protection. Similarly, the functionality of a mesh substation auto reclose/switching relay differs from substation to substation depending upon the number of connected feeders, transformers, etc. It is therefore usual with software-based relays to subject the generic routine tested relay to a range of simulation tests to verify that it is suitably set for the circuit(s) relevant to the project. Such tests are often referred to as 'factory acceptance tests'.

21.3.1.4 Commissioning tests

Commissioning tests, self-evidently, are not factory based but are carried out on site. They are mentioned here solely to illustrate the full range of equipment tests. Commissioning tests must of course comply with the manufacturers commissioning requirements – they are discussed in more detail in Chapters 15 and 23.

Chapter 22
Site installation procedure

22.1 Site installation procedure – overview

This chapter will examine the procedural requirements relevant to the site installation stage of a project, as depicted in Figure 1.4. The term 'site installation' as used in this text is a blanket term covering all project work, except commissioning, that is undertaken on site. The site may be a green field site or an operational site. Civil engineers may prefer the term 'site construction' to site installation since much of the civil work involves constructing buildings, roads, structures, temporary works, etc., whereas many items of power equipment arrive on site as complete (i.e. factory assembled). Within this context, the terms site installation or site construction are reasonably interchangeable.

Site installation essentially comprises the physical installation (or construction) of the power network equipment and associated infrastructure, on site. It also involves the establishment of the site facilities, and the site management arrangements. In summary, the following will be examined:

- Site installation – stages
- Equipment installation – inputs and outputs
- Site procedures
- Site duty holders
- Site installation – meetings arrangements
- Site health, safety and environmental requirements
- Site records, documentation and filing.

22.2 Site installation – stages

22.2.1 Site installation – the five stages

With reference to Figure 22.1, there are essentially five stages associated with site installation, which are as follows:

- Stage 1: pre-site access meeting
- Stage 2: site establishment
- Stage 3: equipment installation
- Stage 4: site demobilisation
- Stage 5: post-site access close out meeting

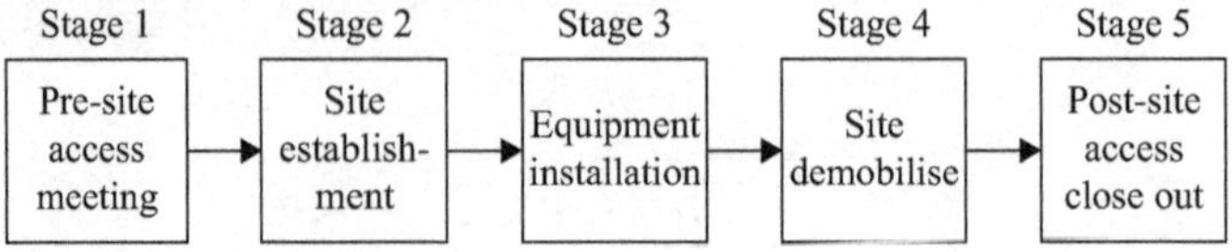

Figure 22.1 Site installation – stages

Pre-site access meeting agenda

1. Site establishment plan – proposal
2. Construction phase health and safety plan – status
3. Site environmental plan – status
4. Site quality plan – status
5. Emergency plan – status
6. CDM zones – arrangements
7. Design/drawings – status
8. Equipment manufacture/procure – status
9. Project programme – status
10. Project stage-by-stage document – status
11. Third-party interface arrangements (e.g. general public) – status
12. Duty holder competence arrangements – status
13. Action log

Figure 22.2 Pre-site access meeting agenda – typical

- **Stage 1: pre-site access meeting**
  Prior to site access, there is a requirement for representatives of both the power network company and the contractor to meet to confirm that all necessary arrangements are in place to enable site access to commence. The meeting would typically take place at least 6 weeks prior to site access and would usually involve the two project managers, the site manager (PC), the site engineer, the senior authorised person (SAP) and others considered essential – see Figures 19.3 and 19.6. Typical subject matter for the meeting is illustrated in Figure 22.2. The outcome of this meeting is finalisation of the site establishment plan.
- **Stage 2: site establishment**
  This stage comprises the physical implementation of the site establishment plan, i.e. the establishment of the work site and making it safe, secure and functional as a place of work. A typical site establishment plan is given in Figure 22.3.
- **Stage 3: equipment installation**
  This comprises the actual physical installation (i.e. construction) of the power network equipment and associated infrastructure – which will be undertaken in accordance with the following:
  - Detail design drawings/schedules
  - Site quality plan (SQP) – which will include the build sequence, equipment installation manuals and handbooks, and inspections and hold points

Site establishment plan

1. Establish all required roadways and access routes, both to/from and within the site
2. Establish site perimeter fence, etc.
3. Establish site security arrangements
4. Establish site cabins/huts (mostly temporary)
5. Implement any site requirements arising out of the CDM construction phase plan and environmental plan
6. Establish site secure storage accommodation (e.g. COSHH)
7. Establish site induction arrangements
8. Establish site documentation and records filing system (see Section 22.8.1)
9. Establish site notice board
    (i) Health and safety law poster
    (ii) CDM F10 form
    (iii) Emergency plan
    (iv) Location map showing nearest hospital
    (v) Emergency telephone and contact details
    (vi) Employers liability insurance certificate
    (vii) Project organisational structure (highlighting duty holder responsibilities), etc.

Figure 22.3 Site establishment plan contents – typical

Site Demobilisation Plan

1. Confirm all CDM zones removed
2. Confirm all equipment commissioned and transferred into the operational control of the power network company
3. Confirm CDM construction phase plan discharged
4. Confirm environmental plan discharged
5. Confirm a set of fully marked up drawings left and filed on site
6. Confirm CDM H&S file complete and transferred to power network company
7. Confirm with all third-party interfaces – no outstanding actions
8. Snagging list and resolution programme prepared – and agreed
9. Programme for the removal of all site facilities and temporary works prepared and agreed.

Figure 22.4 Site demobilisation plan – typical

  - Project stage-by-stage document
  - Site work schedule.
- **Stage 4: site demobilisation**
  This is the reverse of site establishment and also requires a plan to define how it is to be accomplished. It is prepared and executed by the site manager (PC) following completion of all commissioning work but in some instances may be phased following installation completion. The typical content of a site demobilisation plan is illustrated in Figure 22.4.

- **Stage 5: post-site access close out meeting**
  This essentially comprises a meeting, typically held within 6 weeks of Stage 2 commissioning being completed, and involving all those named in the Stage 1 above (i.e. pre-site access meeting). It additionally also usually involves the power network company and contractor's respective commissioning engineers. Typical content of the meeting will comprise the following:
  – Confirmation that the site demobilisation plan has been delivered and actions arising
  – Status of snagging list and actions arising
  – Contractual matters arising
  – Collation of lessons learned for forwarding to the post-project review.

22.3 Equipment installation – inputs and outputs

22.3.1 Site installation inputs and outputs – overview

Figure 22.5 provides an overview of the inputs to, and outputs from, the equipment installation stage. The inputs comprise the following:

1. **Temporary works**
   Site installation is invariably critically dependent upon temporary works. This input comprises the design of the temporary works and associated drawings together with the equipment and materials required to undertake the temporary works (e.g. scaffolding, shuttering, etc.).
2. **Detail design**
   This is the permanent works design comprising the drawings and schedules against which the permanent works will be installed/constructed.

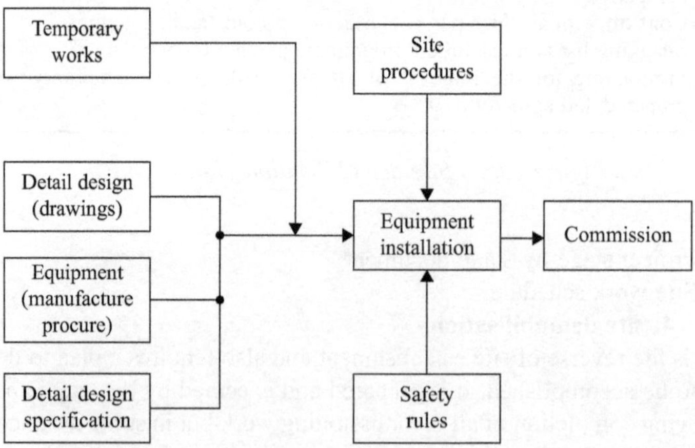

Figure 22.5 Site installation – inputs and outputs

3. **Equipment**
   This comprises the manufactured/procured equipment and materials which are delivered to site to become the newly installed equipment and associated infrastructure.
4. **Detail design specification**
   This document contains detail design, site installation and commissioning requirements. It should be used as a reference document against which site installation should be cross-checked and confirmed.
5. **Site procedures**
   These comprise the procedures and supporting documentation for defining
   (i) How the work site will be established and managed and
   (ii) How and when the equipment and the associated infrastructure will be installed/constructed.
6. **Safety rules**
   Safety rules, covering safe ways of working, are generally in two forms, see Section 18.6.1:
   (i) Safety from the system. These are provided by the power network company.
   (ii) General Safety. These are usually provided by the contractor.

The output from site installation is equipment which has been installed/constructed in accordance with procedural and technical requirements (i.e. a compliant installation) and is available for commissioning to commence.

22.4 Site installation procedures

22.4.1 Site installation procedures – summary

Figure 22.6 provides an overview of the main procedures (and associated inputs) relevant to site installation. In summary, they comprise the following:

1. **CDM construction phase plan**
   The CDM construction phase plan was described in Section 18.9.10.3. It takes account of the health and safety considerations specified in both the pre-construction information document and the site establishment plan (see Section 22.2.1). It addresses the following site requirements:
   (i) The site rules and arrangements for managing the construction site.
   (ii) Arrangements for identifying and controlling hazards and risks.
2. **Risk assessment and method statements (RAMS)**
   An integral part of the CDM construction phase plan is the preparation of risk assessment and method statements (RAMS). Section 18.3.2 defined a risk which is as follows:

   $$\text{Risk} = \text{hazard severity} \times \text{hazard likelihood}$$

   where both the hazard severity and likelihood are ranked in a range, e.g. a range of 1 to 5. A typical hazard categorisation matrix is given in Figure 22.7.

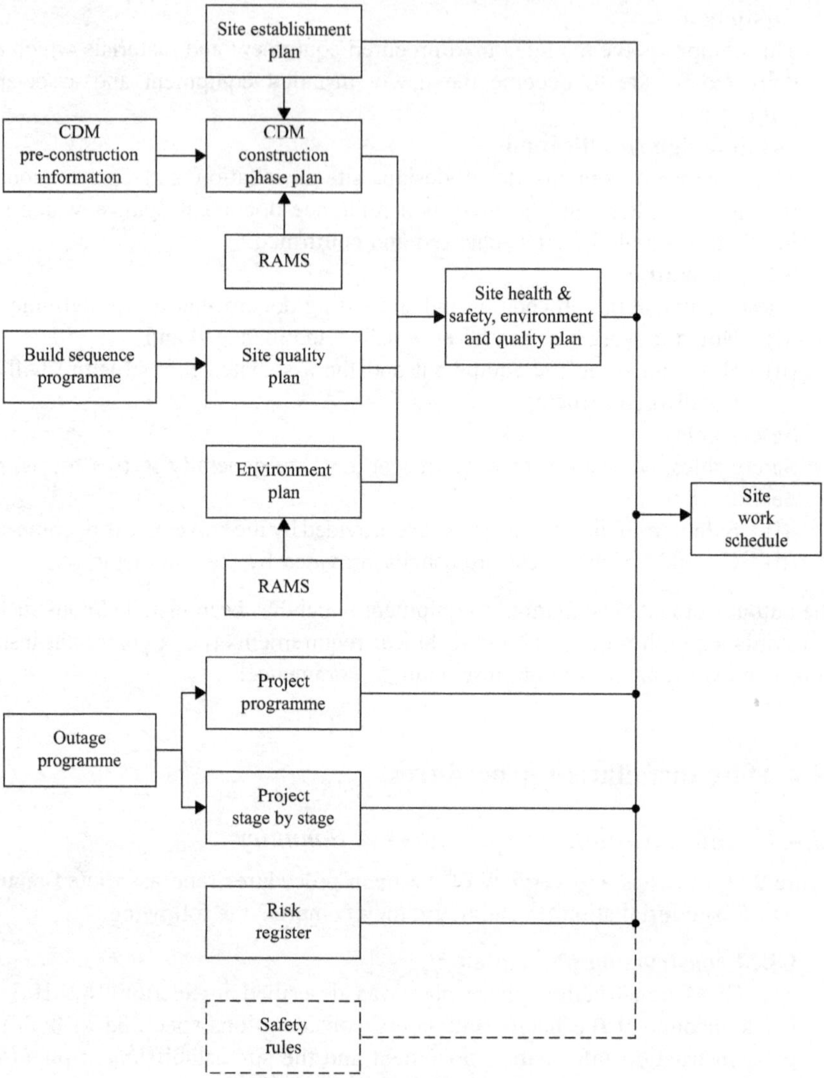

Figure 22.6 Main site procedures

From Figure 22.7, a risk score matrix can be prepared by multiplying together all the permutations. This is given in Figure 22.8.

From the risk score matrix, a risk categorisation can be produced. Figure 22.9 provides one possible risk categorisation.

NB: The categorisation ranges are a matter of judgement.

Figure 22.10 illustrates the typical format of a RAMS that, following preparation, may be issued to a duty holder to commence work.

Hazard Categorisation Matrix

Hazard severity	**Hazard likelihood**
1 = Minor injury	1 = Highly unlikely
2 = First aid required	2 = Unlikely
3 = Hospital attendance required	3 = Possibly
4 = Lost time incident	4 = Highly likely
5 = Major injury/death	5 = Almost certain

Figure 22.7 Risk categorisation matrix – typical

Risk Score Matrix

Hazard severity ↓					
5	5	10	15	20	25
4	4	8	12	16	20
3	3	6	9	12	15
2	2	4	6	8	10
1	1	2	3	4	5
Hazard likelihood →	***1***	***2***	***3***	***4***	***5***

Figure 22.8 Risk score matrix

Risk Categorisation

Risk score	**Risk categorisation**
1 to 4	Low risk (work can proceed)
5 to 12	Medium risk (work only to proceed after suitable controls put in place)
15 to 25	High risk (consider alternative work method or work only to proceed after suitable controls put in place)

Figure 22.9 Risk categorisation

3. **Site quality plan**
   The SQP is the site element of the project quality plan. It is prepared by the contractor (PC) and is a requirement of ISO 9001. The content of a typical SQP is summarised in Figure 22.11.

RISK ASSESSMENT & METHOD STATEMENT (RAMS)

Project No:

Location:

Risk assessment

Hazard description	Risk without control measure			Risk with control measure		
	Severity 1–5 (a)	Likelihood 1–5 (b)	Risk (a) × (b)	Severity 1–5 (a)	Likelihood 1–5 (b)	Risk (a) × (b)

Risk control measure description	

Method statement

Description of work to be undertaken	
Sequence of work (including risk control measures)	
Requirements to make work area safe	
Requirements to make environment safe (e.g. lighting, ventilation)	
PPE requirements	

Figure 22.10 RAMS – typical format

Site Quality Plan

1. Project description and scope
2. Stages of work
3. Site programme and build sequence
4. Site roles and responsibilities
5. Site quality arrangements (audits/inspections)
6. Schedule of installation instructions, manuals, etc., for each item of equipment/site infrastructure
7. Hold points
8. Site procedures
9. Change control arrangements
10. Competency arrangements
11. Site filing and records

Figure 22.11 Site quality plan – typical

The SQP is a key document defining the means and standard by which site installation is undertaken. An input into the SQP will be the build sequence programme.

4. **Environmental plan**

   An environmental plan is the environmental equivalent of the CDM construction phase plan, targeted at ensuring that the construction site and construction activities satisfy legal requirements (as described in Section 2.1). An environmental plan is additionally required to satisfy ISO 14001 (see Section 2.8.3). Typical environmental plan contents are given in Figure 22.12.

   NB: The RAMS would have a similar format to that illustrated in Figure 22.10.

   The environmental aspects to be managed on site typically comprise:

   (i) Emissions to air
   (ii) Releases to water
   (iii) Releases to land
   (iv) Use of raw materials and natural resources
   (v) Use of energy emitted (e.g. heat, vibration)
   (vi) Waste
   (vii) Physical impact (e.g. size, shape, colour)
   (viii) Wildlife and diversity etc.

   A requirement integral to an environmental plan, as required by the site waste management plan (SWMP) regulations, is the production of a SWMP. This should be prepared prior to site access and implemented once on site. The aim of a SWMP is to:

   (i) Improve efficiency and profitability through:
      (a) Reuse
      (b) Recycle
      (c) Recovery
      (d) Disposal

Environmental Plan

1. Project description
2. Roles and responsibilities
3. Reporting arrangements
4. Legislation/procedures to be complied with
5. Environmental risks to be managed
6. Environmental permits to work
7. Programme and key dates
8. Environmental registers
9. Communications (internal and external)
10. Emergency plan
11. Audit
12. Training
13. Documentation/records/filing

Figure 22.12 Environmental plan – typical

(ii) Reduce fly tipping through a full audit trail of waste removal
(iii) Increase environmental awareness in the workforce and management.

5. **Site health and safety, environmental and quality plan**
The requirements of the construction phase H&S plan, the SQP and environmental plan contain areas of duplication (e.g. project description and roles/responsibilities, programme, etc.). Therefore, some contractors in an effort to remove unnecessary bureaucracy integrate these three documents into a single document, with a title similar to the heading above.
6. **Project programme**
The project programme, as described in Section 19.4.1, will also define the events and key dates relevant to site installation. The outage programme would feed into the project programme.
7. **Project stage by stage**
The project stage-by-stage document as described in Chapter 24 becomes operational at the site installation stage. This document is concerned with the stages by which equipment is constructed, becomes subject to the safety rules of the power network company and connected to the power system via Stage 2 commissioning. This document usually encompasses all interacting sites where installation and commissioning is being undertaken simultaneously. The outage programme would usually feed into the project stage-by-stage document.
8. **Risk register**
The risk register is described in Section 19.7.2. Although it would be hoped and expected that many of the project risks would either have not materialised or been resolved by the commencement of site installation, some may still be present. As such the risk register must be kept active with ongoing evaluation and reduction of risk.
9. **Site work schedule**
The site work schedule (or other similar title) is the definitive document for identifying each item of work to be undertaken together with associated requirements. This would include the date(s) of the work together with the required health, safety, environmental and quality procedures. All other documents feed into this document. No work should commence unless entered onto this document and agreed by the site manager, the site engineer and where relevant the SAP. Figure 22.13 illustrates a typical site work schedule. Week 1, the forthcoming week, is firm, and Weeks 2, 3 and 4 in draft, with increasing detail and certainty as Week 1 is approached.

22.5 Site duty holders and responsibilities

22.5.1 Site duty holders

With reference to Figures 19.3 and 19.6, the responsibilities of the following major duty holders, who hold key positions during the site installation period, will be examined:

- Site manager
- Site engineer
- SAP

Site Work Schedule
Week 1 = firm

Task description	Planned dates	Quality procedure Ref. No.	RAMS Ref.No.	Safety rule document (type)	Circuit outage requirements

Week 2 = draft

Headings as Week 1

Week 3 = draft

Headings as Week 1

Week 4 = draft

Headings as Week 1

Figure 22.13 Site work schedule – typical

22.5.1.1 Site manager

The contractor's site manager (i.e. PCs representative on site) is accountable for both the management of the construction site and the installation/construction of the assets in accordance with the requirements of the contract. Specific requirements are summarised as follows:

- Establishing the construction site in accordance with the site establishment plan, and later demobilising the site in accordance with the site demobilisation plan.
- Managing the construction site in accordance with the requirements of Figure 22.6.
- Installing/constructing the power network assets (equipment) and associated site infrastructure in accordance with the requirements of Figures 22.5 and 22.6.
- Managing the site construction workforce including all subcontractors.
- Managing the interface with adjacent projects and third parties.
- Handing over the installed/constructed power network assets and associated site infrastructure to the commissioning panel for subsequent commissioning.

22.5.1.2 Site engineer

The power network company's site engineer would usually have delegated authority from the project manager to deliver site-based responsibilities. In brief, such responsibilities would comprise:

- Contract management (on site) and monitoring/progressing site installation
- Engineering assurance of the site installation stage of the work with reference to health and safety management, environmental management and quality management
- Ensuring that the necessary consents and wayleaves are in place, and adhered to, during the site installation stage.

Different site engineers may be required for overseeing substation electrical, civil engineering, OHL and HV cable site installation requirements, respectively, each possessing specialist capability in the field in question. Detailed responsibilities of a site engineer would comprise the following:

1. **Site manager interface**
   (i) Ongoing liaison and communication with the site manager with reference to contract delivery, work coordination and site installation progress
   (ii) Out of hours working coordination
   (iii) Site security arrangements
   (iv) Third-party interfaces
   (v) Site access coordination
2. **Contract management**
   Maintenance of a site diary with reference to:
   (i) Daily progress
   (ii) Deliveries to site
   (iii) Personnel on site
   (iv) Site incidents
   (v) Site contract claims
   (vi) Site additional work/variations
3. **Health and safety and environmental assurance and management**
   (i) Sensible monitoring, see Section 18.9.12
   (ii) Ensure that site incidents are investigated and progressed
   (iii) Interface with SAP for safety rule requirements
   (iv) SHES audit
   (v) Temporary works monitoring
   (vi) IV hazards evaluation and progressing
   (vii) Monitor development and implementation of the construction phase health and safety plan
   (viii) Monitor development and implementation of the environmental plan
   (ix) Monitor general safety and safety from the system on site
   (x) Ensure the CDM health and safety file is handed to the occupier

4. **Engineering (work quality) assurance**
   (i) Monitor that the SQP is prepared, acceptable and implemented
   (ii) Ensure site establishment plan and site demobilisation plan appropriately executed
   (iii) Monitor that site drawings are appropriately managed, marked up and site installed copy retained on site
   (iv) Checks on HV safety clearances, see Section 18.6.3
   (v) Monitor design change process
   (vi) Ensure all planned outages are booked
   (vii) Ensure snagging list prepared and implementation monitored
   (viii) Confirm installed assets are suitable for transfer to the commissioning panel.

A comprehensive monitoring schedule of site engineer responsibilities should be prepared containing the above – which should be completed as the work progress.

22.5.1.3 Senior authorised person (SAP)

The SAP is usually the most senior person appointed under the power network company safety rules, see Section 18.6.4. The SAP would usually also discharge the role of site occupier (for operational sites), as described in Sections 18.4.1 and 18.4.3. The SAP undertakes a critical role in the smooth running of site installation, requiring an ongoing dialogue with both the site manager and the site engineer. Key SAP responsibilities would usually include the following:

1. **Safety from the system**
   This is achieved through application of the power network company safety rules, see Section 18.6.1.
2. **Sensible monitoring of safety from the system**
   This comprises monitoring that the arrangements put in place to achieve safety from the system are maintained throughout the duration of the work, see Section 18.9.12.
3. **CDM zone – transfer arrangements**
   This consists of transfer of part of an operational site into the safety management control of the PC (site manager), and subsequent transfer back, see Section 18.9.11.
4. **Adding/removing equipment to/from the HV safety rules**
   The SAP is usually the power network company site representative for agreeing the timing of, and completing, the Safety Rules Clearance Certificate, see Figure 18.7. NB: This certificate is sometimes alternatively completed during the commissioning stage.
5. **Impressed voltage safety management**
   The SAP will usually have an authoritative understanding of both the nature of, and risks arising from, IV, see Chapter 11. In addition to controlling these risks when the equipment is under the safety rules, the SAP can also advise the PC on IV when the equipment is under the safety management control of the PC.

6. **General advice**
   The SAP will usually have a good understanding of the functioning of an operational site, including equipment performance, site infrastructure and the general environment surrounding the site. As such the SAP is well positioned to provide guidance and advice on a whole range of matters and issues that may arise during site installation. NB: On some sites, the SAP and site engineer may be one and the same person.

22.6 Site installation meeting arrangements

22.6.1 Site installation key meetings

Generally, two key meetings are essential for successfully progressing site installation, these are:

- Site progress meeting and
- Site health and safety and environmental meeting.

These meetings would usually be held weekly and on site. Their purpose and content are summarised below.

22.6.1.1 Site progress meeting

The site progress meeting serves the following purposes:

- Progressing the site work via the site work schedule and supporting documentation
- Progressing matters arising under the contract
- Monitoring and assessing work quality
- All other matters not covered by the site health and safety and environmental meeting.

Typical attendance at this meeting would comprise the following:

- Site engineer (chair)
- Site manager (PC)
- SAP
- Temporary works coordinator
- Site manager direct reports (as required)
- Subcontractors (as required)
- Project manager/project engineer (as required).

A typical agenda is provided in Figure 22.14.

22.6.1.2 Site health and safety and environmental meeting

This meeting serves the following purpose:

- To plan, progress and coordinate all health, safety and environmental requirements
- To provide assurance that safety, health and environmental requirements are being satisfactorily executed

---

Site Progress Meeting Agenda

1. Site title
2. Project no.
3. Date of meeting
4. Meeting attendance
5. Project stage-by-stage document – status update
6. Outage programme – status update
7. Project programme – status update
8. SQP – status update
9. Drawings management – status update
10. Design change control – status update
11. Manpower levels and resources – status update
12. Site work schedule
    (i) Confirmation of work achieved
    (ii) Confirmation of work scheduled for Week 1 (firm)
    (iii) Confirmation of work scheduled for Weeks 2–4 (draft)
13. Contract matters and issues arising – status update
    (i) Work variations
    (ii) Work instructions
    (iii) Snagging/defects
14. Equipment transferred to commissioning stage – status update
15. Risk register – status update
16. Third-party interfaces – status update
17. Lessons learned and improvement plan
18. Outstanding actions log – status update
19. AOB
20. Date of next meeting

Figure 22.14 Site progress meeting agenda – typical

- To consider, plan and action future health, safety and environmental improvements
- To ensure that all legal requirements are discharged, with particular emphasis on the CDM regulations.

Typical attendance at this meeting will usually comprise the following:

- PC (site manager) (chair)
- Site engineer
- SAP
- Temporary works coordinator
- PC direct reports
- Subcontractor representatives
- Site health and safety officer
- Site environmental officer
- Project managers/engineers (as required)
- Principal designer (as required)
- Commissioning engineers (as required).

Site Health and Safety and Environmental Meeting Agenda

1. Site location
2. Project no.
3. Date of meeting
4. Meeting attendance
5. Construction site – status update (e.g. accommodation, welfare, notices, records, etc.)
6. Project stage-by-stage document – status update
7. CDM requirements – status update
   (i) Construction phase plan
   (ii) Demarcation zones
   (iii) Health and safety file
8. Environmental plan – status update
9. IV management
10. Temporary works – status update
11. Commissioning – status update
12. Third-party interfaces – status update
    (i) Adjacent projects and other PCs
    (ii) General public
13. Sensible monitoring
    (i) General safety
    (ii) Safety from the system
    (iii) Environmental compliance
14. Site work schedule
    (i) Confirmation of work achieved
    (ii) Confirmation of work in Week 1 – firm
    (iii) Confirmation of work in Weeks 2, 3 4 – draft
15. Hazard reviews
16. Accidents and incidents – status update
17. Audit programme – status update
18. Performance improvement initiatives
19. Competency and training – status update
20. Lessons learned
21. AOB
22. Date of next meeting

Figure 22.15 Site health and safety and environmental meeting agenda – typical

NB: This meeting would usually extend into commissioning activities. A typical agenda is given in Figure 22.15.

As can be seen, both the site progress meeting and the site health and safety and environmental meeting have similar attendees, and slightly overlapping agenda items (albeit with a different focus). For reasons of efficient use of time, these meetings are often held back to back.

22.7 Site health and safety requirements

22.7.1 Site health and safety – considerations

Self-evidently the site installation stage of a project provides the greatest health and safety risk to those involved with the project. The risks are managed through the

Site Induction Procedure

1. Senior management statement of commitment to H&S
2. Project outline
3. Site geography and areas of responsibility
4. Site organisational structure and key duty holders
5. Site hazards and risks
6. Management arrangements
   - (i) Site safety rules and procedures
   - (ii) Site access rules
   - (iii) Access/egress routes
   - (iv) Security arrangements
   - (v) Welfare facilities
   - (vi) Site storage arrangements
   - (vii) PPE arrangements
7. First aid arrangements
8. Accident/incident reporting arrangements
9. Arrangements for consulting and briefing the workforce
10. Individual responsibilities for H&S (e.g. first aid).

Figure 22.16 Site induction procedure contents – typical

application of a safety management system as illustrated in Figure 18.1. With reference to site construction work, health and safety management arrangements may be categorised as comprising:

- Management of the construction site
- Installation/construction of the new assets and related temporary works
- Management of third-party interfaces
- Ensuring that duty holders are competent for the work to be undertaken.

The above will be briefly examined.

22.7.1.1 Management of the construction site

The health and safety management arrangements for the construction site itself are covered by, and are a requirement of, the CDM regulations and defined in the construction phase plan, as outlined in Section 18.9.10.3. This should also include site establishment as defined in the site establishment plan, see Figure 22.3. A key requirement of health and safety on site is 'site induction'. This should be in accordance with a site induction procedure, the requirements of which should apply to all those with access to the construction site. Figure 22.16 outlines the contents of a typical site induction procedure. A register of all those who have successfully completed the induction procedure should be maintained.

22.7.1.2 New asset installation/construction

All work on site should be specified on a site work schedule, see Figure 22.13, which defines all H&S requirements necessary to undertake the work. The inputs to the site work schedule are illustrated in Figure 22.6 and comprise either site procedures or safety rules (shown dotted to indicate they are distinct and separate from site procedures).

With references to the site procedures, it is worthy of note that health and safety regulations such as COSHH, working at height, etc. (see Chapter 18) fall under the umbrella of the CDM regulations and where applicable must be considered when preparing the RAMS.

Safety rules are concerned with safety from the electrical power system and are aligned to the requirements of the electricity at work regulations, see Section 18.5.1 (although contractors may also have safety rules applicable to general safety). Both the power network company and the PC (site manager) will have their own safety rules, although those of power network companies are likely to be more comprehensive than those of the contractors since their spectrum of work will be wider. The contractor's safety rules will apply in the CDM zone, when the PC is accountable for the safety management of electrical systems within the zone.

22.7.1.3 Management of third-party interfaces

Third-party interfaces during site installation essentially fall into three areas, which are as follows:

- Interfaces with another project (i.e. another PC). This is covered in Section 18.9.13.
- Interfaces with the general public, including major users of electricity. This must be considered in the CDM construction phase plan, see Figure 18.10.
- Interfaces with either another power network company, generator company or major consumer involving both equipment interfaces and safety rule interfaces.

The safety rule interface arrangements between a power network company and another power network company, or generation company or major consumer may include:

- A permanent connection made between equipment constructed under interfacing projects and the determination of which safety rules apply during the connection process and
- Equipment being constructed under the safety rules of one organisation and then transferred into the safety rules of another organisation.

Advance agreement of how the above will be managed, must be undertaken.

22.7.1.4 Duty holder competence

All those undertaking work on a construction site will require the necessary competencies to undertake the work. Site-based competencies generally fall into three categories which are follows:

- Basic electrical competence

- Safety rule competence
- Task-specific competence

These will be briefly examined below:

- **Basic electrical competence and BESC AME**
  Many UK power network companies require all those working on a construction site to be in possession of a certificate of basic electrical competence. The most commonly requested certificate is BESC AME (basic electricity safety competence – access, movement and egress). This certificate is awarded by the energy and utility skills register (EUSR). EUSR is an employer-led membership organisation that helps ensure that the power, gas and water industries have the skills and competencies that they require. BESC AME covers basic requirements for safely entering, moving around and exiting electrical work sites. There are separate certificates for substations, OHL and HV cables. Requirements of the competency include the following:
  - Safe access, move around and exit a work area, e.g. a substation
  - Comply with company procedures, health and safety regulations and environmental practices
  - Identify risks associated with equipment and excavations
  - Use of, and communicate, relevant data and information
  - Resolve problems efficiently and effectively.

  Successful completion of the competency results in the individual being issued with a card and entered on the EUSR register, i.e. website. Registration usually exists for 3 years thereafter requiring reassessment, see also Section 26.1.5.4.
- **Safety rule competence**
  Most power network companies have their own arrangements for assessing competence under the safety rules for the roles of person, competent person, authorised person and SAP. Some power network companies have detailed safety procedures for specific types of work, under the umbrella of the safety rules, e.g. work on cables or OHL – which require both their own and contractor's personnel to be subject to assessment to achieve a certificate of competence prior to starting the work. Similarly competence also needs to be determined under the contractor's safety rules.
- **Task-specific competence**
  Competence is also required to be demonstrated on specific activities, usually linked to the HASAWA regulations, examples include:
  - Scaffolding
  - Construction plant operation (e.g. excavation)
  - Construction demolition.

  Widely recognised training and assessment of such competencies is provided by the Construction Industry Training Board which is the government recognised training board for all of the UK construction industry, see Section 26.1.5.6. Again successful demonstration of a competency results in the issue of a (competency) card to the person in question and entry on the competency register. These competences are in the domain of the contractor.

22.8 Site records and documentation filing

22.8.1 Records and documentation filing – requirements

An essential requirement of construction site management is the methodical and comprehensive filing of site records and documentation by the contractor. A typical list of files is as follows:

1. **Drawings**
   (i) Schedule of drawings received on site
   (ii) Drawings identified as being subject to design change or error correction
   (iii) Drawings returned to the construction drawing office, either for revision or final mark up
2. **Site work procedures**
   (i) Quality procedures
   (ii) Health and safety procedures
   (iii) Environmental management procedures
3. **Health and safety and environmental management**
   (i) CDM construction phase plan
   (ii) Accident book
   (iii) Fire management arrangements
   (iv) Site electrical wiring certificate
   (v) Site COSHH management arrangements
   (vi) Site waste management arrangements and control documentation
   (vii) Environmental plan
   (viii) Site emergency plan (including hospital details)
4. **Site health and safety and environmental records**
   (i) First-aid equipment inspections
   (ii) Fire extinguisher inspections
   (iii) Scaffolding inspections
   (iv) Environmental licences
   (v) Insurance test certificates (e.g. cranes, excavators, hoists, etc.), etc.
5. **Competency training records**
   (i) Safety rule appointments
   (ii) Site inductions
   (iii) Mobile plant operators
   (iv) Scaffolders, etc.
6. **Equipment inspection procedures**
   (i) E.g. cranes, hoists, lifting equipment, scaffolding, mobile plant, electrical equipment testing
7. **Project documentation**
   (i) Site establishment plan
   (ii) Site demobilisation plan
   (iii) SQP
   (iv) Project stage by stage
   (v) Outage programme

(vi) Risk register
(vii) Project programme, etc.

8. **Site work schedule**
   (i) RAMS
9. **Minutes of meetings**
   (i) Site progress meeting
   (ii) Site H&S and environmental meeting
   (iii) Commissioning panel meetings, etc.

22.9 Site installation – summary

The preceding sections have summarised many of the key requirements associated with the site installation stage of a project. Prior to site installation, virtually all tasks were associated with the preparation and management of documentation. Once site installation commences, not only does the preparation and management of documentation continue but also the enactment of that documentation together with the physical installation/construction of the new assets. It is therefore a highly complex and demanding stage of any project – demanding a highly competent team of site practitioners.

Chapter 23
Equipment commissioning procedure

23.1 Introduction

Commissioning is the final stage in a scheme/project prior to closure, see Figure 1.4. It is a critical and important stage since it is the final opportunity to correct any errors/omissions which thus far have not been detected. In summary, the commissioning process takes the assets that have been assembled/constructed during the site installation stage and subjects them to tests and inspections to verify that they are acceptable for commercial service. Figure 23.1 summarises the inputs and outputs to/from the equipment commissioning stage. This chapter will examine requirements which should be included in a commissioning procedure, as follows:

- Commissioning objectives
- Management arrangements for commissioning
- Concept of Stages 1 and 2 commissioning
- Commissioning programme
- Commissioning inspections
- Commissioning switching programme
- Commissioning certificates
- Commissioning technical documentation
- OHL and HV cable commissioning
- Health and safety
- Equipment decommissioning
- Key commissioning roles and responsibilities
- Commissioning panel meeting agenda – and documentation filing

23.2 Commissioning objectives

23.2.1 Commissioning process objectives

The objective of the commissioning process is to prove that the individual items of equipment, which are interconnected to form electrical systems, circuits and substations accord with the contract specification, are suitable for their intended

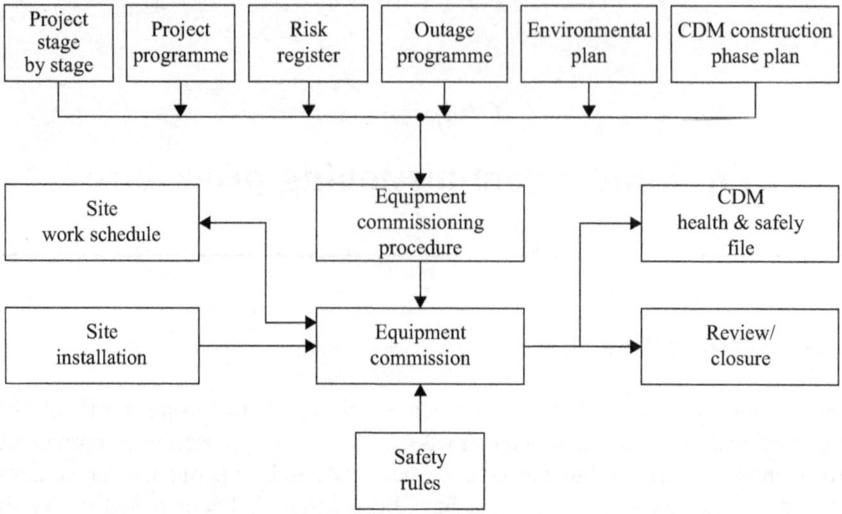

Figure 23.1 Equipment commission – inputs and outputs

purpose and are acceptable for entering commercial service. This umbrella objective can be subdivided into the following:

- Verifying that the equipment has not been damaged in transit, that it has been correctly installed and that it performs as specified
- Obtaining test data as the basis for comparison with future maintenance and fault investigation results.

The above has concentrated on the prime purpose of commissioning which is that associated with proving equipment which will form part of the power system. However, commissioning also extends into power system infrastructure to include assets such as oil containment systems, building engineering, site security, site lighting, etc. and these assets must also be encompassed by the commissioning process.

Commissioning is concerned with the proving of equipment and other assets via inspections and performance tests. However, some assets cannot be subject to testing (e.g. buildings) and therefore must be subject to comprehensive inspections during the site installation stage. As such, these assets are not encompassed by the commissioning process.

23.3 Management arrangements

23.3.1 Commissioning management arrangements

It is usual to form a team with responsibility for managing the commissioning process. This team is frequently termed the 'commissioning panel' or other similar title. Typical composition of a commissioning panel is given in Figure 23.2.

Commissioning Panel Attendees

1. **Power network company representatives**
   - (i) Chairman
   - (ii) Commissioning engineer(s)
   - (iii) Lead design engineer
   - (iv) System operations department representative
   - (v) SAP
   - (vi) Operational safety management representative
   - (vii) Site engineer
   - (viii) Project engineer (as required)
2. **Contractor representatives**
   - (i) Commissioning engineer(s)
   - (ii) Site manager (PC)
   - (iii) Site health and safety officer
   - (iv) Project engineer (as required)
3. **Third-party representatives**
   (e.g. interfacing projects/power station/other power network company/major consumer, etc.)
   - (i) Commissioning engineer
   - (ii) SAP
   - (iii) Project manager

Figure 23.2 Commissioning panel composition – typical

The prime purpose of the commissioning panel is to collectively ensure that arrangements are in place for ensuring the equipment is correctly commissioned, with particular reference to the requirements of the contract. Specific requirements include:

- The preparation, agreement and execution of a commissioning programme
- The management of health and safety and environmental safeguards
- Adherence to quality management procedural arrangements
- Taking all necessary steps to ensure the commissioning activities do not jeopardise the reliability and performance of the existing power system
- Ensuring that all commissioning duty holders are competent for the tasks to be undertaken
- Ensuring full cooperation, coordination and communication with other commission panels concerned with an interfacing part of the network (e.g. the substation at the remote end of the circuit) or with other relevant third parties.

With reference to Figure 23.2, not all possible attendees would be required at every meeting. Those attending should be selectively determined by the chairman. Commissioning panel meetings should ideally take place weekly. They are usually held on-site.

23.4 Commissioning Stages 1 and 2

23.4.1 Stages 1 and 2 commissioning

This section will examine the meaning of four terms which are both useful and commonly used, in defining commissioning activities. The terms are:

1. Off-load commissioning test
2. On-load commissioning test
3. Stage 1 commissioning
4. Stage 2 commissioning

1. **Off-load commissioning test**
   These are commission tests which are undertaken on equipment which is not connected to the in-service power system – with the objective of confirming that the equipment is suitable for connecting to the in-service power system.
2. **On-load commissioning tests**
   These are commissioning tests which are undertaken on equipment which is connected to the in-service (i.e. energised) power system – with the objective of using power system quantities (i.e. voltage, current, frequency, etc.) to confirm that the equipment is suitable for commercial service (i.e. fully commissioned).
3. **Stage 1 commissioning**
   Stage 1 commissioning comprises off-load tests and inspections to confirm that the equipment is ready for Stage 2 commissioning. Stage 1 is usually under full control of the contractors' commissioning engineer and subject to engineering assurance (witnessing) by the power network company's commissioning engineer.
4. **Stage 2 commissioning**
   Stage 2 commissioning comprises the following:
   (i) The method by which equipment that has completed Stage 1 commissioning is connected to the existing power system, and execution of the same
   (ii) Following (i) above, the method of energising the equipment from the existing power system, and execution of the same
   (iii) On-load commissioning tests and inspections.
   Stage 2 commissioning is usually carried out under the direction of the power network company's commissioning engineer [and senior authorised person (SAP)], but with the agreement of the contractor's commissioning engineer who will undertake the on-load commissioning tests and inspections.
   NB: Points (ii) and (iii) above are usually undertaken in accordance with a commissioning switching programme (CSP).

It is to be noted that the above terms apply not only to HV equipment but also LV equipment, including LV supplies, battery systems, protection & control, etc. In the case, for example, of a substation LV generator or 110 V battery system, these will not be connected to the existing power system but are stand-alone systems integral to the total power system.

23.5 Commissioning programme

23.5.1 Commissioning programme – purpose and content

A commissioning programme defines the arrangements by which equipment is tested and inspected on-site. It covers Stages 1 and 2 commissioning requirements and is usually part of the site quality plan. On completion of the commissioning programme, the equipment is available for commercial service. In general, a commissioning programme comprises two complementary documents as described in Figure 23.3.

The commissioning test and inspection schedule would typically be configured as shown in Figure 23.4.

23.6 Commissioning inspections

23.6.1 Commissioning inspection schedules

The commissioning programme outlined in Section 23.5.1 listed three formal inspections schedules namely:

Commissioning Programme

1. **Commissioning timing and staging programme**
   This comprises the dates for undertaking the range of commissioning activities (e.g. when to commence Stage 1 commissioning of a transformer) and the duration of the activity. The activities will be categorised into Stages 1 and 2 and in a logical sequence, e.g. CT tests, followed by primary injection, followed by secondary injection. The programme must also align with the availability of the equipment and the commissioning engineer resource available.
2. **Commissioning tests and inspections schedule**
   This comprises a full and complete listing of all test and inspection documentation that is required, and encompasses the following:
   (i) **Stage 1 commissioning**
      (a) Initial commissioning inspection schedule
      (b) Off-load commissioning test schedule
      (c) Pre-Stage 2 commissioning inspection schedule
   (ii) **Stage 2 commissioning**
      (a) Method of connecting new equipment to the existing power system (via RAMS, and PTW for HV equipment)
      (b) Commissioning switching programme (including method of HV equipment energisation and on-load commissioning tests schedule)
      (c) Final commissioning inspection schedule

Figure 23.3 Commissioning programme – typical

Commissioning Test and Inspection Schedule

Stage 1 commissioning

Activity	Test/ Inspect document ref. no.	Outage requirement/ risk of trip	Safety document type	Test hold point	Date completed	Confirmed by
Initial Comm. inspection schedule						
IR tests						
CT tests 1. Flick test 2. Mag curve, etc.						
Primary injections						
P&C Sec. injections 1. First MP 2. Second MP 3. Back up prot, etc.						
DC logic tests 1. MP1 circuit diagram 2. MP2 circuit diagram 3. Transformer prot circuit diagram, etc.						
Battery tests						
Equipment tests 1. Circuit breaker 2. Transformer 3. Disconnector, etc.						
Comm. method statements						
Pre-Stage 2 comm. inspection schedule						

Figure 23.4 Continued

Stage 2 commissioning

Activity	Test/ Inspect document ref. no.	Outage requirement/ risk of trip	Safety document type	Test hold point	Date completed	Confirmed by
LV diesel generator Comm. method statement no. XYZ						
HV circuit ABC comm. switching programme no. ZYX						
Final inspection schedule						

Figure 23.4 Commissioning test and inspection schedule – typical

1. Initial commissioning inspection schedule
2. Pre-Stage 2 commissioning inspection schedule
3. Final commissioning inspection schedule.

These will be briefly examined below.

1. **Initial commissioning inspection schedule**
   This should comprise a list of the equipment to be inspected to determine whether the equipment has been satisfactorily installed to enable Stage 1 commissioning tests to begin. Typical items to inspect and check comprise:
   (a) All equipment identified and accounted for
   (b) All equipment correctly labelled
   (c) Site installation is complete in accordance with the SQP, and equipment not damaged
   (d) There are no risks to the health and safety of personnel
   (e) All relevant equipment satisfactorily earthed
   (f) Confirm no obvious infringement of safety clearances
   (g) Satisfactory access and egress.
2. **Pre-Stage 2 commissioning inspection schedule**
   This schedule should comprise of a list of items to be inspected to provide assurance that the equipment is in a condition suitable for Stage 2 commissioning. Typical requirements include:
   (a) Confirmation that the Stage 1 commissioning test and inspection schedule is complete and signed off.
   (b) Confirmation that all valves, cubicles, kiosks, etc., are suitably locked.
   (c) Confirmation that all pressure gauges are at the correct pressure.
   (d) Confirmation that all CT links are normal.

   (e) Confirmation that all test equipment has been removed.
   (f) Confirmation that all wiring terminations are made and are tight.
   (g) Confirmation that all relays are set in accordance with relay settings sheets.
   (h) Confirm the status of all alarms.

3. **Final commissioning inspection schedule**

   The purpose of this inspection is to ensure that on completion of the energisation and on-load commissioning tests as required by the CSP, that the equipment has been left in a position suitable for commercial service. It is a last examination to check all is satisfactory. Typical requirements include:

   (a) All temporary commissioning protection systems have been removed/disabled and protection settings are normal
   (b) No item of equipment seems to be in a state of distress
   (c) All in/out switches and tap changer positions are normal
   (d) Confirm status of all auto reclose/switching equipment
   (e) Confirm that all alarms and indications both local and remote are normal
   (f) Confirm that the Stage 2 commissioning test and inspection schedule is complete and signed off.

23.7 Commissioning switching programme

23.7.1 Commissioning switching programme – purpose

As stated in previous sections, a CSP (or other similar title) is a document that defines the operational switching sequence and methodology for Stage 2 commissioning. It is concerned with HV equipment energisation and on-load commissioning tests and comprises a detailed step-by-step procedure.

The CSP will usually be concerned with the Stage 2 commissioning of a circuit or a number of circuits. Therefore, usually more than one substation and commissioning panel would be included in the CSP.

23.7.2 Commissioning switching programme – content

Figure 23.5 summarises the typical content of a CSP.

23.8 Commissioning certificates

23.8.1 Commissioning certificates – types

The commissioning process usually involves a number of certificates to formalise certain key stages and activities. Some of the most commonly used include the following:

- Safety rules clearance certificate
- Commissioning acceptance certificate
- SCADA system commissioning certificate
- Commissioned equipment defect report.

These will be briefly examined below.

Commissioning Switching Programme

1. Project title and number
2. Location(s) of equipment to be commissioned
3. Implementation date
4. List of equipment to be commissioned
5. Names of duty holders and location
6. Communication arrangements
7. Diagrams to be referenced
8. Protection status
   (i) Unproven protection to be commissioned
   (ii) Proven protection to be used for commissioning (location plus temporary settings)
9. Auto reclose/switching equipment – confirm status (usually switched out of service)
10. Documentation
    (i) Confirm that Part 1 of commissioning acceptance certificate is complete
    (ii) Confirm that ratings schedules, protection schedules and operations diagrams are complete and distributed
11. Equipment initial conditions
    (i) Confirm the status of all equipment to be commissioned (e.g. open/closed) NB: Protection systems to be commissioned are usually in service and normal, at the point of circuit energisation
    (ii) Confirm the status of all operational equipment that will be involved with the CSP (e.g. open/closed)
    (iii) Confirm the tap position of all transformers.
    (iv) Confirm the status of all alarms.
12. Equipment energisation and on-load tests. NB: This comprises the actual sequence for both energising the equipment and undertaking the on-load commissioning tests. The following is required:
    (i) A precisely worded step-by-step sequence of activities comprising
        (a) Switching sequence
        (b) Phasing out tests
        (c) Visual inspections
    (ii) On-load commissioning tests typically comprising
        (a) CT ratio and polarity (for polarity sensitive equipment) for each item of protection and control equipment
        (b) Protection polarity (where relevant)
        (c) VT connections and phase rotation
        (d) Metering direction and magnitude
        (e) Auto reclose/switching tests
        (f) AVC tests, etc.
13. Equipment reconfiguration (for commercial service)
14. CSP signatures. Typically: Both the power network company's and contractor's commissioning engineers, and the system operations representative

Figure 23.5 Commissioning switching programme contents – typical

23.8.1.1 Safety rules clearance certificate

The safety rules clearance certificate (SRCC) is the means by which equipment (usually HV equipment) becomes subject to, or is removed from, the power network company's safety rules. From the instant the certificate is issued, no work can be undertaken on, or near to, the equipment without the agreement of the power network company (usually the SAP), see Section 18.7.2 and Figure 18.7. The SRCC may be issued either during the site installation or commissioning stages as determined by the commissioning panel.

23.8.1.2 Commissioning acceptance certificate

The commissioning acceptance certificate (or other similar title) is used to formalise the progress of commissioning to confirm:

- Part 1: Completion of Stage 1 commissioning – and the equipment ready to progress to Stage 2 commissioning
- Part 2: Completion of Stage 2 commissioning – and the equipment available for commercial service

An example of a commissioning acceptance certificate is provided in Figure 23.6.

23.8.1.3 SCADA system commissioning certificate

With reference to Section 20.8.1, it is essential that before a new SCS system, or changes to an existing SCS system, becomes operational, that there is formal confirmation that the databases in both the SCS on-site and the machine in the control centre are identical, and have been subject to commissioning tests to confirm this requirement. Confirmation that this is the case is often declared by a SCADA system commissioning certificate (or other similar title). Typical requirements of such a certificate would include the following:

- Statement of the database reference and revision numbers for the equipment at both site and at the control centre, confirming they are identical.
- Statement that both local and end to end tests have been satisfactorily undertaken, and the system is available for commercial service.
- Formal sign off by both the power network company's and contractor's commissioning engineer and the system operations representative.

23.8.1.4 Commissioned equipment defect report

Following equipment commissioning, any defects with the equipment during the warranty period (as defined in the contract) must be notified to the contractor by the power network company, for purposes of correction. This is usually undertaken via a commissioned equipment defect report, typical requirements of which are as follows:

- Contract reference, equipment location, equipment type, details of defect and date of notification. Signed by both the power network company's project manager, and commercial representative – and forwarded to the contractor.

Commissioning Acceptance Certificate

Project No: **Certificate No:**

Location:

Part 1

I certify that the following equipment, apart from any exceptions noted below, has satisfactorily completed Stage 1 commissioning, and is now available to commence Stage 2 commissioning

Time .... Date .... Signed ....

(Commissioning Engineer, power network company)

Time .... Date .... Signed ....

(Commissioning Engineer, contractor)

Time .... Date .... Signed ....

(System Operations Representative)

Description of Equipment:

Exceptions/Limitations:

Part 2

I certify that the equipment specified in Part 1 above has satisfactorily completed Stage 2 commissioning tests and is now available for commercial service, with the following exceptions/limitations

Exceptions/Limitations

Time .... Date .... Signed ....

(Commissioning Engineer, power network company)

Time .... Date .... Signed ....

(Commissioning Engineer, contractor)

Time .... Date .... Signed ....

(System Operations Representative)

Figure 23.6 Commissioning acceptance certificate – typical

- Contract reference, equipment location, equipment type, details of remedial action and date undertaken. Signed by both the contractor's person responsible for remedial action, and commercial representative – and forwarded to the power network company.

23.9 Commissioning technical documentation

23.9.1 Commissioning technical documentation – types

Commissioning technical documentation may be subdivided into three types, which are as follows:

- Equipment commissioning documentation, i.e. that documentation against which the equipment is commissioned, and provides a record of the commissioning.
- Site-specific commissioning documentation, i.e. commissioning method statement or CSP.
- Technical documentation which needs to be in place before the equipment can enter commercial service (usually prior to Stage 2 commissioning) typically termed as operational and technical data.

The commission panel has responsibility for all of the above. Essential requirements are summarised below.

23.9.2 Equipment commissioning documentation

Equipment commissioning is usually carried out in accordance with two types of documents: Equipment test procedures and drawings. The requirements of each will be briefly examined.

1. **Equipment test procedures**
   Equipment test procedures usually define the step-by-step requirements for inspecting and testing an item of equipment (e.g. circuit breaker, transformer, HV cable, protection relay), see Section 15.2. They usually cover both Stages 1 and 2 commissioning requirements. The test and inspection format is often agreed between a power network company and a contractor/manufacturer as part of the type test package of documentation, see Section 21.3.1.1. They usually form part of a comprehensive suite of power network company (and contractor) commissioning test documentation.
2. **Drawings**
   Drawings generally fall into the following categories:
   (i) Interconnections between equipment, either HV, LV or protection and control. They may be either electrical or mechanical. The greatest volume of such drawings is usually associated with protection and control.
   (ii) Electrical safety clearances.
   (iii) Physical structures such as buildings, towers, support structures, etc.

(iv) Cable and OHL routes.
(v) OHL sag drawings.
(vi) Earthing drawings.

During commissioning, both equipment and site infrastructure must be confirmed as being in accordance with the drawings. In some instances, this will require testing (e.g. proving equipment inputs, outputs and wiring against a protection circuit diagram), and in other instances, inspections and measuring of dimensions. The management of drawings on-site is critically important, particularly with reference to those who are nominated to mark up a drawing, and the requirement for clear identification of those drawings that have been fully proven.

23.9.3 Site-specific commissioning documentation

Site-specific commissioning documentation may be subdivided into commissioning method statements or CSPs, as follows (see Section 15.5.1.3):

1. **Commissioning method statement**
   A commissioning method statement (or other similar title) may be used both for site-specific Stage 1 commissioning and/or Stage 2 commissioning of LV equipment (i.e. equipment not energised from the HV network). It should comprise a step-by-step sequence of activities, and usually involve a risk assessment. It may include a commissioning test procedure within it. Typical examples of use include:
   (i) Primary injections
   (ii) Adding a new circuit of busbar protection to an existing system
   (iii) Replacement of a battery and charger system
2. **Commissioning switching programme**
   This is described in Sections 23.7.1 and 23.7.2 and is essentially concerned with the Stage 2 commissioning of HV equipment. It may contain commissioning test procedures (concerned with on-load commissioning tests) within it (see also Section 15.2.1).

23.9.4 Technical and operational data

The commissioning process usually takes responsibility for confirming completion (and actioning where necessary) of a number of project requirements which fall under the (typical) title of 'technical and operational data'. These are summarised below:

- Protection and control relay settings (see Chapter 20)
- Thermal rating schedules (see Chapter 20)
- Protection and automatic reclose/switching schedules (see Chapter 20)
- Operations diagrams (see Chapter 20)
- SCADA systems (see Chapter 20)
- Asset register technical data (see Chapter 25)
- Equipment technical and operational descriptions.

23.10 OHL and HV cable commissioning

23.10.1 OHL and HV cable commissioning – requirements

The workload concerned with the commissioning of OHL and HV cables is substantially less than that associated with substations. It is therefore usually the case that where commissioning is taking place at a substation connected with an OHL or HV cable, that the OHL or HV cable commissioning is incorporated into the commissioning panel and commissioning programme for the substation. Specific requirements for OHL and HV cables are as follows:

23.10.2 OHL commissioning

Requirements are typically as follows (see also Section 15.11.1.4):

1. **Initial commissioning inspection**
   This is often not undertaken. Unlike substations, there is little to be inspected.
2. **Stage 1 commissioning off-load tests**
   This would usually comprise
   (i) Tower earthing resistance
   (ii) Conductor joint resistance
3. **Pre-Stage 2 commissioning inspection**
   This would usually comprise
   (i) OHL conductors fully de-earthed
   (ii) Correct down-lead phasing colours
   (iii) No apparent infringement of statutory and safety clearances
   (iv) Notices/colour plates/climbing guards in place
   (v) No visible equipment damage (e.g. insulators)
   (vi) All temporary works removed
4. **Stage 2 commissioning switching programme**
   If no related substation work is being undertaken, a CSP will need to be prepared solely for the OHL.
5. **Final commissioning inspection**
   This is frequently dispensed with, on the basis that an inspection will be carried out as part of the maintenance regime, say within 1 year of commissioning.

23.10.3 HV cable commissioning

Requirements for HV cable commissioning are typically as follows (see also Section 15.11.1.3):

1. **Initial commissioning inspection**
   As with OHL, this is often not undertaken, since there is little to inspect
2. **Stage 1 commissioning off-load tests**
   This would typically include:
   (i) High-voltage pressure test

- (ii) Partial discharge test
- (iii) Sheath insulation test
- (iv) DC resistance test
- (v) Cross-bonding cable test
- (vi) Symmetrical components impedance measurement (at higher voltages)
- (vii) SVL tests
- (viii) Calibration of any pressure gauges
- (ix) Confirmation of correct phasing

3. **Pre-Stage 2 commissioning inspection**
The following would typically be included:
   - (i) All labels correct
   - (ii) All locks applied
   - (iii) All pressure gauges within limits
   - (iv) All temporary works removed
4. **Stage 2 commissioning switching programme**
If no related substation work is being undertaken, a CSP will be required solely for the HV cable.
5. **Final commissioning inspection**
This would usually involve a visual inspection of any exposed cable terminations, if not too geographically remote. A check should be made to ensure that all major temporary works have been made good.

23.11 Health and safety

23.11.1 Commissioning health and safety – considerations

Although the commissioning stage of a project does not pose as many health and safety hazards as those of site installation, they must be identified and managed. Commissioning, of course, falls within the ambit of the CDM regulations, see Chapter 18, within the accountability of the PC (site manager). Accordingly, the site health, safety and environment meeting which commenced in the site installation stage, usually extends into the commissioning stage with relevant commissioning activities included in the site work schedule, see Figure 22.13.

As work proceeds through the site installation to the commissioning stage, equipment will be progressively transferred from the PC's (site manager) safety management control to that of the SAP, i.e. the power network company. The transfer of HV equipment is via a SRCC as shown in Figure 18.7. It is worthy of note that commissioning may be undertaken in a CDM zone totally under the contractors health and safety control, or alternatively, under a safety document issued by the power network company.

As with the site installation stage, the identification and control of IV, as outlined in Chapter 11 is critical to the management of health and safety. The SAP and both of the commissioning engineers should have the capability to resolve IV risks and hazards.

23.12 Equipment decommissioning

23.12.1 Decommissioning – considerations

Many projects also require equipment to be decommissioned as well as commissioned. Decommissioning results in equipment being physically disconnected from the power system. It is usually the case that when equipment is decommissioned that it is removed from site, but instances may arise of it being left on-site, at least for a time. In some instances, the decommissioned equipment may be left within safety distance of the energised and operational power network and as such remains subject to the safety rules for HV equipment. More usually, decommissioned equipment is outside safety distance and is removed from the safety rules for HV equipment via use of SRCC, see Section 18.7.2 and Figure 18.7.

Decommissioning is usually under the control of the power network company, and not the contractor. However, once the equipment is decommissioned, the contractor will usually be required to remove the decommissioned equipment from site. A contractor may also undertake the work associated with decommissioning, usually under a safety document issued by the power network company. Control of the decommissioning process is usually vested in a decommissioning engineer, who is often one and the same as the commissioning engineer.

23.12.2 Decommissioning certificate

Notification of equipment being decommissioned is often achieved via a decommissioning certificate (or other similar title). The decommissioning certificate would typically comprise:

- Decommissioned equipment location
- Description and identification of equipment
- Reasons for equipment being decommissioned
- Disconnection point details
- Whether the equipment remains within safety distance and subject to the safety rules for HV equipment.
- Associated SRCC.

Notification of decommissioned equipment may be useful to those responsible for holding equipment spares.

23.13 Commissioning roles and responsibilities

23.13.1 Key roles and responsibilities

With reference to Figures 19.3 and 19.6, the roles and responsibilities of the following key commissioning duty holders will be examined:

- Commissioning panel chairman
- Commissioning engineer – contractor
- Commissioning engineer – power network company

- SAP
- Site Manager (PC)

Typical responsibilities are summarised below.

23.13.1.1 Commissioning panel chairman

The commissioning panel chairman's role is usually undertaken by one of the following: the power network company's project manager/engineer, or the site engineer, or the commissioning engineer or SAP. Alternatively, it could be undertaken by the contractor's project manager/project engineer or commissioning engineer or site manager (PC). However, in the author's experience, the role is most beneficially vested in the power network company's project manager/engineer for the following reasons:

- Good communications back into the power network company for purposes of Stage 2 commissioning arrangements.
- Has lived and breathed the project from the outset, so best informed about the project nuances.
- Good project planning and organising skills – an essential requirement.

Specific duties of the chairman are as follows:

- Accountable, through the commissioning panel for ensuring that the equipment is satisfactorily commissioned, in accordance with both the conditions of contract and power network company and contractor QMS arrangements.
- Calling and arranging meetings, preparing agendas and minutes/action schedules.
- Ensuring the preparation and completion of a commissioning programme.
- Allocating tasks and responsibilities to members of the commissioning panel.
- Formally agreeing with the site manager (PC) when equipment can be transferred from the site installation to the commissioning stage.
- Ensuring that all duty holders are competent for the work to be undertaken.
- Ensuring satisfactory coordination of activities with other commissioning panels and third parties.

23.13.1.2 Commissioning engineer – contractor

The contractor's commissioning engineer will form part of the site manager's (PC) team. This is a major commissioning role in terms of the volume of work and tasks to be delivered. It is essential that this commissioning engineer is not just proficient at equipment performance testing – but can also manage the commissioning process, such that there is a clear and concise audit trail to confirm the satisfactory status of each item of equipment. Specific responsibilities include:

- Tabling the commissioning programme [prepared by the contractor as part of the SQP] and obtaining agreement from the commissioning panel
- Undertaking and managing all Stage 1 commissioning
- Undertaking Stage 2 commissioning under the direction of the power network company commissioning engineer

- Managing the commissioning team, should the scale of the work require additional commissioning engineers
- Managing health and safety during the commissioning stage
- Agreeing the timing and issue of the SRCC and the Commissioning Acceptance Certificate
- Ensuring that IV hazards and risks are identified and managed
- Ensuring coordination with other commissioning panel's personnel
- Providing a complete and presentable set of commissioning documentation and drawings in accordance with the commissioning tests and inspection schedule (see Figure 23.4), on finalisation of commissioning to the commissioning panel.

23.13.1.3 Commissioning engineer – power network company

The power network company's commissioning engineer is primarily responsible for accepting the contractor's commissioning programme and the subsequent commissioning results (i.e. in effect, an engineering assurance role) – in addition to directing Stage 2 commissioning, i.e. equipment energisation from the power system. Specific responsibilities include:

- Formal inspections, see Section 23.6.1
- Witnessing and accepting key Stage 1 commissioning activities
  - CT tests (flick test, magnetisation curves, etc.)
  - Primary injections
  - Secondary injections
  - Confirming correct relay settings applied
  - Transformer vector group tests
  - Circuit tripping tests, etc.
- Circuit loadability test
- Agreeing the timing and issue of the EAC
- Accepting the completed commissioning test and inspection schedule
- Coordination with other commission panels
- Directing the CSP
- Ensuring that IV hazard and risks are identified and managed.

23.13.1.4 Senior authorised person

As stated previously, the SAP may frequently be one and the same as the power network company commissioning engineer, and there is great advantage with one person discharging both roles. The SAP is responsible for all activities that are encompassed by the power network company safety rules. Specific responsibilities include:

- Define and implement arrangements for achieving safety from the system (e.g. identifying when a safety document is required).
- Undertaking the role of site occupier (i.e. controller of the premises), see Sections 18.4.1–18.4.3.

- To provide the required safe access and egress for commissioning.
- To agree the timing and issue of the SRCC.
- To ensure that IV hazards and risks are managed and identified.
- Switching responsibilities associated with the CSP.

23.13.1.5 Site manager (PC)

At first sight, it may be thought that the site manager should also be the chair of the commissioning panel. However, it is usually the case that the site installation skill set is not one and the same as that of commissioning – so the two roles are usually left separate. However, with reference to accountability for health and safety aspects of the commissioning process (and in effect the commissioning process itself), these remain vested via the CDM regulations in the PC (i.e. site manager). Therefore, all activities undertaken by the commissioning panel must be to the satisfaction of the PC, and for this reason the site manager is an important member of the commissioning panel. In addition, the site manager is required to formally declare to the commissioning panel when equipment may satisfactorily be transferred from the site installation to commissioning stages.

23.14 Commissioning panel meeting agenda – and filing

23.14.1 Commissioning panel meeting agenda

A typical commissioning panel agenda is provided in Figure 23.7.

23.14.2 Commissioning documentation filing

As with other stages of a scheme/project, the filing arrangements for commissioning documentation is an essential requirement and requires methodical consideration. Typical filing arrangements are suggested below. These will be under the management of the contractor's commissioning engineer and integral to the site installation files outlined in Section 22.8.1, and under the management umbrella of the site manager:

1. Commissioning management
   (i) Commissioning panel meeting's minutes and actions
   (ii) Other commissioning panel/third-party's minutes/actions
2. Scheme/project specifications
   (i) SDS
   (ii) Detail design specification
3. Commissioning certificates
   (i) SRCC
   (ii) EAC
   (iii) SCADA commissioning certificates
   (iv) Defect report
4. Commissioning programme
   (i) Commissioning timing and staging programme
   (ii) Commissioning test and inspection schedule

Commissioning Panel Meeting Agenda

1. Project no. and title
2. Location
3. Circuits to be commissioned
4. Date/location of meetings
5. List of attendees
6. Equipment to be commissioned (confirm against SDS/DDS)
7. Equipment transferred from site installation to commissioning (confirmed by site manager)
8. Commissioning programme – status
9. Decommissioning programme – status
10. Commissioning/decommissioning responsibilities
11. Commissioning certificates – status
12. Civil, structural and building engineering commissioning – status
13. Heath, safety and environmental
    (i) Site work schedule – status
    (ii) CDM zones and interfaces – status
    (iii) CDM construction phase plan – status
    (iv) Safety rule arrangements and interfaces – status
    (v) CDM file – status
    (vi) Pressure system regulations WSEs – status
    (vii) IV hazards and risks
14. Project stage-by-stage document – status
15. Drawings management and control – status
16. Emergency return to service arrangements
17. Equipment non-interference arrangements
    (i) Safeguards for operational equipment (barriers, screens, etc.)
    (ii) Safeguards for Stage 1 commissioned equipment
18. Technical and operational data – status
19. Design change control – status
20. Commissioning resourcing, and training requirements
21. Equipment spares arrangements
22. Defects status
23. Action log – status
24. Filing arrangements – status
25. Date of next meeting

Figure 23.7 Commissioning panel meeting agenda – typical

    (iii) Commissioning method statements
    (iv) CSPs

5. Technical and operational data
    (i) P&C settings sheets
    (ii) Thermal rating schedules
    (iii) Protection and auto reclose/switching schedules
    (iv) Operations diagrams
    (v) Equipment technical and operational descriptions

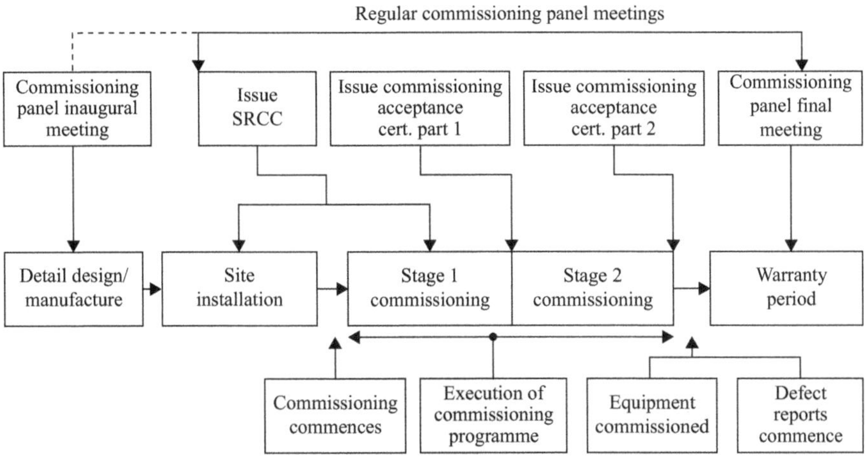

Figure 23.8 Commissioning process – overview

6. Health, safety and environmental
   (i) CDM construction phase plan
   (ii) Environmental plan
   (iii) CDM H&S file
   (iv) RAMS
   (v) IV management arrangements
7. Drawings (categorised into technologies)
   (i) Drawings yet to be commissioned
   (ii) Drawings commissioned (with site markup)
   (iii) Drawings returned to drawing office
8. Procedural documents
   - Project stage-by-stage document
   - Project programme
   - Outage programme
   - Risk register, etc.
9. Suite of generic commissioning test procedures
10. OHL commissioning
11. HV cable commissioning

23.14.3 Commissioning process – overview

Figure 23.8 provides an overview of key aspects of the commissioning process.

Chapter 24
Project stage-by-stage procedure

24.1 Introduction

The term 'project stage by stage' refers to the stage-by-stage high-level sequence of work activities relevant to the site installation and commissioning of HV equipment, specifically including the status of the equipment with reference to the power network company safety rules. Figure 24.1 illustrates the interaction between a project stage-by-stage document and the key scheme/project stages as shown in Figure 1.4. During the scheme development stage, an outline stage-by-stage document would usually be prepared to determine the optimum sequence of work stages by which the equipment is to be installed and commissioned. Following contract release, the project stage-by-stage document would be prepared in detail and subsequently implemented during site installation and commissioning.

24.2 Project stage by stage – procedural requirements

24.2.1 Project stage-by-stage document – scope

The scope of a project stage-by-stage document covers all project high-level site installation and commissioning activities which interact. It therefore usually covers all sites which are encompassed by the same project, since the timing of site installation is usually coordinated across all sites (relevant to the project) and commissioning invariably involves all ends of a circuit (i.e. all sites). The document may also include adjacent projects if project site installation and commissioning activities interact. Alternatively, with large infrastructure projects which may cover multiple sites and circuits over a long period of time, a number of project stage-by-stage documents may be required to match the key stages of the project. Within this context, careful consideration must be given by the project managers concerned to the appropriate number of sites to be included in a project stage-by-stage document, and the number of project stage-by-stage documents to be prepared. In general, a project stage-by-stage document would encompass the following requirements:

- High level, key stages of site installation and commissioning.
- Maintenance work which interacts with site installation and commissioning.
- Points of circuit isolation and position of circuit earths, during site installation and commissioning (i.e. safety from the system arrangements).

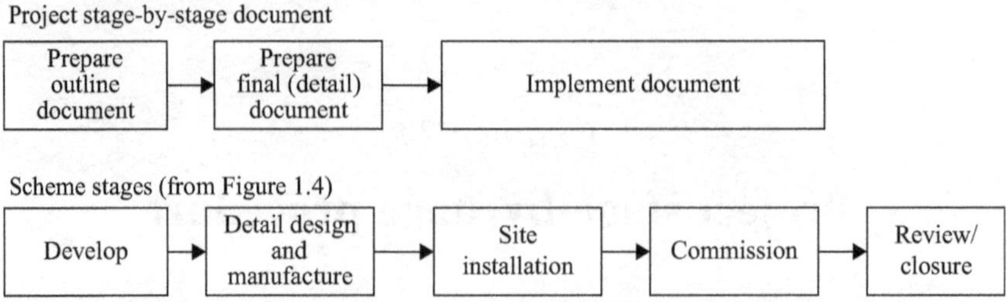

Figure 24.1 Project stage by stage – linkage with scheme stages

- The issue of SRCCs and PTWs (see Sections 18.6.2 and 18.7.1) for adding/removing equipment to/from the system and as such to/from the power network company safety rules.
- Circuit nomenclature change and OHL route colour changes.
- New SCADA systems becoming operational and/or changes to existing SCADA systems.

24.2.2 Project stage-by-stage document – management arrangements

During the development stage of a project, an outline project stage-by-stage document would usually be prepared by the scheme team, see Figure 17.6. After contract release, the development and update of the project stage-by-stage document is usually via a team comprising all relevant parties. The following provides a list of typical team members:

- Project stage-by-stage chairman
- Commissioning panel chairman (all sites)
- Site manager (all sites)
- Commissioning engineer (power network company – all sites)
- Commissioning engineer (contractor – all sites)
- SAP (all sites)
- Site engineer (all sites)
- Project manager (power network company – all sites)
- Project manager (contractor – all sites)
- System operations department representative
- Operational safety department representative
- Third-party representatives (e.g. power station, other power network company, major customer, etc.).

The project stage-by-stage chairman would normally be associated with the major site and would usually be the commissioning panel chairman, or project manager, for that site. From the above large list of team members, the project stage-by-stage chairman would need to ensure that at any team meetings that are convened, only those essential to progressing the meeting are invited, otherwise the numbers would make the meeting too unwieldy. However, all those not attending would need to be

Project Stage-by-Stage Meeting

Agenda

1. Project title(s) and no(s)
2. Sites and circuits involved
3. Date/location of meeting
4. List of attendees
5. Project stage-by-stage interface boundaries – status
6. SRCC strategy – status
7. Safety rules interface strategy – status
8. Circuit nomenclature change strategy – status
9. SCADA system operational date strategy (including changes) – status
10. Equipment maintenance strategy and interfaces – status
11. Project stage-by-stage document – status
12. Outage programme – status
13. Emergency return to service plan – status
14. Third-party/others communications arrangements – status
15. Action log – status
16. Date of next meeting

Figure 24.2 Project stage-by-stage meeting agenda – typical

kept apprised by the chairman. Figure 24.2 summarises a typical meeting agenda. The major deliverable in a project stage-by-stage meeting is the project stage-by-stage document itself. Although the content would be specified through the consensus of the meeting, the document would usually be prepared by the contractor's drawing office.

24.2.3 Project stage-by-stage document preparation and implementation

Project stage-by-stage meetings should be convened as required to produce the proposed project stage-by-stage document as soon as practicable, during the detail design and manufacture stage of the project. Successful implementation of the document, or otherwise, during site installation and commissioning, should be fed back to the project stage-by-stage chairman and meeting attendees, to enable satisfactory progress to be confirmed or otherwise modify and reissue the document if changes are required.

24.3 Project stage-by-stage document

24.3.1 Project stage-by-stage document – content

As described in Figure 24.1, the project stage-by-stage document is initially prepared as an outline and later developed in detail (i.e. the final). The final document

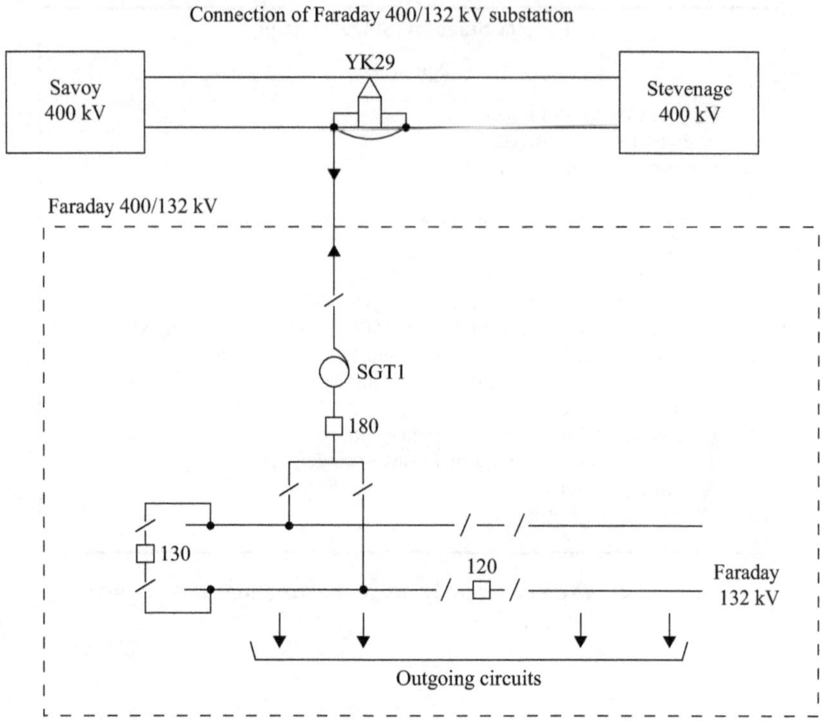

Figure 24.3 Overview of work to be undertaken

would usually incorporate the outline since it provides a clear summary of the high-level stages of the project. The final project stage-by-stage document would typically be structured as follows:

- Summary of the work to be undertaken
- High-level outline stage by stage
- Detailed stage by stage.

An example of a project stage-by-stage document will be provided by examining a typical project as shown in Figure 24.3, requiring the addition of a 'T' into an existing OHL (between Savoy and Stevenage substations) to feed a new substation (i.e. Faraday 400/132 kV substation).

24.3.2 Summary of work to be undertaken

This would comprise a narrative summary of the work to be undertaken, referencing the completed network, as shown in Figure 24.3, and linking/referencing any associated project stage-by-stage documents.

24.3.3 High-level outline stage by stage

The high-level outline stage by stage is given in Figure 24.4 with supporting commentary, which are as follows:

1. **Existing: 1 January 20XY**
   Commence construction work on site on the new Faraday 400/132 kV substation and 1.5 km of 400 kV connecting cable [time duration = 15 months]. NB: Savoy and Stevenage substations are both double busbar
2. **Phase 1: 1 April 20X(Y+1)**
   Commence work on diverting the Savoy-Stevenage No. 1 circuit onto a temporary OHL, so disconnecting from tower YK29 [time duration = 3 weeks]
3. **Phase 2: 22 April 20X(Y+1)**
   Remove Savoy-Stevenage No. 2 circuit from service, and commence work to replace existing tower YK29 with 'T' off tower. On completion, connect to OHL and cable [time duration = 6 weeks]
4. **Phase 3: 3 June 20X(Y+1)**
   Commission new Savoy-Faraday-Stevenage circuit, and Faraday 400/132 kV substation [time duration = 3 weeks]
5. **Phase 4: 24 June 20X(Y+1)**
   Commence work to remove temporary OHL on Savoy-Stevenage circuit and reinstate permanent OHL [time duration = 2 weeks]
6. **8 July 20X(Y+1)**
   All work complete

24.3.4 Detailed stage by stage

The following will be examined:

1. Detailed stage by stage – generic equipment status colour coding – typical
2. Detailed stage-by-stage generic format – typical
3. Detailed stage by stage – summary of typical stages relating to Figure 24.4
4. Detailed stage by stage – example of detailed format for one typical stage

1. **Detailed stage by stage – generic equipment colour coding – typical**
   To ensure maximum clarity of, and removal of ambiguity from, detailed stage-by-stage diagrams, it is preferable to denote the status of relevant items of equipment with an individual colour. Typical colour coding is as follows:
   (i) **Solid black:**
   Equipment energised
   (ii) **Dotted red:**
   Equipment under construction – but not yet under the power network company HV safety rules
   (iii) **Solid red:**
   Equipment constructed and added to the power network company HV safety rules via issue of an SRCC

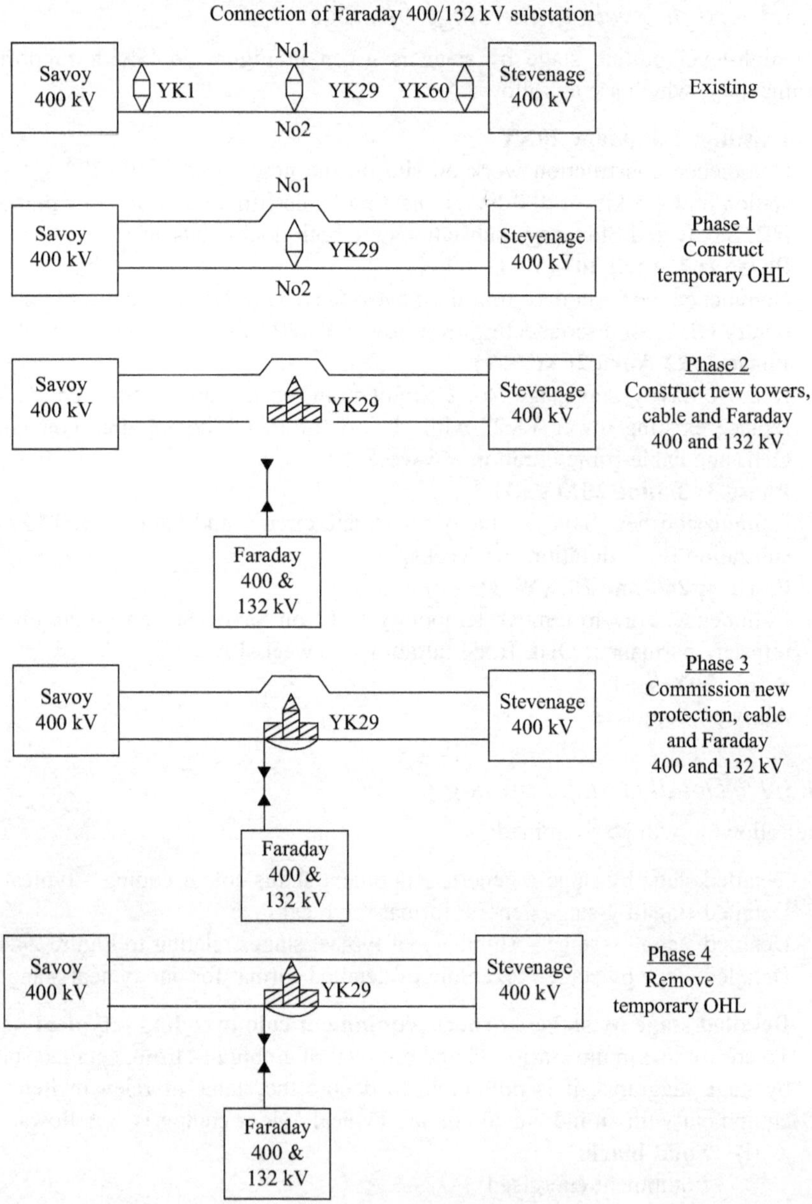

Figure 24.4 High-level outline stage by stage – typical

(iv) **Solid green**
Equipment de-energised and planned to be removed from the power network company HV safety rules

(v) **Dotted green**
Equipment de-energised and removed from the power network company HV safety rules via issue of an HVSCC

(vi) **Solid light blue**
Equipment de-energised and under the power network company HV safety rules

(vii) **Solid dark blue**
Equipment being commissioned

(viii) **Dotted black**
Equipment de-energised and subject to asset replacement.

2. **Detailed stage-by-stage generic format – typical**
Figure 24.5 provides a typical generic format for providing the detailed part of a project stage-by-stage document.

Title (e.g. connection of Faraday 400/132 kV substation)

Stage no. ______ Rev. no. _______

1. **Summary of key activities during this stage**
2. **Key dates**

Activity no.	Date	Activity description	Responsibility

3. **Maintenance to be undertaken**
4. **Points of isolation and circuit earths locations**

Location (site)	Point of isolation designation	Circuit earth designation

5. **Circuit nomenclature changes**
6. **SCADA operational dates**
7. **Associated diagrams (e.g. operations diagrams)**
8. **Corresponding stage-by-stage diagram (to be attached)**

Figure 24.5 Detailed stage-by-stage generic format – typical

3. **Detail stage by stage – summary of typical stages**
   The following provides a summary of one possible detailed stage-by-stage solution, relating to Figures 24.3, 24.4 and 24.7. A corresponding stage-by-stage diagram is required for each individual stage.
   (i) **Stage 0: Existing arrangements**
   This corresponds to the first diagram (i.e. existing) in Figure 24.4
   (ii) **Stage 1**
   Commence construction of new Faraday 400/132 kV substation
   (iii) **Stage 2**
   (a) Construct temporary towers YK29T1 and YK29T2 on Savoy–Stevenage No. 1 circuit
   (b) Ongoing construction of Faraday 400/132 kV substation and 400 kV cable
   (iv) **Stage 3**
   (a) Switch out of service Savoy–Stevenage No. 1 circuit
   (b) Issue PTW to remove conductors between YK28 and YK30 on the No. 1 circuit
   (c) Ongoing construction of Faraday 132 kV substation and 400 kV cable
   (v) **Stage 4**
   (a) Issue SRCC to remove conductors between YK28 and YK30 on the No. 1 circuit from the HV system and safety rules
   (b) Issue SRCC to add towers YK29T1, YK29T2 and conductors from YK28–YK29T1–YK29T2–YK30 to the HV system and safety rules
   (c) Issue PTW to connect conductors from YK28–YK29T1–YK29T2–YK30
   (d) Ongoing construction of Faraday 400/132 kV substation and 400 kV cable
   (vi) **Stage 5**
   (a) Switch in to service and Stage 2 commission Savoy–Stevenage No. 1 circuit – now partly on temporary towers
   (b) Ongoing construction of Faraday 400/132 kV substation and 400 kV cable
   (vii) **Stage 6**
   (a) Switch out of service the Savoy–Stevenage No. 2 circuit
   (b) Ongoing construction of Faraday 400/132 kV substation and 400 kV cable
   (viii) **Stage 7**
   (a) Issue PTW to remove conductors between YK28–YK29–YK30 on No. 2 circuit, to asset replace tower YK29 (and enable T off), and refit conductors between YK28–YK29–YK30 to form a continuous Savoy–Stevenage No. 2 circuit

(b) Ongoing construction of Faraday 400/132 kV substation and 400 kV cable

(ix) **Stage 8**

(a) Issue SRCC to add the following to the HV system and bring under safety rules

(i) Newly constructed Faraday 400/132 kV substation

(ii) Newly constructed 400 kV cable

(iii) Connections from 400 kV cable to tower YK29

(iv) Connections from 400 kV cable to SGT1 at Faraday 400 kV substation

(b) Issue PTW to affect steps (iii) and (iv)

(x) **Stage 9**

(a) Switch into service and Stage 2 commission the following

(i) New Savoy–Faraday–Stevenage circuit

(ii) New Faraday 400/132 kV substation. NB: the new Faraday 132 kV substation will need to provide enough load current to prove the protection systems

(xi) **Stage 10**

(a) Switch out of service the Savoy–Stevenage No. 1 circuit

(b) Issue PTW to remove conductors from YK28–YK29T1–YK29T2–YK30 and temporary towers YK29T1 and YK29T2

(xii) **Stage 11**

(a) Issue SRCC to remove temporary towers YK29T1, YK29T2 and conductors from YK28–YK29T1–YK29T2–YK30 from the HV system and the safety rules

(b) Issue SRCC to reinstate conductors between YK28–YK29–YK30, to add to HV power system and bring under safety rules

(c) Issue PTW to connect conductors from YK28–YK29–YK30 to make the Savoy–Stevenage circuit continuous

(xiii) **Stage 12**

Stage 2 commission the Savoy–Stevenage circuit (no longer termed no. 1) following reinstatement of circuit on tower YK29.

NB: It is assumed (for simplicity) that towers YK28 and YK30 remain as suspension towers when the OHL connects to the temporary towers. For reasons of brevity, the above is a summary and omits some key activities such as nomenclature changes and SCADA requirements.

4. **Detailed stage by stage – example**

To provide an example of the use of the project stage-by-stage generic format as given in Figure 24.5, Stage 4 of the above 'Detail stage-by-stage summary' will be illustrated, see Figure 24.6 and refer to Figure 24.7. NB: Figure 24.7 illustrates the situation on completion of stage 4.

Connection of Faraday 400/132 kV Substation

Stage no.: 4 Rev. no.: 0

1. **Summary of key activities during this stage**
   On Savoy–Stevenage No. 1 circuit erect conductors on temporary towers
2. **Key dates**

Activity no.	Date	Activity description	Responsibility
4.1	1 Apr Year	Issue SRCC to remove conductors between YK28 and YK30 on the no. 1 circuit	Power network company
4.2	1 Apr Year	Issue SRCC to add towers YK29T1 and YK29T2 and conductors from YK28–YK29T1–YK29T2–YK30	Power network company
4.3	1 Apr Year	Issue PTW to connect conductors from YK28–YK29T1–YK29T2–YK30	1 Power network company (issue PTW) 2 Contractor (undertake work)

3. **Maintenance to be undertaken**
   None
4. **Points of isolation and circuit earths locations**

Location (site)	Point of isolation designation	Circuit earth designation
Savoy 400 kV S/S	1. X203 open 2. Stevenage No. 1 400 kV circuit VT fuses removed	OHL E/S X201A closed
Stevenage 400 kV S/S	1. X403 open 2. Savoy No. 1 400 kV circuit VT fuses removed	OHL E/S X401A closed

5. Circuit nomenclature changes
   None
6. SCADA operational dates
   None
7. **Associated diagrams (e.g. operations diagrams)**
   Operations diagram Ref: XYZ
8. **Corresponding stage-by-stage diagram (to be attached)**
   Connection of Faraday 400/132 kV substation: stage no. 4, rev no.: 0

Figure 24.6 Example of detailed stage by stage – format and content

Connection of Faraday 400/132 kV substation

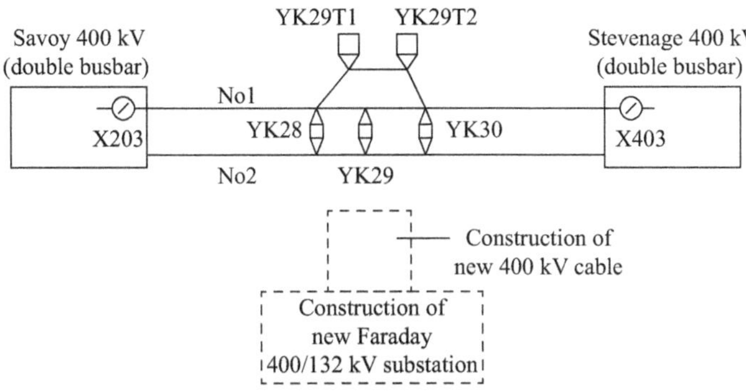

<u>Colour coding of above drawing</u>

- OHL between X203 and YK28, and, YK30 and X403 (No.1 circuit) <u>colour</u> = <u>solid light blue</u> (Equipment de-energised but still under safety rules)

- OHL between YK28 and YK30, <u>colour</u> = <u>dotted green</u> (Equipment removed from safety rules)

- OHL between YK28 and YK30, including YK29T1 and YK29T2 <u>colour</u> = <u>solid red</u> (Equipment under construction not under safety rules)

- Savoy - Stevenage No2 OHL <u>colour</u> = Solid <u>black</u> (Equipment energised)

<u>Stage 4 : Rev 0</u>

Figure 24.7 Example of detailed stage-by-stage diagram

24.3.5 Summary

As illustrated in the preceding sections, the preparation of a project stage-by-stage document is a not insignificant task, requiring careful consideration and wide consultation and agreement. Its preparation and update is however critical to the successful outcome of a project.

Chapter 25
Data management

25.1 Data management – introduction

The term 'data' has a variety of meanings but in the context of this publication, it is simply information that an organisation, either has to, or wants to retain. Power network companies in particular maintain extensive databases of information, which are essential for the running of the company. Much of this information is originated during the construction process, and much is additionally held by contractors and manufacturers, for their own purposes. This chapter will review salient items of data which arise during power network construction, with a view to examining:

- The purpose of data management
- The range of data retained
- Reasons for retaining the data
- Data management procedural requirements

25.2 Data management – purpose

Depending upon the literature consulted, there are numerous definitions of data management, and the following is representative of the many:

> Data management is the development and execution of architecture, policies, practices and procedures in order to manage the information lifecycle needs of an enterprise in an effective manner

Data management is a significant subject in its own right and critical to the management of a company. The above definition identifies an information lifecycle in which power network construction is mostly involved at the outset of the lifecycle, i.e. a source of data input. Within this context, an essential feature of data management is the procedural means by which the data is not only collected, but also assured as being correct.

25.3 Data – range

With reference to power network companies, the range of data to be collected and processed may be generally classified as follows:

- Financial asset register
- Technical asset register

- Technical and operational data
- Drawings and records
- Construction (scheme/project) records
- Equipment subject to statutory inspections

The above will be briefly reviewed.

25.3.1 Financial asset register

A requirement on project completion is the entry into the power network company's financial register of the value of the equipment constructed, i.e. following commissioning. The financial data will have been gathered throughout the lifetime of the project in a format similar to that shown in Figure 19.14. This task will usually be undertaken by the project manager, or project engineer, in conjunction with the project accountant. It is usual to break down the costs on a bay or circuit basis, for example:

- A substation bay overall cost – with the costs further broken down into:
  - HV equipment
  - Protection and control equipment
  - Civil costs (if not absorbed into HV equipment)
- OHL feeder circuit
- HV cable feeder circuit
- Substation infrastructure (i.e. civil) with the cost of large identifiable items (e.g. oil containment, buildings, roadways) separately identified.

Post-construction cost analysis is also frequently undertaken to provide a generic handbook of cost estimating data, to assist in the estimating of future construction work.

25.3.2 Technical asset register

The power network company's technical asset register is a schedule of all installed equipment covering all substation equipment, OHL and HV cables, etc. usually identified in terms of

- Location
- Equipment type specification (including OHL tower and pole types)
- Voltage and current ratings
- Serial number
- Manufacturer
- Date of commissioning (or contractual taking over)
- Equipment registration/certification details (see Chapter 21)
- Equipment nomenclature or title, etc.

This data is required for the following purposes:

- Preparation of maintenance schedules and subsequent maintenance planning (to which maintenance history will be added)

- Equipment data intelligence for:
  - Type defects tracking
  - Condition monitoring
  - Planned future replacement (i.e. location of a particular type of equipment, together with its age and performance/condition history)
  - Power system studies

It is of critical importance that the technical asset register aligns with the content of the financial asset register. Much of the data will be supplied by the contractors and manufacturers.

25.3.3 Technical and operational data

Technical and operational data is essentially required to facilitate operational and future design decisions and generally includes:

1. **Impedance database**
   An impedance database is one that lists the impedance values of all equipment (that contain impedance), that form part of the power network i.e.:
   (i) Transformers, reactors, quadrature boosters
   (ii) OHL
   (iii) HV cables, GIL
   (iv) MSCs, SVCs (usually specified in terms of reactive power parameters)
   (v) Generators (data obtained from generation companies)
   This data is required for both power system modelling and in the determination of relay settings. It is critically important that this data is accurate and kept up to date – to avoid incorrect design, incorrect operational decisions and potential loss of electricity supply.
2. **Protection and control relay settings database**
   This database contains all of the protection and control relay settings for the network as described in Chapters 10 and 20. The number of settings in a typical database may range from 50,000 to in excess of one million. Again, for network reliability reasons, it is critically important that this data is accurate.
3. **Thermal rating schedules**
   As described in Chapters 10 and 20, thermal rating schedules are prepared to specify the thermal (load current) rating of each circuit in the network. They are essential to the operation and control of the power network.
4. **Fault rating schedules**
   Fault rating schedules define the limiting fault current (short-circuit) capability of each circuit in the network, and the limiting item of equipment. They are used by control engineers to ensure that short-circuit ratings are not exceeded.
5. **Protection and auto-reclose switching schedules**
   As described in Chapters 10 and 20, these schedules provide operational data on protection and auto-reclose equipment on a circuit basis.

6. **Site boundary schedules**
   These comprise schedules of responsibility and operational control for a specific substation when two or more power network companies are present. Such schedules identify boundaries of responsibility for busbars, relay panels, busbar protection, AC and DC supplies, etc.
7. **Operations diagrams**
   As described in Chapter 20, these are single line diagrams of substations, OHL, HV cables, etc. with relevant nomenclature which is used as a reference source for both operational, and safety management purposes.

25.3.4 Drawings and records

Drawings and records comprise the suite of drawings lodged in the company drawings' management system, against which equipment was installed and commissioned, see Chapter 20. The drawings act as a general reference source, to aid maintenance, fault investigations and define the starting point for future construction work.

25.3.5 Construction scheme/project records

On completion of a scheme/project, all associated files and records must be retained and stored for a period of time. This will normally be defined in the power network company's record retention policies. Financial records are usually required to be stored for 6 years as required by 'Her Majesties Revenue and Customs' (HMRC) criterion, with general project files typically for up to 10 years (National Archives criterion). With paper files, the task is often one of allocating space for the files, and then filing in logical order, such that they can easily be retrieved, if required. At the outset of a scheme/project, careful consideration needs to be given to the file titles, content, structure and reference number – to facilitate future access. Today, most files are held in computer-based system, with some commercially available systems designed to accommodate all scheme/project records, see Section 19.11.1.

25.3.6 Equipment subject to statutory inspections

Numerous items of equipment are subject to periodic statutory inspection. If this equipment is acquired through the construction process, relevant information pertaining to the equipment needs to be recorded, usually on a register, such that it is routinely called up for inspection. Typical equipment in this category includes:

- Pressure vessels requiring written schemes of examination under the PSSR regulations, see Section 18.3.4.
- Lifting equipment (e.g. cranes, hoists, ropes, winches, etc.) under the LOLER regulations, see Section 18.3.
- Certain portable electrical equipment as required by the Electricity at Work Regulations, see Section 18.5.
- Equipment covered by the PUWER regulations, such as machinery, hammers, drilling machines, see Section 18.3.10.

25.4 Data retention – purpose

Data is usually retained for the following reasons:

1. **Legal**
   Certain data is required to be retained for specified periods of time for statutory reasons. This particularly applies to:
   (i) Financial records
   (ii) Project records
   (iii) Health, safety and environmental records
2. **Corporate management**
   The corporate management of any business is self-evidently dependent upon an understanding of the data that defines the status of the business, as encapsulated by the numerous databases previously described.
3. **Capital investment planning**
   The proposed future capital programme is partly dependent upon an understanding of the condition of the existing assets. This in turn requires data systems that can identify the type of assets on the network and their location and status. Similar data is required for both maintenance planning, and operational planning.
4. **Technical and operational**
   Much of the data is required to enable technical and operational decisions to be made for both the long and short-term operational management of the network. This data also defines the design parameters of the network.

25.5 Data management – procedural requirements

Specialist texts on data management usually point to four distinct requirements in the data management processes as follows:

- A decision on which data to collect and the information details to be recorded
- The process by which the data is collected
- Verification of the data, i.e.
  - That the correct data has been collected
  - That the data is accurate
- A decision on the type of storage medium and confirmation that the data has been correctly entered.

In addition, all of the above needs to be subject to periodic audit.

1. **Which data to collect**
   Consideration needs to be given to each category of data to be collected, and within each category the detailed data requirements to be recorded. A procedure needs to be prepared which fully defines the complete data lifecycle for each category.

2. **Data collection process**
   The procedure alluded to above will need to specify the data collection process defining:
   (i) When the data will be collected
   (ii) The format by which it is recorded (i.e. form, drawing, etc.)
   (iii) Who is responsible for the collection
   (iv) Where it is sent to following collection completion
3. **Data verification**
   It is essential that the required data, once collected, is verified as being the correct data, and free from error. In many instances, because of the technical nature of the data, the verification must be undertaken by a person of appropriate technical competency. However, because of the importance of the data process, some organisations additionally employ a 'data unit' typically charged with ensuring the following:
   (i) That all required data has been collected within specified timescales – as procedural requirements dictate
   (ii) That all collected data has been verified by a person of appropriate competency
   (iii) That the data has been correctly entered into the appropriate database
   (iv) Undertaking a periodic audit of data quality
   (v) Data management process review and performance improvements
4. **Data storage medium**
   Almost all data is now stored in a computer-based storage medium. As with many computer-based systems, these risk becoming very large with reference to the volume of data held, relatively complex in structure and very specialist to use. When designing such systems, a number of considerations arise:
   (i) Ease of data entry
   (ii) Ease of data retrieval
   (iii) User friendliness of the system
   (iv) System life span (and the frequency of updating/replacing)
   (v) Track record of the system used (i.e. is it bespoke or has it been used elsewhere)
   (vi) Requirement for specialist operators
   (vii) Impact on the business of system failure – and recovery arrangements
   (viii) Value for money

25.6 Data management – importance

The collection and processing of a wide raft of data during power network construction is of critical importance. The importance is not always visible to the success of the construction work itself – but may have significant impact on the well-being of the business, perhaps years after the construction work is completed. As such, both an absence of data or incorrect data can be a significant risk both to the well-being of the company, and the security of electricity supply.

Part 3

Engineering competency

Chapter 26
Engineering competency

26.1 Competency requirements

There would appear to be no hard and fast definition of competency but most literature on the subject would broadly agree that it comprises:

> The ability of an individual (duty holder) to independently perform a task, or a role, to a specified standard.

Within this context, successful power network construction projects are highly dependent on engineers with the appropriate competencies, an absence of which creates the following significant risks:

- Health and safety incidents – which in the extreme could endanger life. Or environmental incidents which could cause significant environmental damage.
- Unintentional tripping of circuits either during construction or following completion (i.e. latent deficiencies) – leading to network unreliability and possible loss of electricity supply.
- Inadequate standard of work resulting in eventual re-work and negatively impacting upon cost, time, resources, network availability and reputation.

The standard of engineering competency is therefore of critical importance and leads to five key questions, which are as follows:

- What are the dimensions and components of competency for a specific task or role?
- How is competency acquired?
- How is competency assessed?
- How is competency recorded?
- How is competency maintained?

In addressing the above, the following will be examined:

- Competency levels and components
- Professional competency
- Competency – range of roles
- Competency definition, provision and award organisations
- Engineering competency model
- Competency assessment

- Technician and site workforce competency
- Site operational safety competency
- Management aspects of competency
- Competency and society

26.1.1 Competency considerations

During many years of service with the electrical power industry, the author was involved in numerous incident investigations in which competency was a salient factor. As such the following observations may be worthy of note:

- Well-qualified, knowledgeable and highly experienced engineers may be very competent operating on specific power networks (in the world) but may not be familiar with the nuances relevant to (for example) the UK power networks (and vice versa). Therefore, competence needs to be aligned to the network in question – and must not be assumed.
- An element of competence is being aware of when the boundary of competence has been reached, and more expert advice needs to be sought. This applies particularly to an understanding of impressed voltages (IV) and their danger. Custom and practice (i.e. we have always done it this way) are no substitute for knowledge and understanding.
- Even the most experienced and competent engineers do make mistakes – for a wide range of reasons (mostly concerned with human factors). Therefore, the attributes of competency are not a panacea – but they do very significantly minimise the risk of both unwanted performance and incidents.
- The amount of effort (cost, time, resource) required by an organisation to establish and maintain appropriate levels of competency is usually significant and must not be underestimated.
- Competent performance depends upon a combination of the competency of the individual and the quality of the QMS procedure to be followed. There is an optimum balance (expertise vs bureaucracy) between the two.

26.1.2 Competency – levels and components

Contemporary literature often evaluates competency in two dimensions, namely the level of expertise which defines competency, and the components of competency. A recent Construction Industry Training Board (CITB)/health and safety executive (HSE) report into competency (see Section 26.1.5.6) cited the following:

1. **Competency – levels of expertise**
   Five levels of expertise are considered as follows:
   (i) Novice
   (ii) Advanced novice
   (iii) Competent
   (iv) Proficient
   (v) Expert

The above model identifies the 'competent' level as one in which the duty holder satisfactorily performs the role, independently – and in so doing recognises the boundary of his/her competence such that those at the proficient/expert level are called upon should the task demand (as recognised in Section 26.1.1). This would appear to be a pragmatic and logical evaluation of competency with reference to its position in the spectrum of capabilities.

2. **Competency – components**
   The components of competency are considered to be as follows:
   (i) **Individual competence**
      (a) Knowledge (core qualifications and understanding of the task/role)
      (b) Skills (the practical application of knowledge)
      (c) Human factors (e.g. self-awareness, risk awareness, etc.)
   (ii) **Managerial factors**
      (i) Policy and procedures
      (ii) Ergonomics
      (iii) Environment
      (iv) Communications

   This model identifies key individual attributes of competence but recognises that satisfactory delivery of the task/role is dependent upon a range of additional factors outside the control of the individual that must be provided by the management/organisation.

The above criterion will be used as the basis of a competency model in Section 26.1.8.

26.1.3 Professional competency

The previous section identified levels and components of competency targeted at an individual task or role. However, another dimension of competency is that of competency at a defined 'professional' level. Within this context, the Institution of Engineering and Technology (IET) (and similarly other engineering institutions) defines five generic areas of competency, for all registrants wishing to achieve chartered engineer, incorporated engineer or engineering technician status. The generic areas are common to all three engineering categories, but requiring different levels of attainment for each category. The generic areas of competency comprise:

- Knowledge and understanding
- Design and development of processes, systems, services and products
- Responsibility, management and leadership
- Communications and interpersonal skills
- Professional commitment

In addition, a minimum number of years of experience at an appropriate level must have been achieved.

The above criterion is used to evaluate whether an engineer is capable (i.e. competent) of operating at a specified professional engineering level (e.g. chartered engineer). However, professional engineering status of itself does not provide assurance of competency to deliver a certain task or role, e.g. to discharge the role of a commissioning engineer on the UK 132 kV network. Although it does signify that the requisite competence is attainable with the appropriate training and experience, and it does indicate the level of role that the engineer in question is capable of discharging.

26.1.4 Competency – range of rolls

Clearly, all who are involved in power network construction need to be competent for the role they discharge. Within this context, the list below identifies some of the more significant roles, as an indication of the range of required competency.

1. **Engineering roles**

   In the following PNC = power network company and C = contractor

   (i) Scheme team leader (PNC)
   (ii) Power system design engineers (PNC/C)
   (iii) Development engineers (PNC)
   (iv) System operations engineers (PNC)
   (v) Project managers (PNC/C)
   (vi) Project engineers (PNC/C)
   (vii) Design team leaders (PNC/C)
   (viii) Detail design engineers – all technologies (C)
   (ix) Design assurance engineers – all technologies (PNC)
   (x) Protection and control settings engineers (PNC/C)
   (xi) Site manager (C)
   (xii) Site engineer (PNC)
   (xiii) Commissioning panel chairman (PNC)
   (xiv) Commissioning engineers (PNC/C)
   (xv) SAP (PNC)
   (xvi) Competent persons (under the safety rules) (C)
   (xvii) Health, safety and environmental officers (advisors) (PNC/C)
2. **Technician and site workforce roles**

   (i) Engineering draughtsmen and drawings production (C)
   (ii) Substation electrical and mechanical erection (C)
   (iii) Substation civil and structural engineering construction (C)
   (iv) Substation building engineering workforce (C)
   (v) OHL foundations/towers, poles erection/conductor stringing, etc. workforce (C)
   (vi) Cable excavation/laying/jointing/terminating, etc. workforce (C)

From the above (list not exhaustive), it can be seen that the number of roles is very extensive – and invariably involves all power network companies requiring construction work to be undertaken, and a wide range of contractors who will deliver the work. Each will need to define competency arrangements relevant to their organisation. Smaller organisations with a small number of duty holders may have difficulty in managing and resourcing the arrangements in-house and in the ultimate may need to buy in competent resources. Within this context, the duty holder would require a certificate or ticket of competency, etc. from a recognised body thereby certifying the competency.

26.1.5 Competency definition, provision and award organisations

Numerous UK-government-created bodies and other organisations and institutions (many with worldwide impact and connections) discharge responsibilities relating to competency. Most are concerned with maintaining and raising the standards of competency – and in doing so undertake the following activities:

- Definition of competency framework requirements
- Specification of competency requirements for specific tasks and roles
- Task and role-specific training in competency
- Preparation of methods of assessing competency, e.g. practical demonstration, examination papers, structured interview, etc.
- Competency assessment
- Award of competency tickets, certificates, etc.
- Maintenance of registers of competent people.

Generally, the subject matter can be categorised into health and safety competency and vocational competency although the two are intertwined. Within this context, the following major contributors to the field of competency will be reviewed:

- Health and Safety Executive
- Institution of Occupational Health (IOSH)
- National Examining Board in Occupational Safety and Health (NEBOSH)
- Energy and Utility Skills Register (EUSR)
- Royal Society for the Prevention of Accidents (ROSPA)
- Construction Industry Training Board (CITB)
- Engineering Construction Industry Training Board (ECITB)
- Professional institutions.

26.1.5.1 Health and safety executive

Chapter 2 discussed the role of the HSE, and many of their duties and publications are discussed in Chapter18.

From a health and safety perspective, numerous legal provisions require that certain types of work particularly those relating to testing, inspections and assessments, etc. must be undertaken by a 'competent person'. Examples of such a requirement include the Pressure System Regulations and the LOLER Regulations – see Chapter 18. However, very little guidance is provided on the abilities and expertise required excepting that in principle a competent person is taken to be 'one who has the necessary theoretical and practical knowledge and the hands on experience to carry out the task – and that such experience is matched to the complexity of the work and the degree of risk' – i.e. consistent with many other definitions of competency but with an emphasis on risk.

Two areas of legislation which are of specific relevance to power network construction are as follows:

1. **CDM Regulations 2015**
   In Regulation 8 – general duties, it is stated, 'anyone appointing a designer or contractor to work on a project must take reasonable steps to satisfy themselves that those who will carry out the work have the skills, knowledge, experience .... to carry out the work in a way that secures health and safety'. The word competency is not directly used – but the attributes described are those of a competent person.
2. **Electricity at Work Regulations 1989**
   In Regulation 16 under the heading of 'People to be Competent to Prevent Danger and Injury', it stated:

   No one shall be engaged in any work activity where technical knowledge and experience is necessary to prevent danger, or where appropriate, injury, unless they possess such knowledge and experience or are under such degree of supervision as may be appropriate having regard to the nature of the work

   The regulations go on to state that that technical knowledge and experience may include:
   (i) Adequate knowledge of electricity and experience of electrical work
   (ii) Adequate knowledge of the system to be worked upon and practical experience of the class of system, including an understanding of the hazards that may arise and precautions to be undertaken
   (iii) The ability to recognise at all times whether it is safe for the work to continue.

   A requirement for competency also arises with reference to a permit to work, which is usually required for work on HV equipment (and specified in the electricity at work guidance document). The guidance document stipulates, 'a permit to work (PTW) should be issued by a designated competent person' – where a competent person is 'someone appointed by an employer to undertake specific responsibilities and duties'. It goes on to say that the person 'must be competent by way of training and qualifications and/or experience'. NB: The competent person referred to in the regulations is usually the senior authorised person discussed in Chapter 18.

In addition to the above, the HSE guidance document 'Managing Competence for Safety-related Systems' lists 15 principles for the planning, design, operation and review of a competency system. These are summarised below:

1. Define purpose and scope
2. Establish competence criteria
3. Decide processes and methods
4. Select and recruit staff
5. Assess competence

6. Develop competence
7. Assign responsibilities
8. Monitor competence
9. Deal with failure to perform competently
10. Manage assessors and managers competence
11. Manage supplier competence
12. Manage information
13. Manage change
14. Audit
15. Review

Thus, the requirement for competency in some salient activities relevant to power network construction is mandated by health and safety law.

26.1.5.2 Institution of occupational safety and health (IOSH)

IOSH is a UK-based chartered body for health and safety professionals but with a worldwide membership. It provides support, advice and training to professional health and safety officers, managers and other professionals who require defined levels of competence in health and safety. One of the many courses designed and delivered by IOSH is 'managing safely' – which is often accepted as a minimum level of health and safety competency for engineering professionals working in power network construction.

26.1.5.3 National examination board in occupational safety and health (NEBOSH)

NEBOSH was founded in 1929 as a UK-based awarding body, with charitable status. It offers a range of globally recognised qualifications designed to meet the health, safety and environmental needs of the workplace.

NEBOSH does not deliver courses but develops the syllabus for its qualifications and sets methods of assessments such as examinations and practical course work. The courses are delivered by NEBOSH-accredited course providers. NEBOSH qualifications are considered to be some of the best in class and recognised by professional membership bodies such as IOSH. NEBOSH qualifications such as the 'certificate level' qualification and particularly the 'NEBOSH Diploma' are recognised as benchmark qualifications signifying a high level of attainment, and competency, in health and safety, particularly relevant to the construction industry.

26.1.5.4 Energy and utility skills register (EUSR)

As pointed out in Chapter 22, EUSR is a UK-employer-led organisation that ensures that the UK power, gas and water industries have the skills and experience they require to maintain/improve standards of service. EUSR provides market intelligence, develops national occupational standards (NOS) and benchmarks standards of performance. They also provide organisation support for apprentice training frameworks, etc. Importantly, EUSR also provides cards certifying competence in certain activities and listing on a skills register. A key qualification relating to an electrical work on site is the 'Basic Electrical Safety Competency for Access, Movement and Egress', i.e. BESC AME – successful completion of which results in issue of a card by EUSR and listing on the register.

26.1.5.5 Royal society for the prevention of accidents (ROSPA)

ROSPA dates back to 1917 and is a registered charity. Its objective is to influence legislation and attitudes towards accidents, not only in construction, but in general. Although ROSPA is UK based, it has international connections and impact. ROSPA among its many activities runs numerous courses in health and safety and construction-related activities, e.g. NEBOSH, IOSH courses, manual handling, etc. It produces numerous authoritative publications on health and safety management and health and safety law.

26.1.5.6 Construction industry training board (CITB)

The CITB is a UK-government-supported skills and industry training board for the construction industry. Its activities range from standards, training, qualifications and the award of certificates and cards of competency. In conjunction with the HSE, and supported by government, it has recently produced a report entitled 'Competency in Construction' (undertaken by Pye Tait Consulting). The report comprises a strategic evaluation of competency and the means by which standards of competency can be improved. The recommendations are (at the time of writing) being considered but may ultimately lead to a consolidated engineering-based framework for defining, awarding and recording competency.

26.1.5.7 Engineering construction industry training board (ECITB)

Established in 1991, the ECITB is accountable to the UK government for skills, standards and qualifications in engineering construction. It has a statutory responsibility for the development of the engineering construction workforce in the United Kingdom. It also operates an international arm. It essentially targets the provision of vocational qualifications and prepares 'National Occupational Standards' (NOS). These are employer-agreed statements defining content and levels of knowledge, understanding and performance. Individuals may be assessed against these standards to demonstrate competency to undertake certain defined tasks in the workplace. Examples of NOS standards include:

- Inspection, testing and commissioning of electrical installations (plant)
- Develop and agree detailed project designs in construction contracting operations management
- Assemble components of electrical plant and equipment.

NOS are largely targeted at engineering technicians.

26.1.5.8 Professional institutions

As stated in Section 26.1.3, the professional institutions define levels of competence in accordance with chartered status and membership. With reference to power network construction, the most relevant institutions are the IET, the Institution of Civil Engineers (ICE) and the Institution of Mechanical Engineers. These institutions also provide a range of professional training course in such tasks as project management, specification writing and contract law. For the most part, chartered status does not confer competency to discharge a specific task or role (e.g. a power system design engineer) – but signifies the general level of competency at which an engineer is capable of working. This is evidenced by defined attainments, experience of operating at the defined level and supported by referees usually those in positions of authority in the same technical field.

26.1.6 Technical standards

An increasing number of technical standards (e.g. IEC) also stipulate that those who undertake defined tasks should possess the relevant competency – often with a specification of what the competency comprises. Such competency is increasingly considered as being mandatory.

26.1.7 Competency system

From the previous sections, it can be seen that the United Kingdom has much in place in terms of formal competency definition, assessment and recording – which extends to many other countries worldwide. Organisations that drive competency standards broadly divide into health and safety focused (and environmental) and as such are driven by statutory requirements – and vocationally focused. Much of the vocational is targeted at the technician (or craftsman) level and is mostly task related (e.g. scaffolding). However, all vocational tasks carry an element of health and safety so the two are intrinsically interlinked. In addition, the professional institutions award titles of professional status, thereby signifying a general level of competency at which an engineer/technician is capable of operating, and in doing so they also acknowledge an affiliation with an institution of professional standing.

However, responsibility for ensuring that individuals with the appropriate competencies are matched to the role/task in question lies with the employer (i.e. at the level of the individual organisation), albeit that many of the competencies could be acquired through the external organisations previously discussed. As such the employing organisation must, as part of QMS procedural requirements, have a competency system in place to achieve this end, e.g. typically based upon the managing competency system articulated by the HSE as outlined in Section 26.1.5.1. Within this system, there is a requirement for a model defining the route to achieving the required competency. The following sections will consider a possible model and associated arrangements.

26.1.8 Engineering competency model

Based upon the criterion in Section 26.1.2, Figure 26.1 suggests an engineering competency model suitable for power network construction that may be utilised by

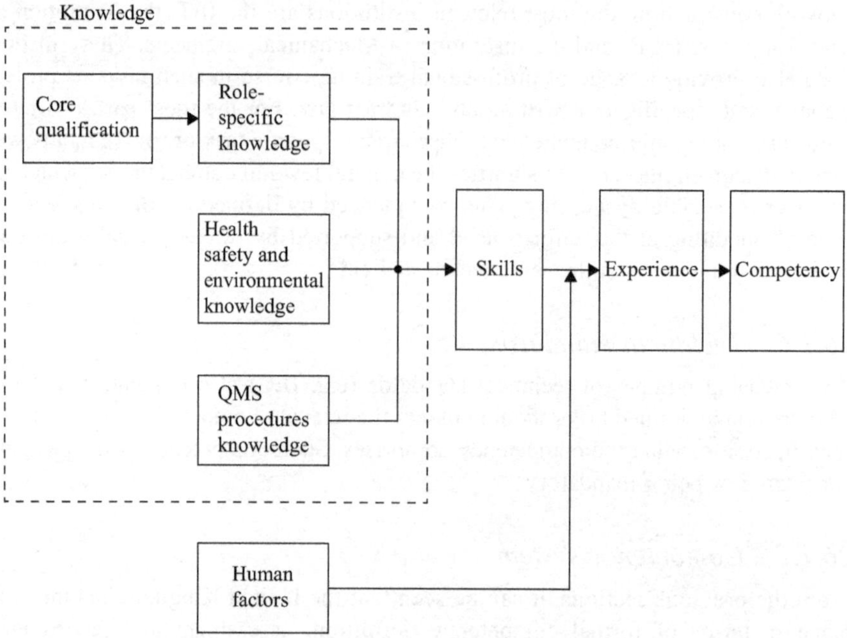

Figure 26.1 An engineering competency model

an employing organisation. It essentially targets the engineering roles listed in Section 26.1.4 (1). The salient components of this model are as follows:

1. **Core qualification**
   The core qualification is the relevant academic qualification for the role to be undertaken, e.g. a first degree (or master's degree) in power engineering or civil engineering, etc. as appropriate.
2. **Role-specific technical knowledge**
   Role-specific technical knowledge comprises the additional knowledge over and above the core qualification that is required to competently discharge the role, e.g. the specific technical knowledge required by an OHL design engineer.
3. **Health and safety (and environmental) knowledge**
   This comprises the health and safety (and environmental) knowledge, to competently discharge the role. This is essentially required by the safety management system (SMS), e.g. CDM regulations, company safety rules (and equivalent environmental procedures), etc.
4. **QMS procedures knowledge**
   This comprises knowledge of the requirements of the QMS procedures, i.e. who does what and when, related to the role in question. It may also relate to the QMS procedures of another organisation if two organisations are to work together seamlessly.

5. **Skills**
   The term 'skills' refers to the practical application of acquired knowledge. Skills may be acquired by practical training, case studies, etc. or on the job. If the latter, it overlaps with experience.
6. **Human factors**
   As stated in Section 26.1.2, human factors include personal skills, situational awareness, risk awareness, the ability to communicate with others and where applicable the ability to work as part of a team. These are often salient requirements in the obtaining of chartered status.
7. **Experience**
   Experience is the repeated execution of a skill on a regular basis, particularly on different types of projects in varying circumstances – thereby acquiring wide-ranging experience. Competency in the context of experience is often acquired in the following way:
   (i) Observe competent practitioners undertaking the role/task and learn through observation
   (ii) Undertake the work under the supervision by a competent practitioner
   (iii) Undertake the work independently as a competent practitioner.
   As a generalisation, the more time the work is undertaken, the greater the experience gained, and the greater the level of expertise.
8. **Competency**
   Competency is achieved when the role/task is undertaken independently to the right technical standard, and in accordance with both SMS requirements and QMS procedural requirements.

26.1.8.1 Engineering competency model – implementation

With reference to the model in Figure 26.1 (or any similar model), organisations need to define the arrangements for implementing the model, which are as follows:

1. **Knowledge specification**
   A specification of knowledge requirements (as encompassed by the dotted box in Figure 26.1) with reference to the role to be undertaken needs to be prepared. In some instances, the depth of knowledge will need to be defined, e.g. just an awareness of a QMS procedure or a comprehensive understanding.
2. **Knowledge acquisition**
   Second, the means by which knowledge is to be acquired needs to be defined. Clearly, the core qualification will be externally awarded by an educational establishment. The remainder will need to be acquired through lectures, case studies, practical exercises, etc. Health and safety is often outsourced to an external specialist provider.
3. **Skills and experience**
   Arrangements for acquiring the requisite skills and level of experience need to be specified. For example, a commissioning engineer may need to be sent to a training school (or on site) to acquire the practical skills of commissioning, prior to gaining experience by undertaking the role himself/herself.

4. **Assessment**
   A key decision to be taken is the degree of formality associated with assessment. This will be discussed further in Section 26.1.9.
5. **Record of competency**
   A record of competency for each duty holder needs to be maintained. This is especially important in larger organisations with large numbers of practitioners, with possibly a relatively large turnover of personnel. An easily accessible and central database record is often preferable to issuing a certificate of competency. It is important at the outset of construction work that the competency of all duty holders is known, and as such a database facilitates confirmation. A database also enables the date of expiry of competency to be identified.
6. **Competency duration**
   Another key decision is the period of time for which the assessed competency remains valid, particularly in changing circumstances. For most roles, a period of 3 to 5 years is a pragmatic and optimum period.
7. **Reassessment of competency**
   Following expiry of a period of competency, a decision needs to be taken as to how competency is reassessed, for a further period. This could comprise a complete or partial reassessment depending upon the role. Consideration needs to be given as to how competency is updated with changing technology, health and safety law or QMS procedures.

26.1.9 Competency assessment

The means by which competency is assessed is a prime consideration for an organisation. Assessment by an external organisation of recognised standing greatly alleviates the workload, but this is usually only available in certain subject areas such as health and safety and some vocational tasks. With many roles, the organisation has to make its own assessment of competency – and as such a decision has to be taken on the degree of formality. Options are as follows:

1. **Job observation**
   Following the necessary training in knowledge and skills, this would comprise observation of performance on a number of jobs (usually partially supervised). To assist assessment, a scoring matrix may be used covering all aspects of the model in Figure 26.1 and where for example:
   - 3 = High
   - 2 = Satisfactory
   - 1 = Questionable
   - 0 = Unsatisfactory
2. **Structured interview**
   This would comprise a reasonably formalised interview with a battery of structured questions, again covering all aspects of Figure 26.1. A scoring system is usually helpful, perhaps using a matrix approach as in (1) above.

3. **Formal assessment**
   Formal assessment comprises a test paper similar to that in an examination, again covering the content in Figure 26.1. The paper would need to be supported by a comprehensive syllabus so the candidate can adequately prepare. This method has the advantage that a wide range of subject matter can be assessed, and in preparing for the paper, the candidate will usually acquire significant relevant knowledge. The formal assessment also usually provides a consistent standard of evaluation (i.e. less subjective) and drives high standards of competency. However, the work involved in establishing and maintaining formal assessment arrangements cannot be underestimated – but once established, it is an efficient and disciplined approach to competency. Formal assessment is probably more applicable to larger organisations with an ongoing throughput of candidates. The assessment format may be multi-choice questionnaire or narrative response. A reasonable time limit should be allowed when taking the assessment.
4. **Mixed**
   A mix of all of the above could also be employed.

In all instances, the engineer being assessed should be provided with feedback on their performance.

26.1.9.1 Organisation competency

One school of thought argues that in order to provide organisations with resource flexibility and speed of movement, competency should be awarded to the organisation and not to the individual. The rationale is that the organisation as a whole would possess the competency (e.g. in substation design, installation and commissioning), and the organisation would manage individuals within the organisations with sub-competencies, as required, to undertake the work. Clearly, some engineers within the organisation would require the full competency, and these individuals would coordinate the work of those with the sub-competencies undertaking elements of the work. Such an approach may benefit organisations with resource constraints and allow individuals with a more limited range of skills to be utilised. However, the following must be considered:

- That health and safety legal requirements must be satisfactorily discharged
- That an audit trail is always put in place indicating how the full spectrum of competencies is applied to the work in question
- QMS procedural arrangements defining the competency and sub-competency arrangements would need to be defined and the method of execution.

26.1.9.2 Specialist competencies

Situations invariably arise in practice where specialists whose competency is limited to a specific technical subject are required for a limited part of a project, e.g. a commissioning engineer on a specialist piece of equipment. Under such instances, consideration will need to be given both to how the competency of the specialist engineer is assured and the arrangements for integrating the work of the specialist into the overall work – e.g. under the control of a competent commissioning engineer.

26.1.10 Technician and site workforce competency

The discussion thus far has primarily focused on competency with reference to engineering roles, although the requirement for competency applies equally to the technician and site workforce referenced in Section 26.1.4. Within this context, thc model described in Figure 26.1 can equally be applied, but with the following considerations:

- In many instances, elements of the workforce may be under the personnel supervision of an individual who possesses the necessary competency. Even so most, if not all, of the workforce on site would usually be required to be in possession of at least a BESC ASM ticket of competency.
- Many site-related competencies are limited to a specific task e.g. scaffolding erection or crane operator, in such instances, the ticket of competency can be provided by one of the organisations listed in Section 26.1.5.

26.1.11 Site operational competency

As stated in Chapters 18 and 22, a specific area of competency relevant to power network construction is competency under the power network company safety rules (and equally any contractor safety rules) – as required by the electricity at work regulations, see also Section 26.1.5.1. The responsibilities are largely concerned with the following:

- Access to and moving around an operational site
- Working under a safety document on an operational site (including OHL and HV cables)
- Issuing/cancelling a safety document on an operational site (including OHL and HV cables).

The arrangements for defining and assessing competency for these circumstances are long established both in the United Kingdom and in many countries worldwide. They are invariably very rigorous and represent very best practice.

26.1.12 Management aspects of competency

It was stated in Section 26.1.2 that competent delivery of work depended not only on individual competency but also a number of management factors. Thus, management has to create a set of wider environmental arrangements in order for the work to be competently undertaken by the individual. A simple example would be the adequacy of lighting and heating in a work area. Another specific area is that of QMS procedures. Within this context, there are two fundamental factors affecting the way in which work can be competently delivered, which are as follows:

1. **QMS procedure**
   By fully defining how a task is to be discharged in a QMS procedure, the requirement for a competent person to understand every facet of the work is diminished. By way of simple analogy, little expertise is required to make a quality cake as long as a comprehensive and well-written recipe can be followed.

2. **Individual competency**
   Alternatively, an individual who is fully competent in an area of work is not dependent upon the stringent direction of a QMS procedure and can apply professional judgement and flexibility to meet the work-specific demands.

A long-standing discussion in many organisations is where the balance should lie between the imposition of procedures (i.e. bureaucracy) or alternatively 'leave it to the professional'. The merit of QMS procedures is that they bring common best practice, standardisation and efficiency and incorporate the lessons of the past. However, they are not always (and for understandable reasons) concisely written nor tailored to match all sizes of projects. The task for the organisation is to get this balance right – since both are critical to competent project delivery and organisation success.

26.1.13 Competency and society

The opening paragraph of this chapter gave three significant reasons why those discharging power network construction must be competent for the work undertaken. Within this context, power networks differ from most other activities that underpin modern society – in as much as the impact of any loss of electricity supply, particularly on a large scale, risks an inability of society to function. As such, society has an understandable expectation that those who have responsibility for power networks, including all those concerned with construction, are both competent in discharging the work they undertake and able to show how that competency was achieved. This is one further and salient reason why engineering competency is the vital and essential third pillar, in addition to technology and QMS procedures, in the definition of a construction execution model.

References and bibliography

It is recognised that many of the publications listed below, although current at the time of preparing the text, are invariably updated with time, although the core principles and core content are usually retained. In some instances, an older or discontinued publication may be referenced because of its ongoing relevance and explanatory text.

1. **Health and Safety Publications**

   Acknowledgement is made to the Open Government Licence for Public Sector Information for permission to reference the following documents. Most documents can be obtained from www.hse.gov.uk

   1.1 Health and Safety at Work Act etc. 1974
   1.2 Management of Health and Safety at Work Regulations 1989
   1.3 Construction (Design and Management) Regulations 2015
   1.4 Electricity at Work Regulations 1999
   1.5 COSHH Regulations 2002
   1.6 Pressure System Regulations 2000
   1.7 Work at Height Regulations 2005
   1.8 Lifting Operations and Lifting Equipment Regulations1998
   1.9 Reporting of Injuries, Diseases and Dangerous Occurrences Regulations 2013
   1.10 Manual Handing Regulations 1992
   1.11 Personal Protective Equipment at Work Regulations 1992
   1.12 Provision and Use of Work Equipment Regulations 1998
   1.13 HSG 47 Avoiding Danger from Underground Services 2014
   1.14 HSG 65 Managing for Health and Safety 2013
   1.15 HSG 85 Electricity at Work: Safe Working Practices 2013
   1.16 HSG 150 Health and Safety in Construction 2006
   1.17 HSG 168 Fire Safety in Construction 2010
   1.18 Managing Health and Safety in Construction – Guidance on CDM Regulations 2015
   1.19 ACOP L5 Control of Substances Hazardous to Health 2013
   1.20 ACOP L22 Pressure System Safety Regulations 2014
   1.21 L25 Personal Protective Equipment at Work 2015
   1.22 L74 First Aid at Work 2013
   1.23 Electricity Safety, Quality and Continuity Regulations 2002
   1.24 HMSO/HSE Tolerability of Risk from Nuclear Power Stations 1992

2. **Environmental Publications**

Acknowledgement is made to the Open Government Licence for Public Sector Information for permission to reference the following documents. Most documents can be obtained from www.legislation.gov.uk

- 2.1 Control Waste Regulations 1992/2012
- 2.2 Hazardous Waste Regulations 2005
- 2.3 Site Waste Management Plan Regulations 2007
- 2.4 Environmental Permitting Regulations 2007
- 2.5 Groundwater Regulations 2009
- 2.6 Land Drainage Act 1991
- 2.7 Control of Pollution (Oil Storage) Regulations 2001
- 2.8 Environmental Permitting Regulations 2001
- 2.9 Road Traffic (Vehicle Emissions) Regulations 1997
- 2.10 Fluorinated Greenhouse Gas Regulations 2009
- 2.11 Environmental Protection Regulations 2011
- 2.12 Clean Air Act 1993
- 2.13 Pollution Prevention and Control Act 2009
- 2.14 Contaminated Land Regulations 2006
- 2.15 Building Regulations 2010
- 2.16 Energy Performance of Building Regulations 2007
- 2.17 Ancient Monuments and Archaeology Areas Act 1979
- 2.18 Hedgerows Regulations 1997
- 2.19 Environmental Protection Act 1990
- 2.20 Town and Country Planning (Environmental Impact Assessment) Regulations 1999
- 2.21 Wildlife and Countryside Act 1971
- 2.22 Conservation of Habitats and Species Regulations 2010
- 2.23 Wild Mammals Protection Act 1996
- 2.24 Environmental Protection Act 1990
- 2.25 Road Traffic Act 1998
- 2.26 Noise and Statutory Nuisance Act 1996
- 2.27 Control of Noise at Work Regulations 1988
- 2.28 Clean Neighbour and Environment Act 2005
- 2.29 Control of Pollution (Oil Storage) (England) Regulations 2005

3. **Government-Related Publications – General**

Acknowledgement is made to the Open Government Licence for Public Sector Information for permission to reference the following documents. Most documents can be obtained from www.legislation.gov.uk

- 3.1 UK Electricity Act 1999
- 3.2 UK Planning Act 2008
- 3.3 Town and Country Planning Order 2005
- 3.4 Department of Energy and Climate Change. Power Lines: Control of Micro-shocks and other Indirect Effects of Public Exposure to Electric Fields 2013

4. **National Grid Publications**
Acknowledgement is made to National Grid Company plc for permission to reference the following documents. Most documents can be obtained from www2.nationalgrid.com
4.1 Grid Code (subject to ongoing update)
4.2 National Electricity Transmission System Security and Quality of Supply Standard (SQSS) 2017
4.3 Holford Rules
4.4 Horlock Rules
5. **British Standards/IEC Publications**
Acknowledgement is made to BSI Group for permission to reference the following documents. Most documents can be obtained from www.bsigroup.com/Shop
5.1 BS 8100 Lattice Towers and Masts 1999
5.2 BS EN 193 Design of Steel Structures 2005
5.3 IEC 60826 Design Criterion for Overhead Transmission Lines 2003
5.4 BS 7354 Design of HV Open Terminal Substations 1990
5.5 IEC 62271 HV Switchgear and Control Gear 2007
5.6 BS EN 61936 Power Installations Exceeding 1 kV AC 2014
5.7 BS EN 60071 Insulation Coordination 2015
5.8 IEC 60060-1 Guide on High Voltage Testing Techniques – General 2010
5.9 BS EN 60694 Common Specifications for High Voltage Switchgear and Control Standards 1996
5.10 IEC 60287 Calculation of the Continuous Current Rating of Cables 2015
5.11 IEC 60853 Calculation of the Cyclic and Emergency Current Ratings of Cables 2015
5.12 IEC 60076 Power Transformers 2011
5.13 IEC 60214 On Load Tap Changers 2014
5.14 IEC 60289 Reactors 1988
5.15 IEC 60354 Loading Guide for Oil Immersed Transformers 1991
5.16 IEC 60270 Partial Discharge Measurements 2000
5.17 IEC 60044 Instrument Transformers 1996
5.18 IEC 255-1 All or Nothing Relays 1975
5.19 IEC 255-3 Single Input Energising Quantity Measuring Relays with Dependent or Independent Time 1989
5.20 IEC 255-6 Measuring Relays and Protection Equipment 1980
5.21 IEC 255-13 Biased Differential Relays 1980
5.22 IEC 60255 Measuring Relays and Protection Equipment 2013
5.23 IEC 60834 Performance and Testing of Teleprotection Equipment of Power Systems 1999
5.24 IEC 60050 International Electrotechnical Vocabulary 2015
5.25 BS 7626 Specification for Current Transformers 1993
5.26 BSEN 600896-22 Stationary Lead Acid Batteries 2015

- 5.27 BS 7671 IET Wiring Regulations 2015
- 5.28 BS EN 50522 Earthing of Power Installations Exceeding 1 kV AC 2010
- 5.29 BS EN 7430 Code of Practice for Earthing 2015
- 5.30 IEC TS 60479 Effects of Current on Human Beings and Livestock 2007
- 5.31 IEC TS 61201 Use of Conventional Touch Voltage Limits – Application Guide 2007–08
- 5.32 IEC 60865 Short Circuit Currents – Calculation of Effects 2011
- 5.33 BSEN 1992 Design of Concrete Structures 2014
- 5.34 BS 1722 Fences 2011
- 5.35 BS 5795 Temporary Works 2011
- 5.36 BS EN 12811-1 Temporary Works Equipment – Scaffolds 2003
- 5.37 BS EN 1990 Basis of Structural Design 2002
- 5.38 BS EN 60617 Graphical Symbols for Diagrams 1996
- 5.39 BS EN ISO 14617 Graphical Symbols for Diagrams 2002
- 5.40 BS 3939 Graphical Symbols for Electrical Power, Telecommunications, and Electronics Diagrams 1966
- 5.41 BS EN ISO 9001 Quality Management Systems 2015
- 5.42 BS EN ISO 14001 Environmental Management Systems 2015
- 5.43 PAS 55 (ISO 5501) Asset Management 2008
- 5.44 PAS 91 Construction Related Procurement 2013
- 5.45 PAS 99 Specification of Common Management System Requirements as a Framework for Integration 2012
- 5.46 BS OHSAS 18001, Occupational Health and Safety 2007
- 5.47 BS OHAS 18002 Guidelines for Implementing OHAS 18001 2007

6. **Electrical Network Association (ENA) Publications**

Acknowledgement is given to ENA for permission to reference the following documents. Most documents can be obtained from http://www.dcode.org.uk or www.energynetworks.org

- 6.1 Distribution Code (subject to ongoing update)
- 6.2 ENA Engineering Recommendation P2/6, Security of Supply 2014
- 6.3 ENA TS 43-40 Specification for Single Circuit OHL on Wood Poles for use up to and Including 33 kV 2004
- 6.4 ENA TS 09-02 Supply Delivery and inspection of Power Cables with Operating Voltage in the Range 33 kV to 400 kV
- 6.5 EA TS 41-24 Guidelines for the Design, Installation, Testing and Maintenance of Main Earthing Systems in Substations 1992
- 6.6 Electricity Association, Engineering Recommendation S34 – A Guide for Assessing the Rise of Earth Potential at Substation Sites 1986
- 6.7 Electricity Council, Engineering Recommendation S5/1, Earthing Installations in Substations 1966
- 6.8 EA TS 36-1 Procedure to Identify and Record Hot Substations 2007
- 6.9 EA TS 41-37 Part 2, GIS Switchgear for use on 66 kV to 132 kV Distribution Systems
- 6.10 EATS 41-15 Part 1, Standard Circuit Diagrams for Equipment in 132 kV Substations (Transformer Feeder, Transformer End)

7. **Other Publications**
 7.1 ASCE Manual 74 Guidelines for Electrical Transmission Line Structural Loading 2010
 7.2 IEEE 45 Battery Sizing 2010
 7.3 IEEE C37.112 Inverse-Time Characteristic for Overcurrent Relays 1996
 7.4 ITU-T K.33 Limits for People Safety Relating to Coupling into Telecommunications Systems from AC Electric Power Installations (Previously CCITT Recommendation) 1996
 7.5 Cigre: 283: Special Bonding of High Voltage Power Cables 2005
 7.6 Cigre:44: Earthing of GIS – An Application Guide 1993
 7.7 Competence in Construction. Report by Pye-Tait Consulting for Citb 2014
 7.8 FIDIC – Conditions of Contract for Electrical and Mechanical Works Including Erection 1987
 7.9 New Engineering Contract (NEC4) – The Institution of Civil Engineers 2017

Further Reading

- Akass R: "What every manager needs to know about health and safety" (Aldershot: Gower 1995).
- Bayliss B and Hardy B: "Transmission and distribution electrical engineering" (Oxford: Newnes 2012).
- BICC Cables Ltd: "Electrical cables handbook" (Oxford: BS Professional Books 1997).
- El-Sharkawi M: "Electric safety – practice and standards" (Boca Raton, FL: CRC Press 2014).
- Giles R L: "Layout of E.H.V Substations" (London: Cambridge University Press/IEE 1970).
- Haddad A and Warne D: "Advances in high voltage engineering" (London: The IET 2004).
- Harker K: "Power system commissioning and maintenance practice" (London: The IET 1998).
- Heathcote M J: "J &P Transformer Book" (London: Newnes 2005).
- Juan M G and Holmes E J: "Protection of electricity distribution networks" (London: The IET 2004).
- Lavervi E and Holmes E J: "Electricity distribution network design" (London: The IET 1995).
- Lythall R T and Lythall P: "J & P Switchgear Book" (London: Butterworth 1976).
- Machowski J, Bialek J W and Bumby J R: "Power system dynamics and stability" (Chichester: Wiley 1997).
- Moor R: "The law of health and safety at work" (London: Croner 2016).
- Ryan H M: "High voltage engineering and testing" (London: The IET 1998).

- Sclabbach J and Rofalski K: "Power system engineering" (Weinheim: Wiley 2008).
- Solymar L: "Lectures on electromagnetic theory" (London: Oxford University Press 1976).
- Sclabbach J: "Short circuit currents" (London: The IET 1999).
- Song Y H and Johns A T: "Flexible ac transmission systems (FACTS)" (London: The IET 1999).
- Stranks J: "A Manager's guide to health and safety at work" (London: Kogan Page 1990).
- Tleis N: "Power system modelling and fault analysis" (Oxford: Newnes 2008).
- Various Authors: "Power system protection" – four volumes (Stevenage: ETA/IET 1995).
- Various Authors: "Protective relay application guide" (Stafford: Alstom 2002).
- Waring B: "Wood pole overhead lines" (London: The IET 2005).

Index